Nature-Based Solutions for Urban Sustainability

AF631540

Nature-Based Solutions for Urban Sustainability

Edited by
Piet N. L. Lens and Xuan-Thanh Bui

IWA PUBLISHING

Published by **IWA Publishing**
Unit 104–105, Export Building
1 Clove Crescent
London E14 2BA, UK
Telephone: +44 (0)20 7654 5500
Fax: +44 (0)20 7654 5555
Email: publications@iwap.co.uk
Web: www.iwaponline.com

First published 2025
© 2025 IWA Publishing

Apart from any fair dealing for the purposes of research or private study, or criticism or review, as permitted under the UK Copyright, Designs and Patents Act (1998), no part of this publication may be reproduced, stored or transmitted in any form or by any means, without the prior permission in writing of the publisher, or, in the case of photographic reproduction, in accordance with the terms of licenses issued by the Copyright Licensing Agency in the UK, or in accordance with the terms of licenses issued by the appropriate reproduction rights organization outside the UK. Enquiries concerning reproduction outside the terms stated here should be sent to IWA Publishing at the address printed above.

The publisher makes no representation, express or implied, with regard to the accuracy of the information contained in this book and cannot accept any legal responsibility or liability for errors or omissions that may be made.

Disclaimer
The information provided and the opinions given in this publication are not necessarily those of IWA and should not be acted upon without independent consideration and professional advice. IWA and the Editors and Authors will not accept responsibility for any loss or damage suffered by any person acting or refraining from acting upon any material contained in this publication.

British Library Cataloguing in Publication Data
A CIP catalogue record for this book is available from the British Library

ISBN: 9781789065008 (paperback)
ISBN: 9781789065015 (eBook)
ISBN: 9781789065022 (ePub)

Doi: 10.2166/9781789065015

This eBook was made Open Access in March 2025.

© 2025 IWA Publishing
This is an Open Access book distributed under the terms of the Creative Commons Attribution Licence (CC BY-NC-ND 4.0), which permits copying and redistribution for non-commercial purposes with no derivatives, provided the original work is properly cited (https://creativecommons.org/licenses/by-nc-nd/4.0/). This does not affect the rights licensed or assigned from any third party in this book.

Contents

List of contributors

Dan A
Zhongkai University of Agriculture and Engineering, College of Resources and Environment, Guangzhou 510225, China

Simone Amadori
Università di Bologna, Via Zamboni 33, 40126 Bologna, Italy

Tamara Avellan
University of Oulu, Water, Energy and Environmental Engineering Department, Pentti Kaiteran katu 1, Linnanmaa, Finland
and
UNU-CRIS, Potterierei 72, 8000 Brugge, Belgium

Sebrian M. Beselly
Coastal and Urban Risk & Resilience, IHE Delft Institute for Water Education, Westvest 7, 2611 AX Delft, the Netherlands
and
Water Resources Engineering, Brawijaya University, Jl. MT Haryono 167, 65142, Malang, Indonesia

Marc Breulmann
Department Systemische Umweltbiotechnologie, Helmholtz-Zentrum für Umweltforschung GmbH - UFZ, Permoserstraße 15, 04318 Leipzig, Germany

Thi-Hang Bui
VNU Vietnam Japan University, Master's Program in Environmental Engineering, Luu Huu Phuoc St., Nam Tu Liem Dist., Hanoi 130000, Vietnam

Xuan-Thanh Bui
Key Laboratory of Advanced Waste Treatment Technology, Ho Chi Minh University of Technology (HCMUT), 1a street No 4, Ward 26, Binh Thanh District, Ho Chi Minh City, Vietnam

Alba Canet-Martí
BOKU University, Universität für Bodenkultur Wien, Gregor-Mendel-Straße 33, 1180 Vienna, Austria

Eriona Canga
Alchemia-nova research & innovation gemeinnützige GmbH, Linzer Str. 284A, 1140, Vienna, Austria

Chawalit Chaiwong
Environmental Engineering and Management, Asian Institute of Technology (AIT), PO Box 4, Klong Luang, Pathum Thani 12120, Thailand

Chart Chiemchaisri
Department of Environmental Engineering, Faculty of Engineering, Kasetsart University, 50 Ngam Wong Wan Road, Chatuchak, Bangkok 10900, Thailand

Wilai Chiemchaisri
Department of Environmental Engineering, Faculty of Engineering, Kasetsart University, 50 Ngam Wong Wan Road, Chatuchak, Bangkok 10900, Thailand

Tjandra Chrismadha
Research Centre for Limnology and Water Resources, National Research and Innovation Agency – Republic of Indonesia (BRIN), KST Soekarno, Jl. Raya Bogor-Raya KM46, Cibinong, Bogor, 16911, Indonesia

Chau-Thi Da
Faculty of Applied Sciences, Ton Duc Thang University, 19 Nguyen Huu Tho Street, District 7, Ho Chi Minh City, Vietnam

Bishal Dahal
University of Oulu, Water, Energy and Environmental Engineering Department, Pentti Kaiteran katu 1, Linnanmaa, Finland

Thanh-Son Dao
Ho Chi Minh University of Technology (HCMUT), 142 To Hien Thanh Street, Ward 14, District 10, Ho Chi Minh City, Vietnam

Thi-Viet-Huong Dao
Ho Chi Minh University of Technology (HCMUT), 142 To Hien Thanh Street, Ward 14, District 10, Ho Chi Minh City, Vietnam

Daneish Despot
Department Systemische Umweltbiotechnologie, Helmholtz-Zentrum für Umweltforschung GmbH - UFZ, Permoserstraße 15, 04318 Leipzig, Germany

Yoshitaka Ebie
Planning Division, National Institute for Environmental Studies (NIES), 16-2 Onogawa, City of Tsukuba, Ibaraki 305-8506, Japan

Jan Friesen
Department Systemische Umweltbiotechnologie, Helmholtz-Zentrum für Umweltforschung GmbH - UFZ, Permoserstraße 15, 04318 Leipzig, Germany

Kedar Ghag
University of Oulu, Water, Energy and Environmental Engineering Department, Pentti Kaiteran katu 1, Linnanmaa, Finland

Valentina Guerrieri
Università di Bologna, Via Zamboni 33, 40126 Bologna, Italy

Harish Harini
Department of Chemical and Environmental Engineering, Royal Melbourne Institute of Technology (RMIT) University, Melbourne, Victoria 3000, Australia

Marco Hartl
Alchemia-nova research & innovation gemeinnützige GmbH, Linzer Str. 284A, 1140, Vienna, Austria

Yuni P. Hastuti
Aquaculture Department Fisheries and Marine Sciences Faculty, IPB University, Bogor, 16680, Indonesia

Mai-Nhu Hoang
Key Laboratory of Advanced Waste Treatment Technology, Ho Chi Minh University of Technology (HCMUT), 142 To Hien Thanh Street, Ward 14, District 10, Ho Chi Minh City, Vietnam

Ky-Phuong-Ha Huynh
Ho Chi Minh University of Technology (HCMUT), 142 To Hien Thanh Street, Ward 14, District 10, Ho Chi Minh City, Vietnam

Tomonori Ishigaki
Material Cycles Division, National Institute for Environmental Studies (NIES) 16-2 Onogawa, City of Tsukuba, Ibaraki 305-8506, Japan

Jeanger P. Juanga-Labayen
College of Industrial Technology, Carlos Hilado Memorial State University, Mabini St., Brgy. Zone 2, Talisay City, 6115 Negros Occidental, Philippines

Ganbaatar Khurelbaatar
Department Systemische Umweltbiotechnologie, Helmholtz-Zentrum für Umweltforschung GmbH - UFZ, Permoserstraße 15, 04318 Leipzig, Germany

Björn Klöve
University of Oulu, Water, Energy and Environmental Engineering Department, Pentti Kaiteran katu 1, Linnanmaa, Finland

Thammarat Koottatep
Environmental Engineering and Management, Asian Institute of Technology (AIT), PO Box 4, Klong Luang, Pathum Thani 12120, Thailand

Ildefonso V. Labayen
Manufacturing Engineering Technology Department, Technological University of the Philippines Visayas, Capt. Sabi St., Brgy. Zone 12, Talisay City, 6115 Negros Occidental, Philippines

Stevo Lavrnic
Università di Bologna, Via Zamboni 33, 40126 Bologna, Italy

Linh-Thy Le
Faculty of Public Health, University of Medicine and Pharmacy at Ho Chi Minh City (UMP), 217 Hong Bang Street, Ward 11, District 5, Ho Chi Minh City, Vietnam

Piet N.L. Lens
Department of Microbiology, National University of Ireland, University Road, Galway, Ireland

Novi Mayasari
Research Centre for Limnology and Water Resources, National Research and Innovation Agency – Republic of Indonesia (BRIN), KST Soekarno, Jl. Raya Bogor-Raya KM46, Cibinong, Bogor, 16911, Indonesia

Roland A. Müller
Department Systemische Umweltbiotechnologie, Helmholtz-Zentrum für Umweltforschung GmbH - UFZ, Permoserstraße 15, 04318 Leipzig, Germany

Syed Mustafa
University of Oulu, Water, Energy and Environmental Engineering Department, Pentti Kaiteran katu 1, Linnanmaa, Finland
and
Wageningen University & Research, PO Box 9101, 6700 HB Wageningen, The Netherlands

Nidhi Nagabhatla
UNU-CRIS, Potterierei 72, 8000 Brugge, Belgium

Kukuh Nirmala
Aquaculture Department Fisheries and Marine Sciences Faculty, IPB University, Bogor, 16680, Indonesia

Hoang-Dung Nguyen
Faculty of Environment and Natural Resources, Ho Chi Minh City University of Technology, Vietnam National University, Ho Chi Minh City, Vietnam

Ngoc-Tuan Nguyen
Faculty of Applied Sciences, Ton Duc Thang University, 19 Nguyen Huu Tho Street, District 7, Ho Chi Minh City, Vietnam

Phuoc-Dan Nguyen
Ho Chi Minh University of Technology (HCMUT), 142 To Hien Thanh Street, Ward 14, District 10, Ho Chi Minh City, Vietnam

Thi-An-Hang Nguyen
Faculty of Advanced Technologies and Engineering, Vietnam Japan University, Luu Huu Phuoc St., Nam Tu Liem Dist., Hanoi 130000, Vietnam

Thi-Kim-Ngan Nguyen
Key Laboratory of Advanced Waste Treatment Technology, Ho Chi Minh University of Technology (HCMUT), 142 To Hien Thanh Street, Ward 14, District 10, Ho Chi Minh City, Vietnam

Thi-Yen-Phuong Nguyen
Key Laboratory of Advanced Waste Treatment Technology, Ho Chi Minh University of Technology (HCMUT), 142 To Hien Thanh Street, Ward 14, District 10, Ho Chi Minh City, Vietnam

Thuong-Thi Nguyen
Ritsumeikan University, Asia-Japan Research Institute, Asia-Japan Research Institute, Ritsumeikan University, 2-150 Iwakura-cho, Ibaraki-shi, Osaka Prefecture 567-8570, Japan

Yuga Ogata
Material Cycles Division, National Institute for Environmental Studies (NIES), 16-2 Onogawa, City of Tsukuba, Ibaraki 305-8506, Japan

Mai-Duy-Thong Pham
Key Laboratory of Advanced Waste Treatment Technology, Ho Chi Minh University of Technology (HCMUT), 142 To Hien Thanh Street, Ward 14, District 10, Ho Chi Minh City, Vietnam

Nhu-Nguyet Phan
Faculty of Environment, University of Science, Vietnam National University, Ho Chi Minh City, Vietnam, 227 Nguyen Van Cu Street, Ward 4, District 5, Ho Chi Minh City, Vietnam

Bernd Plaul
Leipziger Wohnungs- und Baugesellschaft mbH, Wintergartenstraße 4, 04103 Leipzig, Germany

Chongrak Polprasert
Department of Civil Engineering, Faculty of Engineering, Thammasat University, 99 Moo 18, Phahonyothin Road, Khlong Nueng, Khlong Luang, Pathum Thani 12121, Thailand

Tatchai Pussayanavin
Faculty of Science, Ramkhamhaeng University, 2086 Ramkhamhaeng Road, Hua Mak, Bangkapi, Bangkok 10240, Thailand

Dano Roelvink
Coastal and Urban Risk & Resilience, IHE Delft Institute for Water Education, Westvest 7, 2611 AX Delft, the Netherlands
and
Deltares, Boussinesqweg 1, 2629 HV Delft, the Netherlands
and
Department of Hydraulic Engineering, Delft University of Technology, Building 23, Stevinweg 1, 2628 CN Delft, the Netherlands

Awalina Satya
Research Centre for Limnology and Water Resources, National Research and Innovation Agency – Republic of Indonesia (BRIN), KST Soekarno, Jl. Raya Bogor-Raya KM46, Cibinong, Bogor, 16911, Indonesia

Azalea Dyah Maysarah Satya
Department of Chemical and Environmental Engineering, Faculty of Science and Engineering, University of Nottingham Malaysia, Jalan Broga, Semenyih 43500, Selangor Darul Ehsan, Malaysia

Ika Atman Satya
Research Centre for Computation, National Research and Innovation Agency – Republic of Indonesia (BRIN), Komplek Cisitu, Jl. Sangkuriang, Coblong, Bandung, 40135, Indonesia

Tjandra Setiadi
Department of Chemical Engineering, Faculty of Industrial Technology, Institut Teknologi Bandung, Jl. Ganesha No. 10 Bandung 40132, Indonesia

Pau-Loke Show
Department of Chemical Engineering, Khalifa University, Shakhbout Bin Sultan Street – Zone 1, Al Zafranah 127788, Abu Dhabi, United Arab Emirates

Satoshi Soda
Department of Civil and Environmental Engineering, College of Science and Engineering, Ritsumeikan University, 1-1-1 Noji-higashi, Kusatsu, Shiga, 525-8577, Japan

Sirintornthep Towprayoon
The Joint Graduate School of Energy and Environment (JGSEE), King Mongkut's University of Technology Thonburi, 126 Pracha Uthit Road, Bang Mod, Thung Khru, Bangkok 10140, Thailand

Cong-Sac Tran
Key Laboratory of Advanced Waste Treatment Technology, Ho Chi Minh University of Technology (HCMUT), 142 To Hien Thanh Street, Ward 14, District 10, Ho Chi Minh City, Vietnam

Hau-Vuong Tran
Ho Chi Minh City University of Natural Resources and Environment (HCMUNRE), 236B Le Van Sy Street, Tan Binh District, Ho Chi Minh City, Vietnam

Pham-Yen-Nhi Tran
Key Laboratory of Advanced Waste Treatment Technology, Ho Chi Minh University of Technology (HCMUT), 142 To Hien Thanh Street, Ward 14, District 10, Ho Chi Minh City, Vietnam

Thanh-Dai Tran
Faculty of Technology, Dong Nai Technology University, Bien Hoa City, Vietnam, 4 Le Quy Don Street, Tan Hiep Ward, Bien Hoa City, Dong Nai, Vietnam

Manfred van Afferden
Department Systemische Umweltbiotechnologie, Helmholtz-Zentrum für Umweltforschung GmbH - UFZ, Permoserstraße 15, 04318 Leipzig, Germany

Mick van Der Wegen
Coastal and Urban Risk & Resilience, IHE Delft Institute for Water Education, Westvest 7, 2611 AX Delft, the Netherlands
and
Deltares, Boussinesqweg 1, 2629 HV Delft, the Netherlands

Tran Van Dung
Faculty of Soil Science, College of Agriculture, Can Tho University, Can Tho, Vietnam, 3/2 Street, Hung Loi Ward, Ninh Kieu District, Can Tho City, Vietnam

Jegatheesan Veeriah
Department of Chemical and Environmental Engineering, Royal Melbourne Institute of Technology (RMIT) University, Melbourne, Victoria 3000, Australia

Nguyen-Xuan-Phuong Vo
Faculty of Applied Sciences, Ton Duc Thang University, Ho Chi Minh City, Vietnam

Nguyen-Xuan-Que Vo
Faculty of Environment and Natural Resources, Ho Chi Minh City University of Technology, Vietnam National University, Ho Chi Minh City, Vietnam

Thi-Kim-Quyen Vo
Faculty of Biology and Environment, Ho Chi Minh City University of Industry and Trade (HUIT), 140 Le Trong Tan street, Tay Thanh ward, Tan Phu district, Ho Chi Minh City, 700000, Vietnam, 142 To Hien Thanh Street, Ward 14, District 10, Ho Chi Minh City, Vietnam

Agus Waluyo
Aquaculture Department Fisheries and Marine Sciences Faculty, IPB University, Bogor, 16680, Indonesia

Sara Kazemi Yazdi
Department of Chemical and Environmental Engineering, Faculty of Science and Engineering, University of Nottingham Malaysia, Jalan Broga, Semenyih 43500, Selangor Darul Ehsan, Malaysia

Qiuyan Yuan
Department of Civil Engineering, University of Manitoba, E1-368, EITC, 15 Gillson Street, Winnipeg, MB, R3T 2N2, Canada

doi: 10.2166/9781789065015_xix

Preface

Rapid urbanization has brought unprecedented opportunities for economic growth and innovation, yet it has also exacerbated challenges, including the effect of climate change, resource depletion, and environmental pollution. Along with the development of society, population growth has led to an increasing demand for land, water, and energy. As cities expand and urban populations grow, the pressure on natural resources significantly increases, leading to higher levels of consumption and waste generation. They create urgent needs for sustainable and regenerative urban solutions that can balance development with environmental preservation.

This book, *Nature-based Solutions for Urban Sustainability*, is a comprehensive exploration of how nature can play a crucial role in transforming cities into sustainable and resilient ecosystems. Nature-based solutions (NBS) offer innovative approaches to urban challenges by integrating natural processes into urban systems, providing a range of co-benefits, including biodiversity enhancement, climate resilience, and improved quality of life - from urban wetlands that filter pollutants and mitigate flooding, to green infrastructure that improves air quality and cools urban heat islands. By shifting from traditional methods to regenerative systems that restore and enhance natural cycles, NBS provide cities with pathways to become not just sustainable but regenerative. Furthermore, this book is structured to guide readers through the principles, practices, and impacts of NBS, addressing a range of themes critical to sustainable urban development, including pollution control and resource management, water-centric development, urban biorefineries, urban farming, regenerative urban planning, and sustainability assessment. Based on theoretical insights, practical applications, and real-world examples, this book aims to inspire a new paradigm shift toward cities that are sustainable and regenerative.

Authored by experts and researchers at the forefront of ecological engineering and environmental technology, as well as environmental science and pollution research, this comprehensive volume aims to provide an authoritative resource for academics, researchers, industry professionals, and policymakers. It will empower readers with knowledge about the latest advances, challenges, and breakthroughs in the use of nature-based solutions for the treatment of diffuse pollution and climate change mitigation. Each contributed chapter is presented on a stand-alone basis, so that the reader can consider the theme of each chapter. There are, nevertheless, many connections between what may at first seem to be quite different topics. As in all the books of the *Integrated Environmental Technology*

© 2025 IWAP. This is an Open Access book chapter distributed under the terms of the Creative Commons Attribution License (CC BY-NC-ND 4.0) which permits copying and redistribution for non-commercial purposes with no derivatives, provided the work is properly cited (https://creativecommons.org/licenses/by-nc-nd/4.0/). The chapter is from the book *Nature-Based Solutions for Urban Sustainability*, Piet N.L. Lens and Xuan-Thanh Bui (Editors).

series, one of our purposes was to draw out and emphasize these interdisciplinary linkages, so a comprehensive index is included to facilitate cross-referencing. We hope that the work described in this book will inspire those working in the field and will encourage those who are beginning to investigate it.

We wish to thank all contributors to this book for their valuable contributions by sharing their expertise in the various chapters included. We also thank all past and present co-workers as well as collaborators who joined in unravelling different areas of the application of nature-based solutions for urban sustainability as described in this book, especially those at Vietnam National University Ho Chi Minh, National University Ireland Galway and IHE Delft. We would also like to thank the reviewers who put a lot of effort into improving the quality of the publication.

We gratefully acknowledge the national and international granting agencies who supported our work on various aspects of diffuse pollution and ecological engineering over the years, in particular the research funding provided by Vietnam National University Ho Chi Minh (VNU-HCM) under grant number B2024-20-23 and the Science Foundation Ireland (SFI), who financially supported the open access publication of this book through the SFI Research Professorship Programme *Innovative Energy Technologies for Biofuels, Bioenergy and a Sustainable Irish Bioeconomy* (IETSBIO^3; grant number 15/RP/2763) and the Department of Foreign Affairs (DFA) under the SDG Challenge project *Floating Treatment Wetland* (grant number SFI/21/FIP/SDG/9933).

We express our appreciation to Ms Mai-Nhu Hoang, Ms Pham-Yen-Nhi Tran, Ms Thi Yen Phuong Nguyen and Dr Thi-Kim-Quyen Vo for technical support. We are also grateful to the editorial team of IWA Publishing, in particular Katharine Allenby and Andrew Peart, for their help and editorial support in realizing this book.

Piet N. L. Lens
University of Galway, Ireland

Xuan-Thanh Bui
Vietnam National University Ho Chi Minh, VNU-HCM University of Technology, Vietnam

Part 1

Role of nature-based solutions in urban sustainable development

doi: 10.2166/9781789065015_0003

Chapter 1

Role of nature-based solutions in urban sustainable development

*Mai-Nhu Hoang[1,2], Mai-Duy-Thong Pham[1,2], Phuoc-Dan Nguyen[1,2], Piet N. L. Lens[3] and Xuan-Thanh Bui[1,2]**

[1]Key Laboratory of Advanced Waste Treatment Technology, Ho Chi Minh City University of Technology (HCMUT), 268 Ly Thuong Kiet, District 10, Ho Chi Minh City, Vietnam
[2]Vietnam National University Ho Chi Minh (VNU-HCM), Ho Chi Minh City 700000, Vietnam
[3]National University of Ireland Galway, University Road, Galway H91 TK33, Ireland
**Email: bxthanh@hcmut.edu.vn*

ABSTRACT

Urbanization poses significant challenges, including pollution, flooding, resource depletion, and climate vulnerability, demanding innovative solutions for sustainable development. Nature-based solutions (NBS) offer adaptable and multifunctional approaches that integrate natural processes to mitigate these issues while enhancing ecosystem services, biodiversity, and urban resilience. This chapter explores the potential of NBS, in conjunction with Urban Ecosystem Service (EUS) and Green Infrastructure (GI), to achieve zero pollution in urban areas. It highlights practical applications (e.g. constructed wetlands, green roofs, and floating treatment wetlands), emphasizing their roles in managing water quality, reducing flood risks, and promoting ecological balance. Additionally, it provides a shift from traditional sustainable design to regenerative architecture, which not only minimizes harm but actively restores ecosystems. By combining theoretical insights and practical case studies, this chapter presents the vital strategy of NBS in creating zero-pollution cites that can be in harmony with nature.

Keywords: nature-based solution, urban sustainable development, pollution control, climate resilience, regenerative architecture

1.1 NATURE-BASED SOLUTIONS FOR ZERO POLLUTION IN URBAN AREAS

In an era of rapid urbanization, the sustainability of cities has become a critical global concern. Urbanization is closely linked to economic development and population growth in cities. By 2050, the urban population is projected to increase to 6.7 billion, accounting for 68% of the global population (Bai *et al.*, 2016; Fang *et al.*, 2023). Alongside, urban expansion leads to many environmental and social issues, including air pollution, urban flooding, heat waves, resource depletion, biodiversity loss, and vulnerability to climate change (Fang *et al.*, 2023, 2024; Fink, 2016). Urban air pollution, for instance, caused nearly 1 million deaths annually in China, while in South Asia, floods in 2017 affected more than 45 million people, resulting in over 1000 deaths and leaving many homeless (Bai

© 2025 IWAP. This is an Open Access book chapter distributed under the terms of the Creative Commons Attribution License (CC BY-NC-ND 4.0) which permits copying and redistribution for non-commercial purposes with no derivatives, provided the work is properly cited (https://creativecommons.org/licenses/by-nc-nd/4.0/). The chapter is from the book *Nature-Based Solutions for Urban Sustainability*, Piet N.L. Lens and Xuan-Thanh Bui (Editors).

et al., 2018; Yue *et al.*, 2020). To achieve sustainable urban development, it is crucial to mitigate these environmental challenges and build resilience against future risks.

Traditionally, cities have relied on technological and engineering-based solutions to address environmental risks. These approaches, such as expanding rainwater sewer systems to mitigate flooding, building sound barriers to reduce noise pollution and constructing seawalls to protect coastlines, offer predictable and straightforward solutions (Fang *et al.*, 2023, 2024). However, these methods tend to be expensive, both environmentally and economically, and often serve a single function. Moreover, they are often effective only in the short term, as they may not be adaptable to changing environmental conditions (Fang *et al.*, 2024). As in cities environmental pressures are growing, there is a need for more sustainable and multifunctional solutions.

To address the drawbacks of traditional pollution control methods and global societal challenges, nature-based solutions (NBS) can be applied (Kabisch *et al.*, 2022). NBS is defined as actions that are not only nature-inspired but also help build resilience, making them valuable for tackling environmental, social, and economic problems (Fink, 2016). For instance, a constructed wetland (CW) not only mitigates flood risks but also enhances biodiversity, improves water quality, and serves as a recreational space for urban dwellers (Alikhani *et al.*, 2021). This multifunctionality makes NBS a cost-effective and sustainable alternative to conventional urban infrastructure (Fang *et al.*, 2024). A key advantage of NBS is their adaptability to changing environmental conditions, which is critical to address climate change and urbanization. Furthermore, NBS are unique for their problem-driven and solution-oriented approach, focusing on long-term, adaptable solutions to environmental and social challenges. Another key aspect of NBS is their integrative and governance-based approach. NBS take various forms, including natural, managed, restored, or designed green spaces, and can also incorporate hybrid elements that combine natural and engineered components, such as green roofs (GFs). This flexibility allows NBS to effectively address urban sustainability, climate resilience, and biodiversity conservation (Fang *et al.*, 2023).

1.2 URBAN ECOSYSTEM SERVICES AND GREEN INFRASTRUCTURE FOR URBAN SUSTAINABLE DEVELOPMENT

Urban ecosystem services (UES) and green infrastructure (GI) are also related to NBS and urban sustainability. UES is the benefits derived from natural ecosystems within cities, including provisioning (food, water, and raw materials), regulating (climate, air, and water purification), cultural (recreational, aesthetic, and mental health), and supporting (soil formation, nutrient cycling, and biodiversity) services (Fang *et al.*, 2023). NBS are designed to maximize UES in urban environments. By preserving or restoring natural systems within cities, NBS enhance the delivery of these essential services (Fang *et al.*, 2024). In terms of GI, it provides the physical network, such as GFs, wetlands, and parks that supports the implementation of NBS, enabling cities to deliver ecosystem services more efficiently (Dorst *et al.*, 2019). All three approaches are multifunctional, involve interdisciplinary research, and rely on participatory governance, making them crucial to sustainable urban planning (Figure 1.1).

NBS plays a vital role in managing urban water pollution within sustainable urban drainage systems (SuDS). Urban water management (UWM) involves the planning, development, and oversight of water infrastructure, including water supply, wastewater treatment, stormwater drainage, and conservation measures. It helps to ensure access to safe and sufficient clean water and sanitation in urban areas.

Table 1.1 illustrates a range of NBS applications in water management, showing their adaptability across diverse environmental contexts. These solutions contribute to improving water quality, mitigating flood risks, and fostering biodiversity, demonstrating their capacity to address multiple urban water management challenges effectively.

Based on the extensive literature that highlights the significance of NBS, this book explores the fundamental role of NBS in achieving sustainable urban development through five distinct parts, each covering a critical aspect of NBS applications. This introductory chapter outlines the increasing

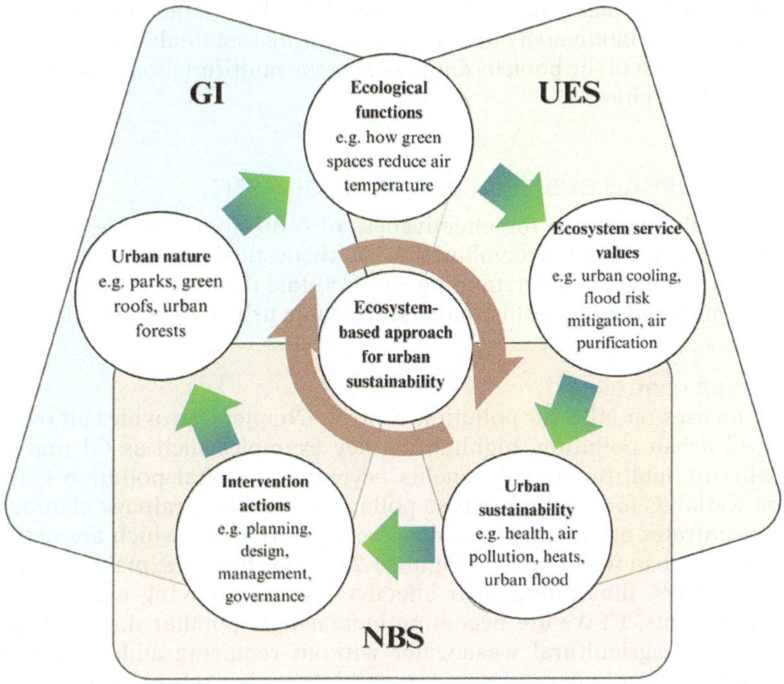

Figure 1.1 Correlation between urban ecosystem services (UES), green infrastructure (GI) and nature-based solutions (NBS) for urban sustainable development (Fang *et al.*, 2023).

Table 1.1 Nature-based treatment technologies and their applications in water management.

No.	Technology	Type of Water	Location	Issues Addressed	References
1	Hybrid aerated vertical/horizontal treatment wetland	Raw wastewater	France	Online control of aeration for nitrogen removal	Miyazaki *et al.* (2024)
2	Aerated horizontal subsurface treatment wetland	Primary treated municipal wastewater	Brazil	Maximize nitrification process due to the high presence of dissolved oxygen available; enhance the removal efficiency of organic matter and nutrients	Stopa *et al.* (2022)
3	GF	Rainwater	Germany	Increase rainwater retention time, evaporative cooling, and biodiversity; reduce flood risk from heavy rain events	Richter and Dickhaut (2023)
4	GF	Rainwater	Italy	Richter and Dickhaut (2023)	Piro *et al.* (2018)
5	CWs	Urban runoff & sewer overflow	Italy	Prevent flood, improve water quality, and increase biodiversity	Rizzo *et al.* (2018)

importance of NBS in urban planning. It details how NBS help cities manage challenges such as pollution, water scarcity, and biodiversity loss while promoting sustainable development. This chapter sets the foundation for the rest of the book by emphasizing the multifunctionality and cost-effectiveness of NBS in creating resilient cities.

1.3 ROLE OF NBS IN URBAN SUSTAINABLE DEVELOPMENT

All chapters of this book emphasize the effectiveness of NBS in addressing a wide array of urban sustainability challenges. Through a combination of theoretical explorations and practical case studies, this book aims to enhance understanding and facilitate the implementation of NBS, ultimately promoting the development of more resilient and sustainable urban environments.

1.3.1 NBS for pollution control

Part 2 of this book focuses on NBS for pollution control. Chapter 2 provides an overview of various NBS used to control urban pollution, highlighting key examples such as GI that mitigate air and water pollution, offering multifunctional benefits beyond traditional pollution control techniques. Floating treatment wetlands for treating diffuse pollution in urban drainage channels are discussed in Chapter 3. It concentrates on floating treatment wetlands (FTWs), which are natural systems that remove pollutants from urban water bodies (Figure 1.2a). The chapter explains the design, operation, and maintenance of FTWs, illustrating their effectiveness in removing organics, nutrients, heavy metals, and micropollutants. FTWs are becoming increasingly popular due to their ability to treat domestic, industrial, and agricultural wastewater without requiring additional land. In addition, Chapter 4 presents CWs as a potential solution for antibiotic removal from wastewater. It explores the mechanisms of antibiotic degradation in CWs and discusses the challenges and future perspectives for scaling these systems for widespread use. Continuing the theme of contamination control, Chapter 5 examines biological filtration techniques to mitigate methane emissions from solid waste and wastewater treatment. The chapter highlights the role of methanotrophic bacteria in converting methane into less harmful carbon dioxide which contributes to greenhouse gas reduction and offers insights into the environmental factors influencing methanotrophic activities.

1.3.2 Re-naturing cities through water-centric development

In Part 3 of this book on re-naturing cities, Chapter 6 presents regenerative living walls as an innovative solution for addressing urban sanitation and wastewater management issues, particularly in developing countries. These systems provide a low-cost, space-efficient solution for treating wastewater while contributing to climate resilience and improving urban biodiversity. Similarly, Chapter 7 discusses the potential of GFs as NBS that not only treat wastewater but also enhance urban green spaces (Figure 1.2b). The chapter covers the fundamentals of GF systems, their benefits for pollutant removal, and the technical considerations for implementing them in densely populated cities like those in Vietnam.

1.3.3 Integration of NBS into urban biorefineries

Part 4 focuses on the integration of NBS into urban biorefineries and material flows. A case study from Indonesia is presented in Chapter 11. It explores the integration of aquaculture into urban areas for water purification and food production. This chapter also shows the socio-economic and environmental benefits of urban aquaculture systems. Following this, Chapter 12 analyzes the successful implementation of NBS in solid waste management, highlighting the city's innovative

Figure 1.2 A module of (a) Floating Treatment Wetland (FTW) for treating diffuse pollutants in a canal and (b) Green roofs (GRs) for domestic wastewater treatment

practices in using vermicomposting and landfill gas recovery. The chapter emphasizes the potential for expanding green technologies to enhance sustainability. Chapter 13 evaluates innovative technologies for converting agricultural by-products into bio-fertilizers in Vietnam, demonstrating how these fertilizers reduce reliance on inorganic inputs and support sustainable agriculture.

1.3.4 Sustainability assessment of urban NBS

Part 5 of this book focuses on the sustainability assessment of urban NBS. It provides a comprehensive framework for assessing the environmental impacts of CWs in agriculture and integrated multi-trophic aquaculture (IMTA) systems through life cycle assessment (LCA). This part discusses the challenges of urban water management and outlines the importance of sustainable, low-cost technologies in urban environments. It offers practical case studies and trends for integrating wetlands and aquaculture systems into urban wastewater solutions as part of the broader NBS strategy.

1.4 FROM SUSTAINABLE TO REGENERATIVE ARCHITECTURE AND URBAN DESIGN

This book illustrates well that in the past decades, sustainability aspects have received ample attention toward design, urban planning, and architect profession. In the realm of environmental consciousness, the term sustainable design is becoming broadened toward regenerative design (Reed, 2007). The future actually demands to explore sustainability, and consider the whole ecosystem, and how the infrastructure we build (and our houses and buildings in particular) can be integrated in the ecosphere we live in (Figure 1.3). Shifting from a sustainability lens to a regenerative one means that architects need to question how we can design structures that not only use resources, but also restore them.

The problem with sustainable design is that it ignores the crisis that mankind has created in the natural environment. To compensate for this existing damage, we need to push beyond a 'neutral' or sustainable approach and instead target a positive impact on our surroundings (Table 1.2). Regenerative architecture engages the natural world as the medium for and generator of architecture. Living systems on the site become the building blocks of the structure built in harmony with the overall ecosystem. In contrast to sustainably designed buildings, regenerative buildings are designed and operated to reverse ecological damage and have a net-positive impact on the natural environment.

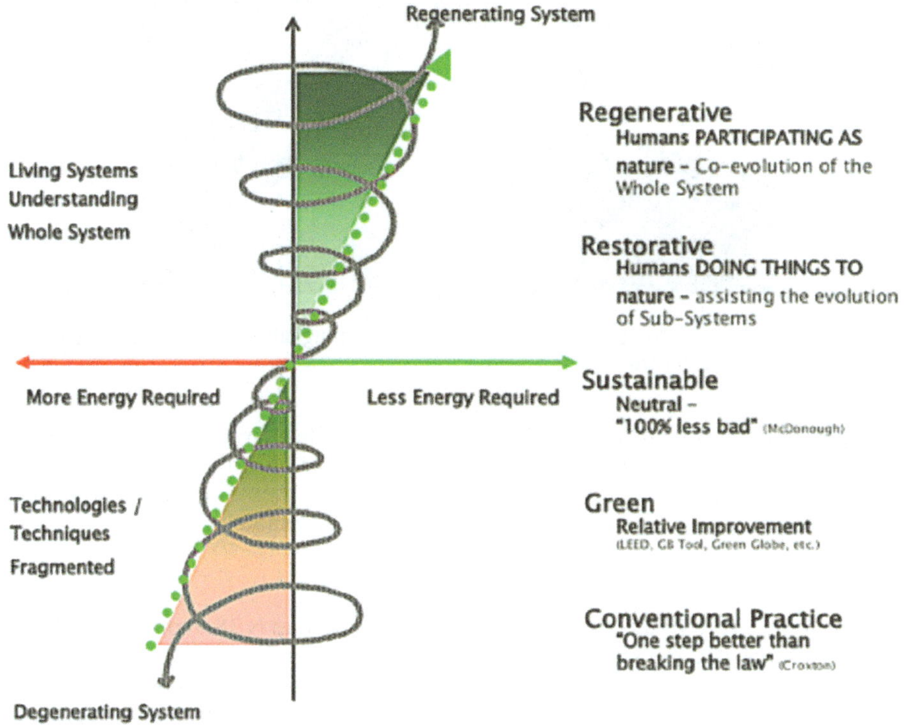

Figure 1.3 Trajectory of environmentally responsible design (from Reed, 2007).

Table 1.2 Differences between sustainable design and regenerative design.

Sustainable design	Regenerative design
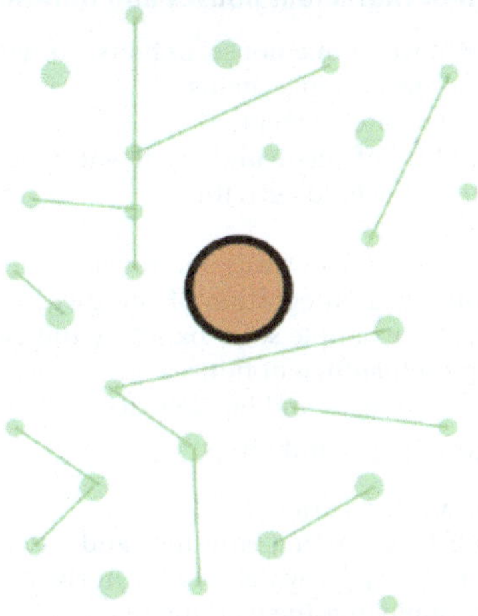	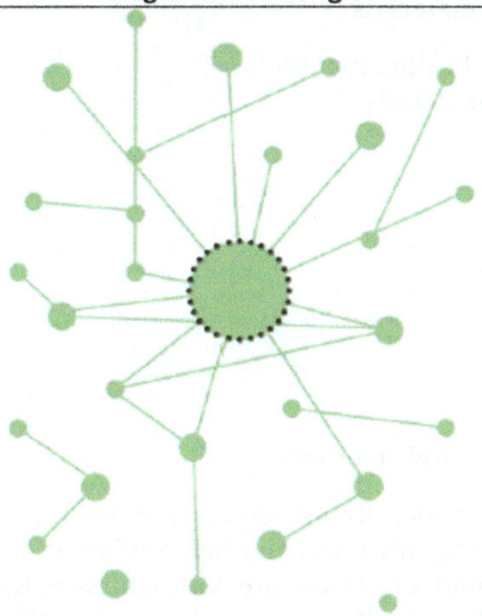
Creating a clear site boundary and reducing impact within it.	Removing the assumption of a site boundary and creating a positive impact to the surrounding environment
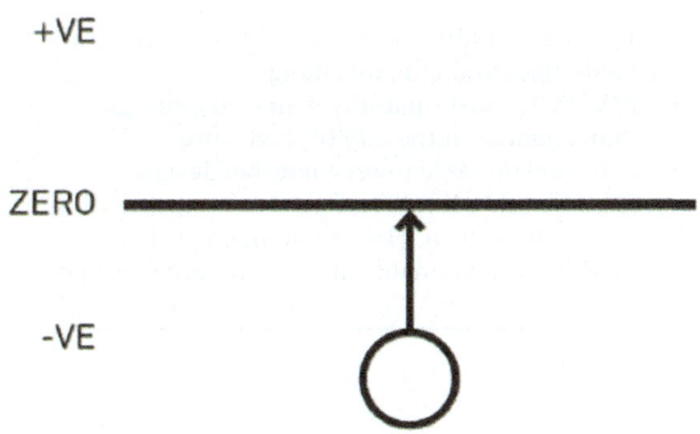	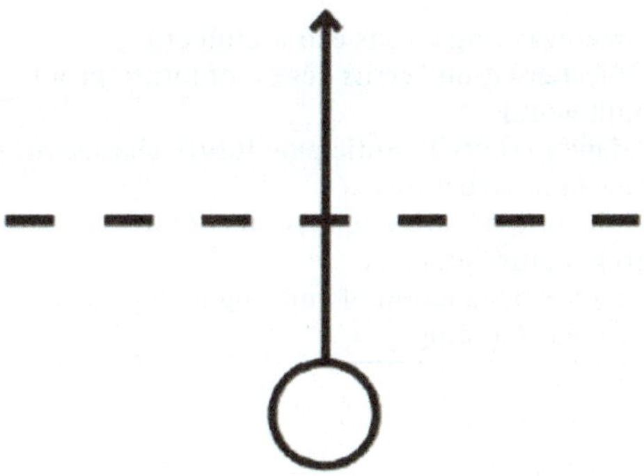
Reducing the negative impacts as much as possible	Designing beyond "less negative", aiming at positive impact

The built world exists within social, environmental, and business networks that are changing at a rate that traditional architecture cannot support anymore. Regenerative design demands architecture to be an extension of the place, the site, the flora and fauna, and the ecosystem (Venter *et al.*, 2019). Buildings thus become a part of a larger system, helping to produce and share resources like clean water, energy, and food, next to the traditional functions architecture provides (Table 1.3).

Table 1.3 Possible contributions of environmental technologies for pollution prevention and resource recovery that support the development of sustainable design and regenerative architecture.

Living materials	**Redesign of individual houses and buildings**
• Self-healing materials • Biomaterials	• Reuse of water in a house, in house or within garden • On-site sanitation systems • Separate water systems • Integration of waste and wastewater treatment within household activities • Wetroofs • Green walls/algal walls/bio-reactive façades • Architectural integration of energy generation in a house, for example solar panels, wind energy, … • Energy self-sufficient houses • Autarkic houses and houseboats
Interior architecture	**Redesign villages and cities**
• Integration living devices in interior design, for example microbial fuel cell lamp • Biomimicry to imitate nature for air-cleaning building walls, water-purifying structures, or carbon-capturing architecture.	• Stormwater treatment • Retrofitting existing buildings and city districts • NBS in a city for water quality in rivers and canals, for example floating wetlands • NBS for coastal protection, for example mangroves • Urban planning for biodiversity • Urban farming
Sustainability of the architecture industry	**City of the future**
• Greenwashing versus eco-architecture • Object creation versus design of future proof built world • Adaptive reuse to anticipate future change in new built structures • Reduction of carbon emissions of the architecture industry • Greener production of building materials • Circular building	• Resilient buildings that can cope with natural challenges and climate change • SWITCH: Sustainability water and energy management in the city of the future • From sustainable to regenerative design: sustainability limits resource use; regeneration replenishes them. How to implement this? • Scaling of sustainable and regenerative design

REFERENCES

Alikhani S., Nummi P. and Ojala A. (2021). Urban wetlands: a review on ecological and cultural values. *Water*, **13**(22), 3301, https://doi.org/10.3390/w13223301

Bai X., Surveyer A., Elmqvist T., Gatzweiler F. W., Güneralp B., Parnell S., Prieur-Richard A.-H., Shrivastava P., Siri J. G., Stafford-Smith M., Toussaint J.-P. and Webb R. (2016). Defining and advancing a systems approach for sustainable cities. *Curr. Opin. Environ. Sustain*, **23**, 69–78, https://doi.org/10.1016/j.cosust.2016.11.010

Bai X., Dawson R. J., Ürge-Vorsatz D., Delgado G. C., Barau A. S., Dhakal S., Dodman D., Leonardsen L., Masson-Delmotte V., Roberts D.C., Schultz S. (2018). Six research priorities for cities and climate change. *Nature*, ***555***(7694), 23–25, https://doi.org/10.1038/d41586-018-02409-z

Dorst H., Jagt A., Raven R. and Runhaar H. (2019). Urban greening through nature-based solutions – Key characteristics of an emerging concept. *Sustainable Cities and Society*, **49**, 101620, https://doi.org/10.1016/j.scs.2019.101620

Fang X., Li J. and Ma Q. (2023). Integrating green infrastructure, ecosystem services and nature-based solutions for urban sustainability: a comprehensive literature review. *Sustainable Cities and Society*, **98**, 104843, https://doi.org/10.1016/j.scs.2023.104843

Fang X., Li J., Ma Q., Zhou R. and Du S. (2024). A quantitative review of nature-based solutions for urban sustainability (2016–2022): from science to implementation. *Science of the Total Environment*, **927**, 172219, https://doi.org/10.1016/j.scitotenv.2024.172219

Fink H. (2016). Human-nature for climate action: nature-based solutions for urban sustainability. *Sustainability*, **8**(3), 254, https://doi.org/10.3390/su8030254

Kabisch N., Frantzeskaki N. and Hansen R. (2022). Principles for urban nature-based solutions. *Ambio*, **51**(6), 1388–1401.

Miyazaki C. K., Morvannou A., Petitjean A., Nivala J. and Molle P. (2024). Hydraulic characterization of a hybrid aerated vertical and horizontal treatment wetland. *Ecological Engineering*, **206**, 107301, https://doi.org/10.1016/j.ecoleng.2024.107301

Piro P., Carbone M., De Simone M., Maiolo M., Bevilacqua P. and Arcuri N. (2018). Energy and hydraulic performance of a vegetated roof in sub-Mediterranean climate. *Sustainability*, **10**(10), 3473, https://doi.org/10.3390/su10103473

Reed B. (2007). Shifting from 'sustainability' to regeneration. *Building Research and Information*, **35**(6), 674–680, https://doi.org/10.1080/09613210701475753

Richter M. and Dickhaut W. (2023). Long-term performance of blue-green roof systems–results of a building-scale monitoring study in Hamburg, Germany. *Water*, **15**(15), 2806, https://doi.org/10.3390/w15152806

Rizzo A., Bresciani R., Masi F., Boano F., Revelli R. and Ridolfi L. (2018). Flood reduction as an ecosystem service of constructed wetlands for combined sewer overflow. *Journal of Hydrology*, **560**, 150–159, https://doi.org/10.1016/j.jhydrol.2018.03.020

Stopa J. M., Benassi R. F., Bueno F., & Coelho R. and G L. H. (2022). Horizontal subsurface flow constructed wetlands as post-treatment of aerated pond effluent. *International Journal of Environmental Science and Technology*, **19**(5), 4447–4458, https://doi.org/10.1007/s13762-021-03418-8

Venter A., Marais L. and Morgan H. (2019). Informal settlement upgrading in South Africa: a preliminary regenerative perspective. *Sustainability*, **11**, 2685. https://doi.org/10.3390/su11092685

Yue H., He C., Huang Q., Yin D. and Bryan B. (2020). Stronger policy required to substantially reduce deaths from PM2.5 pollution in China. *Nature Communications*, **11**(1), 1462, https://doi.org/10.1038/s41467-020-15319-4

Part 2

Nature-based solutions for pollution control

doi: 10.2166/9781789065015_0015

Chapter 2

Nature-based solutions in water management: pathways to resource recovery

Harini Harish[1] and Veeriah Jegatheesan[1]*

Royal Melbourne Institute of Technology (RMIT), Department of Chemical and Environmental Engineering, Melbourne, Australia
**Corresponding author: jega.jegatheesan@rmit.edu.au*

ABSTRACT

Natural resources extracted or generated during the treatment process by nature-based solutions (NBS) offer valuable raw materials. If these resources can be reclaimed from NBS after treatment and reintegrated into the environment, they pave the way for a circular economy and sustainable resource management, reducing stress on global resources. The strength of natural systems lies in their ability to connect various environmental factors, creating a closed-loop process that benefits the environment. By limiting the consumption and wastage of non-renewable resources such as water, carbon, food, and soil nutrients, NBS can help stabilize ecosystems. This chapter explores the role of NBS in resource recovery, covering different types of NBS, their contributions to recovery, and the methods and technologies used to extract resources from the waste they generate. These practices can guide the transition toward circularity in urban areas. By adopting NBS for treatment and resource recovery, cities and rural regions can foster economic growth while enhancing quality of life. Integrating NBS into resource management promotes more resilient, efficient systems, contributing to long-term sustainability and improved living conditions across diverse regions.

Keywords: Bioenergy production, circular economy, constructed wetlands, environmental sustainability, green infrastructure, nature-based solutions, nutrient recycling, resource recovery, sustainable water management, wastewater treatment

2.1 INTRODUCTION

Nature-based solutions (NBS) for water sustainability encompass a wide array of strategies that leverage natural processes to enhance water environments and recover valuable resources. Resource recovery can be seamlessly integrated into NBS, allowing for the sustainable reuse, recycling, or regeneration of resources. This approach offers a sustainable way of managing resources while protecting the environment for the benefit of society. NBS can be implemented at micro, meso, and macro levels, providing a range of environmental services (Langergraber *et al.*, 2020). By harnessing NBS in modern urban lifestyles and combining them with resource recovery, the environmental and economic benefits can be multiplied, fostering more sustainable resource management and improved ecosystem health.

© 2025 IWAP. This is an Open Access book chapter distributed under the terms of the Creative Commons Attribution License (CC BY-NC-ND 4.0) which permits copying and redistribution for non-commercial purposes with no derivatives, provided the work is properly cited (https://creativecommons.org/licenses/by-nc-nd/4.0/). The chapter is from the book *Nature-Based Solutions for Urban Sustainability*, Piet N.L. Lens and Xuan-Thanh Bui (Editors).

Urban areas store and generate resources from development activities, and urban treatment systems act as resource sinks from which these resources can be reclaimed and reused, thereby reducing the burden on natural resources (Ellen Mac Arthur Foundation, 2017). Utilizing sustainability management solutions serves as a highly effective resource recovery strategy. The following sections provide a detailed overview of the components of NBS, the categories of recoverable resources, and the methods used in the recovery process, all of which contribute to urban sustainability. They focus on the application of NBS in urban water management, as urban environments require greater attention compared to other settings.

2.2 COMPONENTS OF NBS USED IN WATER MANAGEMENT

NBS serve two primary purposes: conservation and restoration of natural resources. These solutions can help eliminate pollutants in the environment and improve local habitats. Key components that contribute to the effectiveness of NBS include various types of vegetation, naturally available materials to enhance system efficiency, naturally occurring or induced microbiomes, and well-designed systems (Figure 2.1). The environmental beneficiaries of NBS – such as past and present habitats or communities – benefit from the resource recovery opportunities these systems provide. The following discussion will explore the functions of these components and how they contribute to efficient resource recovery.

2.2.1 Contribution from vegetation

The selection of plants is crucial for the efficiency of NBS such as green roofs and walls (GRW), constructed wetlands, swales, and rain gardens (Figure 2.2). For example, plants must be resilient to both drought conditions and heavy rainfall (Raimondo *et al.*, 2015). Therefore, considering key parameters during plant selection is essential. Important factors include growth characteristics such as vegetation height and coverage area, as well as sunlight requirements, cold tolerance, irrigation

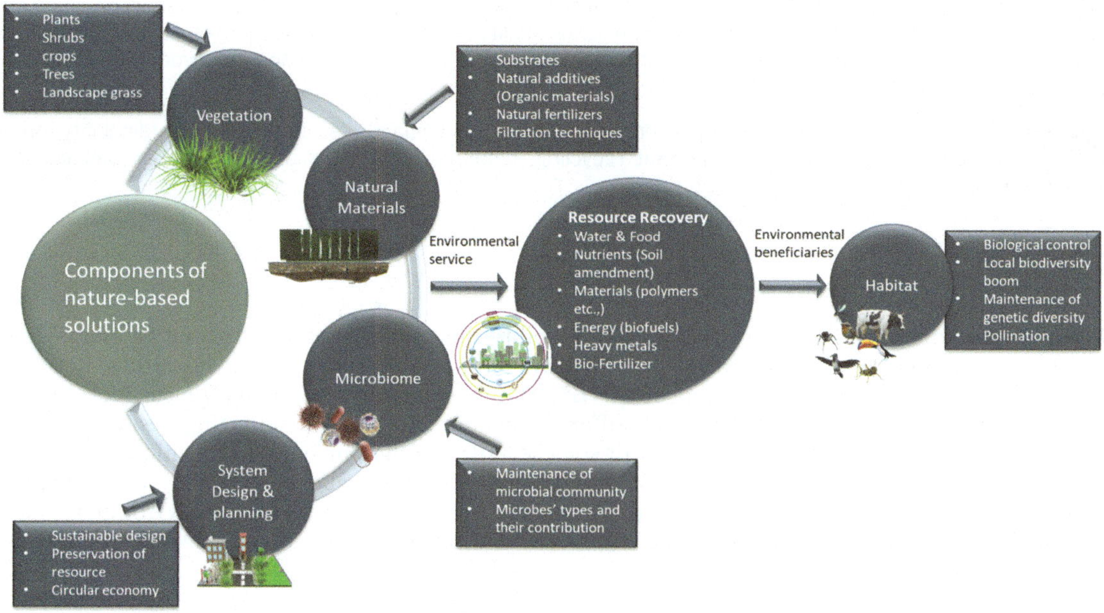

Figure 2.1 Schematic representation of various components of NBS and environmental services delivered by NBS.

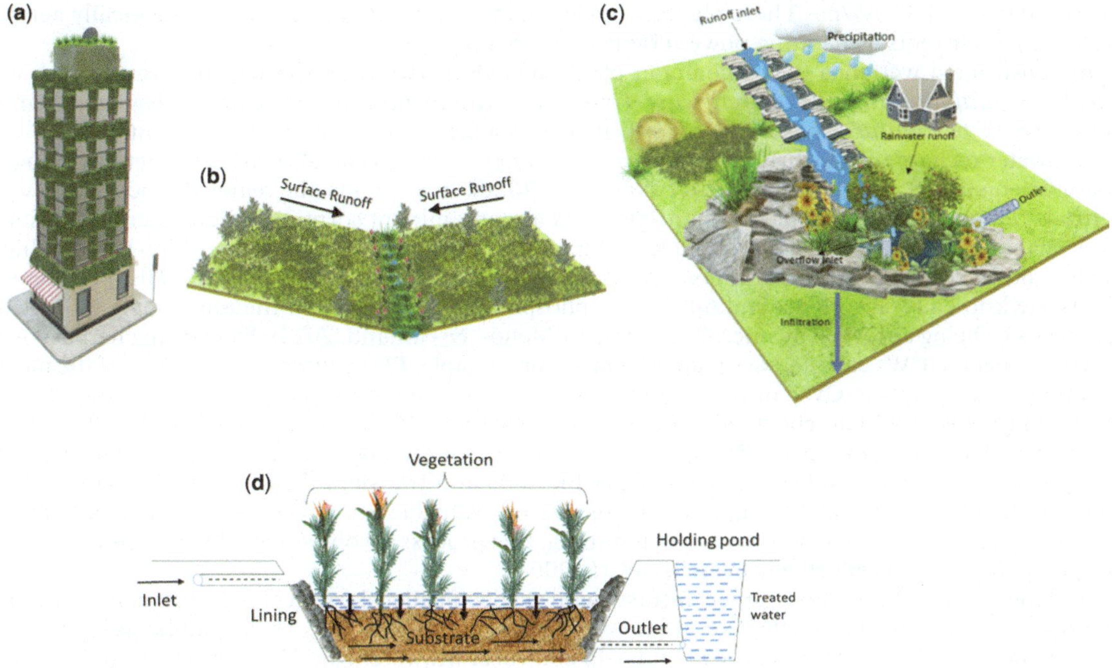

Figure 2.2 Examples of NBS: (a) green roofs and walls, (b) swales, (c) rain gardens, and (d) constructed wetlands.

needs, nutrient absorption capabilities, and the ability to survive in extreme soil conditions (e.g., high metal concentrations) (Gong *et al.*, 2021). Other factors include low stomatal resistance, leaf area index (LAI) (Cascone *et al.*, 2019; Seyedabadi *et al.*, 2021; Skuza *et al.*, 2022; Whittinghill *et al.*, 2014; Zhou *et al.*, 2018), barrier dimensions, leaf reflectivity, humidity levels, evapotranspiration, wind speed, and direction – all of which play a significant role in NBS performance (Morakinyo *et al.*, 2016; Poë *et al.*, 2015; Weerakkody *et al.*, 2017; Zhang *et al.*, 2021).

2.2.1.1 Sedum and herbaceous plants

The most common plant types seen on building roofs include sedum herbs (succulent leaves and stems), and herbaceous perennial plants valued for their high photosynthesis rate, wide availability range, and low cost (Durhman *et al.*, 2006; Seyedabadi *et al.*, 2021). A few sedum varieties and herbaceous perennial plant species are *Callisia repens, Portulaca grandiflora, Plectranthus prostrates, Sedum lineare*, and *Danthonia spicata, Coreopsis rosa, Heuchera Americana 'Raspberry Regal'*, which are used in the GRW vegetation (Hu *et al.*, 2022; Lundholm *et al.*, 2009). When sedum varieties are mixed with herbaceous plants for rainwater treatment, the accumulation of total nitrogen (TN), ammonium nitrogen (NH_4^+-N), and total phosphorus (TP) is significantly higher compared to systems with a single species. About 90% of nutrient accumulation is observed in sedum species, and the survival rate of herbaceous plants increases during drought or low-rainfall conditions (Brandão *et al.*, 2017; Gong *et al.*, 2021; Nagase & Dunnett, 2010).

One of the primary goals of NBS is to mitigate climate change, making carbon reduction a key factor in selecting vegetation for urban areas. Research has shown that sedum plants when planted on roofs and walls, have a high potential for carbon absorption (Klein *et al.*, 2017). The sedum species *Sedum hybridum* has also been identified as an effective driver in plant microbial fuel cells (PMFC), achieving a power density of 92 $\mu W/m^2$. These plants provide organic matter that feeds electrochemically active bacteria in their roots, generating power (Tapia *et al.*, 2017).

In constructed wetlands (CW), common plants include herbaceous species and grasses such as *Baccharis latifolia, Bromus catharticus, Hypericum humboldtiana, Plantago orbignyana,* and *Potamogeton pectinatus.* These plants are primarily used for water and soil treatment, as well as groundwater recharge. They thrive in highly contaminated soils and can absorb heavy metals through phytoremediation (Leguizamo *et al.*, 2017; Nan *et al.*, 2020). In wetlands, the removal process involves natural attenuation and multiple mechanisms, such as contaminant sorption, where microorganisms grow on the plant surface and assist in biodegradation. Plant exudates from decomposition and biofilms further aid in the removal of polar contaminants (Nivala *et al.*, 2013; Sharif *et al.*, 2014). Wetland plants are known to absorb around 60–70% of phosphorus in water, with the remainder settling in sediments or being absorbed by microbial biomass (Menon & Holland, 2013). By planting herbaceous or woody species, CWs can increase plant biomass. For example, *Phragmites australis*, one of the most commonly used plants in CWs, produces 19 ($\pm$ 13) tonnes of biomass/ha/year (Avellán & Gremillion, 2019), and helps eliminate chemicals of emerging concerns (CECs) as identified by the EU Water Framework Directive (EU, 2013; Zhang *et al.*, 2012). One of the major aspects of phytoremediation is phytoextraction which takes up metals from the soils and transfers them into their shoots. This method can offer a sustainable approach to recovering valuable resources, especially metals from contaminated soils. It is also known as phytomining, this process involves several key steps to extract and reuse these metals effectively (Vassilev *et al.*, 2004).

A study of New York City's green infrastructure, specifically bioswales, found that herbaceous plants and grasses are more effective at removing nutrients from water. These plants also promote microbiome growth for water treatment and help control stormwater runoff. For instance, the species *Nepeta* $\times$ *faassenii* has shown maximum conductance and high TN storage, which can be recovered (Brodsky *et al.*, 2019).

2.2.1.2 Landscaping grass and shrubs

Grass planting requires minimal effort but is highly effective in mitigating non-point source pollution. Various native grass species, readily available depending on geographic location, contribute to this. A case study in the UK found that species such as *Anthoxanthum odoratum, Festuca ovina, Koeleria macrantha,* and *Trisetum flavescens,* which are taller and have larger diameters with more root biomass, effectively reduce surface water runoff (Nagase & Dunnett, 2012). Similarly, in Taiwan, native C4 grass species, known for their efficient photosynthesis, have been used to control erosion in a reservoir. These grasses are categorized as M-type, meaning they are thick and highly fibrous (Lee *et al.*, 2022), supporting the restoration and conservation goals of NBS.

In ornamental landscaping, native grasses are commonly selected. A study comparing herbaceous grasses, deciduous shrubs, broad-leaf evergreen shrubs (*Buxus sempervirens*), and narrow-leaf evergreen shrubs (*Pinus mugo, Juniperus chinensis*) found that grasses and shrubs both contribute to carbon sequestration. Shrubs, however, are more effective, sequestering 65.27–66.53 kg of carbon/m^2, compared to 55.59–61.33 kg of carbon/m^2 for grasses. Additionally, in-ground landscaping systems were found to sequester more carbon than green roof systems (Whittinghill *et al.*, 2014). Grasses in tropical humid regions with clayey soils also demonstrate a high carbon storage potential (Mathew *et al.*, 2017). A study in the USA on the *Poaceae, Napiergrass,* and *Cyperaceae* families showed an annual silica uptake of 2.68%, making these species promising for biorefinery applications (Bichot *et al.*, 2018; Xu *et al.*, 2020).

Ornamental plants such as *Canna indica, Pontederia cordata, Thalia dealbata* Fraser ex Roscoe, *Acorus calamus, Juncus effusus, Iris laevigata, Mentha aquatica, Oenanthe javanica* (Blume) DC, *Calthapalustris,* and *Sparganium erectum* are commonly found in floating wetlands and are effective in soil and water phytoremediation. These species have a nitrogen uptake potential of 15 g/m^2 and exhibit 100% survivability (Barco & Borin, 2020). In California, the shrub species *Rhaphiolepis umbellata* was planted as part of an 8-year stormwater management system. It achieved nutrient uptake efficiencies of 99.1% for nitrogen, 99.5% for phosphate, and 99.4% for total organic carbon, highlighting its effectiveness after rainfall events in urban areas (Xiao *et al.*, 2017).

2.2.1.3 Urban agriculture and forests

Green infrastructure, including urban agriculture and forestry, is developed to provide benefits for both indoor and outdoor environments (Russo *et al.*, 2017). A study in Bolzano (Italy) found that urban trees remove 2.42 tonnes of air pollutants per year, with the highest removal of ozone (1.2 tonnes/year) and the lowest removal of carbon monoxide (0.03 tonnes/year). The study also observed a 2°C decrease in temperature during the summer, highlighting the benefits of urban green infrastructure for cooling (Russo *et al.*, 2016). This encourages the use of urban streetscapes and green spaces.

In Slovenia, an evapo-transpirative willow system (*Salix alba* varieties 'V 093', 'V 051', and 'V 160') was found to uptake primary treated wastewater while producing 34–38 tonnes of dry mass/ha. These trees were highly efficient in taking up nitrogen (up to 48%) and phosphorus (up to 45%), along with trace amounts of heavy metals like iron, manganese, chromium, and zinc within their woody biomass. When the harvested willows are converted into pellets and chips, they can be used in heating or used as insulation material during construction thus reducing household energy consumption (Istenič & Božič, 2021; Topić Božič *et al.*, 2024). This demonstrates their potential for efficiently utilizing urban resources.

Urban spaces, such as allotments, parks, community and private gardens, urban farms, forests, and even indoor farming, can be used to grow food crops like tomatoes, potatoes, cucumbers, and lettuce using techniques like hydroponics or soil-based systems (Payen *et al.*, 2022). The aim of urban green systems goes beyond food production; they are intended to bridge the gap between society and urban sustainability. A study by Payen *et al.* (2022) found that crops such as legumes, vegetables, and tubers grown in urban spaces can produce four times more than conventional soil-based agriculture. Urban agriculture also sequesters between 100,000 and 170,000 tonnes of nitrogen, promoting a more sustainable environment. It is estimated that urban agriculture has the potential to produce 100–180 million tonnes of food annually, with countries such as China, Japan, Germany, and the United States being top producers (Clinton *et al.*, 2018).

Cereal crops, due to their high carbon sequestration capacity, should be promoted in urban agriculture. Legumes also contribute significantly, with a high biomass production and a root-to-carbon stock ratio of 3.29 ($\pm$ 0.63) tonnes/ha/annum and 0.86 ($\pm$ 0.04), respectively. By integrating urban agriculture with green infrastructure, various ecosystem services, including climate regulation, air quality improvement, biological control, and food production, can be provided. Additionally, edible vegetation in urban areas offers fuel and fiber resources (Evans *et al.*, 2022).

Urban agriculture also plays a role in regulating global climate through carbon sequestration. A case study in Indonesia found that food grain crops were more efficient at sequestering carbon than vegetable crops. Proper management, considering economic and social values, could yield better results. The process involves calculating carbon stocks (C) from biomass, which are then used to estimate absorbed CO_2 and finally, the amount of oxygen (O_2) released (Hastuti *et al.*, 2022):

$$\text{Sequestered}\,CO_2 = C \times \left(\text{Molecular weight of}\,O_2/\text{Molecular weight of}\,CO_2\right) \tag{2.1}$$

One significant benefit of urban agriculture is its economic contribution. Through carbon sequestration, both farming and non-farming lands were found to release oxygen worth approximately 10,000 USD/ha (Hastuti *et al.*, 2022). Greater emphasis should be placed on urban agriculture in gray infrastructure – areas covered by artificial, impermeable materials – where food production is still possible. The agronomic suitability of these spaces is under-researched but holds potential for sustainable urban development (Evans *et al.*, 2022; Walsh *et al.*, 2022). With better policies focused on environmental and economic benefits, urban agriculture can elevate urban spatial management.

2.2.2 Natural materials as support matrices

Natural materials such as substrates and filters in NBS can store absorbed nutrients (Buffam & Mitchell, 2015). Additives like activated carbon and organic materials enhance system performance by acting as media for nutrient adsorption (Chetri *et al.*, 2022). Using natural media fertilizers improve plant growth and soil quality. Rather than relying on synthetic nutrients, recycling resources recovered from natural systems can help preserve these resources. NBS solutions like permeable pavements and soil aquifer treatment (SAT) act as natural filters that recharge groundwater aquifers, increasing the capacity of water systems (Sharma & Kennedy, 2017).

However, NBS does have a few disadvantages, including the need for larger infiltration areas, changes in groundwater hydrology, and the lack of sufficient pretreatment methods, which can affect water quality (Ayuso-Gabella *et al.*, 2011). A case study at the Shafdan treatment plant in Israel demonstrated that it could infiltrate 130 million cubic meters of biological secondary effluent per year. The plant removed 90% of biological oxygen demand (BOD), total suspended solids (TSS), ammonia, nitrogen, and phosphorus. With a series of open ponds for pretreatment, the plant eliminated nitrogen and achieved a recharge rate of 70–210 m/year (Juanico, 2008).

Substrates are key components in NBS such as CW, GRW, and rain gardens. They are designed to mimic natural conditions and can be modified according to geographical location. For example, in coastal regions with salty air, substrates are prepared using mineral-based materials to tolerate salinity and support plant growth (Farrell *et al.*, 2012). GRW systems are installations of vegetation layers on urban rooftops or walls, commonly referred to as living walls or vertical gardens (Bauduceau *et al.*, 2015). GRW can be retrofitted to existing buildings, making it a preferred NBS in cities with limited land availability (Manso & Castro-Gomes, 2015).

The first green roof was established in Sydney (Australia) using a timber frame design, while another was constructed in Rio de Janeiro (Brazil) using blockwork techniques (Wilkinson *et al.*, 2017). GRW offers multiple ecosystem services, including mitigating urban heat islands (Razzaghmanesh *et al.*, 2016), improving air quality (both indoor and outdoor), managing stormwater (Vijayaraghavan & Raja, 2014), enhancing biodiversity (Mayrand & Clergeau, 2018), improving energy efficiency (Sproul *et al.*, 2014), reducing noise pollution (Van Renterghem *et al.*, 2013), and supporting urban agriculture and food production (Dorr *et al.*, 2021). GRW can also contribute to resource recovery by storing water, carbon, and energy.

The potential of GRW varies based on factors such as precipitation, evapotranspiration, vegetation type, and substrate design depth (Hu *et al.*, 2022; Oberndorfer *et al.*, 2007; Paraskevopoulou *et al.*, 2021). In regions with high temperatures and arid conditions, water-retention materials such as peat or organic materials are used to improve the water-retention capacity (Stewart-Wade, 2020). Substrates less than 10 cm or exceeding 100 cm in depth can reduce aeration and hinder plant growth (De Boodt & Verdonck, 1971). However, recent technological advances have shown that substrates as shallow as 7.62 cm can support sedum plants for better functionality. This highlights the importance of designing regulated substrates and selecting appropriate vegetation to maximize resource recovery.

2.2.2.1 Stormwater retention and recovery

Substrates serve as the primary retention hub for absorbed resources, followed by the plant's roots and leaves. As substrate depth increases, so does its water-retention capacity, which is largely dependent on the types of vegetation present (Dvorak & Volder, 2010). To assess the performance of GRWs, factors such as the moisture content in the substrate and water levels in the storage layer are measured. Another important factor influencing stormwater recovery is evapotranspiration, which relates to the plant's transpiration capacity and can negatively impact rainwater recovery from the storage layer during dry seasons (Poë *et al.*, 2015; Zhang *et al.*, 2021).

By determining substrate depth and the difference between its field capacity and initial moisture content, the retention capacity of green roofs can be calculated (Zhang & Guo, 2013). During periods of low rainfall, the water content in the storage area declines, and by calculating the retention capacity of the storage layer, the overall recovery process of green roofs can be estimated. The variation in this process is also tied to the retention performance of the vegetation (Hu *et al.*, 2022). Studies on various GRW systems show retention percentages of 80–99% (Gong *et al.*, 2021; Zhang *et al.*, 2021).

In dry seasons, while plants can be self-reliant based on their retention capacity, thicker substrates–though providing higher retention–can sometimes lead to issues like roof collapse or ponding if the bulk density of the soil is too high and the hydraulic conductivity is too low. During rainfall, the stored water leads to elevated nutrient concentrations in the vegetation, reducing nutrient runoff pollution (Gong *et al.*, 2021).

In one study, green roof systems were modified into vertical flow CW to treat graywater. By increasing the substrate depth from 15 cm to 25 cm with vermiculite, the system achieved an average removal efficiency of 91% for COD and BOD, 93% for TSS and turbidity, and 87% for TN. The depth of the substrate was found to be directly proportional to the system's removal efficiency, with performance decreasing by 20% when the depth was reduced. To enhance the system's self-reliance, 40% of the treated water was recirculated into the makeshift CW (Thomaidi *et al.*, 2022). This approach supports water security, plant growth (food security), and soil resource protection.

2.2.3 Benefits of microbiome

One of the most integral components of NBS is the naturally developing microbiome community. Microbes, particularly those in soil, play a significant role in degrading pollutants, making the system more efficient (Vimal *et al.*, 2022). Beyond soil, aquatic biomass also benefits from microorganisms, which contribute about 70% of the energy recycling and nutrient supply, ensuring ecosystem services are maintained (Bar-On *et al.*, 2018).

The primary factor for healthy NBS maintenance is maintaining microbial growth at the required level. Second, selecting the appropriate microbial species for specific pollutant treatment and resource recovery is crucial. Microbes contribute to various aspects of resource recovery, including nutrient cycling (van Hullebusch *et al.*, 2021), bioremediation (Kisser *et al.*, 2020), energy production (Rathour *et al.*, 2024), water treatment (Gupta *et al.*, 2022), and recovery of valuable materials (Vítová & Mezricky, 2024).

One method to improve soil quality in areas polluted by heavy metals is phytomining, a type of phytoremediation. By intercropping plants, the microbial community is stimulated, enhancing soil quality. Using improved plant strains can also help increase element extraction through agroecological processes (Bani *et al.*, 2015). Microorganisms like algae, bacteria, and cyanobacteria have been found to stock CO_2, mitigating emissions by 23.4 ppm of CO_2, 0.11 ppm of ozone, and 0.01 ppm of nitrogen dioxide, according to a study in Ontario (Canada) (Anderson & Gough, 2020).

A case study involving urban buildings explored CO_2 mitigation using algae-based systems. In this example, photobioreactors installed on broiler exhausts captured heat and converted it into electricity, a solution found to be more cost-effective than solar energy (van Hullebusch *et al.*, 2021). Similarly, in Sweden, photobioreactors have been used to treat flue gas emissions, aiding in organic carbon uptake (Sörenson *et al.*, 2021).

Bacterial–microalgal consortia have been studied for their ability to treat textile dyes, where algae absorb pollutants due to their large surface area, and bacteria degrade the absorbed pollutants (Raza *et al.*, 2024; Samo *et al.*, 2018). This symbiotic system addresses the metabolic limitations of individual species, enhancing the overall degradation process (Sörenson *et al.*, 2021). Resources recovered through this process include renewable biofuels (Biswas *et al.*, 2021; Tighiri & Erkurt, 2019), biomaterials (polyphosphate) (Majee *et al.*, 2017), biofertilizer from landfills (Das *et al.*, 2019), dietary supplements (animal feed) (Abd El-Hack *et al.*, 2023), and CO_2 to O_2 exchange through carbon fixation (Yang *et al.*, 2017).

Combining plant biomass with microalgae can produce bio-oils through photofermentation (Li *et al.*, 2003). While energy supplements are recovered through these methods, further optimization is needed to ensure the purity of the resources obtained (Logan & Visvanathan, 2019). Research is focusing on recovering nutrients from treatment systems using photosynthetic microorganisms such as purple phototrophic bacteria, cyanobacteria, fungi, and green microalgae, which generate value-added products like biofuels, biomethane, bioethanol, carotenoids, pigments, and single-cell protein (Gupta *et al.*, 2022).

Enzymes like laccase, alkane hydroxylase (Elumalai *et al.*, 2017), and azo reductase (Sarkar *et al.*, 2017) produced by microorganisms can degrade polymeric substances and pollutants (Agrawal & Verma, 2020). By adopting new microbial electrochemical systems, nutrients like ammonium and phosphate can be recovered from agricultural waste effluent (Li *et al.*, 2020). Furthermore, wastewater containing organic and inorganic compounds (e.g. heavy metals) can be treated using phytoremediation, which relies heavily on the microbiome in the rhizosphere to efficiently remediate wastewater and recycle nutrients (Romanis *et al.*, 2021; Sood *et al.*, 2015).

2.2.4 Role of NBS system design in resource recovery

The most critical component of NBS is the design element, which ensures the system's effectiveness and sustainability. Key design considerations include source separation, resource flow analysis, technology readiness (scalability and flexibility), integrated environmental services, community engagement, operations and maintenance, and compatibility with existing urban systems. Incorporating sustainable urban drainage systems (SuDS) (Figure 2.5), water-sensitive urban design (WSUD) (Figures 2.3 and 2.4), green infrastructure for resource management, the use of microorganisms, and a focus on circularity in design can address various environmental challenges (Kisser *et al.*, 2020; Robinson *et al.*, 2003; Tsatsou *et al.*, 2023; van Hullebusch *et al.*, 2021).

Urban circularity and environmental challenges can be addressed through vertical greening systems, green roofs, urban food, and biomass production, public green spaces, rainwater and graywater

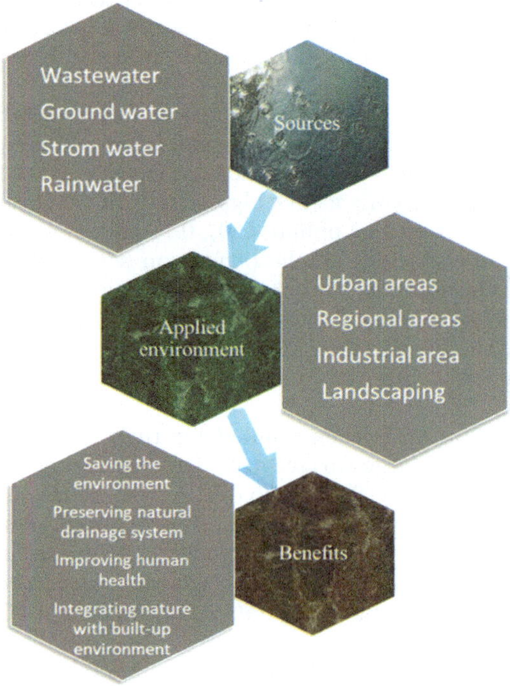

Figure 2.3 Diagrammatic representation of the elements of water-sensitive urban design.

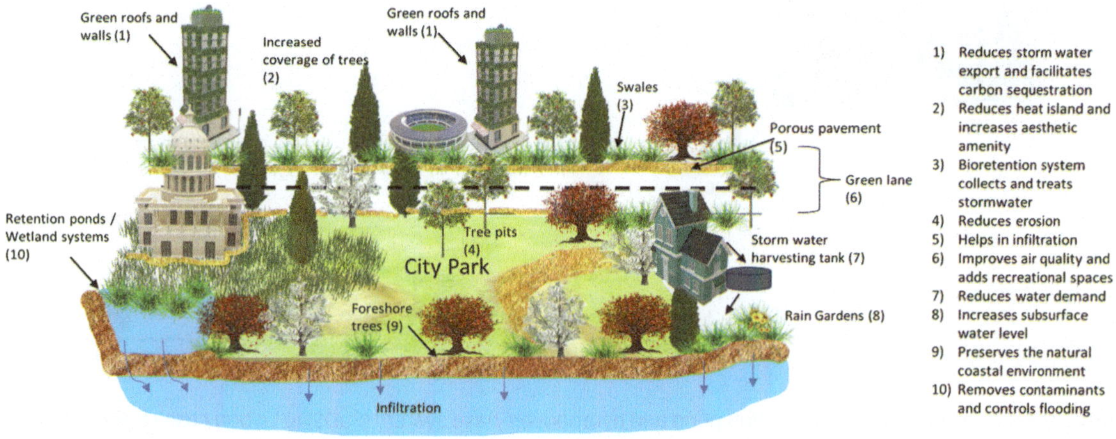

Figure 2.4 Benefits of an urban ecosystem with water-sensitive urban design.

management, soil and water bioengineering, flood mitigation, and river restoration (Almenar *et al.*, 2021; Ferreira *et al.*, 2022; Langergraber *et al.*, 2021). These solutions support the creation of a resilient and efficient resource recovery system that contributes to overall environmental well-being.

In a study conducted in a small town near Lisbon, sustainable urban drainage systems design (Figure 2.5) elements were explored for energy recovery. Retention ponds, combined with hydropower converters, produced renewable energy while also mitigating floods represented in Figure 2.6. With flow rates ranging from 50 to 420 L/s and a pond volume of 10,000 m^3, a maximum of 800 W or 210 MWh/year was generated (Ramos *et al.*, 2013). This NBS design can be integrated into existing urban street layouts, enhancing biodiversity and upgrading green spaces.

A notable example from Greener Grangetown (Cardiff, UK) illustrates the benefits of rain gardens as part of NBS for stormwater management. Around 100 rain gardens, planted with native shrubs, grasses, and trees, helped prevent flooding, reduce water pollution, and minimized sewer discharges in an aging infrastructure. Additionally, this design reduced the cost of pumping stormwater for treatment, saving energy and financial resources (Nature-based Solutions Initiative, 2018).

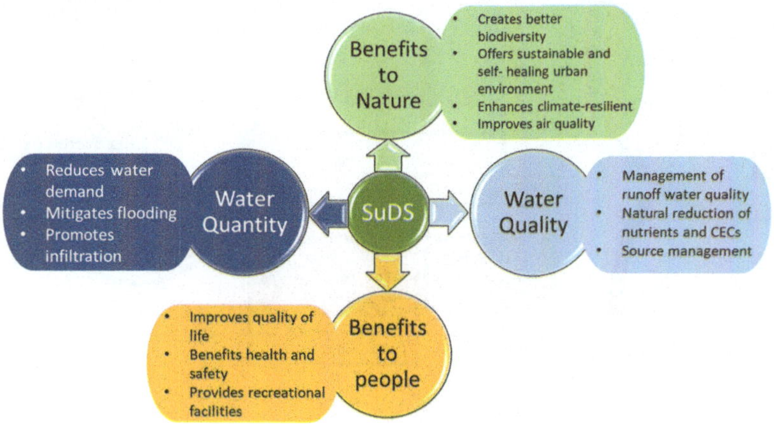

Figure 2.5 Schematic representation of pillars of sustainable urban drainage systems.

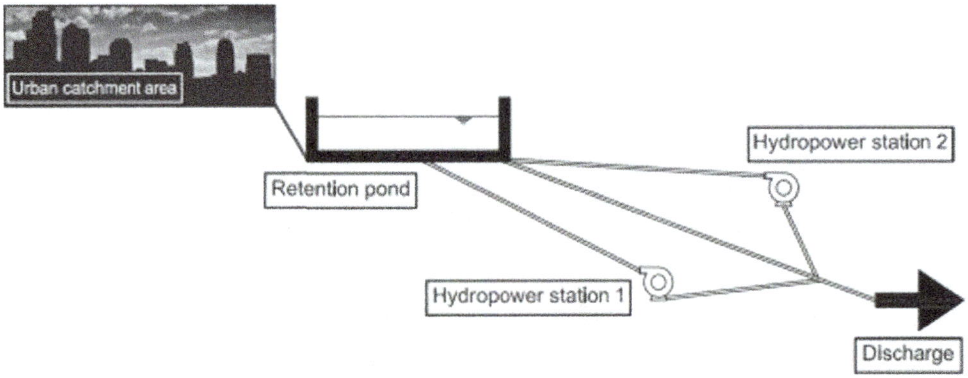

Figure 2.6 Representation of a retention pond and hydropower-base system (SuDS) in Lisbon case study (extracted from Ramos *et al.*, 2013).

In Melbourne (Australia) the Fishermen's Bend project applied WSUD principles to manage water resources sustainably. The goal was to transform an industrial region into a sustainable community, leading to a 45% reduction in freshwater consumption and a 50% decrease in wastewater discharge (Furlong *et al.*, 2019; Jegatheesan, 2020). For example, car wash facilities in Australia use 17.5 billion liters of water annually, costing AUD 75 million. Implementing WSUD in such sectors could significantly reduce water use, making the process more sustainable (Jegatheesan, 2020). If extended

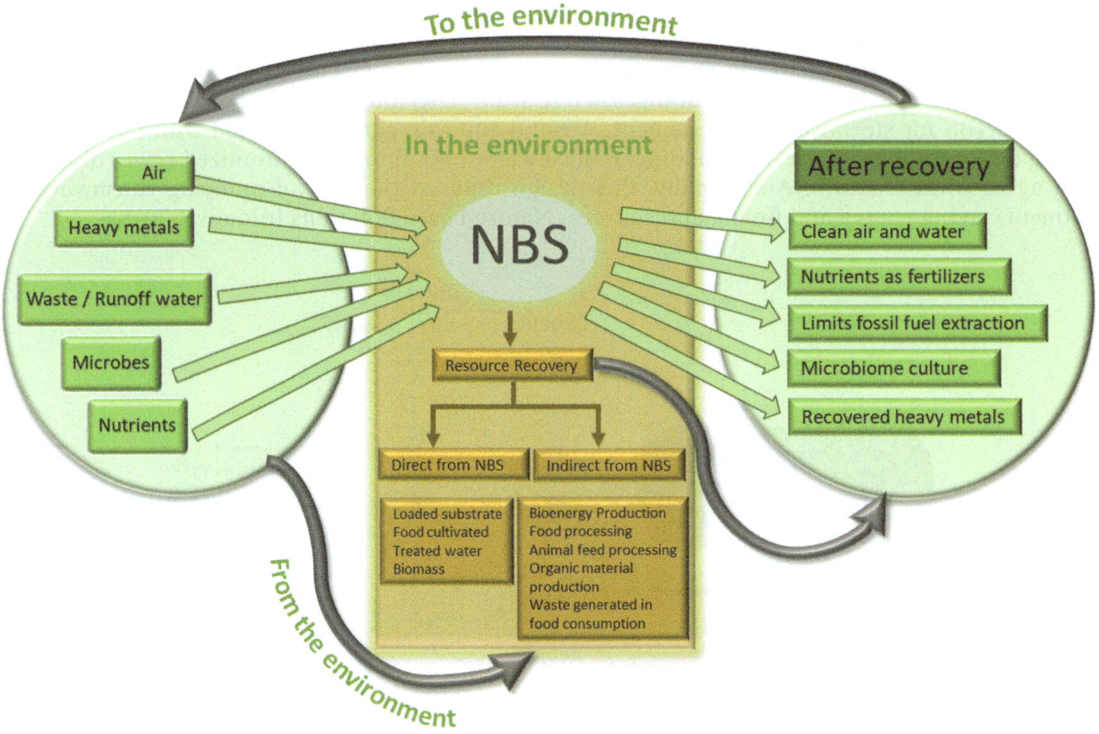

Figure 2.7 Diagrammatic representation of closing the environmental resource loop through resource recovery.

across various services, such designs could have a larger impact on society, the environment, and the economy.

WSUD can also aid in stormwater management, graywater recycling, flood mitigation as well as energy and nutrient recovery. These examples demonstrate how NBS can integrate with urban processes to recover resources and provide multiple environmental and social benefits. These components contribute to urban metabolism, a framework for quantifying how cities consume resources, generate waste, and recycle materials. By understanding urban metabolism, cities can develop strategies to enhance sustainability, reduce waste, and close the loop on resource flows. Figure 2.7 visually demonstrates how NBS addresses and treats various environmental resources. Additionally, the figure highlights the resource recovery stage, detailing the range of products that can be reintroduced into the environment. This process not only facilitates the recycling of valuable materials but also contributes to closing the environmental loop, thereby enhancing sustainability and supporting a more resilient ecosystem.

2.3 RESOURCE RECOVERY

2.3.1 Types of resources

Resources that can be recovered from NBS are broadly classified into four categories: nutrients, water, energy, and materials. Table 2.1 consolidates the types of resources that can be recovered from various NBS. Different NBS provide various recovery opportunities. For example, a CW uses 6.8% of the energy required for operations compared to an activated sludge plant and plays a significant role in maintaining the urban water cycle by providing reclaimed water for irrigation or fertigation (Muduli *et al.*, 2023; Shao *et al.*, 2013). The plant biomass from CW systems can be composted, contributing to nutrient recovery and energy production (Waly *et al.*, 2022). It has been noted that biomass from CW systems worldwide can produce 170–360 LN (liters of methane normalized to standard conditions)/kg, potentially supporting the energy needs of a village with 200 residents and leading to a 55% reduction in fuel consumption (Avellán & Gremillion, 2019). Additionally, biomass can be used as construction material and to produce composite panels or insulation material (Koop & van Leeuwen, 2017; Lisowski & Glinicki, 2023; Ryłko-Polak *et al.*, 2022). In China, it is estimated that using 20% of its fallow land could treat all generated wastewater and produce 8.2×107 GJ of bioenergy annually (Zheng *et al.*, 2014).

When floating treatment wetlands are combined with microbial fuel cells, bioenergy is generated, with peak voltage and power of 225 mV and 0.93 mW/m^2, respectively, in batch flow. Plant species with larger root systems release more oxygen, leading to higher voltage values compared to plants with smaller roots (Colares *et al.*, 2021).

Resource recovery from NBS is not always confined to the classified categories. Some NBS provide indirect benefits by improving the overall environment, which can reduce long-term maintenance and repair costs. For instance, living shorelines (nature-based coastal defences) offer multiple environmental benefits, including coastal restoration, biogenic reef development, carbon sequestration, and hazard mitigation. Countries such as Australia, the USA, and the Netherlands have significantly benefited from living shorelines (Morris *et al.*, 2018). For example, in Australia, living shoreline systems are implemented along 178 kilometers of coastline (Morris *et al.*, 2018). There are several other projects including sand nourishment, coral transplanting on artificial reefs, oyster shell bags for natural oyster recruitment, mangrove planting, and juvenile kelp transplants that benefit the coastal restoration (Morris *et al.*, 2024).

Eco-industrial parks represent innovative strategies to enhance the synergy between environmental and industrial activities. These parks integrate sustainable practices and resource efficiency to optimize environmental performance and industrial productivity. They focus on soil restoration, carbon sequestration, biodiversity conservation, water management, and eco-efficiency concerning energy use. In essence, energy produced within eco-industrial parks is used by the industries, making the parks self-resilient and promoting resource cycle optimization (Domenech *et al.*, 2019; Ehrenfeld & Gertler, 1997). Australia has also developed sustainable green energy innovations, such as smart energy benches in urban parks that allow individuals to sit, relax, and charge their devices.

Table 2.1 Possible resources that can be recovered from NBS.

Nature-based Systems	Nutrient Recovery	Water Reuse	Energy Recovery	Materials Recovery
Green roofs and walls	Yes (TN, TP, COD, BOD)	Yes (gray water reuse as irrigating GRW system)	Yes (plant biomass and bioenergy)	Yes (eco-friendly construction materials)
Urban wetlands and constructed wetlands	Yes (TN, TP, and used as biofertilizers)	Yes (irrigation or industrial reuse)	Yes (plant biomass and bioenergy)	Yes (eco-friendly construction materials)
Green infrastructure for stormwater management	Yes (TN, TP)	Yes (groundwater recharge)	Yes (indirect ways by integrating with renewable energy technologies)	Yes (pollutant trap)
Urban forests and street trees	Yes (minimal)	Yes (infiltration)	Yes (plant biomass and bioenergy)	Yes (carbon, improves air quality)
Permeable pavements	No	Yes (groundwater recharge)	No	No
Riparian buffers and stream restoration	Yes (plant uptake and biomass used as fertilizers)	Yes (infiltration improves water quality)	Yes (by reducing energy consumption for water treatment)	Yes (habitat restoration)
Algae-based systems	Yes (TN, TP)	Yes (acts as filter removes suspended particles, produces oxygen)	Yes (plant biomass and bioenergy)	No
Soil aquifer treatment	Yes (filtration system)	Yes (filtration system recharges aquifer)	No	No
Urban agriculture and community gardens	Yes (soil remediation)	Yes (minimal irrigation)	Yes (biomass and bioenergy)	Yes (food security)
Floating treatment wetlands	Yes (TN, TP)	Yes (better quality water)	Yes (bioelectricity)	Yes (heavy metals)
Blue-green infrastructure	Yes (minimal)	Yes (stormwater reuse)	Yes (reduce energy demand)	No
Phytoremediation	Yes (TN, TP, COD, BOD)	Yes (minimal)	Yes (plant biomass and bioenergy)	Yes (heavy metals)
Ecological parks and natural reserves in urban areas	Yes (minimal)	Yes (irrigation, GW recharge)	Yes (plant biomass and bioenergy)	Yes (native species restoration)
Sustainable Urban Drainage Systems (SUDS)	Yes (reduction of nutrients in surface runoff)	Yes (reduction in surface runoff, improves water quality, flood mitigation)	Yes (retention ponds with hydropower converter)	Yes (native species restoration, improves air quality)
Living shorelines	Yes (plant uptake and biomass used as fertilizers)	Yes (improved water quality)	No	Yes (habitat recovery, biogenic reefs, limits soil erosion)
Bioretention systems	Yes (hydrocarbon)	Yes (improved water quality through filtration)	Yes (by reducing energy consumption for water treatment)	Yes (heavy metals, soil remediation)

Abandoned railways, highways, and underutilized public spaces are being converted into sources of renewable energy (Atrash, 2024). The Australian Recycling Investment Fund, with $100 million in Clean Energy Finance Corporation (CEFC) finance, was established in December 2019 to invest in projects utilizing clean energy technologies to support recycling and resource recovery (CEFC Australian Government, 2024).

The effectiveness and cost benefits of various NBS can be categorized as shown in Table 2.1. NBS systems such as urban wetlands, constructed wetlands, green roofs, and walls, green infrastructure for stormwater management, floating treatment wetlands, and bioretention systems are efficient across all four categories of resource recovery: nutrients, water, energy, and materials. In contrast, systems such as living shorelines, soil aquifer treatment, and permeable pavements may not be equally effective in all these categories. However, these latter systems offer significant indirect ecological benefits. In summary, while not all NBS focus directly on resource recovery across all categories, they collectively contribute to overall environmental sustainability and provide indirect benefits related to ecosystem efficiency and climate mitigation. Table 2.2 quantifies the various types of resource recovery methods discussed, providing an overview of their performance effectiveness in extracting resources from generated waste.

Table 2.2 Quantification of resource recovery by different methods.

Methods for Resource Recovery	Recovered Sources	Quantity Recovered	Reference
Enzymes assisted extraction	VFAs lead to polyhydroxyalkanoates (PHA) production	0.3 g PHA/g VFA in 24 h reaction time from food waste	Strazzera *et al.* (2018)
Pyrolysis	Bio-oils, heavy metals	Bio oils – 24.12 wt% from sugarcane bagasse at 450°C, Ni – 13,400 to 24,000 mg/kg in DW of (synthetic hyperaccumulator) *corn stover*, Zn – 24,700 to 34,700 mg/kg in DW of *Sedum alfredii Hance*, Cu – 17,600 to 33,000 mg/kg in DW of *Sedum alfredii Hance*, Cr- 17,400 to 21,300 mg/kg in DW of (synthetic hyperaccumulator) *corn stover*, Cd – 89 (± 3.2) mg/kg in DW of *Sedum alfredii Hance*, Pb – 174 (± 8) mg/kg in DW of *Sedum alfredii Hance*	Chen *et al.* (2019); Koppolu and Clements (2003); Parihar *et al.* (2007)
Anaerobic digestion	Methane, bioenergy	1400 MW, 40 tonnes of maize per hectare can produce 0.35 m^3 of CH_4 per kg of volatile solids from agricultural biomass	Battista *et al.* (2019)
Solvent extraction	Lipids for biofuels	106.2–125.4 mg/g dry weight of algal biomass	Kim *et al.* (2012); Ranjith Kumar *et al.* (2015)
Neoteric solvent extraction	Phytochemicals (Proanthocyanidins) phenolic compounds	114.86 mg/g using 1-butyl–3-methylimidazolium bromide and 88.3% extraction of phenolic compounds	Thakker *et al.* (2014)
Urban agriculture	Nitrogen sequestration, food	170,000 tonnes of nitrogen sequestration, 100–180 million tonnes of food	Clinton *et al.* (2018)
Ash leaching	Sodium, Chlorine, potassium	93.6% – Na, 90%- Cl, 82% – K	Gonçalves *et al.* (2008); Mei *et al.* (2020)

2.3.2 Methods used in resource recovery

Understanding the categories of resource recovery is crucial, but focusing on the specific processes involved in effectively recovering these resources is equally important. By examining detailed methods and technologies such as enzymes, pyrolysis, nutrient leaching, ash leaching, anaerobic digestion, chemical precipitation, and innovative solvent extraction, we can optimize the recovery of valuable materials and energy, minimize environmental impact, and promote sustainability. These recovery methods also discuss the role of microbes in resource recovery and the fate of biomass after harvest. The processes primarily address bioenergy recovery from the end of the pipeline of NBS. The subsequent sections aim to demonstrate how these methods can be integrated with the end products of NBS, such as biomass and substrate. Water treatment and the provision of clean water are performed by the NBS. The components of the NBS, namely biomass and substrate, play a key role in treating water and soil, from which resources are subsequently recovered.

2.3.2.1 Enzyme extraction

Enzymes play a crucial role in extracting valuable bioactive compounds by disrupting the cell walls of source materials, which enhances the yield of these compounds (Marathe *et al.*, 2017). Biocatalysts, such as microbial enzymes, facilitate the degradation of cellulose and, aid in the hemicellulose solubilization process. This enhances the extraction of volatile fatty acids (VFAs) from food waste, which are highly valued in the bioenergy and bioproduct production industries. VFAs are instrumental in producing biohydrogen, a promising renewable fuel. Other methods for producing biohydrogen from biomass include dark fermentation and photofermentation. In dark fermentation, enzymes like cellulases, amylases, and proteases break down complex organic materials, such as amino acids and fatty acids. These simpler molecules are then fermented by bacteria to produce biohydrogen and other by-products. The enzymatic process in photofermentation is similar; however, in this process, photosynthetic bacteria utilize the by-products from dark fermentation to produce biohydrogen (Elbeshbishy *et al.*, 2017; Ghimire *et al.*, 2015; Strazzera *et al.*, 2018).

2.3.2.2 Solvent extraction technique

The Bligh and Dyer solvent extraction method is highly preferred for lipid extraction. Lipids, which are embedded within the cell walls of microalgae, require the cell walls to be broken down for extraction. This can be accomplished through mechanical methods (e.g., bead beating, ultrasonication), chemical methods (e.g., using solvents), or biological methods (e.g., enzyme treatment). Solvents such as hexane, chloroform, and methanol are used to dissolve lipids from the plant biomass. These lipids are crucial for the production of biofuels from microalgae (Ranjith Kumar *et al.*, 2015).

Lipids from *C. vulgaris* can also be effectively extracted using ionic liquids with a volume ratio of 1:1 methanol, achieving a high lipid yield of 19.0% with 1-butyl-3-methylimidazolium. Extraction yields in this method range from 106.2 to 125.4 mg/g of dry weight (Kim *et al.*, 2012). Additionally, subcritical solvent extraction, which employs solvents like dimethyl ether, can eliminate the need to dry the biomass, allowing for direct lipid extraction (Wang *et al.*, 2021). The extracted lipids are then purified and converted into biodiesel through a chemical reaction called transesterification. In this process, lipids react with alcohol (usually methanol) in the presence of a catalyst to produce biodiesel and glycerol (Guldhe *et al.*, 2016).

Phenolic compounds present in food and its byproducts are used in the nutraceutical, pharmaceutical, and cosmetic industries. These compounds can be extracted from the bio-waste generated from gardens, eco-parks, and urban green spaces. There are various methods to extract these compounds; however, extraction using deep eutectic solvents is common. Choline chloride butane-1,4-diol is a solvent used in the extraction of anthocyanins while 16 different solvent mixtures were used in flavonoid extraction. Various flavonoids and anthocyanins were extracted to produce 110.95 ($\pm$ 5.55) – 73.25 ($\pm$ 3.66) mg/g and 16.0 ($\pm$ 0.80) mg/g, respectively, using the solvent extraction method (Masala *et al.*, 2024). A mixture of hydrogen bond acceptor and donor is required in this extraction process. The choline chloride is a hydrogen bond acceptor while cheap and safe hydrogen bond donors are urea, ethylene glycol, and glycerol, but other alcohols, amino acids, carboxylic acids, and sugars have also been used (Ruesgas-Ramón *et al.*, 2017).

Neoteric solvents such as surfactants, hydrotropes, and ionic liquids are used in the extraction of phytochemicals, and phenolic compounds from the aqueous state (Thakker *et al.*, 2014). These solvents can replace the conventional organic solvents for extraction. They offer medicinal and health benefits such as antioxidants and anti-inflammatory compounds. 96.67% extraction of a few components such as phenol, 4-nitrophenol, 2,4-dinitrophenol, 2,6-dinitrophenol, 1-naphthol, 2-naphthol, and 4-chlorophenol has been achieved from an aqueous solution with hydrophobic solvents such as imidazolium (Zhuang *et al.*, 2014; Khachatryan *et al.*, 2005).

2.3.2.3 Pyrolysis

Pyrolysis is a promising method for recovering valuable resources from biomass through a thermochemical process that decomposes organic material at high temperatures (400–800°C) (Bañuelos *et al.*, 2015). There are two main types of pyrolysis: fast and slow (Bridgwater, 2012; Parihar *et al.*, 2007). In the bio-oil production process, a slow pyrolysis method typically operating at temperatures between 400 and 500°C, with a nitrogen flow rate of 50–200 ml/min, producing a maximum yield of 24.12 wt%. The bio-oil produced generally has a calorific value ranging from 17 to 20 GJ/ton, which is lower than that of fuel oil, around 40 GJ/ton (Parihar *et al.*, 2007). Pyrolysis also yields other products, such as biochar, biochemicals, and syngas, with production of many sustainable products still only demonstrated at lab-scale (Amenaghawon *et al.*, 2021). Despite the advantages of pyrolysis, several challenges remain, including feedstock variability, heat and mass transfer issues, product quality, scaling up, and economic viability (Mettler *et al.*, 2012; Rasaq *et al.*, 2021; Wang *et al.*, 2022).

Pyrolysis after phytoremediation, or phytomining, is a common process used to extract heavy metals like nickel, zinc, cadmium, copper, and chromium (Muthusaravanan *et al.*, 2018). Plants that can accumulate high concentrations of these metals are known as "hyperaccumulators." Heavy metal accumulation in biomass ranges from 1000 to 50,000 mg/kg for the metals mentioned (Reeves *et al.*, 1996). Therefore, it adds up to the source for the resource recovery process.

The fixed-bed reactor is a commonly used pyrolysis reactor. It is effective for converting biomass into biofuels and valuable products, characterized by long biomass residence time, high carbon conservation, reduced ash content, and low sweeping gas velocity, which is suitable for small-scale energy production (Amenaghawon *et al.*, 2021; Chen *et al.*, 2016). Trees, including species like eucalyptus, grown in urban and rural areas are used in various applications such as manufacturing composite wood products, construction, and landscape management (Bañuelos *et al.*, 2015; Przewoźna *et al.*, 2022). However, one significant challenge in pyrolysis is tar removal, which can affect the quality of the gas produced.

2.3.2.4 Leaching

The two most commonly used methods for resource recovery are nutrient leaching and ash leaching. Nutrient leaching affects groundwater and surface water quality, posing challenges such as eutrophication. However, controlled leaching can manage nutrient application in soil and water bodies more effectively (Meisinger & Delgado, 2002). Riparian buffer strips along coastal areas help control the nutrients reaching water bodies. By regularly monitoring nutrient levels in soil and water, the benefits of nutrient leaching can be maximized while mitigating its negative aquatic impacts (NaturalEngland, 2022; Rashmi *et al.*, 2017). In-depth research and understanding of controlled nutrient leaching can help strategize the reintroduction of resources into the environment.

Ash leaching is a pretreatment process used in pyrolysis to remove agricultural residues, wood chips, and other organic matter. This process enhances the efficiency of gasification and combustion in producing biofuels and bio-oil. Elements such as potassium (K), sodium (Na), and chlorine (Cl) are extracted or removed during ash leaching (Wang *et al.*, 2024). These elements are purified and used in various industrial and agricultural applications. For instance, potassium, extracted as potash, is used in soaps, glass, and certain types of batteries. Sodium, in the form of sodium hydroxide, is utilized in paper and textile manufacturing (Capodaglio *et al.*, 2017; Mei *et al.*, 2020). Similarly, chlorine is important as a disinfectant and is used in polyvinyl chloride, solvents, and various other chemical production processes (Gonçalves *et al.*, 2008; Pan *et al.*, 2021).

2.3.2.5 Anaerobic digestion

A key technology for converting waste into usable resources is anaerobic digestion. These systems are commonly found in municipal wastewater treatment plants, agricultural waste management, and industrial waste management. Anaerobic digesters break down organic waste in the absence of oxygen, converting it primarily into biogas. The digestate, which is rich in nutrients, can be used as fertilizer (Kandagatla *et al.*, 2024). However, some studies raise concerns about using digestate as fertilizer due to trace element loading, environmental risks such as odor, and high content of VFAs (Monlau *et al.*, 2015). Additionally, the presence of weeds, pathogens, pharmaceutical residues, and antibiotic-resistant genes in digestate can affect the environment. Stabilizing the digestate through composting can make it more usable and reduce these risk elements by up to 20–30% after application to soil (Bai *et al.*, 2020; Teglia *et al.*, 2011). Therefore, integrating composting with anaerobic digestion can help stabilize the digestate and enhance its usability.

A biogas plant in Italy was studied to assess the need for plant upgrades. The feedstock (maize) was quantified, and its biogas and energy production were calculated. For 40 tons of maize per hectare, the plant produced 0.35 m^3 of CH_4 per kg of volatile solids, generating 1400 MW of bioenergy (Battista *et al.*, 2019). Carbohydrates from vegetable waste were broken down faster and more easily compared to proteins and fats, resulting in higher biogas yields (Dhanalakshmi Sridevi & Ramanujam, 2012). This case study highlights the potential of anaerobic digestion in agricultural settings, showcasing its benefits and challenges. Further research is needed to minimize digester costs and simplify the process.

2.3.2.6 Other recovery methods

Apart from the discussed recovery methods, several other techniques can be used to recover resources from NBS. Chemical precipitation is one such process that aids in phosphorus recovery. Multivalent metal ions, such as calcium, aluminum, and iron, are used to form precipitates of sparingly soluble phosphates. Struvite, for example, is precipitated using magnesium and is controlled by factors such as temperature, mixing energy, pH, and the magnesium concentration (Hallas *et al.*, 2019).

Composting transforms organic waste into valuable products like mulch and compost. Plant matter, food scraps, and paper products can be composted to create resources. Composting reduces the amount of waste sent to landfills, thereby addressing associated environmental impacts (Lacovidou *et al.*, 2017). It also offers other environmental benefits, such as reducing the need for chemical fertilizers, enriching soil, and recovering energy (Qi *et al.*, 2022; Robinson, 2021).

Photobioreactors use light and CO_2 to cultivate microalgae or phototrophic bacteria, which can then be harvested to recover valuable resources. The recovered nutrients can be used as biofertilizers, and the harvested biomass can be utilized as animal feedstock or for bioenergy production (Ferrera *et al.*, 2024; Lindeboom *et al.*, 2022). Various types of photobioreactors, such as raceway ponds, tubular reactors, and flat-plate reactors, are designed to produce different types of end products. These reactors also offer environmental benefits such as CO_2 capture and water reclamation (Fradinho *et al.*, 2021; Kim *et al.*, 2022).

The integration of plant biology with microbial fuel cell principles has led to the development of plant microbial fuel cells. These cells have potential applications in sustainable agriculture, water and wastewater treatment, and renewable energy production in both urban and rural settings. The ultimate goal of these systems is to generate electricity using vegetation and the microbes in the roots.

Microbes degrade organic materials, releasing electrons that lead to electricity generation (Paucar & Sato, 2021; Singuru *et al.*, 2023; Zhang *et al.*, 2024). Overall, the resource recovery methods discussed are essential for achieving sustainable development goals, minimizing environmental impact, and generating economic value from waste.

2.4 CONCLUSIONS

The integration of NBS in water management represents a transformative approach to resource recovery, utilizing natural processes to address critical environmental challenges. This chapter explores the diverse pathways through which NBS contributes to sustainable water management and resource recovery, while highlighting both its strengths and limitations. Various NBS methods, such as green roofs, constructed wetlands, and floating treatment wetlands, exemplify how these systems enhance resource recovery. The chapter categorizes the recovered resources into four key areas: nutrients, water, energy, and materials. Understanding the transition of these resources within the environment, including their uptake, demand, and waste generation, is crucial for optimizing recovery processes. While NBS offers significant potential for resource recovery and environmental sustainability, its success hinges on thoughtful design, implementation, and management. Overcoming the associated challenges and optimizing these systems will be essential to maximize their benefits and support the achievement of sustainable development goals. Moving forward, continued innovation and the integration of these solutions will be vital in promoting resilient and resource-efficient water management practices.

ACKNOWLEDGEMENTS

The financial assistance provided to the first author by the Royal Melbourne Institute of Technology (RMIT) University through the 'Australian Government Research Training Program' Scholarship, for her PhD study to which this chapter will contribute is greatly acknowledged.

REFERENCES

Abd El-Hack M. E., Abdel-Moneim A.-M. E., Shehata A. M., Mesalam N. M., Salem H. M., El-Saadony M. T. and El-Tarabily K. A. (2023). Chapter 34 – Microalgae applications in poultry feed. In: Handbook of Food and Feed From Microalgae, E. Jacob-Lopes, M. I. Queiroz, M. M. Maroneze and L. Q. Zepka (eds), Academic Press, pp. 435–450, https://doi.org/10.1016/B978-0-323-99196-4.00008-5

Agrawal K. and Verma P. (2020). Potential removal of hazardous wastes using white laccase purified by ATPS–PEG–salt system: an operational study. *Environmental Technology & Innovation*, **17**, 100556, https://doi.org/10.1016/j.eti.2019.100556

Almenar J. B., Elliot T., Rugani B., Philippe B., Gutierrez T. N., Sonnemann G. and Geneletti D. (2021). Nexus between nature-based solutions, ecosystem services and urban challenges. *Land Use Policy*, **100**, 104898, https://doi.org/10.1016/j.landusepol.2020.104898

Amenaghawon A. N., Anyalewechi C. L., Okieimen C. O. and Kusuma H. S. (2021). Biomass pyrolysis technologies for value-added products: a state-of-the-art review. *Environment, Development and Sustainability*, **23**(10), 14324–14378, https://doi.org/10.1007/s10668-021-01276-5

Anderson V. and Gough W. A. (2020). Evaluating the potential of nature-based solutions to reduce ozone, nitrogen dioxide, and carbon dioxide through a multi-type of green infrastructure study in Ontario, Canada. *City and Environment Interactions*, **6**, 100043, https://doi.org/10.1016/j.cacint.2020.100043

Atrash D. (2024). Why green energy innovations matters in parks and open spaces. *Ecorenewable Energy*. https://www.ecorenewableenergy.com.au/articles/green-energy-innovations/

Avellán T. and Gremillion P. (2019). Constructed wetlands for resource recovery in developing countries. *Renewable and Sustainable Energy Reviews*, **99**, 42–57, https://doi.org/10.1016/j.rser.2018.09.024

Ayuso-Gabella N., Page D., Masciopinto C., Aharoni A., Salgot M. and Wintgens T. (2011). Quantifying the effect of managed aquifer recharge on the microbiological human health risks of irrigating crops with recycled water. *Agricultural Water Management*, **99**(1), 93–102, https://doi.org/10.1016/j.agwat.2011.07.014

Bai L., Deng Y., Li J., Ji M. and Ruan W. (2020). Role of the proportion of cattle manure and biogas residue on the degradation of lignocellulose and humification during composting. *Bioresource Technology*, **307**, 122941, https://doi.org/10.1016/j.biortech.2020.122941

Bani A., Echevarria G., Sulçe S. and Morel J. L. (2015). Improving the agronomy of *Alyssum murale* for extensive phytomining: a five-year field study. *International Journal of Phytoremediation*, **17**(2), 117–127, https://doi.org/10.1080/15226514.2013.862204

Bañuelos G. S., Arroyo I., Pickering I. J., Yang S. I. and Freeman J. L. (2015). Selenium biofortification of broccoli and carrots grown in soil amended with Se-enriched hyperaccumulator *Stanleya pinnata*. *Food Chemistry*, **166**, 603–608, https://doi.org/10.1016/j.foodchem.2014.06.071

Barco A. and Borin M. (2020). Ornamental plants for floating treatment wetlands: preliminary results. *Italian Journal of Agronomy*, **15**(2), 109, https://doi.org/10.4081/ija.2020.1602

Bar-On Y. M., Phillips R. and Milo R. (2018). The biomass distribution on earth. *Proceedings of the National Academy of Sciences*, **115**(25), 6506–6511, https://doi.org/10.1073/pnas.1711842115

Battista F., Frison N. and Bolzonella D. (2019). Energy and nutrients' recovery in anaerobic digestion of agricultural biomass: an Italian perspective for future applications. *Energies*, **12**(17), 3287, https://www.mdpi.com/1996-1073/12/17/3287, https://doi.org/10.3390/en12173287

Bauduceau N., Berry P., Cecchi C., Elmqvist T., Fernandez M., Hartig T., Krull W., Mayerhofer E., Sandra N. and Noring L. (2015). Towards an EU research and innovation policy agenda for nature-based solutions and re-naturing cities. Final report of the horizon 2020 expert group on nature-based solutions and re-naturing cities. 1–70.

Bichot A., Delgenès J.-P., Méchin V., Carrère H., Bernet N. and García-Bernet D. (2018). Understanding biomass recalcitrance in grasses for their efficient utilization as biorefinery feedstock. *Reviews in Environmental Science and Bio/Technology*, **17**, 707–748, https://doi.org/10.1007/s11157-018-9485-y

Biswas T., Bhushan S., Prajapati S. K. and Chaudhuri S. R. (2021). An eco-friendly strategy for dairy wastewater remediation with high lipid microalgae-bacterial biomass production. *Journal of Environmental Management*, **286**, 112196, https://doi.org/10.1016/j.jenvman.2021.112196

Brandão C., do Rosário Cameira M., Valente F., de Carvalho R. C. and Paço T. A. (2017). Wet season hydrological performance of green roofs using native species under Mediterranean climate. *Ecological Engineering*, **102**, 596–611, https://doi.org/10.1016/j.ecoleng.2017.02.025

Bridgwater A. V. (2012). Review of fast pyrolysis of biomass and product upgrading. *Biomass and Bioenergy*, **38**, 68–94, https://doi.org/10.1016/j.biombioe.2011.01.048

Brodsky O. L., Shek K. L., Dinwiddie D., Bruner S. G., Gill A. S., Hoch J. M., Palmer M. I. and McGuire K. L. (2019). Microbial communities in bioswale soils and their relationships to soil properties, plant species, and plant physiology. *Frontiers in Microbiology*, **10**, 2368, https://doi.org/10.3389/fmicb.2019.02368

Buffam I. and Mitchell M. E. (2015). Nutrient cycling in green roof ecosystems. In: Green Roof Ecosystems, Ecological Studies, vol **223**, R. Sutton (ed.), Springer International Publishing, Springer, Cham., pp. 107–137, https://doi.org/10.1007/978-3-319-14983-7.

Capodaglio A. G., Callegari A., Cecconet D. and Molognoni D. (2017). Sustainability of decentralized wastewater treatment technologies. *Water Practice and Technology*, **12**(2), 463–477, https://doi.org/10.2166/wpt.2017.055

Cascone S., Gagliano A., Poli T. and Sciuto G. (2019). Thermal performance assessment of extensive green roofs investigating realistic vegetation-substrate configurations. *Building Simulation*, **12**, 379–393, https://doi.org/10.1007/s12273-018-0488-y

CEFC Australian Government (2024). Clean energy finance corporation. https://www.cefc.com.au/where-we-invest/about-our-finance/investment-programs/australian-recycling-investment-fund/

Chen D., Li Y., Cen K., Luo M., Li H. and Lu B. (2016). Pyrolysis polygeneration of poplar wood: effect of heating rate and pyrolysis temperature. *Bioresource Technology*, **218**, 780–788, https://doi.org/10.1016/j.biortech.2016.07.049

Chen H., Wang X., Lyu X., Xu L., Wang J. and Lu X. (2019). Hydrothermal conversion of the hyperaccumulator *Sedum alfredii* Hance for efficiently recovering heavy metals and bio-oil. *Journal of Environmental Chemical Engineering*, **7**(5), 103321, https://doi.org/10.1016/j.jece.2019.103321

Chetri J. K., Reddy K. R. and Green S. J. (2022). Use of methanotrophically activated biochar in novel biogeochemical cover system for carbon sequestration: microbial characterization. *Science of the Total Environment*, **821**, 153429, https://doi.org/10.1016/j.scitotenv.2022.153429

Clinton N., Stuhlmacher M., Miles A., Uludere Aragon N., Wagner M., Georgescu M., Herwig C. and Gong P. (2018). A global geospatial ecosystem services estimate of urban agriculture. *Earth's Future*, **6**(1), 40–60, https://doi.org/10.1002/2017EF000536

Colares G. S., Dell'Osbel N., Barbosa C. V., Lutterbeck C., Oliveira G. A., Rodrigues L. R., Bergmann C. P., Lopez D. R., Rodriguez A. L., Vymazal J. and Machado E. L. (2021). Floating treatment wetlands integrated with microbial fuel cell for the treatment of urban wastewaters and bioenergy generation. *Science of the Total Environment*, **766**, 142474, https://doi.org/10.1016/j.scitotenv.2020.142474

Das S., Das S. and Ghangrekar M. (2019). Quorum-sensing mediated signals: a promising multi-functional modulator for separately enhancing algal yield and power generation in microbial fuel cell. *Bioresource Technology*, **294**, 122138, https://doi.org/10.1016/j.biortech.2019.122138

De Boodt M. A. V. O. and Verdonck O. (1971). The physical properties of the substrates in horticulture. III Symposium on Peat in Horticulture, **26**, pp. 37–44.

Dhanalakshmi Sridevi V. and Ramanujam R. A. (2012). Biogas generation in a vegetable waste anaerobic digester: an analytical approach. *Research Journal of Recent Sciences*, **2277**, 2502.

Domenech T., Bleischwitz R., Doranova A., Panayotopoulos D. and Roman L. (2019). Mapping industrial symbiosis development in Europe typologies of networks, characteristics, performance and contribution to the circular economy. *Resources, Conservation and Recycling*, **141**, 76–98, https://doi.org/10.1016/j.resconrec.2018.09.016

Dorr E., Goldstein B., Horvath A., Aubry C. and Gabrielle B. (2021). Environmental impacts and resource use of urban agriculture: a systematic review and meta-analysis. *Environmental Research Letters*, **16**(9), 093002, https://doi.org/10.1088/1748-9326/ac1a39

Durhman A. K., Rowe D. B. and Rugh C. L. (2006). Effect of watering regimen on chlorophyll fluorescence and growth of selected green roof plant taxa. *HortScience*, **41**, 1623–1628, https://doi.org/10.21273/HORTSCI.41.7.1623

Dvorak B. and Volder A. (2010). Corrigendum to 'Green roof vegetation findings for North American ecoregions: a literature review' [Landscape Urban Plan. (2010) 96(4), 197–213]. *Landscape and Urban Planning*, **97**(2), 146, https://doi.org/10.1016/j.landurbplan.2010.06.008

Ehrenfeld J. and Gertler N. (1997). Industrial ecology in practice: The evolution of interdependence at Kalundborg. *Journal of Industrial Ecology*, **1**(1), 67–79, https://doi.org/10.1162/jiec.1997.1.1.67

Elbeshbishy E., Dhar B. R., Nakhla G. and Lee H.-S. (2017). A critical review on inhibition of dark biohydrogen fermentation. *Renewable and Sustainable Energy Reviews*, **79**, 656–668, https://doi.org/10.1016/j.rser.2017.05.075

Ellen Mac Arthur Foundation (2017). Urban Biocycles. https://www.ellenmacarthurfoundation.org/cities-urban-biocycles

Elumalai P., Parthipan P., Karthikeyan O. P. and Rajasekar A. (2017). Enzyme-mediated biodegradation of long-chain n-alkanes (C32 and C40) by thermophilic bacteria. *3 Biotech*, **7**(2), 116, https://doi.org/10.1007/s13205-017-0773-y

EU, Directive (2013). 39/EU OF THE EUROPEAN PARLIAMENT AND OF THE COUNCIL of 12 August 2013 amending directives 2000/60/EC and 2008/105/EC as regards priority substances in the field of water policy. L226/1, EPaCotE Union, Editor.

Evans D. L., Falagán N., Hardman C. A., Kourmpetli S., Liu L., Mead B. R. and Davies J. A. C. (2022). Ecosystem service delivery by urban agriculture and green infrastructure – a systematic review. *Ecosystem Services*, **54**, 101405, https://doi.org/10.1016/j.ecoser.2022.101405

Farrell C., Mitchell R., Szota C., Rayner J. and Williams N. (2012). Green roofs for hot and dry climates: interacting effects of plant water use, succulence and substrate. *Ecological Engineering*, **49**, 270–276, https://doi.org/10.1016/j.ecoleng.2012.08.036

Ferreira C. S. S., Potočki K., Kapović-Solomun M. and Kalantari Z. (2022). Nature-based solutions for flood mitigation and resilience in urban areas. In: Nature-Based Solutions for Flood Mitigation, Handbook of Environmental Chemistry, vol **107**, C. S. S. Ferreira, Z. Kalantari, T. Hartmann and P. Pereira (eds), Springer International Publishing, Springer, Cham., pp. 59–78, https://doi.org/10.1007/698_2021_758

Ferrera E., Ruigómez I., Vela-Bastos C., Ferreira A., Gouveia L. and Vera L. (2024). Resources recovery from domestic wastewater by a combined process: anaerobic digestion and membrane photobioreactor. *Environmental Science and Pollution Research*, **31**(37), 49560–49573, https://doi.org/10.1007/s11356-024-34468-3

Fradinho J. C., Carvalho V. C. F. and Reis M. A. M. (2021). New phototrophic factories for resource recovery. In: Enzymes for Solving Humankind's Problems: Natural and Artificial Systems in Health, Agriculture, Environment and Energy, J. J. G. Moura, I. Moura and L. B. Maia (eds), Springer International Publishing, Springer, Cham., pp. 413–438, https://doi.org/10.1007/978-3-030-58315-6_13

Furlong C., Dobbie M., Morison P., Dodson J. and Pendergast M. (2019). Chapter 16 – infrastructure and urban planning context for achieving the visions of integrated urban water management and water sensitive urban design: the case of Melbourne. In: Approaches to Water Sensitive Urban Design, A. K. Sharma, T. Gardner and D. Begbie (eds), Elsevier Science & Technology, United States, pp. 329–350, https://doi.org/10.1016/B978-0-12-812843-5.00016-2

Ghimire A., Valentino S., Frunzo L., Trably E., Escudié R., Pirozzi F., Lens P. N. L. and Esposito G. (2015). Biohydrogen production from food waste by coupling semi-continuous dark-photofermentation and residue post-treatment to anaerobic digestion: a synergy for energy recovery. *International Journal of Hydrogen Energy*, **40**(46), 16045–16055, https://doi.org/10.1016/j.ijhydene.2015.09.117

Gonçalves C., Tran H., Braz S., Puig F. and Shenassa R. (2008). Chloride and potassium removal efficiency of an ash leaching system. *Pulp and Paper Canada – Ontario*, **109**(3), p.33.

Gong Y., Zhang X., Li H., Zhang X., He S. and Miao Y. (2021). A comparison of the growth status, rainfall retention and purification effects of four green roof plant species. *Journal of Environmental Management*, **278**, 111451, https://doi.org/10.1016/j.jenvman.2020.111451

Guldhe A., Singh B., Ansari F. A., Sharma Y. and Bux F. (2016). Extraction and conversion of microalgal lipids. In: Algae Biotechnology: Products and Processes, F. Bux and Y. Chisti (eds), Springer International Publishing, Springer, Cham., pp. 91–110, https://doi.org/10.1007/978-3-319-12334-9_6

Gupta K., Bardhan P., Rather M. A., Saikia D., Loying S., Mandal M. and Kataki R. (2022). Microbes in resource and nutrient recovery via wastewater treatment: In: Industrial Microbiology and Biotechnology, P. Verma (ed.), Springer, Singapore, pp. 643–665, https://doi.org/10.1007/978-981-16-5214-1_22

Hallas J. F., Mackowiak C. L., Wilkie A. C. and Harris W. G. (2019). Struvite phosphorus recovery from aerobically digested municipal wastewater. *Sustainability*, **11**(2), 376, https://www.mdpi.com/2071-1050/11/2/376, https://doi.org/10.3390/su11020376

Hastuti D. R. D., Darma R., Salman D., Santoso S. and Rahim A. (2022). Carbon sequestration of city agriculture: between farming and non-farming land. *IOP Conference Series. Earth and Environmental Science*, **1041**(1), 012009, https://doi.org/10.1088/1755-1315/1041/1/012009

Hu Y., Qin H., Ouyang Y. and Yu S. L. (2022). Seasonal variation in recovery process of rainwater retention capacity for green roofs. *Water*, **14**(18), 2799, https://doi.org/10.3390/w14182799

Istenič D. and Božič G. (2021). Short-rotation willows as a wastewater treatment plant: biomass production and the fate of macronutrients and metals. *Forests*, **12**(5), 554, https://doi.org/10.3390/f12050554

Jegatheesan J. (2020). Resource recovery and concentrate management in environmental quality management. *Environmental Quality Management*, **30**, 5–8, https://doi.org/10.1002/tqem.21721

Juanico M. (2008). Israel As a case study – international survey of wastewater reclamation and reuse practice. In: International Survey of Wastewater Reclamation and Reuse Practice, B. Jimenez and T. Asano (eds), IWA Publishing, pp. 483–502, https://www.researchgate.net/publication/292320772

Kandagatla N., Sridhar P., Rao P. V. and Tyagi R. D. (2024). Anaerobic digestion (AD) and resource recovery. In: Biological and Hybrid Wastewater Treatment Technology: Recent Developments in India, M. M. Ghangrekar, S. Yadav and R. N. Yadava (eds), Springer Nature, Switzerland, pp. 197–222, https://doi.org/10.1007/978-3-031-63046-0_8

Khachatryan K. S., Smirnova S. V., Torocheshnikova I. I., Shvedene N. V., Formanovsky A. A. and Pletnev I. V. (2005). Solvent extraction and extraction–voltammetric determination of phenols using room temperature ionic liquid. *Analytical and Bioanalytical Chemistry*, **381**(2), 464–470, https://doi.org/10.1007/s00216-004-2872-y

Kim Y.-H., Choi Y.-K., Park J., Lee S., Yang Y.-H., Kim H. J., Park T.-J., Kim Y. H. and Lee S. H. (2012). Ionic liquid-mediated extraction of lipids from algal biomass. *Bioresource Technology*, **109**, 312–315, https://doi.org/10.1016/j.biortech.2011.04.064

Kim K., Yoo J. and Lee W. G. (2022). Recent advances in photobioreactor systems for sustainable and enhanced microalgal biofuel production. *Sustainable Energy and Fuels*, **6**(24), 5459–5473, https://doi.org/10.1039/D2SE01345B

Kisser J., Wirth M., De Gusseme B., Van Eekert M., Zeeman G., Schoenborn A., Vinnerås B., Finger D. C., Kolbl Repinc S., Bulc T. G., Bani A., Pavlova D., Staicu L. C., Atasoy M., Cetecioglu Z., Kokko M., Haznedaroglu B. Z., Hansen J., Istenič D., Canga E., Malamis S., Camilleri-Fenech M. and Beesley L. (2020). A review of nature-based solutions for resource recovery in cities. *Blue-Green Systems*, **2**(1), 138–172, https://doi.org/10.2166/bgs.2020.930

Klein T., Vasl A., Shalom H., Kadas G. and Blaustein L. (2017). Sedum-dominated green-roofs in a semi-arid region increase CO_2 concentrations during the dry season. *Science of the Total Environment*, **584**, 1147–1151.

Koop S. H. and van Leeuwen C. J. (2017). The challenges of water, waste and climate change in cities. *Environment, Development and Sustainability*, **19**(2), 385–418, https://doi.org/10.1007/s10668-016-9760-4

Koppolu L. and Clements L. D. (2003). Pyrolysis as a technique for separating heavy metals from hyperaccumulators. Part I: Preparation of synthetic hyperaccumulator biomass. *Biomass and Bioenergy*, **24**(1), 69–79, https://doi.org/10.1016/S0961-9534(02)00074-0

Lacovidou E., Millward-Hopkins J., Busch J., Purnell P., Velis C. A., Hahladakis J. N., Zwirner O. and Brown A. (2017). A pathway to circular economy: developing a conceptual framework for complex value assessment of resources recovered from waste. *Journal of Cleaner Production*, **168**, 1279–1288, https://doi.org/10.1016/j.jclepro.2017.09.002

Langergraber G., Pucher B., Simperler L., Kisser J., Katsou E., Buehler D., Mateo M. C. G. and Atanasova N. (2020). Implementing nature-based solutions for creating a resourceful circular city. *Blue-Green Systems*, **2**(1), 173–185, https://doi.org/10.2166/bgs.2020.933

Langergraber G., Castellar J. A., Andersen T. R., Andreucci M.-B., Baganz G. F., Buttiglieri G., Canet-Martí A., Carvalho P. N., Finger D. C. and Griessler Bulc T. (2021). Towards a cross-sectoral view of nature-based solutions for enabling circular cities. *Water*, **13**(17), 2352, https://doi.org/10.3390/w13172352

Lee J. T., Lin Y. S., Shih C. Y. and Lee M. J. (2022). Root functional traits and water erosion-reducing potential of two indigenous C4 grass species for erosion control of mudstone badlands in Taiwan. *Water*, **14**(9), 1342, https://doi.org/10.3390/w14091342

Leguizamo M. A. O., Gómez W. D. F. and Sarmiento M. C. G. (2017). Native herbaceous plant species with potential use in phytoremediation of heavy metals, spotlight on wetlands – a review. *Chemosphere*, **168**, 1230–1247, https://doi.org/10.1016/j.chemosphere.2016.10.075

Li Y.-M., Chaney R., Brewer E., Roseberg R., Angle J. S., Baker A., Reeves R. and Nelkin J. (2003). Development of a technology for commercial phytoextraction of nickel: economic and technical considerations. *Plant and Soil*, **249**(1), 107–115, https://doi.org/10.1023/A:1022527330401

Li N., Wan Y. and Wang X. (2020). Nutrient conversion and recovery from wastewater using electroactive bacteria. *Science of the Total Environment*, **706**, 135690, https://doi.org/10.1016/j.scitotenv.2019.135690

Lindeboom R., Poór-Pócsi E., Kim J. E. and Hofman J. (2022). Factsheet – photobioreactors (V5), next gen circular water solutions, p. 01, https://mp.uwmh.eu/media/factsheets/LA_Factsheet_Photobioreactor.pdf

Lisowski P. and Glinicki M. (2023). Promising biomass waste-derived insulation materials for application in construction and buildings. Biomass Conversion and Biorefinery, 1, https://doi.org/10.1007/s13399-023-05192-8

Logan M. and Visvanathan C. (2019). Management strategies for anaerobic digestate of organic fraction of municipal solid waste: current status and future prospects. *Waste Management & Research*, **37**(1_suppl), 27–39, https://doi.org/10.1177/0734242X18816793

Lundholm J., MacIvor J. and Ranalli M. (2009). Benefits of green roofs on Canada's east coast. Annual Greening Rooftops for Sustainable Communities Conference, Atlanta, GA.

Majee S. B., Avlani D. and Biswas G. R. (2017). Pharmacological, pharmaceutical, cosmetic and diagnostic applications of sulfated polysaccharides from marine algae and bacteria. *African Journal of Pharmacy and Pharmacology*, **11**(5), 68–77, https://doi.org/10.5897/AJPP2016.4695

Manso M. and Castro-Gomes J. (2015). Green wall systems: a review of their characteristics. *Renewable and Sustainable Energy Reviews*, **41**, 863–871, https://doi.org/10.1016/j.rser.2014.07.203

Marathe S. J., Jadhav S. B., Bankar S. B. and Singhal R. S. (2017). Enzyme-assisted extraction of bioactives. In: Food Bioactives: Extraction and Biotechnology Applications, M. Puri (ed.), Springer International Publishing, pp. 171–201, https://doi.org/0.1007/978-3-319-51639-4_8

Masala V., Jokić S., Aladić K., Molnar M. and Tuberoso C. I. G. (2024). Exploring phenolic compounds extraction from saffron (*C. sativus*) floral by-products using ultrasound-assisted extraction, deep eutectic solvent extraction, and subcritical water extraction. *Molecules*, **29**(11), 2600, https://doi.org/10.3390/molecules29112600

Mathew I., Shimelis H., Mutema M. and Chaplot V. (2017). What crop type for atmospheric carbon sequestration: results from a global data analysis. *Agriculture, Ecosystems and Environment*, **243**, 34–46, https://doi.org/10.1016/j.agee.2017.04.008

Mayrand F. and Clergeau P. (2018). Green roofs and green walls for biodiversity conservation: a contribution to urban connectivity? *Sustainability*, **10**(4), 985, https://doi.org/10.3390/su10040985

Mei Y., Wang Z., Zhou X., Ding L., Liu Z., Huang J. and Fang Y. (2020). Pressured carbon dioxide hydrothermal leaching of catalytic gasification ash for dealkalization of solid waste, sodium catalyst recovery and carbon dioxide utilization. *Journal of Cleaner Production*, **247**, 119109, https://doi.org/10.1016/j.jclepro.2019.119109

Meisinger J. J. and Delgado J. A. (2002). Principles for managing nitrogen leaching. *Journal of Soil and Water Conservation*, **57**(6), 485. http://www.jswconline.org/content/57/6/485.abstract

Menon R. and Holland M. M. (2013). Phosphorus retention in constructed wetlands vegetated with *Juncus effusus*, *Carex lurida*, and *Dichanthelium acuminatum* var. acuminatum. *Water, Air, and Soil Pollution*, **224**(7), 1–, https://doi.org/10.1007/s11270-013-1602-5

Mettler M. S., Vlachos D. G. and Dauenhauer P. J. (2012). Top ten fundamental challenges of biomass pyrolysis for biofuels. *Energy & Environmental Science*, **5**(7), 7797–, https://doi.org/10.1039/c2ee21679e

Monlau F., Sambusiti C., Ficara E., Aboulkas A., Barakat A. and Carrère H. (2015). New opportunities for agricultural digestate valorization: current situation and perspectives. *Energy & Environmental Science*, **8**(9), 2600–2621, https://doi.org/10.1039/C5EE01633A

Morakinyo T. E., Lam Y. F. and Hao S. (2016). Evaluating the role of green infrastructures on near-road pollutant dispersion and removal: modelling and measurement. *Journal of Environmental Management*, **182**, 595–605, https://doi.org/10.1016/j.jenvman.2016.07.077

Morris R. L., Konlechner T. M., Ghisalberti M. and Swearer S. E. (2018). From grey to green: efficacy of eco-engineering solutions for nature-based coastal defence. *Global Change Biology*, **24**(5), 1827–1842, https://doi.org/10.1111/gcb.14063

Morris R. L., Campbell-Hooper E., Waters E., Bishop M. J., Lovelock C. E., Lowe R. J., Strain E. M. A., Boon P., Boxshall A., Browne N. K., Carley J. T., Fest B. J., Fraser M. W., Ghisalberti M., Gillanders B. M., Kendrick G. A., Konlechner T. M., Mayer-Pinto M., Pomeroy A. W. M., Rogers A. A., Simpson V., Van Rooijen A. A., Waltham N. J. and Swearer S. E. (2024). Current extent and future opportunities for living shorelines in Australia. *Science of the Total Environment*, **917**, 170363, https://doi.org/10.1016/j.scitotenv.2024.170363

Muduli M., Choudharya M. and Ray S. (2023). A review on constructed wetlands for environmental and emerging contaminants removal from wastewater: traditional and recent developments. Environment, Development and Sustainability, **26**, 30181–30220, https://doi.org/10.1007/s10668-023-04190-0

Muthusaravanan S., Sivarajasekar N., Vivek J. S., Paramasivan T., Naushad M., Prakashmaran J., Gayathri V. and Al-Duaij O. K. (2018). Phytoremediation of heavy metals: mechanisms, methods and enhancements. *Environmental Chemistry Letters*, **16**(4), 1339–1359, https://doi.org/10.1007/s10311-018-0762-3

Nagase A. and Dunnett N. (2010). Drought tolerance in different vegetation types for extensive green roofs: effects of watering and diversity. *Landscape and Urban Planning*, **97**(4), 318–327, https://doi.org/10.1016/j.landurbplan.2010.07.005

Nagase A. and Dunnett N. (2012). Amount of water runoff from different vegetation types on extensive green roofs: effects of plant species, diversity and plant structure. *Landscape and Urban Planning*, **104**(3), 356–363, https://doi.org/10.1016/j.landurbplan.2011.11.001

Nan X., Lavrnić S. and Toscano A. (2020). Potential of constructed wetland treatment systems for, agricultural wastewater reuse under the EU framework. *Journal of Environmental Management*, **275**, 111219, https://doi.org/10.1016/j.jenvman.2020.111219

Natural England (2022). Nutrient Neutrality and Mitigation: a summary guide and frequently asked questions (NE776). Natural England (edition 7). https://publications.naturalengland.org.uk/publication/6248597523005440 (NE776 Edition 7 Nutrient Neutrality and Nutrient Mitigation, PDF, 281.7 kB)

Nature-based Solutions Initiative, U. o. O. (2018). Greener Grangetown SuDS project. https://nbshub.naturebasedsolutionsinitiative.org/case-studies-tool/

Nivala J., Headley T., Wallace S., Bernhard K., Brix H., van Afferden M. and Müller R. A. (2013). Comparative analysis of constructed wetlands: The design and construction of the ecotechnology research facility in langenreichenbach, Germany. *Ecological Engineering*, **61**, 527–543, https://doi.org/10.1016/j.ecoleng.2013.01.035

Oberndorfer E., Lundholm J., Bass B., Coffman R. R., Doshi H., Dunnett N., Gaffin S., Köhler M., Liu K. K. and Rowe B. (2007). Green roofs as urban ecosystems: ecological structures, functions, and services. *BioScience*, **57**(10), 823–833, https://doi.org/10.1641/B571005

Pan J., Hassas B. V., Rezaee M., Zhou C. and Pisupati S. V. (2021). Recovery of rare earth elements from coal fly ash through sequential chemical roasting, water leaching, and acid leaching processes. *Journal of Cleaner Production*, **284**, 124725, https://doi.org/10.1016/j.jclepro.2020.124725

Paraskevopoulou A. T., Zafeiriou S. and Londra P. A. (2021). Plant growth of *Atriplex portulacoides* affected by irrigation amount and substrate type in an extensive green roof system. *Ecological Engineering*, **165**, 106223, https://doi.org/10.1016/j.ecoleng.2021.106223

Parihar M. F., Kamil M., Goyal H. B., Gupta A. K. and Bhatnagar A. K. (2007). An experimental study on pyrolysis of biomass. *Process Safety and Environmental Protection*, **85**(5), 458–465, https://doi.org/10.1205/psep07035

Paucar N. E. and Sato C. (2021). Microbial fuel cell for energy production, nutrient removal and recovery from wastewater: a review. *Processes*, **9**(8), 1318, https://www.mdpi.com/2227-9717/9/8/1318, https://doi.org/10.3390/pr9081318

Payen F. T., Evans D. L., Falagán N., Hardman C. A., Kourmpetli S., Liu L., Marshall R., Mead B. R. and Davies J. A. C. (2022). How much food can we grow in urban areas? Food production and crop yields of urban agriculture: a meta-analysis. *Earth's Future*, **10**(8), e2022EF002748, https://doi.org/10.1029/2022EF002748

Poë S., Stovin V. and Berretta C. (2015). Parameters influencing the regeneration of a green roof's retention capacity via evapotranspiration. *Journal of Hydrology*, **523**, 356–367, https://doi.org/10.1016/j.jhydrol.2015.02.002

Przewoźna P., Mączka K., Mielewczyk M., Inglot A. and Matczak P. (2022). Ranking ecosystem services delivered by trees in urban and rural areas. *Ambio*, **51**(9), 2043–2057, https://doi.org/10.1007/s13280-022-01722-2

Qi C., Yin R., Gao X., Chen J., Wang R., Xu Z., Luo W., Li G. and Li Y. (2022). Development of solid-state anaerobic digestion and aerobic composting hybrid processes for organic solid waste treatment and resource recovery: a review. *Current Pollution Reports*, **8**(3), 221–233, https://doi.org/10.1007/s40726-022-00223-w

Raimondo F., Trifilò P., Lo Gullo M. A., Andri S., Savi T. and Nardini A. (2015). Plant performance on Mediterranean green roofs: interaction of species-specific hydraulic strategies and substrate water relations. *AoB Plants*, **7**, plv007, https://doi.org/10.1093/aobpla/plv007

Ramos H. M., Teyssier C., Samora I. and Schleiss A. J. (2013). Energy recovery in SUDS towards smart water grids: a case study. *Energy Policy*, **62**, 463–472, https://doi.org/10.1016/j.enpol.2013.08.014

Ranjith Kumar R., Hanumantha Rao P. and Arumugam M. (2015). Lipid extraction methods from microalgae: a comprehensive review. *Frontiers in Energy Research*, **2**, 61, https://doi.org/10.3389/fenrg.2014.00061

Rasaq W. A., Golonka M., Scholz M. and Białowiec A. (2021). Opportunities and challenges of high-pressure fast pyrolysis of biomass: a review. *Energies*, **14**(17), 5426, https://www.mdpi.com/1996-1073/14/17/5426, https://doi.org/10.3390/en14175426

Rashmi I., Shirale A., Kartikha K. S., Shinogi K. C., Meena B. P. and Kala S. (2017). Leaching of plant nutrients from agricultural lands. In: Essential Plant Nutrients: Uptake, Use Efficiency, and Management, M. Naeem, A. A. Ansari and S. S. Gill (eds), Springer International Publishing, pp. 465–489, https://doi.org/10.1007/978-3-319-58841-4_19

Rathour R. K., Sharma D., Ullah S., Mahmoud E.-H. M., Sharma N., Kumar P., Bhatt A. K., Ahmad I. and Bhatia R. K. (2024). Bacterial–microalgal consortia for bioremediation of textile industry wastewater and resource recovery for circular economy. *Biotechnology for the Environment*, **1**(1), 6, https://doi.org/10.1186/s44314-024-00005-2

Raza N., Rizwan M. and Mujtaba G. (2024). Bioremediation of real textile wastewater with a microalgal-bacterial consortium: an eco-friendly strategy. *Biomass Conversion and Biorefinery*, **14**(6), 7359–7371, https://doi.org/10.1007/s13399-022-03214-5

Razzaghmanesh M., Beecham S. and Salemi T. (2016). The role of green roofs in mitigating urban heat island effects in the metropolitan area of Adelaide, South Australia. *Urban Forestry and Urban Greening*, **15**, 89–102, https://doi.org/10.1016/j.ufug.2015.11.013

Reeves R. D., Brooks R. R. and Baker A. J. (1996). Abnormal Accumulation of Trace Metals by Plants (No. CONF-960592-). International Business Communications, Southborough, MA (United States).

Robinson D. (2021). How composting can reduce our impact on the planet. UNEP. https://www.unep.org/news-and-stories/story/how-composting-can-reduce-our-impact-planet

Robinson D., Stankovic S., Morel N., Deque F., Rylatt M., Kabele K., Manolakaki E. and Nieminen J. (2003). Integrated resource flow modelling of urban neighbourhoods: project SUNtool. Proc. Building Simulation, Eighth International IBPSA Conference Eindhoven, Netherlands, 1117–1122.

Romanis C. S., Pearson L. A. and Neilan B. A. (2021). Cyanobacterial blooms in wastewater treatment facilities: significance and emerging monitoring strategies. *Journal of Microbiological Methods*, **180**, 106123, https://doi.org/10.1016/j.mimet.2020.106123

Ruesgas-Ramón M., Figueroa-Espinoza M. C. and Durand E. (2017). Application of deep eutectic solvents (DES) for phenolic compounds extraction: overview, challenges, and opportunities. *Journal of Agricultural and Food Chemistry*, **65**(18), 3591–3601, https://doi.org/10.1021/acs.jafc.7b01054

Russo A., Escobedo F. and Zerbe S. (2016). Quantifying the local-scale ecosystem services provided by urban treed streetscapes in Bolzano, Italy. *AIMS Environmental Science*, **3**, 58–76, https://doi.org/10.3934/environsci.2016.1.58

Russo A., Escobedo F. J., Cirella G. T. and Zerbe S. (2017). Edible green infrastructure: an approach and review of provisioning ecosystem services and disservices in urban environments. *Agriculture, Ecosystems & Environment*, **242**, 53–66, https://doi.org/10.1016/j.agee.2017.03.026

Ryłko-Polak I., Komala W. and Białowiec A. (2022). The reuse of biomass and industrial waste in biocomposite construction materials for decreasing natural resource use and mitigating the environmental impact of the construction industry: a review. *Materials*, **15**(12), 4078, https://doi.org/10.3390/ma15124078

Samo T. J., Kimbrel J. A., Nilson D. J., Pett-Ridge J., Weber P. K. and Mayali X. (2018). Attachment between heterotrophic bacteria and microalgae influences symbiotic microscale interactions. *Environmental Microbiology*, **20**(12), 4385–4400, https://doi.org/10.1111/1462-2920.14357

Sarkar S., Banerjee A., Halder U., Biswas R. and Bandopadhyay R. (2017). Degradation of synthetic azo dyes of textile industry: a sustainable approach using microbial enzymes. *Water Conservation Science and Engineering*, **2**, 121–131, https://doi.org/10.1007/s41101-017-0031-5

Seyedabadi M. R., Eicker U. and Karimi S. (2021). Plant selection for green roofs and their impact on carbon sequestration and the building carbon footprint. *Environmental Challenges*, **4**, 100119, https://doi.org/10.1016/j.envc.2021.100119

Shao L., Wu Z., Zeng L., Chen Z., Zhou Y. and Chen G. (2013). Embodied energy assessment for ecological wastewater treatment by a constructed wetland. *Ecological Modelling*, **252**, 63–71, https://doi.org/10.1016/j.ecolmodel.2012.09.004

Sharif F., Westerhoff P. and Herckes P. (2014). Impact of hydraulic and carbon loading rates of constructed wetlands on contaminants of emerging concern (CECs) removal. *Environmental Pollution*, **185**, 107–115, https://doi.org/10.1016/j.envpol.2013.10.001

Sharma S. K. and Kennedy M. D. (2017). Soil aquifer treatment for wastewater treatment and reuse. *International Biodeterioration & Biodegradation*, **119**, 671–677, https://doi.org/10.1016/j.ibiod.2016.09.013

Singuru R., Praveen Kumar G. and Kumawat A. S. (2023). Microbial fuel cell usage in treatment, resource recovery and energy production from bio-refinery wastewater. In: Biorefinery for Water and Wastewater Treatment, M. P. Shah (ed.), Springer International Publishing, pp. 425–443, https://doi.org/10.1007/978-3-031-20822-5_20

Skuza L., Szućko-Kociuba I., Filip E. and Bożek I. (2022). Natural molecular mechanisms of plant hyperaccumulation and hypertolerance towards heavy metals. *International Journal of Molecular Sciences*, **23**(16), 9335, https://doi.org/10.3390/ijms23169335

Sood A., Renuka N., Prasanna R. and Ahluwalia A. S. (2015). Cyanobacteria as potential options for wastewater treatment. *Phytoremediation: Management of Environmental Contaminants*, **2**, 83–93.

Sörenson E., Capo E., Farnelid H., Lindehoff E. and Legrand C. (2021). Temperature stress induces shift from co-existence to competition for organic carbon in microalgae-bacterial photobioreactor community – enabling continuous production of microalgal biomass. *Frontiers in Microbiology*, **12**, 607601, https://doi.org/10.3389/fmicb.2021.607601

Sproul J., Wan M. P., Mandel B. H. and Rosenfeld A. H. (2014). Economic comparison of white, green, and black flat roofs in the United States. *Energy and Buildings*, **71**, 20–27, https://doi.org/10.1016/j.enbuild.2013.11.058

Stewart-Wade S. M. (2020). Efficacy of organic amendments used in containerized plant production: part 1–Compost-based amendments. *Scientia Horticulturae*, **266**, 108856, https://doi.org/10.1016/j.scienta.2019.108856

Strazzera G., Battista F., Garcia N. H., Frison N. and Bolzonella D. (2018). Volatile fatty acids production from food wastes for biorefinery platforms: a review. *Journal of Environmental Management*, **226**, 278–288, https://doi.org/10.1016/j.jenvman.2018.08.039

Tapia N. F., Rojas C., Bonilla C. A. and Vargas I. T. (2017). Evaluation of *Sedum* as driver for plant microbial fuel cells in a semi-arid green roof ecosystem. *Ecological Engineering*, **108**, 203–210, https://doi.org/10.1016/j.ecoleng.2017.08.017

Teglia C., Tremier A. and Martel J.-L. (2011). Characterization of solid digestates: part 2, assessment of the quality and suitability for composting of six digested products. *Waste and Biomass Valorization*, **2**, 113–126, https://doi.org/10.1007/s12649-010-9059-x

Thakker M. R., Desai M. A. and Parikh J. K. (2014). Extraction of phytochemicals using neoteric solvents. *J. Agroecol. Nat. Resour. Manage*, **1**, 153–158.

Thomaidi V., Petousi I., Kotsia D., Kalogerakis N. and Fountoulakis M. S. (2022). Use of green roofs for greywater treatment: role of substrate, depth, plants, and recirculation. *Science of the Total Environment*, **807**(Pt 3), 151004–, https://doi.org/10.1016/j.scitotenv.2021.151004

Tighiri H. O. and Erkurt E. A. (2019). Biotreatment of landfill leachate by microalgae-bacteria consortium in sequencing batch mode and product utilization. *Bioresource Technology*, **286**, 121396, https://doi.org/10.1016/j.biortech.2019.121396

Topić Božič J., Fric U., Čikić A. and Muhič S. (2024). Life cycle assessment of using firewood and wood pellets in Slovenia as two primary wood-based heating systems and their environmental impact. *Sustainability*, **16**(4), 1687, https://www.mdpi.com/2071-1050/16/4/1687, https://doi.org/10.3390/su16041687

Tsatsou A., Frantzeskaki N. and Malamis S. (2023). Nature-based solutions for circular urban water systems: a scoping literature review and a proposal for urban design and planning. *Journal of Cleaner Production*, **394**, 136325, https://doi.org/10.1016/j.jclepro.2023.136325

van Hullebusch E. D., Bani A., Carvalho M., Cetecioglu Z., De Gusseme B., Di Lonardo S., Djolic M., van Eekert M., Griessler Bulc T., Haznedaroglu B. Z., Istenič D., Kisser J., Krzeminski P., Melita S., Pavlova D., Płaza E., Schoenborn A., Thomas G., Vaccari M., Wirth M., Hartl M. and Zeeman G. (2021). Nature-based units as building blocks for resource recovery systems in cities. *Water*, **13**(22), 3153, https://www.mdpi.com/2073-4441/13/22/3153, https://doi.org/10.3390/w13223153

Van Renterghem T., Hornikx M., Forssen J. and Botteldooren D. (2013). The potential of building envelope greening to achieve quietness. *Building and Environment*, **61**, 34–44, https://doi.org/10.1016/j.buildenv.2012.12.001

Vassilev A., Schwitzguébel J.-P., Thewys T., Van Der Lelie D. and Vangronsveld J. (2004). The use of plants for remediation of metal-contaminated soils. *The Scientific World Journal*, **4**(1), 9–34, https://doi.org/10.1100/tsw.2004.2

Vijayaraghavan K. and Raja F. D. (2014). Design and development of green roof substrate to improve runoff water quality: plant growth experiments and adsorption. *Water Research*, **63**, 94–101, https://doi.org/10.1016/j.watres.2014.06.012

Vimal S. R., Singh J. S. and Prasad S. M. (2022). Plant–microbe dynamics as a nature-based solution for sustainable agriculture. *Anthropocene Science*, **1**(4), 428–443, https://doi.org/10.1007/s44177-023-00043-7

Vítová M. and Mezricky D. (2024). Microbial recovery of rare earth elements from various waste sources: a mini review with emphasis on microalgae. *World Journal of Microbiology and Biotechnology*, **40**(6), 189, https://doi.org/10.1007/s11274-024-03974-4

Walsh L. E., Mead B. R., Hardman C. A., Evans D., Liu L., Falagan N., Kourmpetli S. and Davies J. (2022). Potential of urban green spaces for supporting horticultural production: a national scale analysis. *Environmental Research Letters*, **17**(1), 014052, https://doi.org/10.1088/1748-9326/ac4730

Waly M. M., Ahmed T., Abunada Z., Mickovski S. B. and Thomson C. (2022). Constructed wetland for sustainable and low-cost wastewater treatment: review article. *Land*, **11**(9), 1388, https://www.mdpi.com/2073-445X/11/9/1388, https://doi.org/10.3390/land11091388

Wang Q., Oshita K. and Takaoka M. (2021). Effective lipid extraction from undewatered microalgae liquid using subcritical dimethyl ether. *Biotechnology for Biofuels*, **14**(1), 17, https://doi.org/10.1186/s13068-020-01871-0

Wang W., Gu Y., Zhou C. and Hu C. (2022). Current challenges and perspectives for the catalytic pyrolysis of lignocellulosic biomass to high-value products. *Catalysts*, **12**(12), 1524, https://www.mdpi.com/2073-4344/12/12/1524, https://doi.org/10.3390/catal12121524

Wang Y., Guo S., Cao F., He C., Wei Y., Qin Y., He Y., Du X., Vassilev S. V. and Vassileva C. G. (2024). Leaching mechanisms of ash-forming elements during water washing of corn straw. *Biomass Conversion and Biorefinery*, **14**(1), 133–146, https://doi.org/10.1007/s13399-021-02184-4

Weerakkody U., Dover J. W., Mitchell P. and Reiling K. (2017). Particulate matter pollution capture by leaves of seventeen living wall species with special reference to rail-traffic at a metropolitan station. *Urban Forestry and Urban Greening*, **27**, 173–186, https://doi.org/10.1016/j.ufug.2017.07.005

Whittinghill L. J., Rowe D. B., Schutzki R. and Cregg B. M. (2014). Quantifying carbon sequestration of various green roof and ornamental landscape systems. *Landscape and Urban Planning*, **123**, 41–48, https://doi.org/10.1016/j.landurbplan.2013.11.015

Wilkinson S., Feitosa R. C., Kaga I. T. and De Franceschi I. H. (2017). Evaluating the thermal performance of retrofitted lightweight green roofs and walls in Sydney and Rio de Janeiro. *Procedia Engineering*, **180**, 231–240, https://doi.org/10.1016/j.proeng.2017.04.182

Xiao Q., Gregory McPherson E., Zhang Q., Ge X. and Dahlgren R. (2017). Performance of two bioswales on urban runoff management. *Infrastructures*, **2**(4), 12, https://doi.org/10.3390/infrastructures2040012

Xu Y., Porter N., Foster J. L., Muir J. P., Schwab P., Burson B. L. and Jessup R. W. (2020). Silica production across candidate lignocellulosic biorefinery feedstocks. *Agronomy*, **10**(1), 82, https://doi.org/10.3390/agronomy10010082

Yang K., Chen Q., Zhang D., Zhang H., Lei X., Chen Z., Li Y., Hong Y., Ma X., Zheng W., Tian Y., Zheng T. and Xu H. (2017). The algicidal mechanism of prodigiosin from *Hahella* sp. KA22 against *Microcystis aeruginosa*. *Scientific Reports*, **7**(1), 7750, https://doi.org/10.1038/s41598-017-08132-5

Zhang S. and Guo Y. (2013). Analytical probabilistic model for evaluating the hydrologic performance of green roofs. *Journal of Hydrologic Engineering*, **18**(1), 19–28, https://doi.org/10.1061/(ASCE)HE.1943-5584.0000593

Zhang D. Q., Gersberg R. M., Hua T., Zhu J., Tuan N. A. and Tan S. K. (2012). Pharmaceutical removal in tropical subsurface flow constructed wetlands at varying hydraulic loading rates. *Chemosphere*, **87**(3), 273–277, https://doi.org/10.1016/j.chemosphere.2011.12.067

Zhang S., Lin Z., Zhang S. and Ge D. (2021). Stormwater retention and detention performance of green roofs with different substrates: observational data and hydrological simulations. *Journal of Environmental Management*, **291**, 112682, https://doi.org/10.1016/j.jenvman.2021.112682

Zhang H., Duan L., Li S., Gao Q., Li M., Xing F. and Zhao Y. (2024). Simultaneous wastewater treatment and resources recovery by forward osmosis coupled with microbial fuel cell: a review. *Membranes*, **14**(2), 29, https://www.mdpi.com/2077-0375/14/2/29, https://doi.org/10.3390/membranes14020029

Zheng X., Wei C., Qin P., Guo J., Yu Y., Song F. and Chen Z. (2014). Characteristics of residential energy consumption in China: findings from a household survey. *Energy Policy*, **75**, 126–135, https://doi.org/10.1016/j.enpol.2014.07.016

Zhou L. W., Wang Q., Li Y., Liu M. and Wang R. Z. (2018). Green roof simulation with a seasonally variable leaf area index. *Energy and Buildings*, **174**, 156–167, https://doi.org/10.1016/j.enbuild.2018.06.020

Zhuang X. L., Li Z. Q. and Wang E. Q. (2014). Study on the extraction of phenolic compounds from model oil by using ionic liquid. *Advanced Materials Research*, **906**, 137–141, https://doi.org/10.4028/www.scientific.net/AMR.906.137

doi: 10.2166/9781789065015_0041

Chapter 3

Fundamentals of floating treatment wetlands: functions, design, and maintenance

Pham-Yen-Nhi Tran[1], Thi-Viet-Huong Dao[1], Thi-Kim-Quyen Vo[2], Mai-Nhu Hoang[1], Linh-Thy Le[3], Nhu-Nguyet Phan[4], Ky-Phuong-Ha Huynh[1], Thanh-Dai Tran[5], Thanh-Son Dao[1], Piet N. L. Lens[6] and Xuan-Thanh Bui[1]

[1]Key Laboratory of Advanced Waste Treatment Technology & Faculty of Environment and Natural Resources, Ho Chi Minh City University of Technology (HCMUT), 268 Ly Thuong Kiet Street, District 10, Ho Chi Minh City, Vietnam; Vietnam National University Ho Chi Minh (VNU-HCM), Linh Trung ward, Ho Chi Minh City 700000, Vietnam
[2]Faculty of Biology and Environment, Ho Chi Minh City University of Industry and Trade (HUIT), 140 Le Trong Tan street, Tay Thanh ward, Tan Phu district, Ho Chi Minh City, 700000, Vietnam
[3]Faculty of Public Health, University of Medicine and Pharmacy at Ho Chi Minh City (UMP), Ward 11, district 5, Ho Chi Minh City, Vietnam
[4]Faculty of Environment, University of Science, Vietnam National University, Ho Chi Minh City, Vietnam
[5]Faculty of Technology, Dong Nai Technology University, Bien Hoa City, Vietnam
[6]National University of Ireland Galway, University Road, Galway H91 TK33, Ireland
E-mail: Xuan-Thanh Bui bxthanh@hcmut.edu.vn

ABSTRACT

Floating treatment wetlands (FTWs) are among the nature-based solutions for purifying surface water bodies (e.g., rivers, ponds, and lakes) contaminated by diffuse pollution sources. Different from conventional constructed wetlands, FTWs are composed of emergent macrophytes growing on floating mats on the water surface, thus the systems can withstand various water depths minimum at 0.8–1.0 m. The root systems hang underneath the floating mat, creating a vast surface area for biofilm adhesion. FTWs are becoming popular as a treatment method due to their strong capacity to remove organics, nutrients, heavy metals, and contaminants of emerging concerns (CECs) such as antibiotics, pesticides, and hormones without the need for land. FTWs efficiently treat industrial, municipal, and agricultural wastewater, stormwater runoff, and landfill leachate. This chapter overviews the functions, design, operation, and maintenance recorded of FTWs so far, and highlights the operational performance of FTWs applied for various types of wastewater treatment.

Keywords: nature-based solution, contaminant treatment, emergent macrophytes, floating mat, biofilms.

3.1 INTRODUCTION

Freshwater systems, including canals, lakes, rivers, and streams, are important resources providing three-quarters of water for agriculture and industry, and one-third of drinking water for households (Hairom *et al.*, 2021). Although surface water is essential for humans as well as other living organisms,

© 2025 IWAP. This is an Open Access book chapter distributed under the terms of the Creative Commons Attribution License (CC BY-NC-ND 4.0) which permits copying and redistribution for non-commercial purposes with no derivatives, provided the work is properly cited (https://creativecommons.org/licenses/by-nc-nd/4.0/). The chapter is from the book *Nature-Based Solutions for Urban Sustainability*, Piet N.L. Lens and Xuan-Thanh Bui (Editors).

it is easily polluted due to uncontrolled discharge activities (Sikdar, 2021). The discharge of untreated or partially treated wastewater has increased surface water pollution and depleted clean water sources. The main pollution sources are domestic wastewater, industrial wastewater, and agricultural runoff. Some other potential emissions from human activities include clinic wastewater, fishery activities, animal husbandry, forest pest protection, fertilizers, and pesticides consisting of inorganic and organic compounds that can cause water pollution (Chowdhary *et al.*, 2018; Hiremath *et al.*, 2024). Besides, surface water sources may contain pathogenic microorganisms, pharmaceutical waste, heavy metals, antibiotics, body care products, pesticides, herbicides, nitrogen, and phosphorus as well as pathogens (Chowdhary *et al.*, 2018). In addition to human activities, algal blooms, biomass decomposition, animal waste, and flood sediments also contribute to and enhance the pollution of surface water (Sasakova *et al.*, 2018). The increase of contaminants and release of wastes including microplastic into the water system disrupt the food chain and web in both aquatic and terrestrial environments (Hairom *et al.*, 2021; Verma *et al.*, 2024). Surface water pollution has extreme consequences, resulting in the loss of biodiversity because it affects the life of aquatic organisms. In addition to affecting human well-being, poor surface water quality impacts the economy because of the investment in water treatment and water quality restoration measures (Andersson *et al.*, 2016).

Various treatment technologies have been developed to address surface water contamination, though each technology has its advantages and limitations. Adsorption, ion exchange, coagulation/flocculation, filtration and sedimentation as well as disinfection are some of the typical surface water treatment methods. In general, these methods are effective, however, there are some limitations. For example, using the flocculation process to treat contaminated surface water results in a high treatment efficiency, however, high dosage levels of common coagulants (e.g., alum, ferric chloride, poly aluminum chloride and poly ferric sulfate) result in significant amounts of sludge, requiring sludge treatment (Jassim & AlAmeri 2020). The filtration method has several disadvantages, including sluggish filtering speed, sensitivity to algae in the input water, less effectiveness in cold areas due to lower biological activity, easy clogging, and cleaning (Maurya *et al.*, 2020). Furthermore, the single-treatment technology often faces implementation challenges. Alternative approaches or combined technologies are required to solve the requirement for surface water treatment. The treatment of surface water requires a significant investment in innovative technologies. The application of plant filtration systems or filter lagoons for large areas of surface water shows lower treatment efficiency than for small areas (Singh *et al.*, 2019). Some technologies have high energy consumption or high cost and can harm the environment, such as ozonation, membrane filtration, and ultraviolet (UV) disinfection (Palansooriya *et al.*, 2020). Treatment systems, particularly plant filtration systems, can accumulate residues and sludge, requiring frequent treatment to remove accumulated residues (Hairom *et al.*, 2021; Singh *et al.*, 2019). As a result, it is important to implement a sustainable technology that contributes to reducing diffuse pollution in surface water.

Generally, surface water pollution is a complex challenge that requires comprehensive and innovative solutions. While treatment technologies play a crucial role in mitigating the impacts of pollution, it is essential to carefully consider their disadvantages and limitations. A balanced and context-specific approach, taking into account the economic, environmental, and social aspects, is necessary for the successful and sustainable implementation. Therefore, it is essential to supplement an ecosystem with a system that could contribute to reduce pollutants in surface water and fix the limitation of conventional methods. A biological treatment method is the constructed wetlands filter system, which uses grass, plants, and soil to filter the polluted water. Plants aid in the absorption and removal of nutrients and other organic pollutants (Singh *et al.*, 2019). Recently, floating treatment wetlands (FTWs) are considered as a promising technique that can reduce diffuse pollution in surface water resources and improve the environment in numerous ways.

A FTW is a type of constructed wetlands (CWs) which is a highly efficient biological treatment technology that has been applied in wastewater treatment applications (Colares *et al.*, 2021). The contaminant removal is based on phytoremediation, in which the roots of growing plants on a rack contact with the water/wastewater via physicochemical–biological processes to eliminate pollutants. A FTW is a modified system by using naturally floating macrophytes put in floating supports so that their roots are fully in touch with the water or wastewater. FTW systems are primarily used to treat various wastewater sources to improve the water quality. Furthermore, biofilms forming on plant roots, floating mats, media, and racks play the main role in pollution remediation. This is an extremely beneficial water treatment method since it utilizes plant roots and the diversity of bacteria growing on the roots. FTW remove and metabolize contaminants through microbial biodegradation and plant uptake, thereby enhancing water quality (Colares *et al.*, 2021). Therefore, FTWs are considered as a potential technology that improves the diffuse pollution in surface water system in general, and canals or water body resources in urban cities in particular. FTWs have also been used to remove nutrients from wastewater (Pang *et al.*, 2022). Furthermore, FTW systems can be utilized for a variety of applications to treat various kinds of water/wastewater such as pond water, urban lake water, swine wastewater, domestic wastewater, agricultural run-off water, and water supply reservoirs to improve water quality (Barco *et al.*, 2021; Colares *et al.*, 2021; Meng *et al.*, 2023).

3.2 FTWS: GREEN TECHNOLOGY IN POLLUTANT REMOVAL

The interactions of floating aquatic plants, microbes, water and the atmosphere is fundamental for the removal mechanisms in FTWs (Yeh *et al.*, 2015). The removal mechanisms of FTWs are entrapment, sediment accumulation, and suspended particles flocculation, the uptake, and assimilation of contaminants, oxidation–reduction and anammox reactions, processes of nitrification and denitrification, mineralization, immobilization, degradation of organic contaminants, and decomposition (Yeh *et al.*, 2015) (Figure 3.1).

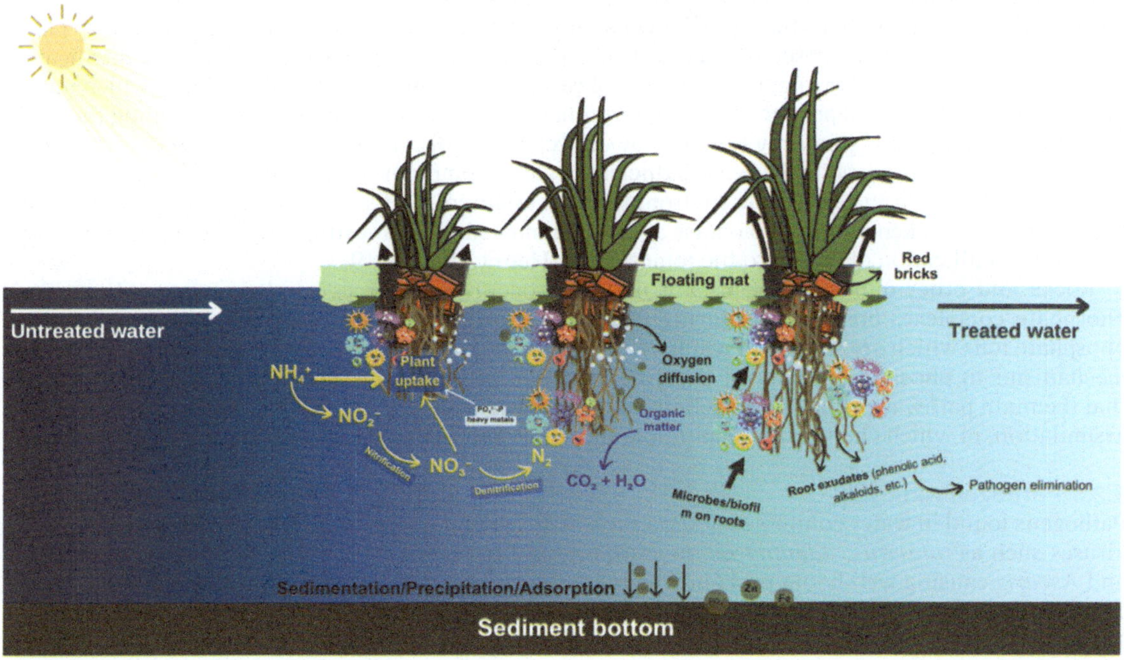

Figure 3.1 Processes involved in phytoremediation by FTWs.

3.2.1 Organic matter removal

The elimination of chemical oxygen demand (COD) depends on both physical and microbiological processes. This can be settling and physical filtering/entrapment because physical root filtration mechanisms may filter and trap suspended particles in rhizomes and roots of plants that allow for better biodegradation of organic solids (Benvenuti *et al.*, 2018). Besides, high COD removal efficiencies are obtained by settling of solid particles and by decomposition processes. Linking to total suspended solids (TSS) removal, the mechanism in FTWs is similar to COD removal (Benvenuti *et al.*, 2018). Suspended solids (SS) are inorganic and organic (e.g., bacteria and algae) particles. The size and characteristics of the SS that are present in wastewater determine the removal procedure. Both biofilm filtration and physical settling can eliminate the coarse suspended solids (CSS or settleable particles with a size larger than 1 μm). Additional methods to improve the removal of fine suspended solids (FSS) removal and colloidal particles are provided by the presence of macrophytes in FTWs (Shen *et al.*, 2021). Organic compounds released by plants can increase the biomass of biofilm or enable coagulation–flocculation processes of FSS which lead to larger floc formation, easy to settle down, followed by sedimentation (Shen *et al.*, 2021).

3.2.2 Nutrients removal

For biochemical reactions and adsorption, the roots of plants offer a high biologically active surface area for biofilm formation (Shen *et al.*, 2021). The nutrient removal efficiency depends on the concentration of nutrients and forms, various types of macrophytes, characteristics of wastewater and environmental factors (e.g., pH, temperature, climate and light intensity) (Shen *et al.*, 2021). This section sheds light on the nitrogen and phosphorus removal mechanisms by FTWs.

Various nitrogen compounds present in wastewater, such as organic nitrogen (amino acids, proteins, and nucleic acids), ammonium-nitrogen (NH_4^+-N), nitrite-nitrogen (NO_2^--N), and nitrate-nitrogen (NO_3^--N), are eliminated through different removal mechanisms (Sharma *et al.*, 2021). The most common removal mechanism is denitrification, which is the conversion of nitrate to nitrogen (in gaseous form) under anoxic conditions by denitrifying bacteria. Nitrogen can be taken up and removed by plants through their growth and subsequent harvest. The presence of macrophytes enhances the nitrification and denitrification by increasing the utilization of carbon compounds from root exudates, which impact the aerobic conditions and produce micro-environments with oxic–anoxic zones surrounding the roots. In addition, ammonia is primarily removed by conversion to nitrate in the presence of oxygen known as the nitrification process. If there are sufficient organic compounds, denitrification may remove the nitrate in both anoxic and anaerobic environments. In addition, plants can uptake nitrogen in inorganic forms, which are taken up in the biomass and then removed via harvesting. High pH levels (>9.0) can cause the volatilization of NH_3 into the atmosphere (Henny *et al.*, 2019).

Rocks and other mineral deposits are naturally occurring sources of phosphorus. Mineralized phosphate complexes break down during the natural weathering process, releasing phosphorus as phosphate ions which are soluble in water. Adsorption, sedimentation, and complexation are the main mechanisms in phosphorus removal (Sharma *et al.*, 2021). However, Zhang *et al.* (2020) concluded that the main pathways of phosphate elimination in wetlands are precipitation, adsorption, and plant assimilation, of which chemical precipitation and adsorption onto the matrix is dominant.

3.2.3 Pathogen removal

Pathogens found in water comprise bacteria such as *E. coli*, *Salmonella*, *Shigella*, and *Vibrio cholerae*, viruses such as *rotavirus*, *adenovirus*, *hepatitis A*, and parasites such as *Giardia*, *Cryptosporidium*, and *Ascaris* causing well-being problems including diarrhea, cholera, typhoid, and worm infections. The mechanisms of eliminating pathogens are the combination of (i) physical processes including sedimentation, filtration, UV radiation inactivation, and aggregation; (ii) chemical processes as adsorption to the plant root, oxidation, and toxin exposure caused by other microbes and plants; and (iii) biological processes such as predation, biotic processes, antibiosis, and natural die-off (Barco

& Borin, 2020; Park *et al.*, 2019). The indicator bacteria most usually reported are fecal coliforms and *Escherichia coli* (Vo *et al.*, 2023), and in certain cases *Enterococci* (Čule *et al.*, 2021) with most investigations achieving 1–2-log reduction (90–99% in pathogen removal). FTW are efficient systems to remove *E. coli* and coliforms, which is similar to Park *et al.* (2019) for non-aerated FTWs (approximately 3-log reduction in vegetated systems). In addition, the study of Kansiime and Van Bruggen (2001) in Uganda showed a contribution of 2-log fecal coliforms removal in a full-scale papyrus (*Cyperus papyrus*) system.

Pathogen removal by physical mechanisms such as sedimentation depends on settling velocities, detention or retention time and attached ability to substrate in water bodies, which is related to the filtration of pathogenic microorganism. The presence of plants in wetlands can increase the rate of sedimentation. Alufasi *et al.* (2017) demonstrated that a higher number of fecal coliforms were adhered to root surfaces of *Cyperus papyrus* compared to those of *Microbotryum violaceum*. This is because *C. papyrus* has a hollow and interwoven rooting mat system leading to a high surface area, whereas the root system of *M. violaceum* is tight and compact which results in a reduced surface area. It shows that in wetlands, the presence of plants can promote the microbial removal efficiency by attachment to the root surface area available. Besides, the mechanical filtration can also treat pathogens as a physical mechanism at high rates with the key factors being medium granularity and plant type. In other words, pathogenic microorganisms with modest settling velocities in wastewater can attach to larger particles and settle out from the suspension.

Chemical mechanisms including oxidation, UV radiation, and exposure to biocides contribute to pathogen removal. The study of Wu *et al.* (2016) indicated that in an un-vegetated system, solar UV light is the most important component in promoting *E. coli* wastewater disinfection by destroying the DNA of the bacterium. However, the contribution of UV rays to FTW disinfection is low because floating elements and vegetative biomass cover the wastewater surface, preventing sun radiation from penetrating into the water column. The increased concentration of DO released from plant roots is directly correlated with the pathogen removal efficiency (Tran *et al.*, 2024). The fundamental mechanism is the delivery of oxygen to the root system, which plays an important role in their activity and the type of metabolism of the microorganisms in the root zone called oxidation. Plant roots transfer oxygen into the rhizosphere of constructed wetlands, raising the system's dissolved oxygen content. Pathogen reductions are directly connected with higher dissolved oxygen concentrations (Wu *et al.*, 2016). Headley *et al.* (2013) reported a greater *E. coli* removal performance of aerated wetlands compared to non-aerated ones. In addition, the penetration of UV radiation into the water column may assist in causing further pathogenic microorganisms mortality. Nevertheless, the vegetation of the wetland may impede penetration of sunlight into the water column by its coverage. The release of oxygen through roots into the rhizosphere may be used to decompose organic waste generated by root decay and macrophytes. Exudates are often released in response to a variety of biotic and abiotic stimulations in the rhizosphere, and this induces interactions between plant roots, soil, and microorganisms.

Generally, vegetation has a critical function in lowering the *E. coli* abundance evidenced by Toscano *et al.* (2009), who found that *E. coli* removal was 55% in CWs vegetated with *P. australis* higher than in un-vegetated (38%) CWs under similar conditions. This was also found in another study (Leto *et al.*, 2013) with the species *Cyperus alternifolious* achieved 86% (nearly 1-log) reduction, which was higher than in the un-planted system. Indeed, macrophytes release several organic compounds directly into wastewater via the root system (e.g., alkaloid, phenolic, tannic and gallic acids) that have an antimicrobial action against *E. coli*, emphasizing the role of plants in microbial pollution abatement (Barco & Borin, 2020). Mahmoud *et al.* (2021) found that the ethanol content of the *Canna generalis* rhizome was 21%, with 98.1 μg/mg extract of phenolic and gallic acid that are toxic to *E. coli* and coliforms. Similarly, organic compounds released from *Phragmites* root (alkaloids) inhibited pathogens (Chen *et al.*, 2022). Also the *Cyperaceae* family including *Cyperus alternifolius* and *Cyperus odoratus* have organic compounds with disinfection capacities in the leaves (Al-mayyahi and Sosa 2017).

3.2.4 Heavy metal removal

Several studies investigated the removal mechanism of heavy metals such as manganese (Mn), iron (Fe), nickel (Ni), copper (Cu), and zinc (Zn) (Bauer *et al.*, 2021; Odinga *et al.*, 2013; Peterson *et al.*, 2021; Shahid *et al.*, 2020). Adsorption and sedimentation are the main mechanisms involved in removing metals in FTWs (Qasem *et al.*, 2021; Shahid *et al.*, 2020; Yu *et al.*, 2021). The study of Shen *et al.* (2021) showed that positively charged metals were more likely to adsorb on small particles and organic matter under favorable conditions at pH levels ranging from 6 to 8 and a temperature of 6–28°C. In the heavy metal removal pathway, plant uptake is not considered to constitute a significant removal mechanism. Aquatic plants can acquire a small amount of dissolved heavy metals, the majority of which accumulate in the plant roots by uptake and physico-chemical adsorption (Shen *et al.*, 2021). Furthermore, the removal of heavy metals is accomplished by sorption, chelation, micro-precipitation, precipitation, ion-exchange, redox processes, trapping of particulate forms into biofilms, binding to clay particles or particulate matter, sorption, and macrophyte absorption (Shahid *et al.*, 2018).

3.3 DESIGN PARAMETERS FOR FTWS

FTWs are known as a cost-effective technology with basic design, simple operation, high efficiency, and saving energy. Besides, this system can provide an attractive sightseeing experience. The buoyant frame, support materials for plant growth, water depth, and environmental factors (e.g., season and temperature) all have a significant impact on the design, operation and maintenance (O&M), and treatment performance of FTW (Vo *et al.*, 2023). The design of FTWs is crucial due to its contribution to the system performance. To design a complete FTW, a variety of parameters need to be considered: types of plant species, plant coverage density, microbial community, water depth, shading, and the installation of artificial methods of achieving buoyancy (Colares *et al.*, 2020; Pavlineri *et al.*, 2017; Samal *et al.*, 2019).

3.3.1 Floating mat buoyancy

Buoyancy is one of the most important design characteristics since it allows FTWs to respond to water level variations and operate continuously under the toughest conditions (Borne *et al.*, 2015; Schwammberger *et al.*, 2020) with a floating mat/raft. Plant species are distinguished by the massive aerenchyma in the rhizomes and roots which increases the floating capacity. In addition, materials used for floating mats/rafts to boost the buoyancy should be hydrophobic. Floating mat materials should be buoyant, durable, flexible, safe, easy to install, environmentally and ecologically friendly, and cost-effective (Vo *et al.*, 2023). For instance, in the artificial FTWs, the most common materials such as foam, artificial matrix-like Biohaven™, rubber mat, tubes or pipes, which are made of PVC, polyester (PE), and polypropylene (PP) or polystyrene (PS) foam pontoons, are used as buoyant mats (Figure 3.2). Furthermore, natural buoyant materials such as bamboo have been tested to make an alternative frame to be more cost-effective in designing FTW.

3.3.2 Vegetation

The plant species used to create FTWs require some basic characteristics. Vegetation is a main component and plays a vital role in FTWs due to its function of enhancing water and wastewater treatment along with the provision of a wide surface area, and other favorable growth conditions for microbial development (Malaviya *et al.*, 2020), scenery aesthetics, have high catabolic profiles, and biomass productivity. FTW vegetation forms a dense, partly integrated root system in the water environment (Mfarrej *et al.*, 2023). For choosing acceptable vegetation in FTWs, fundamental plant species features are important (Sharma *et al.*, 2021). FTWs plants are known for their capacity to thrive in harsh environments, high tolerance to toxicants, ability to absorb high nutrient concentrations and adapt to local climates. Macrophytes should be tall, robust, adapt well to local climatic environments, be aesthetic, non-intrusive, emergent, preferably perennial species, well-grown in aquaculture, and

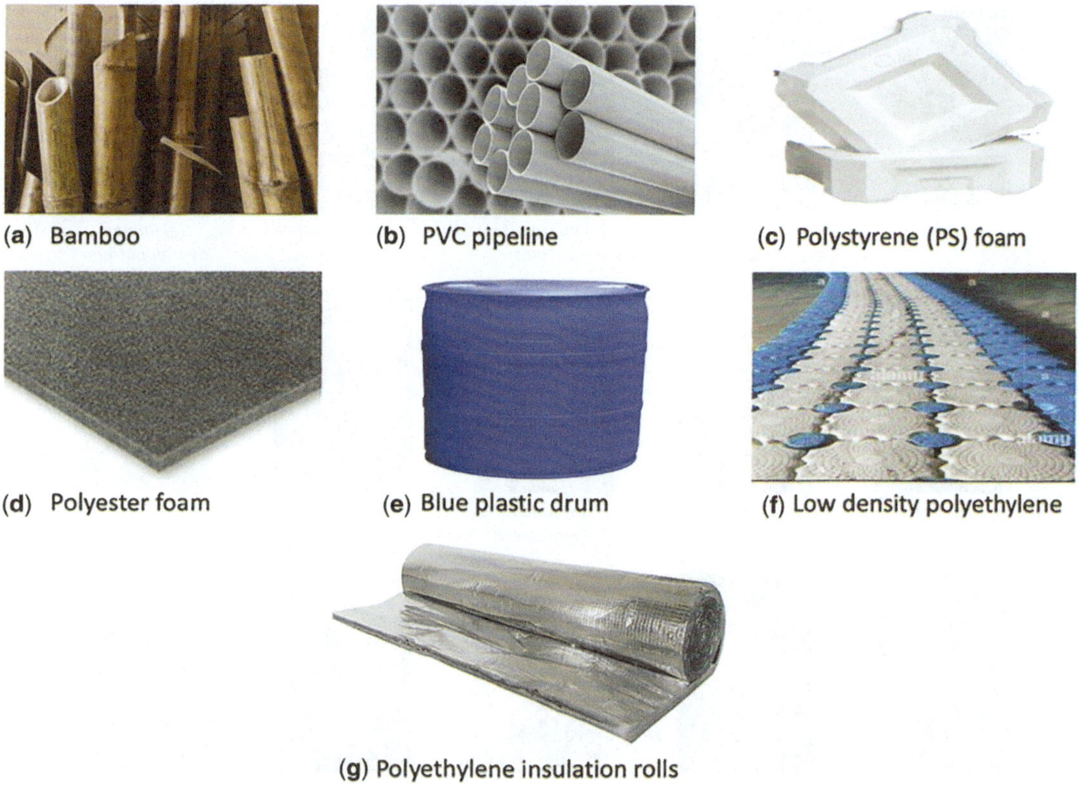

Figure 3.2 Floating materials used in the construction of FTWs.

must have various aerenchyma tissues for higher efficiency of oxygen diffusion from aerial parts to underground parts (Gupta *et al.*, 2020). Based on these criteria, *Canna indica, Phragmites australis, Juncus, Typha domingensis, Elodea* and *Cyperus leavigatus* have been used effectively in FTWs (Vymazal, 2013). Furthermore, some wild plants including *Phragmites australis, Rheum rhabarbarum, Blumea malcolmii, Rumex hydrolapathum,* and *Typhonium flagelliforme* (Figure 3.3) have the ability to treat various types of wastewater such as domestic, textile, and swine wastewater, stormwater, sewage, and brackish water (Barco *et al.*, 2021; Colares *et al.*, 2021; Landaverde *et al.*, 2024; Sasakova *et al.*, 2018). In addition, the study of Kadam *et al.* (2022) used *Fimbristylis dichotoma,* and *Ipomoea aquatic* to treat real textile effluent contaminated with hazardous waste. Plant–plant consortia comprising *I. aquatica,* and *F. dichotoma* were able to degrade Orange 3R more effectively than the individual system. Moreover, to boost the use of FTW, vegetation plays an important role of transporting atmospheric oxygen to the root system, controlling microbial respiration, excreting low molecular weight compounds together with secondary metabolites and ameliorating microhabitat conditions (Park *et al.*, 2019; Sharma *et al.*, 2021). Another study in the USA to treat brackish water using *Distichlis spicata, Juncus roemerianus, Spartina alterniflora,* and *Spartina patens* showed the plants can tolerate salinity stress up to 0.5 and 5.0 g/L salinity (Landaverde *et al.*, 2024).

FTWs can be considered as an innovative path to sustainable water treatment. The pollutant treatment performance of different plant species is presented in Table 3.1. The utilization of various plant species of different sizes is often advantageous. Tall plant species may force FTWs to tip over. As a result, the basis of FTWs must be strong enough to hold the plants in place. Helophytes, reeds, and

Figure 3.3 Plant species used in FTWs.

Table 3.1 Plant species used for FTWs.

Plant Species	Scale	Plant form	Wastewater	Treatment Efficiency	Reference
Colocasia gigantea	Full-scale	Emergent	Domestic	TN: 72–99% TP: 64–98% NH_4^+-N: 75–100%	Kumwimba *et al.* (2017)
Common reed (*Phragmites australis*)	Full-scale	Emergent Height: up to 3.65 m	Domestic livestock wastewater	TN: 58–78% TP: 47–66% NH_4^+-N: 88–98%	Wu *et al.* (2011)
Parrot heliconia (*Heliconia psittacorum*)	Full-scale	Emergent, ornament Tall: 90–180 cm The basal one 8–15 cm long	Commercial Area Shah Alam, Selangor	BOD: 85% COD: 95% NH_4^+-N: 55%	Moktar *et al.* (2018)
Canna lily (*Canna generalis*)	Demo	Emergent, ornament Stem: 1–1.75 m Root: 10–20 cm	Sewage	N: 84%	Leto *et al.* (2013)
Calla lily cell (*Zantedeschia aethiopica*)	Full-scale	Emergent, ornament Height: 60–100 cm Root: able to 16 cm	Municipal wastewater	COD: 97% NH_4^+-N: 59% NO_3^--N: 82% Norg: 71% TN: 65% TP: 82%	Nakase *et al.* (2019)
Napier grass (*Pennisetum purpureum*)	Pilot	Emergent Height: 4–7 m Root length: 25–35 cm	Swine wastewater	COD: 91% BOD: 81%	Sijimol *et al.* (2021)
Fragrant flatsedge (*Cyperus odoratus*)	Pilot	Emergent Height: 10–51 cm	Synthetic industrial wastewater (575 μS/cm EC)	Phase I: COD: 86% Cr: 95% Zn: 96%	de Moya Sánchez *et al.* (2021)

cattails are ideal for artificial and semi-natural systems as they can live in water-logged circumstances, release low molecular weight compounds and secondary metabolites, and control microbial respiration helping the FTW to eliminate pollutants (Sharma *et al.*, 2021). *Canna* macrophytes are more suitable for phytoremediation studies because of their rapid natural development, resilience to unfavorable climatic conditions, hyper-accumulation ability, and long adventitious root systems (Savio *et al.*, 2023). Root growth in FTWs is determined by a number of parameters, including wetland plant species, maturity time, nitrogen intake, redox conditions, and water trophic conditions (Chen *et al.*, 2016; Shahid *et al.*, 2018). These perennial and rhizomatous macrophytes are commonly encountered in nature. The study of Barco *et al.* (2021) in North Italy investigated plant survival, biomass production on mat, nitrogen absorption, root length, and the ratio of root–shoot in five plant species: *Lythrum salicaria, Iris pseudacorus, Carex* spp., *Phragmites australis, and Typha latifolia* (Figure 3.3). Following this investigation, all species demonstrated good survival under aquatic conditions in both the growing season and winter, although there were some cases that effected the survival of *I. pseudacorus* and *P. australis*. In addition, *Typha latifolia, Phragmites australis,* and *Glyceria maxima* were also used in other studies. They were selected due to the high growth rate, capacity to rapidly absorb nutrients, and extensive use as phytoremediation (Fletcher *et al.*, 2024). As a result, selecting each type of plant for each type of wastewater in each climate zone will provide the FWT system with significant pollutant removal efficiency.

3.3.3 Media selection

Another important design parameter in FTWs is the media substrate. It provides a growth medium for macrophytes and facilitates wastewater circulation. The media/substrate selection is based on its hydraulic permeability and ability to absorb contaminants (Wu *et al.*, 2015). The media/substrate is the carrier, and impacts directly the overall treatment efficiency of FTWs by its physical and chemical properties.

Furthermore, materials used as media/substrates should have the patch with a porous medium inside with a larger surface area. It can enhance the hydrodynamics and mechanical characteristics of FTW as well as provide a larger surface area for microbial attachment, thereby increasing the pollutant removal performance (Lu *et al.*, 2016). Poor hydraulic conductivity may cause system clogging, greatly decreasing system efficacy, and insufficient substrate adsorption might also have an impact on the long-term removal efficiency of the FTWs (Wu *et al.*, 2015). The microbial population in FTWs is attached to the substrate and extensive root system, and the substrate type has a substantial influence on microbial species diversity and community structure (Long *et al.*, 2016). The substrate is providing oxygen by diffusion, creating an aerobic environment that favors the activity of microbes (Tian *et al.*, 2017). Large particles and contaminants can be removed by the substrate via physical filtration and interception. Tiny particles can pass through the substrates and be eliminated by subsequent processes.

Some common media used in FTWs include raw (natural) and synthetic media, and by-products from industry, such as coconut fiber, charcoal, carbon fiber, mussel shell, oyster shell, pyrite, volcanic scoria stone (scoria gravel), ceramsite, light expanded clay aggregate boulder (LECA boulder), cobble, clay, zeolite, sand, clay, natural pumice, and coal ash (Bruze, 2017; Geng *et al.*, 2022; Okechukwu *et al.*, 2021; Parjane & Sane, 2011; Sabara *et al.*, 2022; Xu *et al.*, 2021). These materials improve bacterial attachment and nutrient uptake. Some of the sorbent media apart from bricks, stones, sand, and aggregates that may be utilized in the FTW technologies can boost the removal effectiveness of pollutants (Parde *et al.*, 2021). Additionally, rice husk, compost, alum sludge, maerl, and peat are effective substrates for removing nitrogen and organic matter (Okechukwu *et al.*, 2021; Omotade *et al.*, 2019; Wu *et al.*, 2015; Yang *et al.*, 2022). Prochaska and Zouboulis (2006) used a combination of dolomite and sand to eliminate phosphates, while Li *et al.* (2011) used a mixed substrate made of ceramsite, vermiculite, gravel, and calcium silicate hydrate to treat surface water with low nutrient concentrations. Marble, oyster shell, and charcoal were used as substrates to remove phosphorus, heavy metals, and odor (Parde *et al.*, 2021). These are capable of immediately absorbing nutrients from water and wastewater. Based on the type of media/substrate, its physical and chemical features impact variations in contaminant removal effectiveness. However, these substrates must be carefully selected to avoid release of excess nutrients into water.

The use of artificial biofilm carriers can increase the surface area to sustain microbial attachment and growth. Several researches suggested that charcoal can be used for odor or color, and phosphorus removal by adsorption (Ghosh *et al.*, 2020; Omotade *et al.*, 2019; Okechukwu *et al.*, 2021). One of the most frequently used absorbents for treating water and wastewater is activated carbon (Bruze, 2017). Activated carbon effectively absorbs nutrients from wastewater. It has been used extensively in environmental applications in the last decades to reduce pollution in gasses and water. It is efficient in lowering the quantity of organic matter in wastewater, as well as other undesirable substances including chlorinated solvents, herbicides, and phenolics (Bruze, 2017). In the study of Selvaraj and Velvizhi (2021), the organic matter, nitrate, phosphates, and sulfates removal efficiency amounted to 12%, 5%, 6%, and 8%, respectively, by using only activated carbon for treating domestic wastewater taken from sewage treatment plants with a capacity of 1000 cubic meters per day. As a result, activated carbon effectively absorbed organic material due to its sorption and catalytic capabilities. On the other hand, coconut fiber or coconut shell is a prevalent and frequently utilized material with locally available, low-cost, and more novel products in wastewater treatment. Compared to activated carbon, coconut fiber has been less investigated. However, several studies have indicated that the material is highly effective at adsorbing particular chemicals. The result of Sabara *et al.* (2022) showed that the treatment performance using coconut shells without plants was 55.66; 4.04; and 0.76 mg/kg media per day for COD, TN, and TP removal, respectively.

Several studies have been conducted to select media/substrates for removing phosphorous odor and color from wastewater by physicochemical processes, such as adsorption (Ghosh *et al.*, 2020; Liu *et al.*, 2017; Prochaska & Zouboulis, 2006; Selvaraj & Velvizhi, 2021; Wu XiaoFeng *et al.*, 2012). Moreover, oyster shells can be introduced for phosphorus and heavy metal removal. Some ions such as cadmium (Cd), zinc (Zn), and copper (Cu) were adsorbed by raw oyster shells without plants with the removal rate of 0.3; 0.09; and 0.14 mg/kg media per day, respectively (Xu *et al.*, 2021). Approximately 80–95% of the oyster shell's weight is composed of calcium carbonate. This substrate presents the ability to remove COD at >40% (with the input concentration ranging from 19.7 to 78.4 mg/L) (Wu XiaoFeng *et al.*, 2012). The specific surface area of the oyster shells was increased by the clay structure, which facilitated the absorption of organic molecules. Other metal components found in oyster shells, such as iron, and aluminum can precipitate phosphates (Yang *et al.*, 2022). Oyster shells in FTWs were used in a long-term decontamination experiment that reached 88% total phosphorus removal efficiency by adsorption while plants made up for around 7%. The oyster shell ($CaCO_3$ microparticles) absorbs metal ions via ion exchange in three steps: (i) cations (heavy metals) can be absorbed on porous surfaces by dissolving partial calcium carbonate, releasing Ca^{2+} and CO_3^{2-}, (ii) precipitating metal ions and (iii) formation of heavy metal complexes and crystals (Figure 3.4).

Figure 3.4 Types of media substrates used in FTWs.

3.3.4 Enhanced treatment performance for FTWs

The treatment performance of FTWs may be enhanced by system size, input contaminant concentrations, environmental conditions, and coverage ratio (Lucke *et al.*, 2019). Therefore, adding extras and/or integrating additional technologies such as aeration or carriers for biofilm formation can improve the treatment performance of the system (Gao *et al.*, 2017; Garcia Chance & White, 2018). Aeration has a positive impact on the phytoremediation of contaminants in wetlands (Garcia Chance & White, 2018). Aeration has been studied extensively for both surfaces of ponds and constructed wetlands. Supplemental aeration significantly increases the efficiency of pollutant removal including nitrogen, organic matter, and pathogen elimination (Headley *et al.*, 2013; Nivala *et al.*, 2019; Park *et al.*, 2019). However, phosphorus elimination by artificial aeration in FTW remains unclear (Barco & Borin, 2020).

Aeration is critical for boosting the effectiveness of FTW removal operation. Aeration promotes root growth under floating mats as well as creates aerobic zones, which facilitate biofilm creation within a multi-substrate, resulting in an overall improvement in pollutant removal (Sharma *et al.*, 2021). To boost the efficiency of organic matter such as COD, aeration has been employed. The study by Qiao *et al.* (2022) in China concluded that the aeration increased the DO concentration in water, compensated for the reduction in air reoxygenation due to the FTW coverage, achieved the mixing of water body, facilitated the transformation and pollutant removal, and therefore enhanced the purifying capacity of FTW. Aeration speeds up the aerobic nitrification process, resulting in more nitrate generation that plants can absorb (Zhang *et al.*, 2024). However, high dissolved oxygen (DO) levels might impede the denitrification process (Zhang *et al.*, 2024). To solve this, the application of aeration may be managed using strategies such as intermittent aeration or restricted aeration to ensure it is ideal for nutrient removal (Zhang *et al.*, 2024). It may be possible for plants to transfer oxygen internally through their aerenchymous tissues and to leak oxygen into the rhizosphere through the plant roots. The percentage of plant cover, and hence shading, has a significant impact on the concentration of dissolved oxygen (Pavlineri *et al.*, 2017). In general, depending on other factors such as plant species to evaluate the vegetation cover impacts the treatment performance of FTW. For instance, planting distribution and plant coverage ratio influence the reduction of the daily load for *Juncus*, while nitrogen removal decreases at half-density systems. However, neither of these elements has an impact on the deduction concentration or the decreased daily load for *Canna* (Garcia Chance & White, 2018).

The bacterial concentration in biofilms depends on pollutant load and metabolic activities that exist in the water body, root morphology and growth rate as well as root exudates (Shahid *et al.*, 2020; Spangler *et al.*, 2019a). Artificial biofilm carriers like bamboo fibers, polystyrene based, plastic-made carriers, and magnetic particles are provided to FTWs to boost the treatment performance of FTWs that is always higher in these FTWs than by natural biofilms (Zhang *et al.*, 2016).

3.4 OPERATION AND MAINTENANCE

FTWs should be anchored for operation and maintenance. In addition, the anchor prevents FTW from aggressively moving during hydraulic conditions and poor weather which potentially obstructs the overflow system partially or completely (Borne *et al.*, 2015). Materials for anchoring the FTW should not be metal cables because they leach ion metals into the water body.

Anchor selection should be customized based on the dimensions of ponds, lakes or rivers, environmental conditions, and type of sediment in the water body. It is important to install anchor systems that effectively connect to the FTW system. Regular and thorough inspection and maintenance of mooring lines are essential ongoing practices to ensure their sustained tautness and overall security over time. Prioritizing the durability of mooring systems becomes crucial for safeguarding the structural integrity of the entire FTW setup. The design of the anchoring system should incorporate features facilitating seamless adaptation to fluctuations in water levels. Based on the design of a pond, the cable should be long enough for moving from typical flow to overflow. This entails considering the installation of adjustable or flexible mooring options that enable the system to dynamically respond to changes in water depth. On the other hand, FTWs construction materials should be PVC pipes, polyester foam or bamboo, among others (Shahid *et al.*, 2018). Generally, floating mats/rafts provide the buoyancy and the base for plant vegetation. According to Wei *et al.* (2020), to ensure the sustainability in the long term, the buoyant mat should be resilient enough to withstand the plant weight, media/substrate, as well as be resistant to the alteration of atmospheric conditions such as sun, winds, and waves. By anchoring the corners of the floating mat to the side of ponds, lakes, and rivers, the installation of FTWs can stabilize the mats. Moreover, to prevent the mat from sinking or being submerged in high tide, it is important to ensure the anchored ropes are flexible to the adjustment (Borne *et al.*, 2015).

3.5 BENEFITS AND CHALLENGES OF FTWS

The application of FTWs improves the surface water quality, the growth of aquatic plants and aquatic biodiversity which is important for maintaining a well-balanced environment and ecosystem. The operation of FTWs delivers benefits including reducing land requirement and creating an ecosystem.

FTWs have several advantages over constructed wetlands. They can manage the variations in water columns, have an attractive appeal (particularly when using blooming macrophytes), provide natural surroundings for wildlife, and offer sensory green in the city (Barco *et al.*, 2021). FTWs can be integrated with other methods to improve the treatment efficiency and provide numerous benefits. For example, a hybrid system consisting of an anaerobic biofilter and a combination of FTWs with microbial fuel cell (MFC) to treat wastewater from the university campus and bioelectricity generation. The combined system shows the potential to treat wastewater and energy production by integrating FTWs and MFC (Colares *et al.*, 2021).

Pollutant removal by FTWs depends on plant absorption and interception, as well as the presence of microorganisms on the plant root surface (Meng *et al.*, 2023). However, numerous studies have revealed that plants have a limited ability to remove contaminants, particularly during low-temperature seasons. Low temperatures during the cold season may severely impede the growth of floating plants, causing them to wither or decline. Besides, wilted or dead plants release nutrients, creating secondary contamination in water bodies. Furthermore, low temperatures will impact microbial activity, nitrification, and denitrification, reducing the purifying effectiveness of contaminants when utilizing FTWs. As a result, improving the purifying effectiveness of the ecological floating layer at low temperatures is critical (Meng *et al.*, 2023).

To overcome the disadvantages as well as improve the treatment performance of FTWs, the utilization of aeration, synthetic biofilm carriers, and macrophytes are necessary. Substrate supplementation is a good approach to enhance the purifying efficiency of FTWs. The substrate's high surface area can create extra niches for microbes to adhere and grow, raising the microbial biomass. Straw and bamboo are biological substrates. Bio-substrates, ceramic materials, zeolite, sponge iron, and artificial polymers with large specific surface areas can all boost microbial biomass attachment. Iron-rich substrates can release iron ions to improve the phosphate elimination (Meng *et al.*, 2023).

3.6 CASE STUDIES OF FTWS

In recent decades, FTWs have been applied to enhance the quality of rainwater, ponds, lakes, rivers, domestic wastewater, and landfill wastewater (Gaballah *et al.*, 2021; Nivala *et al.*, 2019). Some previous studies on FTWs application indicate that in different scales, including laboratory scale, pilot scale, and full scale, they show very different pollutant treatment effects. A recent study in Italy applied eight FTWs to treat different wastewater types, including aquaculture and canal water, domestic wastewater and irrigation water for more than ten years (Barco & Borin, 2020). *Phragmites australis, Iris pseudacorus* L., *Typha latifolia* L., *Alnus glutinosa* L., and *Carex* spp. were vegetated in an ethylene vinyl acetate (EVA) floating system. The study found that FTW contributed to the removal of *E. coli*, with removal efficiencies ranging from 19% to 95%. This study investigated that UV light was the main removal mechanism in un-planted systems, whereas the *E. coli* elimination was about 90% and 55% for *Cyperus alternifolius* and *P. australis* FTW, respectively. They explained that the release of different organic compounds, antibiotic substances, and H^+ cations from the root system into the aquatic environment interfered with the COD removal. In the studies of Benvenuti *et al.* (2018), Effendi *et al.* (2018), and Tara *et al.* (2019), the COD removal efficiencies amounted to 55.6%, 88.6%, and 92% by using *T. domingensis* Pers., *Chrysopogon zizanioides* L., and *Phragmites australis*, respectively. *Cyperus papyrus* was grown to treat wastewater in the research of Benrahmane *et al.* (2022) under environmental temperature conditions between 17.5°C and 27.0°C and 2 and 4 days of hydraulic retention time (HRT). The removal efficiency of COD, BOD_5, and SS was doubled during the 4-day HRT compared to the 2-day HRT. These findings showed that the yield is more efficient the higher the HRT. From these studies, *Phragmites australis* presented the highest COD removal in FTWs. This species is the most prevalent in FTW (Barco *et al.*, 2021; Li & Guo, 2017; Saeed *et al.*, 2016; Wu *et al.*, 2011).

In addition to reducing COD, and pathogen removal, FTWs have been used for nutrient removal. The agricultural runoff treatment by *Pontederia cordata* reached a TP and TN removal efficiency ranging from 90.3–92.4%, and 84.3–88.9%, respectively (Spangler *et al.*, 2019b). Moreover, the removal of TN and PO_4^{3-}-P from textile industry wastewater were 87% and 61%, respectively, at the HRT of 10 days by *Phragmites australis* (Tara *et al.*, 2019). Another study in Vietnam investigated that some tropical climate aquatic vegetables were also suitable for application in FTW to improve surface water quality. After 7 days of HRT, SS, COD, BOD_5, and NH_4^+-N treatment, the performance amounted to, respectively, 93%, 90%, 94%, and 93.4% for nearly a month. Common plants that have been used to treat fish farm wastewater in An Giang province (Vietnam) include *Eichhornia crassipes*, *Ipomoea aquatica*, and *Commelina diffusa* (Table 3.2).

Table 3.2 FTW studies for remediation of various water/wastewater types representing the type of wastewater, plants used, and average pollutant removal efficiencies.

Wastewater Type	Plant Species	HRT (days)	Removal (%)	Reference
Textile wastewater	*Chrysopogon zizanioides L.*	49	NH_4^+-N: 78.14 BOD: 97.8 COD: 88.6	Effendi *et al.* (2018)
Simulated runoff	*Juncus effusus L Canna flaccida L*	2	TN: 67 TP: 49	Garcia Chance and White (2018)
Textile industry wastewater	*Phragmites australis*	10	COD: 92, BOD: 91 TN: 87 PO_4^{3-}-P: 61	Tara *et al.* (2019)
Residential stormwater runoff	*Carex appressa*	–	NH_4^+-N: 50 TP: 17	Schwammberger *et al.* (2019)
Synthetic samples receiving runoff and sewer	*Typha domingensis*	28	TP: 95% NH_4^+-N: 94-96	Di Luca *et al.* (2019)
Domestic wastewater	*T. domingensis Pers.*	11.5	NH_4^+-N: 40 TP: 37 COD: 55 TSS: 78	Benvenuti *et al.* (2018)
Agricultural runoff		7		Spangler *et al.* (2019b)
River water	*Brachia mutica, Leptochala fusca, P. australis, and T. dominigenesis*	35	Fe, Mn, Ni, Pb, and Cr to 0.53, 0.2, 0.09, 1.04, and 0.07 mg L^{-1}	Shahid *et al.* (2020)

3.7 CONCLUSIONS

FTWs are a nature-based solution for removing diffuse chemical and microbiological pollution in water systems. While organic materials can be eliminated by physical filtration or trapping through the root systems as well as being consumed by the plant root biofilms, FTWs acquire nutrients from water through plant uptake and bioprocesses occurring in the biofilm attaching to the plant roots. Furthermore, sedimentation, attachment/trapping on the roots, UV radiation, and inhibition by high DO concentrations are the primary mechanisms for pathogen elimination. Floating mat buoyancy and vegetation are important in designing FTWs. Selecting suitable native plants can reduce maintenance costs while simultaneously enhancing treatment efficacy. Moreover, FTWs combined with media/substrates could improve the treatment performance. Utilizing materials such as coconut fiber, charcoal, and oyster shells reduces environmental pollution. Besides, additional technologies including aeration, water fountains, and lighting can raise the dissolved oxygen levels in the water body and create an aesthetic landscape attractive to the community.

ACKNOWLEDGEMENTS

This book chapter is funded by Vietnam National University Ho Chi Minh City (VNU-HCM) under Grant C2025 (Thi-Viet-Huong Dao), and Science Foundation Ireland (SFI) and the Department of Foreign Affairs (DFA) under the SDG challenge project Floating Treatment Wetland, supporting teams to develop solutions addressing challenges under UN SDG3: Good Health and Well-being (grant No SFI/21/FIP/SDG/9933). Ms Pham-Yen-Nhi Tran would like to thank the financial support of the MURATA SCIENCE FOUNDATION (grant No 24VH06). Besides, we would like to thank Thi-Anh-Dao Luong, Huynh-Phuong Nguyen, Le-Tuan-Hai Lam, and Quang-Minh Nhat Ho for their support.

REFERENCES

Al-mayyahi T. F. and Sosa A. A. (2017). Chemical study of some species for *Cyperus* L. (Cyperaceae) in Diwaniyah river using gas chromatography – mass spectrometry. *1st Int. Sci. Conf. 2014*, 17–33.

Alufasi R., Gere J., Chakauya E., Lebea P., Parawira W. and Chingwaru W. (2017). Mechanisms of pathogen removal by macrophytes in constructed wetlands. *Environmental Technology Reviews*, **6**, 135–144, https://doi.org/10.1080/21622515.2017.1325940

Andersson K., Lawrence D., Zavaleta J. and Guariguata M. R. (2016). More trees, more poverty? The socioeconomic effects of tree plantations in Chile, 2001–2011. *Environmental Management*, **57**, 123–136, https://doi.org/10.1007/s00267-015-0594-x

Barco A. and Borin M. (2020). Treatment performances of floating wetlands: a decade of studies in North Italy. *Ecological Engineering*, **158**, 106016, https://doi.org/10.1016/j.ecoleng.2020.106016

Barco A., Bona S. and Borin M. (2021). Plant species for floating treatment wetlands: a decade of experiments in North Italy. *Science of the Total Environment*, **751**, 141666, https://doi.org/10.1016/j.scitotenv.2020.141666

Bauer L. H., Arenzon A., Molle N. D., Rigotti J. A., Borges A. C. d. A. C. A. A., Machado N. R. and Rodrigues L. H. R. H. R. (2021). Floating treatment wetland for nutrient removal and acute ecotoxicity improvement of untreated urban wastewater. *International Journal of Environmental Science and Technology*, **18**, 3697–3710, https://doi.org/10.1007/s13762-020-03124-x

Benrahmane L., Mouhir L., Kabbour A., Laaouan M. and El Hafidi M. (2022). Effectiveness of floating treatment wetlands with *Cyperus papyrus* used in sub-humid climate to treat urban wastewater: a case study. *Journal of Ecological Engineering*, **23**(7), 157–168, https://doi.org/10.12911/22998993/149852

Benvenuti T., Hamerski F., Giacobbo A., Bernardes A. M. A. M. A. M., Zoppas-Ferreira J. and Rodrigues M. A. A. S. S. (2018). Constructed floating wetland for the treatment of domestic sewage: a real-scale study. *Journal of Environmental Chemical Engineering*, **6**, 5706–5711, https://doi.org/10.1016/j.jece.2018.08.067

Borne K. E., Fassman-Beck E. A., Winston R. J., Hunt W. F. and Tanner C. C. (2015). Implementation and maintenance of floating treatment wetlands for urban stormwater management. *Journal of Environmental Engineering*, **141**, 1, https://doi.org/10.1061/(ASCE)EE.1943-7870.0000959

Bruze A. (2017). VT 2017 A comparison of nutrient reduction between activated carbon and coconut fibre in wastewater treatment.

Chen Z., Cuervo D. P., Müller J. A., Wiessner A., Köser H., Vymazal J., Kästner M. and Kuschk P. (2016). Hydroponic root mats for wastewater treatment – a review. *Environmental Science and Pollution Research*, **23**, 15911–15928, https://doi.org/10.1007/s11356-016-6801-3

Chen Y., Li L., Jiang L. R., Tan J. Y., Guo L. N., Wang X. L., Dong W., Wang W. B., Sun J. K. and Song B. (2022). Alkaloids constituents from the roots of *Phragmites australis* (Cav.) Trin. ex Steud. with their cytotoxic activities. *Natural Product Research*, **36**, 1454–1459, https://doi.org/10.1080/14786419.2021.1888291

Chowdhary P., Raj A. and Bharagava R. N. (2018). Environmental pollution and health hazards from distillery wastewater and treatment approaches to combat the environmental threats: a review. *Chemosphere*, **194**, 229–246, https://doi.org/10.1016/j.chemosphere.2017.11.163

Colares G. S., Dell'Osbel N., Wiesel P. G., Oliveira G. A., Lemos P. H. Z., da Silva F. P., Lutterbeck C. A., Kist L. T. and Machado Ê. L. (2020). Floating treatment wetlands: a review and bibliometric analysis. *Science of the Total Environment*, **714**, 136776, https://doi.org/10.1016/j.scitotenv.2020.136776

Colares G. S., Dell'Osbel N., Barbosa C. V., Lutterbeck C., Oliveira G. A., Rodrigues L. R., Bergmann C. P., Lopez D. R., Rodriguez A. L., Vymazal J. and Machado E. L. (2021). Floating treatment wetlands integrated with microbial fuel cell for the treatment of urban wastewaters and bioenergy generation. *Science of the Total Environment*, **766**, 142474, https://doi.org/10.1016/j.scitotenv.2020.142474

Čule N., Lučić A., Nešić M., Brašanac-Bosanac L., Mitrović S., Veselinović M. and Ćirković-Mitrović T. (2021). The removal of pathogenic microorganisms in a biological system with floating islands. *Sustainable Forestry: Collection*, 13–25, https://doi.org/10.5937/SustFor2183013C

de Moya Sánchez Á, Martínez H. C., Ramírez X. V. and Osorio A. C. (2021). Chromium and zinc removal from synthetic industrial wastewater in pilot-scale constructed wetlands planted with *Cyperus odoratus* L. *INGE CUC*, **17**(2), 76–88, https://doi.org/10.17981/ingecuc.17.2.2021.08

Di Luca G. A., Mufarrege M. M., Hadad H. R. and Maine M. A. (2019). Nitrogen and phosphorus removal and *Typha domingensis* tolerance in a floating treatment wetland. *Science of the Total Environment*, **650**, 233–240, https://doi.org/10.1016/j.scitotenv.2018.09.042

Effendi H., Margaretha J. and Krisanti M. (2018). Reducing ammonia and chromium concentration in batik wastewater by vetiver (*Chrysopogon zizanioides* L.) grown in floating wetland. *Applied Ecology and Environmental Research*, **16**, 2947–2956, https://doi.org/10.15666/aeer/1603_29472956

Fletcher J., Willby N., Oliver D. and Quilliam R. S. (2024). Engineering aquatic plant community composition on floating treatment wetlands can increase ecosystem multifunctionality. *Environmental Research*, **243**, 117818, https://doi.org/10.1016/j.envres.2023.117818

Gaballah M. S., Ismail K., Aboagye D., Ismail M. M., Sobhi M. and Stefanakis A. I. (2021). Effect of design and operational parameters on nutrients and heavy metal removal in pilot floating treatment wetlands with *Eichhornia crassipes* treating polluted lake water. *Environmental Science and Pollution Research*, **28**, 25664–25678, https://doi.org/10.1007/s11356-021-12442-7

Gao L., Zhou W., Huang J., He S., Yan Y., Zhu W., Wu S. and Zhang X. (2017). Nitrogen removal by the enhanced floating treatment wetlands from the secondary effluent. *Bioresource Technology*, **234**, 243–252, https://doi.org/10.1016/j.biortech.2017.03.036

Garcia Chance L. M. and White S. A. (2018). Aeration and plant coverage influence floating treatment wetland remediation efficacy. *Ecological Engineering*, **122**, 62–68, https://doi.org/10.1016/j.ecoleng.2018.07.011

Geng B., Li Y., Liu X., Ye J. and Guo W. (2022). Effective treatment of aquaculture wastewater with mussel/microalgae/bacteria complex ecosystem: a pilot study. *Scientific Reports*, **12**, 1, https://doi.org/10.1038/s41598-021-99269-x

Ghosh S. K., Saha P. D. and Francesco Di M. (2020). Recent trends in wastewater treatment and water resource management. Springer, Nature, Singapore, pp. 224, https://doi.org/10.1007/978-981-15-0706-9

Gupta V., Courtemanche J., Gunn J. and Mykytczuk N. (2020). Shallow floating treatment wetland capable of sulfate reduction in acid mine drainage impacted waters in a northern climate. *Journal of Environmental Management*, **263**, 110351, https://doi.org/10.1016/j.jenvman.2020.110351

Hairom N. H. H., Soon C. F., Mohamed R. M. S. R., Morsin M., Zainal N., Nayan N., Zulkifli C. Z. and Harun N. H. (2021). A review of nanotechnological applications to detect and control surface water pollution. *Environmental Technology & Innovation*, **24**, 102032.

Headley T., Nivala J., Kassa K., Olsson L., Wallace S., Brix H., Afferden M. V. and Müller R. (2013). *Escherichia coli* removal and internal dynamics in subsurface flow ecotechnologies: effects of design and plants. *Ecological Engineering*, **61**, 564–574, https://doi.org/10.1016/j.ecoleng.2013.07.062

Henny C., Kurniawan R. and Trisuryono (2019). Application of floating treatment wetlands in a highly eutrophic lake, Indonesia: a new tool for lake restoration and provision of microhabitat. *Earth and Environmental Sciences*, **380**, 12003, https://doi.org/10.1088/1755-1315/380/1/012003

Hiremath P. G., Chennabasappa M., Mallik C. and Thejashree V. (2024). Fluoride removal using tartaric acid-modified rice husk biochar: comprehensive batch and column studies. *Sustainable Chemistry One World*, **2**, 100005, https://doi.org/10.1016/j.scowo.2024.100005

Jassim N. and AlAmeri M. (2020). Single objective optimization of surface water coagulation process using inorganic/organic aid formulation by Taguchi method. *Periodicals of Engineering and Natural Sciences*, **8**, 1924–1934.

Kadam S. K., Tamboli A. S., Chandanshive V. V., Govindwar S. P., Choo Y. S. and Pak J. H. (2022). Construction and implementation of floating wetpark as effective constructed wetland for industrial textile wastewater treatment. *Journal of Hazardous Materials*, **424**, 127710, https://doi.org/10.1016/j.jhazmat.2021.127710

Kansiime F. and Van Bruggen J. J. A. (2001). Distribution and retention of faecal coliforms in the Nakivubo wetland in Kampala, Uganda. 199–206.

Kumwimba M. N., Zhu B., Muyembe D. K. and Dzakpasu M. (2017). Growth characteristics and nutrient removal capability of eco-ditch plants in mesocosm sediment receiving primary domestic wastewater. *Environmental Science and Pollution Research*, **24**, 23926–23938, https://doi.org/10.1007/s11356-017-9992-3

Landaverde A. C., Strosnider W. H. J. and White S. A. (2024). Plant suitability for floating treatment wetland applications in brackish waters. *Ecological Engineering*, **200**, 107183, https://doi.org/10.1016/j.ecoleng.2024.107183

Leto C., Tuttolomondo T., La Bella S., Leone R. and Licata M. (2013). Effects of plant species in a horizontal subsurface flow constructed wetland–phytoremediation of treated urban wastewater with *Cyperus alternifolius* L. and *Typha latifolia* L. in the West of Sicily (Italy). *Ecological Engineering*, **61**, 282–291, https://doi.org/10.1016/j.ecoleng.2013.09.014

Li X. and Guo R. C. (2017). Comparison of nitrogen removal in floating treatment wetlands constructed with *Phragmites australis* and *Acorus calamus* in a cold temperate zone. *Water, Air, & Soil Pollution*, **228**, 132, https://doi.org/10.1007/s11270-017-3266-z

Li C. J., Wan M. H., Dong Y., Men Z. Y., Lin Y., Wu D. Y. and Kong H. N. (2011). Treating surface water with low nutrients concentration by mixed substrates constructed wetlands. *Journal of Environmental Science and Health, Part A*, **46**, 771–776, https://doi.org/10.1080/10934529.2011.571632

Liu M., Li B., Xue Y., Wang H. and Yang K. (2017). Constructed wetland using corncob charcoal substrate: pollutants removal and intensification. *Water Science and Technology*, **76**, 1300–1307, https://doi.org/10.2166/wst.2017.305

Long Y., Zhang Z., Pan X., Li B., Xie S. and Guo Q. (2016). Substrate influences on archaeal and bacterial assemblages in constructed wetland microcosms. *Ecological Engineering*, **94**, 437–442, https://doi.org/10.1016/j.ecoleng.2016.06.015

Lu S., Zhang X., Wang J. and Pei L. (2016). Impacts of different media on constructed wetlands for rural household sewage treatment. *Journal of Cleaner Production*, **127**, 325–330, https://doi.org/10.1016/j.jclepro.2016.03.166

Lucke T., Walker C. and Beecham S. (2019). Experimental designs of field-based constructed floating wetland studies: a review. *Science of the Total Environment*, **660**, 199–208, https://doi.org/10.1016/j.scitotenv.2019.01.018

Mahmoud T. N., El-Maadawy W. H., Kandil Z. A., Khalil H., El-fiky N. M. and El Alfy T. S. M. A. (2021). *Canna* × *generalis* L.H. Bailey rhizome extract ameliorates dextran sulfate sodium-induced colitis via modulating intestinal mucosal dysfunction, oxidative stress, inflammation, and TLR4/ NF-κs and NLRP3 inflammasome pathways. *Journal of Ethnopharmacology*, **269**, 113670, https://doi.org/10.1016/j.jep.2020.113670

Malaviya P., Singh A. and Anderson T. A. (2020). Aquatic phytoremediation strategies for chromium removal. *Reviews in Environmental Science and Bio/Technology*, **19**, 897–944, https://doi.org/10.1007/s11157-020-09552-y

Maurya A., Singh M. K. and Kumar S. (2020). Biofiltration technique for removal of waterborne pathogens. In: Waterborne pathogens. Elsevier, Butterworth-Heinemann, pp. 123–141.

Meng T., Cheng W., Ren J. and Wang M. (2023). Effect of zeolite–sponge iron combined substrate on microbial community in ecological floating bed. *Environmental Research Communications*, **5**, 085011, https://doi.org/10.1088/2515-7620/acf02c

Mfarrej M. F. B., Wang X., Fahid M., Saleem M. H., Alatawi A., Ali S., Shabir G., Zafar R., Afzal M. and Fahad S. (2023). Floating treatment wetlands (FTWs) is an innovative approach for the remediation of petroleum hydrocarbons-contaminated water. *Journal of Plant Growth Regulation*, **42**, 1402–1420, https://doi.org/10.1007/s00344-022-10674-6

Moktar K. A., Mazlan W. S. and Tajuddin R. M. (2018). Wastewater treatment using phytoremediation by *Imperata cylindrica* and *Heliconia psittacorum*. *Journal of Built Environment*, **5**, 34–39.

Nakase C., Zurita F., Nani G., Reyes G., Fernández-Lambert G., Cabrera-Hernández A. and Sandoval L. (2019). Nitrogen removal from domestic wastewater and the development of tropical ornamental plants in partially saturated mesocosm-scale constructed wetlands. *International Journal of Environmental Research and Public Health*, **16**, 4800, https://doi.org/10.3390/ijerph16234800

Nivala J., Boog J., Headley T., Aubron T., Wallace S., Brix H., Mothes S., van Afferden M. and Müller R. A. (2019). Side-by-side comparison of 15 pilot-scale conventional and intensified subsurface flow wetlands for treatment of domestic wastewater. *Science of the Total Environment*, **658**, 1500–1513, https://doi.org/10.1016/j.scitotenv.2018.12.165

Odinga C. A., Swalaha F. M., Otieno F. A. O., Ranjith K. R. and Bux F. (2013). Investigating the efficiency of constructed wetlands in the removal of heavy metals and enteric pathogens from wastewater. *Environmental Technology Reviews*, **2**, 1–16, https://doi.org/10.1080/21622515.2013.865086

Okechukwu J. O., Osemeahon S. A. and Dimas B. J. (2021). Development and evaluation of the performance of ceramic water filter prepared from dried duckweed plant. *African Journal of Environment and Natural Science Research*, **4**, 53.

Omotade I. F., Alatise M. O. and Olanrewaju O. O. (2019). Recycling of aquaculture wastewater using charcoal based constructed wetlands. *International Journal of Phytoremediation*, **21**, 399–404, https://doi.org/10.1080/15226514.2018.1537247

Palansooriya K. N., Yang Y., Tsang Y. F., Sarkar B., Hou D., Cao X., Meers E., Rinklebe J., Kim K.-H. and Ok Y. S. (2020). Occurrence of contaminants in drinking water sources and the potential of biochar for water quality improvement: a review. *Critical Reviews in Environmental Science and Technology*, **50**, 549–611, https://doi.org/10.1080/10643389.2019.1629803

Pang Q., Xu W., He F., Peng F., Zhu X., Xu B., Yu J., Jiang Z. and Wang L. (2022). Functional genera for efficient nitrogen removal under low C/N ratio influent at low temperatures in a two-stage tidal flow constructed wetland. *Science of the Total Environment*, **804**, 150142, https://doi.org/10.1016/j.scitotenv.2021.150142

Parde D., Patwa A., Shukla A., Vijay R., Killedar D. J. and Kumar R. (2021). A review of constructed wetland on type, treatment and technology of wastewater. *Environmental Technology & Innovation*, **21**, 101261, https://doi.org/10.1016/j.eti.2020.101261

Parjane S. B. and Sane M. G. (2011). Performance of grey water treatment plant by economical way for Indian rural development. *International Journal of Chemtech Research*, **3**, 1808–1815.

Park J. B. K. K., Sukias J. P. S. S. and Tanner C. C. (2019). Floating treatment wetlands supplemented with aeration and biofilm attachment surfaces for efficient domestic wastewater treatment. *Ecological Engineering*, **139**, 105582, https://doi.org/10.1016/j.ecoleng.2019.105582

Pavlineri N., Skoulikidis N. T. and Tsihrintzis V. A. (2017). Constructed floating wetlands: a review of research, design, operation and management aspects, and data meta-analysis. *Chemical Engineering Journal*, **308**, 1120–1132, https://doi.org/10.1016/j.cej.2016.09.140

Peterson E. W., Nicodemus P., Spooner E. and Heath A. (2021). The effectiveness of an artificial floating wetland to remove nutrients in an urban stream: a pilot-study in the Chicago river, Chicago, IL, USA. *Hydrology*, **8**, 115, https://doi.org/10.3390/hydrology8030115

Prochaska C. A. and Zouboulis A. I. (2006). Removal of phosphates by pilot vertical-flow constructed wetlands using a mixture of sand and dolomite as substrate. *Ecological Engineering*, **26**, 293–303, https://doi.org/10.1016/j.ecoleng.2005.10.009

Qasem N. A. A., Mohammed R. H. and Lawal D. U. (2021). Removal of heavy metal ions from wastewater: a comprehensive and critical review. *NPJ Clean Water* **4**, https://doi.org/10.1038/s41545-021-00127-0

Qiao Y., Ma C. and Xu K. (2022). Nutrient removal by floating treatment wetlands under different spatial arrangement modes: a field study. *Environmental Management*, **70**, 420–430, https://doi.org/10.1007/s00267-022-01677-w

Sabara Z., Anwar A., Yani S., Prianto K., Junaidi R., Umam R. and Prastowo R. (2022). Activated carbon and coconut coir with the incorporation of ABR system as greywater filter: the implications for wastewater treatment. *Sustain*, **14**, https://doi.org/10.3390/su14021026

Saeed T., Paul B., Afrin R., Al-Muyeed A. and Sun G. (2016). Floating constructed wetland for the treatment of polluted river water: a pilot scale study on seasonal variation and shock load. *Chemical Engineering Journal*, **287**, 62–73, https://doi.org/10.1016/j.cej.2015.10.118

Samal K., Kar S. and Trivedi S. (2019). Ecological floating bed (EFB) for decontamination of polluted water bodies: design, mechanism and performance. *Journal of Environmental Management*, **251**, 109550, https://doi.org/10.1016/j.jenvman.2019.109550

Sasakova N., Gregova G., Takacova D., Mojzisova J., Papajova I., Venglovsky J., Szaboova T. and Kovacova S. (2018). Pollution of surface and ground water by sources related to agricultural activities. *Frontiers in Sustainable Food Systems*, **2**, 42, https://doi.org/10.3389/fsufs.2018.00042

Savio N., Pandey D. and Srivastava R. K. (2023). Potentialities of plant based hybrid wetland systems for the treatment of household waste water using *Canna indica, Agave americana, Pistia stratiotes* and *Tagetes erecta*. *Materials Today: Proceedings*, **77**, 217–222, https://doi.org/10.1016/j.matpr.2022.11.264

Schwammberger P. F., Lucke T., Walker C. and Trueman S. J. (2019). Nutrient uptake by constructed floating wetland plants during the construction phase of an urban residential development. *Science of the Total Environment*, **677**, 390–403, https://doi.org/10.1016/j.scitotenv.2019.04.341

Schwammberger P. F., Yule C. M. and Tindale N. W. (2020). Rapid plant responses following relocation of a constructed floating wetland from a construction site into an urban stormwater retention pond. *Science of the Total Environment*, **699**, 134372, https://doi.org/10.1016/j.scitotenv.2019.134372

Selvaraj D. and Velvizhi G. (2021). Sustainable ecological engineering systems for the treatment of domestic wastewater using emerging, floating and submerged macrophytes. *Journal of Environmental Management*, **286**, 112253, https://doi.org/10.1016/j.jenvman.2021.112253

Shahid M. J., Arslan M., Ali S., Siddique M. and Afzal M. (2018). Floating wetlands: a sustainable tool for wastewater treatment. *Clean – Soil, Air, Water*, **46**, 1800120, https://doi.org/10.1002/clen.201800120

Shahid M. J., Ali S., Shabir G., Siddique M., Rizwan M., Seleiman M. F. and Afzal M. (2020). Comparing the performance of four macrophytes in bacterial assisted floating treatment wetlands for the removal of trace metals (Fe, Mn, Ni, Pb, and Cr) from polluted river water. *Chemosphere*, **243**, 125353, https://doi.org/10.1016/j.chemosphere.2019.125353

Sharma R., Vymazal J. and Malaviya P. (2021). Application of floating treatment wetlands for stormwater runoff: a critical review of the recent developments with emphasis on heavy metals and nutrient removal. *Science of the Total Environment*, **777**, 146044, https://doi.org/10.1016/j.scitotenv.2021.146044

Shen S., Li X. and Lu X. (2021). Recent developments and applications of floating treatment wetlands for treating different source waters: a review. *Environmental Science and Pollution Research*, **28**, 62061–62084, https://doi.org/10.1007/s11356-021-16663-8

Sijimol M. R., Mansa M. L. and Joseph S. (2021). Analysis of potential of *Napier grass, Vetiver* and *Equisetum* plants for the treatment of domestic greywater using box-type constructed wetlands. *Water Science and Technology*, **84**, 2913–2922, https://doi.org/10.2166/wst.2021.289

Sikdar P. K. (2021). Environmental management: issues and concerns BT. In: Environmental management: issues and concerns in developing countries, P. K. Sikdar (Ed.), Springer International Publishing, Cham, pp. 1–21, https://doi.org/10.1007/978-3-030-62529-0_1

Singh N. K., Gupta G., Upadhyay A. K. and Rai U. N. (2019). Biological wastewater treatment for prevention of river water pollution and reuse: perspectives and challenges. In: Water conservation, recycling and reuse: issues and challenges, R. P. Singh, A. S. Kolok and S. L. Bartelt-Hunt (eds), Springer Singapore, Singapore, pp. 81–93, https://doi.org/10.1007/978-981-13-3179-4_4

Spangler J. T., Sample D. J., Fox L. J., Albano J. P. and White S. A. (2019a). Assessing nitrogen and phosphorus removal potential of five plant species in floating treatment wetlands receiving simulated nursery runoff. *Environmental Science and Pollution Research*, **26**, 5751–5768, https://doi.org/10.1007/s11356-018-3964-0

Spangler J. T., Sample D. J., Fox L. J., Owen J. S. and White S. A. (2019b). Floating treatment wetland aided nutrient removal from agricultural runoff using two wetland species. *Ecological Engineering*, **127**, 468–479, https://doi.org/10.1016/j.ecoleng.2018.12.017

Tara N., Iqbal M., Mahmood Khan Q. and Afzal M. (2019). Bioaugmentation of floating treatment wetlands for the remediation of textile effluent. *Water and Environment Journal*, **33**, 124–134, https://doi.org/10.1111/wej.12383

Tian T., Tam N. F. Y., Zan Q., Cheung S. G., Shin P. K. S., Wong Y. S., Zhang L. and Chen Z. (2017). Performance and bacterial community structure of a 10-years old constructed mangrove wetland. *Marine Pollution Bulletin*, **124**, 1096–1105, https://doi.org/10.1016/j.marpolbul.2017.07.005

Toscano A., Langergraber G., Consoli S. and Cirelli G. L. (2009). Modelling pollutant removal in a pilot-scale two-stage subsurface flow constructed wetlands. *Ecological Engineering*, **35**, 281–289, https://doi.org/10.1016/j.ecoleng.2008.07.011

Tran P. Y. N., Dao T. V. H., Vo T. K. Q., Nguyen T. A. C., Nguyen T. M. X., Tran C. S., Nguyen T. Y. P., Le L. T., Tra V. T., Phan N. N., Lens P. N. L. and Bui X. T. (2024). Enhanced pollution removal from canal water by coupling aeration to floating treatment wetlands. *International Journal of Phytoremediation*, **0**, 1–12, https://doi.org/10.1080/15226514.2024.2401957

Verma A., Chand N., Upadhyay P., Sharma S. and Prajapati S. K. (2024). Systematic review on microplastics as a threat to terrestrial and aquatic eco-environment. *Sustainable Chemistry One World*, **3**, 100013, https://doi.org/10.1016/j.scowo.2024.100013

Vo T.-K.-Q., Vo T.-D.-H., Ntagia E., Amulya K., Nguyen N.-K.-Q., Tran P.-Y.-N., Ninh N.-T.-T., Le S.-L., Le L.-T., Tran C.-S., Ha T.-L., Pham M.-D.-T., Bui X.-T. and Lens P. N. L. (2023). Pilot and full scale applications of floating treatment wetlands for treating diffuse pollution. *Science of the Total Environment*, **899**, 165595, https://doi.org/10.1016/j.scitotenv.2023.165595

Vymazal J. (2013). Emergent plants used in free water surface constructed wetlands: a review. *Ecological Engineering*, **61**, 582–592, https://doi.org/10.1016/j.ecoleng.2013.06.023

Wei F., Shahid M. J., Alnusairi G. S. H., Afzal M., Khan A., El-Esawi M. A., Abbas Z., Wei K., Zaheer I. E., Rizwan M. and Ali S. (2020). Implementation of floating treatment wetlands for textile wastewater management: a review. *Sustain*, **12**, 1–29, https://doi.org/10.3390/su12145801

Wu H., Zhang J., Li P., Zhang J., Xie H. and Zhang B. (2011). Nutrient removal in constructed microcosm wetlands for treating polluted river water in northern China. *Ecological Engineering*, **37**, 560–568, https://doi.org/10.1016/j.ecoleng.2010.11.020

Wu XiaoFeng W. X., Zou XiaoBiao Z. X. and Li LinFeng L. L. (2012). A study on purification capacity of different substrates on the treatment performance of vertical flow constructed wetland.

Wu H., Zhang J., Ngo H. H., Guo W., Hu Z., Liang S., Fan J. and Liu H. (2015). A review on the sustainability of constructed wetlands for wastewater treatment: design and operation. *Bioresource Technology*, **175**, 594–601, https://doi.org/10.1016/j.biortech.2014.10.068

Wu S., Carvalho P. N., Müller J. A., Manoj V. R. and Dong R. (2016). Sanitation in constructed wetlands: a review on the removal of human pathogens and fecal indicators. *Science of the Total Environment*, **541**, 8–22, https://doi.org/10.1016/j.scitotenv.2015.09.047

Xu Z., Valeo C., Chu A. and Zhao Y. (2021). The efficacy of whole oyster shells for removing copper, zinc, chromium, and cadmium heavy metal ions from stormwater. *Sustainability*, **13**, 4184, https://doi.org/10.3390/su13084184

Yang C., Zhang X., Tang Y., Jiang Y., Xie S., Zhang Y. and Qin Y. (2022). Selection and optimization of the substrate in constructed wetland: a review. *Journal of Water Process Engineering*, **49**, 103140, https://doi.org/10.1016/j.jwpe.2022.103140

Yeh N., Yeh P. and Chang Y.-H. (2015). Artificial floating islands for environmental improvement. *Renewable and Sustainable Energy Reviews*, **47**, 616–622, https://doi.org/10.1016/j.rser.2015.03.090

Yu G., Li P., Wang G., Wang J., Zhang Y., Wang S., Yang K., Du C. and Chen H. (2021). A review on the removal of heavy metals and metalloids by constructed wetlands: bibliometric, removal pathways, and key factors. *World Journal of Microbiology and Biotechnology*, **37**, 1–12, https://doi.org/10.1007/s11274-020-02944-w

Zhang L., Zhao J., Cui N., Dai Y., Kong L., Wu J. and Cheng S. (2016). Enhancing the water purification efficiency of a floating treatment wetland using a biofilm carrier. *Environmental Science and Pollution Research*, **23**, 7437–7443, https://doi.org/10.1007/s11356-015-5873-9

Zhang G., Ma K., Zhang Z., Shang X. and Wu F. (2020). Waste brick as constructed wetland fillers to treat the tail water of sewage treatment plant. *Bulletin of Environmental Contamination and Toxicology*, **104**, 273–281, https://doi.org/10.1007/s00128-020-02782-4

Zhang F., Wang J., Li L., Shen C., Zhang S., Zhang J., Liu R. and Zhao Y. (2024). Technologies for performance intensification of floating treatment wetland – An explicit and comprehensive review. *Chemosphere*, **348**, 140727, https://doi.org/10.1016/j.chemosphere.2023.140727

doi: 10.2166/9781789065015_0061

Chapter 4

Constructed wetlands for sustainable bioremediation of antibiotics from wastewater: potential and challenge

T. A. H. Nguyen[1], *T. H. Bui*[2], *T. T. Nguyen*[3], *Dan A* [4] *and S. Soda*[5]

[1]*VNU Vietnam Japan University, Faculty of Advanced Technologies and Engineering, Hanoi, Vietnam*
[2]*VNU Vietnam Japan University, Master's Program in Environmental Engineering, Hanoi, Vietnam*
[3]*Ritsumeikan University, Asia-Japan Research Institute, Shiga, Japan*
[4]*Zhongkai University of Agriculture and Engineering, College of Resources and Environment, Guangzhou, China*
[5]*Ritsumeikan University, Department of Civil and Environmental Engineering, Shiga, Japan*
Corresponding author: nta.hang@vju.ac.vn

ABSTRACT

Due to a growing concern about antibiotic pollution and resistance, there is an urgent need to seek cost-effective and eco-friendly antibiotic treatment technologies. This book chapter provides a comprehensive overview of constructed wetlands (CWs) as a promising solution for antibiotic abatement in wastewater. It commences by comparing various antibiotic pollution sources and treatment technologies. The focus is placed on evaluating the antibiotic removal in CWs, considering simultaneous effects of multiple influential factors, including antibiotic properties, wastewater characteristics, as well as CWs design and operation parameters. Moreover, it offers insights on antibiotic removal mechanisms. Finally, present obstacles and future perspectives associated with antibiotic removal by CWs are discussed. It was found that livestock was the major source of antibiotic pollution. The antibiotic removal efficacy of CWs can vary, relying on antibiotic properties, CW design and operation parameters. Among various removal pathways of antibiotics in CWs, biodegradation plays a key role, followed by substrate adsorption, plant uptake, and other processes. However, the widespread application of CWs for antibiotic decontamination from wastewater is currently facing challenges, such as poor efficacy of conventional media, excessive cost of emerging substrates, low intolerance of plants to antibiotic-rich wastewater, and limited application of full-scale CWs with real wastewater. Key strategies for boosting antibiotic treatment in CWs include substrate renovation, microbial augmentation, and integration of grey and green technologies. This book chapter aims to facilitate the development of CWs as a viable, affordable, and sustainable technology for intensified purification of antibiotic contaminated wastewater.

Keywords: Antibiotics, constructed wetlands, enhancing strategies, influential factors, mechanisms, pollution sources, removal efficiency, treatment technologies, wastewater

4.1 ANTIBIOTIC POLLUTION SOURCES AND TREATMENT TECHNOLOGIES

4.1.1 Definition and classification of antibiotics

4.1.1.1 Definition

Antibiotics are defined as metabolites produced by biological and artificial means (Ma *et al.*, 2022), which can impede the growth, reproduction or kill bacteria (Button *et al.*, 2019). The antibacterial

© 2025 IWAP. This is an Open Access book chapter distributed under the terms of the Creative Commons Attribution License (CC BY-NC-ND 4.0) which permits copying and redistribution for non-commercial purposes with no derivatives, provided the work is properly cited (https://creativecommons.org/licenses/by-nc-nd/4.0/). The chapter is from the book *Nature-Based Solutions for Urban Sustainability*, Piet N.L. Lens and Xuan-Thanh Bui (Editors).

mechanisms involve damaging the cell structure (e.g., membrane disorganization) or interfering with the synthesis of macromolecules (e.g., proteins, nucleic acids) (Fu *et al.*, 2022). Despite their toxicity for bacteria, antibiotics are metabolized and excreted by the human body (Sosa-Hernández *et al.*, 2021).

4.1.1.2 Classification

There are a wide variety of antibiotic classification methods, which are based on formation method (biological, semi-biological, synthetic), administration route (injectable, oral), bacterial spectrum (broad, narrow), action mechanism (bactericidal, bacteriostatic), and chemical structure. Among them, antibiotic classification based on chemical structure is most commonly used. This book chapter uses the antibiotic classification based on chemical structures as they may affect the removal of antibiotics in wastewater.

It is estimated that tens of thousands of antibiotics are currently available worldwide. Based on their biochemical composition, antibiotics can be classified as diverse groups. The most widely used antibiotic categories include β-lactams (BLAs), fluoroquinolones (FQs), macrolides (MLs), sulfonamides (SAs), and tetracyclines (TCs) (He *et al.*, 2021). BLAs contain penicillin, which is the first discovered antibiotic. TCs include both biosynthetic antibiotics (e.g., tetracycline, oxytetracycline) and semi-synthetic antibiotics (e.g., doxycycline, lymecycline), which are referred to as first generation and second generation, respectively. SAs are considered as broad-spectrum antibiotics as they adversely affect both Gram-positive and Gram-negative bacteria, as well as some protozoa, mainly via bacteriostatic mechanisms (Choi *et al.*, 2016). The properties of common antibiotics are presented in Table 4.1.

4.1.2 Pollution sources of antibiotics

Antibiotics in water medium may derive from major sources, including agricultural, clinical, industrial, and domestic ones (Sosa-Hernández *et al.*, 2021) as shown in Figure 4.1.

4.1.2.1 Aquaculture

Aquaculture is one of the major economic sectors for many countries in the world. The 15 largest aquaculture production countries (2008–2018) consist of China, South Korea, Japan, India, Vietnam, Indonesia, Philippines, Bangladesh, Thailand, Myanmar, Malaysia (Asia), Egypt (Africa), Norway (Europe), Brazil, Chile (South America). China is the leading country, which accounted for 58% of the global aquaculture production in 2016 (Sosa-Hernández *et al.*, 2021).

To prevent infectious diseases from occurring during aquaculture, antibiotics are normally utilized in huge amounts and in uncontrolled manner. It was reported that 67 antibiotics were used in the 15 world leading countries in aquaculture (2008–2018), especially including the prohibited ones (Sosa-Hernández *et al.*, 2021). As a result of long-term exposure to antibiotics at low concentrations and natural selection pressure, some bacteria have developed antibiotic-resistant genes (ARGs). This not only affects the therapeutic efficacy of pathogenic diseases but also increases the number of pathogens in water and sediments (Huang *et al.*, 2019). The antibiotic concentrations in aquaculture wastewater vary in the ranges of 2.57–211.2 ng/L in mainland China (Huang *et al.*, 2019) and 59–217.6 ng/L in Taiwan (Hsieh *et al.*, 2015).

4.1.2.2 Livestock

The livestock sector is regarded as a main consumer of antibiotics for the goals of infectious diseases prevention and treatment, animal growth promotion, and feed additives production (Du *et al.*, 2020; García *et al.* 2020). The most frequently used antibiotics in livestock facilities in the world are SAs and FQs (Choi *et al.*, 2016). The heavy and improper use of veterinary antibiotics leads to substantial residues in the animal faeces and urine (García *et al.*, 2020). Consequently, the direct discharge of the animal faeces and urine into the environment, as well as their direct applications in agricultural land can promote the spread of antibiotics in the receiving environment (Carvalho *et al.*, 2014). The

Table 4.1 Properties and environmental concentrations of commonly used antibiotics.

Group	Antibiotic	Molecular Weight (g/mol)	pK_a	Log K_{ow}	Boiling Point (°C)	Solubility at pH 7; 25°C	Density (g/cm³)	Environmental Concentration (ng/L)	Reference
β-Lactams (BLAs)	AMP	349.4	2.44 ± 0.50	1.35	683.9 ± 55.0	0.87 g/L	1.45 ± 0.1	2–690	He *et al.* (2021)
Fluoroquinolones (FQs)	CIP	331.34	6.43 ± 0.41	0.28	581.8 ± 50.0	1.4 × 10⁻³ mol/L	1.461 ± 0.06	15–3350	Fu *et al.* (2022)
	ENR	359.39	6.43 ± 0.41	1.1	560.5 ± 50.0	0.14 g/L	1.385 ± 0.06	3–100	Sabri *et al.* (2021)
	NOR	319.33	1.344 ± 0.06	0.46	555.8 ± 50.0	1.1 × 10⁻³ mol/L	1.344 ± 0.06	128–2034	Dan *et al.* (2020)
	OFL	361.37	5.19 ± 0.40	−0.39	571.5 ± 50.0	0.18 g/L	1.48 ± 0.1	265–536	Dan *et al.* (2020)
Imidazoles	MTZ	171.15	14.44 ± 0.10		405.4 ± 25.0	405.4 ± 25.0	1.45 ± 0.1		Fu *et al.* (2022)
Lincosamide	LIN	406.54	12.91 ± 0.70	0.2	646.8 ± 55.0	350 g/L	1.29 ± 0.1	5850 ± 428	Chen *et al.* (2019b)
Macrolides (MLs)	AZM	748.98	13.28 ± 0.70	4.02	822.1 ± 65.0	1000 g/L	1.18 ± 0.1	53–6810	Fu *et al.* (2022)
	CLR	747.95	13.08 ± 0.70	3.16	805.5 ± 65.0	460 g/L	1.18 ± 0.1	26–8000	Fu *et al.* (2022)
	ERY	733.93	13.09 ± 0.70	2.48–3.06	818.4 ± 65.0	0.57 mol/L	1.20 ± 0.1	17–320	Ohore *et al.* (2022)
	RTM	837.05	13.00 ± 0.70		864.7 ± 75.0	8.6 × 10⁻³ mol/L	1.25 ± 0.1	109–122	Dan *et al.* (2020)
Polychloro phenoxy phenol	TCL	289.54	7.80 ± 0.35		50–54	1.3 × 10⁻³ g/L	1.490 ± 0.06		Fu *et al.* (2022)
Sulfonamides (SAs)	SDZ	250.28	6.81 ± 0.10		6.81 ± 0.10		1.52		Fu *et al.* (2022)
	SDM	310.33		1.63		0.34 g/L			Oberoi *et al.* (2019)
	SM	280.3	6.58 ± 0.10		539.4 ± 56.0	1.5 g/L	1.465 ± 0.06	4050 ± 47.1	Chen *et al.* (2019b)
	SMZ	278.33	7.89 ± 0.10	0.14	526.2 ± 52.0		1.4655	4550 ± 73.6	Chen *et al.* (2019b)
	SMX	253.28	5.81 ± 0.50	0.89	482.1 ± 55.0	500 in 25°C	1.4895	59–3100	Fu *et al.* (2022)
	SMM	280.3	6.67 ± 0.30		513.2 ± 60.0	16 g/L		5100 ± 200	Chen *et al.* (2019b)
	SP	249.29	8.54 ± 0.30		473.5 ± 51.0	0.85 g/L	1.432 ± 0.06	2.54–493	He *et al.* (2021)
	ST	255.32	7.24 ± 0.10	0.05	479.5 ± 47.0	0.012 mol/L	1.53	0.01–49.3	He *et al.* (2021)
Tetracyclines (TCs)	TMP	290.32	7.04 ± 0.10	0.91	526.0 ± 60.0	3.6 × 10⁻³ mol/L	1.252 ± 0.06	79–6000	Sabri *et al.* (2021)
	OTC	460.43	4.50 ± 1.00	−0.9	839.6 ± 65.0	0.29 g/L	1.634	6.7–132	Ohore *et al.* (2022)
	TET	444.43	4.50 ± 1.00	−1.37	790.6 ± 60.0	4.3 × 10⁻⁴ mol/L	1.64 ± 0.1	34–101	Dan *et al.* (2020)
	TCT	478.88	4.50 ± 1.00		821.1 ± 65.0	0.086 g/L	1.52	4320–5560	Ohore *et al.* (2022)
	DXY	444.4	3.09					0.01–2585	Sabri *et al.* (2021)

**Note:* AMP: Ampicillin. AZM: Azithromycin. CIP: Ciprofloxacin. CLR: Clarithromycin. DXY: Doxycycline. ENR: Enrofloxacin. ERY: Erythromycin. LIN: Lincomycin. MTZ: Metronidazole. NOR: Norfloxacin. OFL: Ofloxacin. OTC: Oxytetracycline. RTM: Roxithromycin. SDM: Sulfadimethoxine. SDZ: Sulfadiazine. SM: Sulfameter. SMM: Sulfamonomethoxine. SMX: Sulfamethoxazole. SMZ: Sulfamethazine. SP: Sulfapyridine. ST: Sulfathiazole. TCL: Triclosan. TCT: Chlortetracycline. TET: Tetracycline. TMP: Trimethoprim.

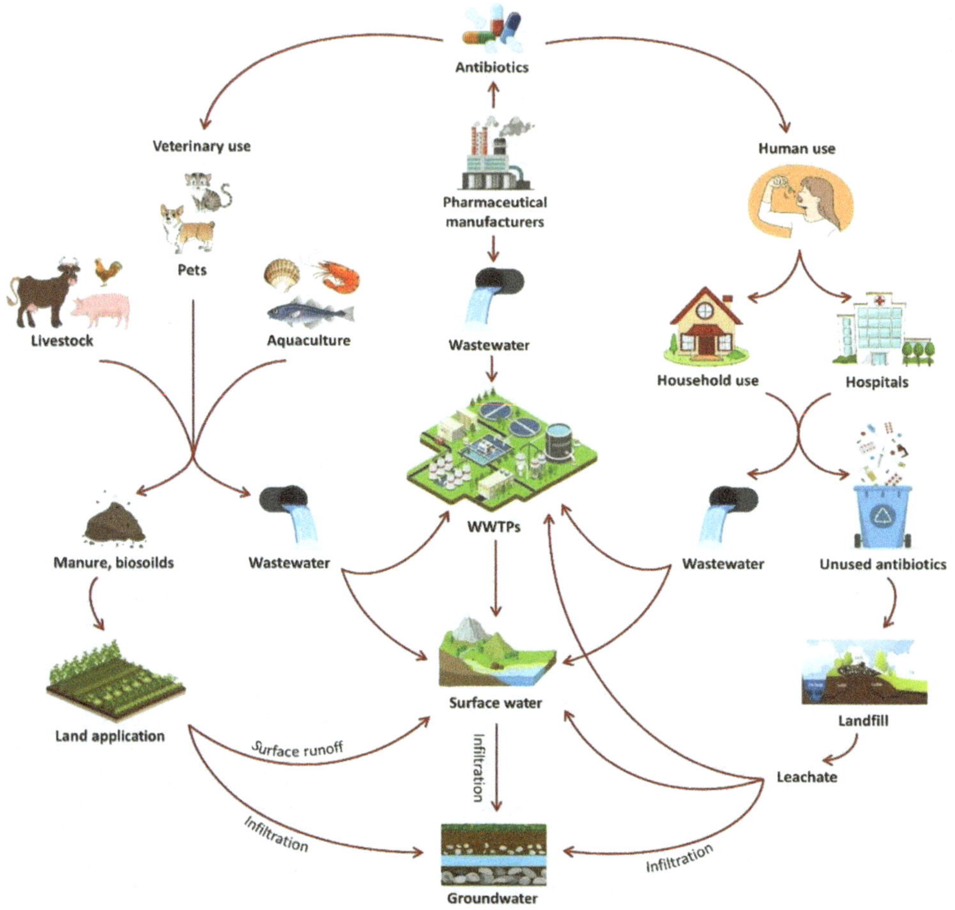

Figure 4.1 Sources and migration paths of antibiotics in the environment (WWTPs – wastewater treatment plants).

big footprint of antibiotics from livestock farms is evidenced by the frequent detection of antibiotics in streams and soils surrounding livestock farms. This poses the consumers at high risks of food contamination and ARGs prevalence (Du *et al.*, 2020).

Livestock wastewater exhibited high antibiotic concentrations as compared to other types of wastewaters (e.g., domestic wastewater, hospital wastewater, wastewater treatment plant (WWTP) effluents), being 4320–14000 ng/L in China (Huang *et al.*, 2015), 100 mg/L in Portugal (Fernandes *et al.*, 2015), 10–300,330 mg/L in Korea (Choi *et al.*, 2016). Although the antibiotic level in livestock wastewater is usually lower than that in pharmaceutical manufacturers, livestock wastewater is one of the most important sources of antibiotic pollution due to its high concentration and great availability.

4.1.2.3 Hospitals

It is reported that antibiotics are overused in medical treatment. This ranges from ineffective use of antibiotics for the cure of viral diseases to arbitrary use of old-generation antibiotics or those prescribed for other people (García *et al.*, 2020). Unfortunately, most antibiotics consumed by humans are released via faeces and urine. For instance, up to 58% of 93,000 tons of antibiotics used in China in 2013 was found to enter the environment (Ma *et al.*, 2022). Since most antibiotics are not effectively removed by hospital WWTPs, antibiotics often occur in the surface water at levels in the range of ng/L to μg/L (Berglund *et al.*, 2014). Therefore, the regular discharge of hospital wastewater is a main pathway of contributing antibiotics into the environment (García *et al.*, 2020). The level of antibiotics in hospital wastewater in Ethiopia ranges from 1.2 to 17 mg/L (Mesenbet *et al.*, 2021).

4.1.2.4 Wastewater treatment plants

Several studies have reported that conventional WWTPs do not effectively remove antibiotics from wastewater. The treatment efficiency of antibiotics from sewage was limited to only 60% (Carvalho *et al.*, 2014). That is because as emerging pollutants, antibiotics are usually difficult to be degraded (Button *et al.*, 2019). More importantly, conventional WWTPs are not specially designed to treat antibiotics (Fu *et al.*, 2022). Consequently, WWTPs can become anthropogenic pollution sources of antibiotics. For example, in a study on the occurrence of antibiotics in 45 WWTPs in China, Gothwal and Shashidhar (2015) discovered that a WWTP effluent in Beijing contained FQs, SAs, and MLs with levels as high as 4916, 2916, and 365 ng/L. Similarly, FQs were dominant in sludge from WWTPs of the Shanxi province with a content up to 29,647 mg/kg. The detection of antibiotics in WWTPs effluent and sludge implied their possible adverse impacts on the environment, including antibiotic resistance (Sosa-Hernández *et al.*, 2021). It is evident that treatment of antibiotics remains a big challenge to conventional WWTPs (Button *et al.*, 2019). Hence, extensive treatment is required to ensure the successful decontamination of antibiotics from wastewater (Sabri *et al.*, 2021). The levels of antibiotics in WWTP effluents are in the range of 0.003–0.933 mg/L in China (Dan *et al.*, 2021) and 0.2–20 mg/L in the Netherlands (Sabri *et al.*, 2021).

4.1.2.5 Pharmaceutical manufacturers

Pharmaceutical production normally generates large amounts of wastewater. It is estimated that 500–6500 m^3 of wastewater can result from the production of one ton of antibiotics (Xue *et al.*, 2022). Moreover, wastewater from pharmaceutical factories usually contains high levels of antibiotics. The concentrations of spiramycin I and neo spiramycin I in wastewater from a pharmaceutical factory in Wuxi (China) were found to be 115.6 and 5.3 mg/L, respectively (Xue *et al.*, 2022). The concentrations of antibiotics in wastewater from pharmaceutical factories (as high as mg/L) are many times larger than those in the WWTPs and livestock effluents (as high as ng/L to μg/L). Hence, pharmaceutical factories can be considered as a major source of antibiotic pollution, significantly contributing to the occurrence of antibiotics in the surface water in Europe, China, and Vietnam (Wang *et al.*, 2021).

4.1.2.6 Household consumption

Consumption of antibiotics includes the use of antibiotics for medical treatment and the utilization of pharmaceutical and personal care products (PPCPs) that contain antibiotics (e.g., soaps, facial cleaners, mouthwashes, sterilizing agents) (Sosa-Hernández *et al.*, 2021). It is reported that the human body can hardly absorb antibiotics. As a result, 30–90% of antibiotics entering the human body are excreted via faeces and urine, subsequently going into domestic sewage treatment systems. Due to ineffective treatment of antibiotics by the domestic septic systems (García *et al.*, 2020), a large proportion of antibiotics (40–90%) in domestic wastewater will enter the soil and water environment in the original form or as primary metabolites (Fu *et al.*, 2022). There is a significant variation in the antibiotic concentration in wastewater, depending on the types of wastewaters, venue, and season. It varies in the range of 1.8–3.7 ng/L in urban wastewater in Spain (Avila *et al.*, 2014), 0–2034 ng/L in domestic wastewater during summer and 0–181,000 ng/L during winter in China, respectively (Dan *et al.*, 2020), 160–255 mg/L in domestic sewage in China (Bai *et al.*, 2022), and 2–193 ng/L in rural wastewater in China (Chen *et al.*, 2015).

4.1.3 Environmental concerns of antibiotics

The occurrence of antibiotics in the surface water may result from the lack of efficient wastewater treatment systems (Sosa-Hernández *et al.*, 2021). The use of wastewater containing antibiotics for irrigation may lead to the accumulation of antibiotics in soil and plants, thus posing a risk to the environment and public health (Cui *et al.*, 2023). Additionally, once entering soil, antibiotics may induce ecological imbalance of microbial communities (Ma *et al.*, 2022), which are responsible for the biogeochemical cycling of elements (Sosa-Hernández *et al.*, 2021). Antibiotics can pose ecosystems at risks even at trace levels (Lv *et al.*, 2022).

4.1.3.1 Antibiotic resistance

Microorganisms can become resistant to antibiotics when exposed to even small amounts of antibiotics (Chen *et al.*, 2019a). This leads to a reduction in therapeutic efficacy of antibiotics (Sosa-Hernández *et al.*, 2021). The abuse of antibiotics can result in the emergence of antibiotic-resistant bacteria (ARB) (Bai *et al.*, 2022; Choi *et al.*, 2016). ARB carrying ARGs are reservoirs for gene transfer between pathogens (Berglund *et al.*, 2014). Even when ARB die, their DNA can be released into the environment, facilitating the assimilation of ARGs by other bacteria (Ma *et al.*, 2022).

Another risk of antibiotic contamination is the spread of ARGs in the environment. It was found that environments with high contents of antibiotics also exhibited abundant levels of ARGs (Berglund *et al.*, 2014). For instance, substantial numbers of ARGs were determined in surface water near Chauvin and Port Fourchon and Bayou Lafourche in Louisiana (García *et al.*, 2020). In addition, the continuous use of antibiotics increases the availability of ARGs in the environment. The level of ARGs in soil in the Netherlands in 2008 was 15-fold higher than that in 1970. Furthermore, ARGs are hard to remove, even after ARB die (Du *et al.*, 2020). According to García *et al.* (2020), ARGs can exist in the soil for two years. Due to their persistence, ARGs can continuously be transmitted between organisms via horizontal gene transfer (HGT) (Ma *et al.*, 2022). It is worth noting that ARGs are also able to be transmitted to humans via polluted water or contaminated food. This was evidenced by the fact that the antibiotic resistance gene *tetW* carried by anaerobic bacteria in ruminants and humans have a similarity of 99.9% in the nucleotide sequence (Ma *et al.*, 2022).

Additionally, the abuse of antibiotics has promoted the occurrence of antibiotic-resistant organisms (e.g., superbugs), inducing incurable diseases and elevated human lethality (Chen *et al.*, 2019a).

4.1.3.2 Adverse impacts on indigenous biota

The presence of antibiotics in agricultural soils can impact both plant growth and soil microbial activity (Carvalho *et al.*, 2014).

Antibiotics can be metabolized by plants via a three-step mechanism: (i) reactivity and polarity of antibiotics are enhanced thanks to oxidation, reduction, or hydrolysis; (ii) antibiotics are conjugated with polar molecules (e.g., sugars, amino acids, malonic acids, and glutathione) of plants; and (iii) the conjugates are stored in cellular vacuoles or bound to insoluble components (e.g., lignin and polysaccharide) (Lv *et al.*, 2022).

However, excessive contents of antibiotics in soils can exhibit toxic effects on plant development (Choi *et al.*, 2016). Antibiotics may result in the disruption of photosynthetic systems of plant cells and a decrease in the rate of CO_2 transformation (Sosa-Hernández *et al.*, 2021). TCs, FQs, and MLs affect protein synthesis in plants. FQs negatively impact plant morphology and photosynthesis. Streptomycin may hinder chlorophyll synthesis. Sulfadimethoxine and enrofloxacin markedly impede plant growth. Ciprofloxacin reduces photosynthesis and plant growth. TCs may induce chromosomal aberrations and inhibit plant growth. Though BLAs are less toxic, they can affect plastid division in lower plants. Tetracycline, ciprofloxacin, and erythromycin may decrease the levels of photosynthetic pigments, chlorophylls, and carotenoids in plants. Penicillin, cephalosporins, and tetracycline control the rate of photosynthetic electron transport. Ciprofloxacin and cephalosporins strongly reduce stomatal conductance, thus inhibiting the assimilation rate (Gothwal & Shashidhar, 2015).

The occurrence of antibiotics can alter the composition and diversity of the soil microbial community (Lv *et al.*, 2022). The number of soil bacteria can be declined due to antibiotics. This may lead to a reduction in the number of soil fauna (e.g., protozoa, nematodes, and micro-arthropods). Consequently, the soil functions (e.g., decomposition of organic matter, denitrification) are hindered. These in turn indirectly impact the plant growth (Carvalho *et al.*, 2014). According to Button *et al.* (2019), tetracycline and sulfamethoxazole ex-situ exposure induced significant effects on microbial functions, whereas their in-situ exposure exhibited marginal impacts.

4.1.4 Antibiotic treatment technologies

Due to persistence of antibiotics and their potential adverse impacts on human and ecological health, a wide variety of technologies have been developed for the removal and degradation of antibiotics from aquatic media (Fu *et al.*, 2022). These technologies can be classified as physical, chemical, biological, and integrated ones as illustrated in Figure 4.2 (Huang *et al.*, 2022).

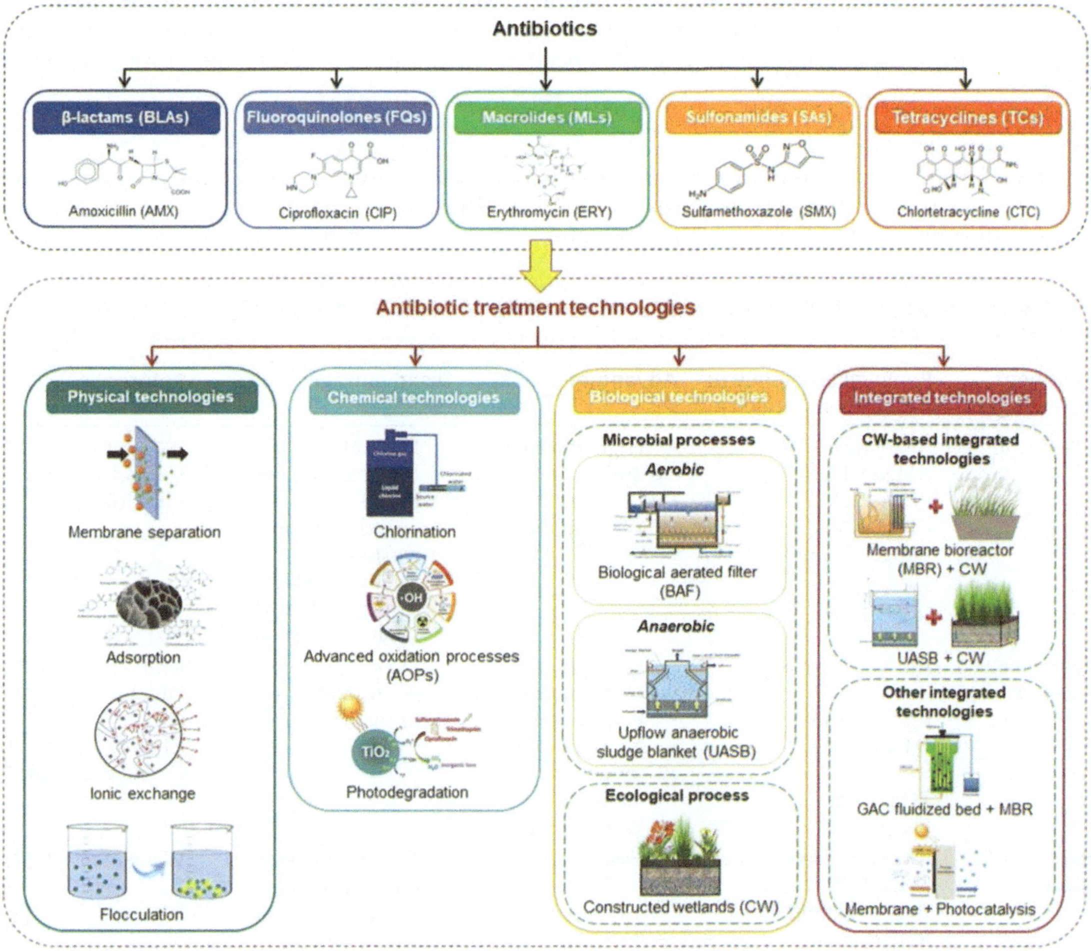

Figure 4.2 Major groups of antibiotics and technologies for antibiotic treatment in wastewater.

4.1.4.1 *Physical technologies*

Physical methods for antibiotic treatment in wastewater are based on concentration and transfer of contaminants by physical means. Physical methods include membrane filtration, adsorption, ion exchange, and flocculation as presented in Table 4.2 (Huang *et al.*, 2022).

4.1.4.1.1 Adsorption

Adsorption is one of the most frequently used methods for eliminating antibiotics from wastewater. That is because it possesses significant advantages, such as high efficiency, simple operation, and low cost. However, this technology also demonstrates some limitations, namely low selectivity, being influenced by pH and co-existing pollutants in wastewater and requiring the disposal of antibiotic loaded adsorbents. The removal percentages of antibiotics by adsorption vary in the range of 53.7–94.6% as shown in Table 4.2 (Hamadeen & Elkhatib, 2022; Weng *et al.*, 2018).

Recently, many types of nanomaterials have been fabricated and applied in decontamination of antibiotics in wastewater. It is acknowledged that nanomaterials can not only eliminate but also degrade antibiotics into environmentally benign by-products. The removal efficiency of antibiotics by nanomaterials is often above 90%. The high removal capability of antibiotics by nanomaterials can be attributed to their extremely tiny size, high porosity, and active surface. However, the adsorption of antibiotics by nanomaterials is strongly influenced by wastewater pH and temperature (Chen *et al.*, 2019a).

4.1.4.1.2 Membrane filtration

Membrane filtration is a technology for water and wastewater treatment that retains pollutants while allowing water to pass through its pores. Membrane filtration can be divided into microfiltration (MF), ultrafiltration (UF), nanofiltration (NF), and reverse osmosis (RO). This technique has merits,

Table 4.2 Advantages and disadvantages of physical technologies for decontamination of antibiotics in wastewater.

Treatment Technologies	Removal Efficiency or Adsorption Capacity	Advantages	Disadvantages	Reference
Adsorption	AMP: 80.6% AMX: 94.6% PEN: 53.7% CIP: 84.74–89.94% Range: 53.7–94.6%	• High efficiency • Simple operation • Low cost • Wide application	• Low selectivity • Being influenced by pH values, co-existing ions in wastewater • Requiring the disposal of antibiotic loaded adsorbents	Weng *et al.* (2018); Hamadeen and Elkhatib (2022)
Membrane filtration	AMX: 52.7–90% CA: 72% ENR: 85–93% OTC: 87.5% SMX: > 95% TOB: 96% Range: 52.7–96%	• High efficiency • Fast process • Simple operation • No requirement for chemical addition • Wide application	• Depending on types of antibiotics and pH of the wastewater	Krall *et al.* (2018); Oulebsir *et al.* (2020); Zhao *et al.* (2018); Cheng *et al.* (2015)
Ionic exchange	AMX: 0.155 mg/L SMX: 0.789 mg/L TET: 0.443 mg/L Range: 0.155–0.789 mg/L	• Fast process	• Depending on pH and coexisting anions in the wastewater	Wang *et al.* (2017)

**Note:* AMP: Ampicillin. AMX: Amoxicillin. CA: Cephalexin. CIP: Ciprofloxacin. ENR: Enrofloxacin. OTC: Oxytetracycline. PEN: Penicillin. SMX: Sulfamethoxazole. TET: Tetracycline. TOB: Tobramycin.

such as high efficiency, fast process, simple operation, and limited chemical use (Huang *et al.*, 2021). Nonetheless, high costs (Hu *et al.*, 2021) and strong dependence on antibiotic categories and wastewater pH have limited the popular application of membranes (Cheng *et al.*, 2015; Krall *et al.*, 2018; Oulebsir *et al.*, 2020; Zhao *et al.*, 2018). Membrane technology is relatively effective in eliminating antibiotics from wastewater with a removal efficiency of 52.7–96%. However, it is rarely applied for decontamination of livestock wastewater (Huang *et al.*, 2021). This is owing to the complicated composition of this type of wastewater.

4.1.4.1.3 Ion exchange

Another process used for removal of antibiotics from wastewater is ion exchange. While this technology is a fast process, its performance is affected by pH and coexisting pollutants present in wastewater. The removal capability of antibiotics by ion exchange resins varies in the range of 0.155–0.789 mg/L (Wang *et al.*, 2017).

4.1.4.2 Chemical technologies

In chemical methods, target contaminants are converted into intermediates or completely degraded via chemical reactions between oxidizing agents or high reactive species and contaminants (Huang *et al.*, 2022). Chemical methods for removal of antibiotics from wastewater can be divided into strong oxidant oxidation, advanced oxidation, photolysis, and photocatalysis as given in Table 4.3.

4.1.4.2.1 Oxidation

Among chemical technologies, oxidants are employed for the decomposition of antibiotics in water. These processes include chlorination, chloramination, and ferrate oxidation. The chlorination has a high removal efficiency, high reaction rate, and wide application (Ye *et al.*, 2021). However, the major limitations of this technique are the formation of toxic by-products and relying on the wastewater pH. The removal efficiency of antibiotics by chlorination is relatively low (0.43–54.26%) (Ye *et al.*, 2021). Another process is ferrate (VI) oxidation. It has a high removal efficiency and fast reaction rate. The removal efficiency of antibiotics by ferrate oxidation varies in a wide range (50–100%). Nevertheless, this technique has drawbacks, such as low stability, limited industrial application, and strong dependence on the pH of the wastewater (Acosta-Rangel *et al.*, 2020; Kovalakova *et al.*, 2021).

4.1.4.2.2 Advanced oxidation processes

Advanced oxidation processes (AOPs) are oxidation processes that decompose or mineralize target pollutants by using highly reactive hydroxyl radicals (·OH) (Carvalho *et al.*, 2014). As a strong oxidant, ·OH breaks chemical bonds or is involved in chemical reactions (e.g., addition, substitution) to partly decompose or completely mineralize pollutants into CO_2 and H_2O (Huang *et al.*, 2021). Typical examples of AOPs are Fenton or photo-Fenton oxidation, ozonation, and photolysis. Thanks to chemical, photochemical or electrochemical processes in AOPs, these processes exhibit a high efficacy for antibiotic removal (70–100%). AOPs show an enormous potential for decontamination of antibiotics in livestock wastewater (Huang *et al.*, 2021). The significant advantages of AOPs are high degradation efficiency and fast reaction rate. Nevertheless, each AOP faces different obstacles. For example, the Fenton process results in the formation of unknown by-products and ferrous sludge (Han *et al.*, 2020; Hang *et al.*, 2022). Meanwhile, prohibitive cost, strong dependence on pH and co-existing pollutants are demerits of ozonation (Alsager *et al.*, 2018; Östman *et al.*, 2019).

4.1.4.2.3 Photolysis

Photolysis is a chemical process for removing antibiotics in wastewater. The removal percentages of antibiotics by photolysis are remarkably high (90–100%). Photolysis is reported to be more effective than adsorption, a fast process with low costs. However, its antibiotic treatment performance is influenced by turbidity of the wastewater. Furthermore, the removal of antibiotics by direct photolysis is rather low. These are regarded as constraints for the broad application of photolysis (Timm *et al.*, 2019).

Table 4.3 Advantages and disadvantages of chemical technologies for decontamination of antibiotics in wastewater.

Treatment Technologies	Removal Efficiency or Adsorption Capacity	Advantages	Disadvantages	Reference
Oxidation processes				
Chlorination	CTC, DXY, OTC, TET 0.43–54.26% Range: 0.43%–54.26%	• High removal efficiency • High reaction rate • Wide application	• Formation of toxic by-products • Depending on pH of wastewater	Ye *et al.* (2021)
Ferrate oxidation	CIP, ERY, OFL, OTC, SMX, TET, TRM > 50% SAs: 100% Range: > 50%–100%	• High removal efficiency • High reaction rate	• Affected by pH of wastewater • Poor stability • Limited practical application	Kovalakova *et al.* (2021); Acosta-Rangel *et al.* (2020)
Advanced oxidation processes (AOPs)				
Fenton and photo-Fenton oxidation	OTC: 94.8%; TET: 94.2% SMX: 100% Range: 94.2%–100%	• High degradation efficiency • Rapid degradation of recalcitrant compounds • Simple operation and maintenance	• Formation of unknown by-products • The effluent contains high levels of anions • Production of iron containing sludge • Feasible at lab-scale • Having technical constraints	Han *et al.* (2020); Hang *et al.* (2022)
Ozonation	AMX: 70% CPR: 100% CPR, DXY, SDZ: 92–98% ERY: 94% MTR: 100% TRM: 94% Range: 70%–100%	• Mild reaction conditions • High reaction rate • Easy adjustment of the flow rate • No secondary pollution	• Low mass transfer efficiency • Low continuous oxidation ability • Affected by pH and co-existing pollutants of wastewater • High costs for ozone production and aeration energy • Easy loss of ozone	Alsager, *et al.* (2018); Östman *et al.* (2019)
Photolysis	AMP: 100% AMX: 100% PNV: 95% PPR: 90% Range: 90–100%	• More effective than adsorption • Fast process • Low maintenance and operation costs	• Limited removal by direct photolysis • Influenced by turbidity of wastewater	Timm *et al.* (2019)

**Note:* AMP: Ampicillin. AMX: Amoxicillin. CIP: Ciprofloxacin. CPR: Cefprozil. CTC: Chlortetracycline. DXY: Doxycycline. ERY: Erythromycin. MTR: mithramycin. OFL: Ofloxacin. OTC: Oxytetracycline. PNV: Penicillin V. PPR: Piperacillin. SAs: Sulfonamides. SDZ: Sulfadiazine. SMX: Sulfamethoxazole. TET: Tetracycline. TRM: Trimethoprim.

4.1.4.3 Biological technologies

Biological methods are processes that can degrade antibiotics in situ under mild conditions by means of living beings (e.g., microbes, plants, animals). Biological methods are more advantageous than physical and chemical processes in terms of the cost and no formation of toxic by-products (Zhou & Wang, 2023). However, the performance of biological methods depends on many factors, such as operation conditions and water quality parameters (Huang *et al.*, 2021). The biological processes can be grouped into microbial and ecological processes. Based on oxygen requirement, the microbial processes can be further classified as aerobic, anaerobic, and aerobic–anaerobic combined processes as given in Table 4.4. In microbial processes, single or multiple strains can be used for biodegradation of antibiotics in wastewater (Zhou & Wang, 2023).

4.1.4.3.1 Aerobic processes

The common aerobic processes are activated sludge (AS) and biological aerated filter (BAF).

AS is the most widely used biological method in WWTPs. Nevertheless, the removal efficiency of antibiotics by AS varies from negative values to 90% (Table 4.4). Sludge retention time (SRT) and HRT

Table 4.4 Advantages and disadvantages of microbial technologies for decontamination of antibiotics in wastewater.

Treatment Technologies	Removal Efficiency or Adsorption Capacity	Advantages	Disadvantages	Reference
Aerobic processes				
Activated sludge (AS)	AZT, CLAR, ERY: 54.8–91.4% CIP: 76.6–99.9% CTC, MIN, OTC, TET: 58.8–80.3% Range: 54.8–99.9%	• Good treatment efficiency • Simple process • Low cost • Small footprint • Wide application range	• Influenced by antibiotic chemical structures, sludge properties, and operation conditions • Requiring long hydraulic retention time (HRT)	Oberoi *et al.* (2019); Tran *et al.* (2016); Zhao *et al.* (2022)
Biological aerated filter (BAF)	LCM, LIN, NFX, OFX, OTC, SCP, SMM, SMZ, TRM: 89–91% Range: 89–91%	• Better degradation than anaerobic treatment • High automation • Small footprint	• More energy consumption • High operation cost • Unable to digest all antibiotics	Chen *et al.* (2017)
Anaerobic processes				
Anaerobic digestion (AD) Upflow anaerobic sludge blanket (UASB) Anaerobic filter (AF) Anaerobic baffled reactor (ABR)	CIP, ENR, OFL: 85% CTC, OTC, TET: 65% Range: 65–85%	• Simple operation • No requirement for aeration or energy • Being applicable in smaller size than aerobic treatment • Methane production • Formation of less sludge	• Higher operation cost • Unable to digest all antibiotics, requiring hybrid systems for better treatment	Huang *et al.* (2021); Wang *et al.* (2019)
Combined aerobic–anaerobic processes				
Sequencing batch reactor (SBR) Membrane bioreactor (MBR) Intermittently aerated membrane bioreactor	TET: > 90% QN: < 70% SA: > 90%	• High removal efficiency	• Requiring longer HRT • Higher cost	Huang *et al.* (2021)

Note: AZT: Azithromycin. CIP: Ciprofloxacin. CLAR: Clarithromycin. CTC: Chlortetracycline. ENR: Enrofloxacin. ERY: Erythromycin. LCM: Leucomycin. LIN: Lincomycin. MIN: Minocycline. NFX: Norfloxacin. OFL: Ofloxacin. OTC: Oxytetracycline. QN: Quinolone. SA: Sulfonamide. SMM: Sulfamonomethoxine. SMZ: Sulfamethazine. TET: Tetracycline. TRM: Trimethoprim.

are the most important factors regulating the removal of antibiotics in AS. With the extended HRT, the removal of sulfamethazine by AS was increased from 45% to 90%. In the AS process, antibiotics are removed via two major pathways, namely adsorption and biodegradation (Fu *et al.*, 2022).

BAF is the combination of biological oxidation and granular filtration (Chen *et al.*, 2017). The BAF has a deep bed of filter media, which is submerged in wastewater. With a large surface area, the packed media favour biofilm formation, thus enhancing the antibiotic degradation in wastewater. Air is provided to intensify the oxidation of antibiotics in BAF. The antibiotic treatment performance of BAF is relatively high, being 89–91% (Chen *et al.*, 2017). BAF is one of the currently used processes for decontamination of antibiotics in livestock wastewater. The technology has merits, such as higher efficiency than AD, small footprint, and high automation. However, limitations of the BAF process include intensive energy consumption, high operation cost, and BAF is unable to degrade all types of antibiotics (Huang *et al.*, 2021).

4.1.4.3.2 Anaerobic processes

The major anaerobic processes include anaerobic digestion (AD), upflow anaerobic sludge blanket (UASB), anaerobic filter (AF), and anaerobic baffled reactor (ABR). The AD process consists of four steps, that is, hydrolysis, acidogenesis, acetogenesis, and methanogenesis (Fu *et al.*, 2022). Compared to AS, AD has more merits, such as less sludge production, better energy saving, and biogas generation. Consequently, AD is frequently used for the treatment of wastewater with elevated levels of antibiotics, such as swine wastewater. It was reported that under conditions of 37°C, HRT of 16 d, and organic loading rate of 1.38–2.16 kg COD/m^3/d, the removal efficiency of tetracycline and quinolone from swine wastewater by AD was 65 and 85%, respectively (Wang *et al.*, 2019). Temperature is the most important operation parameter, influencing the AD performance. The removal efficiencies of SAs and FQs from swine wastewater in summer (66.2% and > 97%) were greater than those in winter (−19.4% and −42.5%). This can be explained by the stronger activity of methanogens at higher temperatures in summer. However, AD is unable to achieve complete removal of antibiotics. As a result, further treatment or integration of AD with aerobic treatment is required (Fu *et al.*, 2022).

4.1.4.3.3 Combined aerobic and anaerobic processes

The typical combined aerobic and anaerobic process is a sequencing batch reactor (SBR). The SBR has one reactor, in which five steps occur including anoxic filling, aerobic reaction, sludge settling, decanting, and idling. While a SBR has advantages of small footprint and energy saving, its operation is complicated. The SBR is widely used for decontamination of antibiotics from swine wastewater. With a HRT of 3–5 d, a lab-scale intermittently aerated SBR could remove 88% tetracycline and 96% sulfonamide from swine wastewater (Huang *et al.*, 2021).

4.1.4.3.4 Ecological methods

Constructed wetlands (CWs) are artificial wastewater treatment systems with main components of substrates, plants, and microorganisms (García *et al.*, 2020). Various pollutants including antibiotics in wastewater can be removed via physical, chemical, and biological processes in CWs (Ma *et al.*, 2022). CWs have two main categories, namely surface water flow CWs (SWF-CWs) and sub-surface flow CWs (SSF-CWs). Based on the wastewater flow direction, SSF-CWs can be further divided into horizontal SSF-CWs (HSSF-CWs) and vertical SSF-CWs (VSSF-CWs). As a result of the combination of different types of CWs, hybrid CWs can be formed (Ilyas & van Hullebusch, 2020). Generally, SWF-CWs possessing larger scale than SSF-CWs are applied for tertiary treatment, whereas SSF-CWs are utilized in decentralized treatment systems (García *et al.*, 2020).

CWs offer many advantages, such as simple design and operation, low energy requirement, low costs (e.g., periodic maintenance requirement) (Choi *et al.*, 2016), some level of selectivity in antibiotic removal (Huang *et al.*, 2019), good compatibility (e.g., applicable to different wastewater or flow rates) (Cui *et al.*, 2023), ecological friendliness (e.g., no chemical addition, marginal sludge generation, habitat creation, landscape enhancement) (Chen *et al.*, 2019a). Because of significant technical, economic, and environmental benefits, CWs are usually preferred by the public over traditional wastewater treatment methods (Omondi & Navalia, 2020). Regarding antibiotic treatment, CWs are efficient in eliminating a wide variety of antibiotics (e.g., TCs, SAs, FQs, MLs) from different types of wastewater (e.g. hospital, livestock, aquaculture, pharmaceutical, municipal) (Cui *et al.*, 2023). Antibiotic treatment efficiencies by CWs vary in the range of 75.8–98.6% (Bai *et al.*, 2022). According to Chen *et al.* (2016a), many mesocosm CWs exhibit greater antibiotic removal efficacies than conventional WWTPs.

However, CWs may suffer from drawbacks, such large land area requirement, seasonal variation in treatment performance, lower stability than conventional treatment methods, and possible secondary pollution generation (e.g., contaminant accumulated plants) (Hu *et al.*, 2021). The CWs' demerit of requiring large land area results from its operation at a low hydraulic loading rate (HLR) (Avila *et al.*, 2014).

4.1.4.4 Integrated technologies

The integrated technologies are those that combine at least two different treatment processes for the goal of complete treatment of antibiotics in wastewater. Integrated technologies are based on several component processes, including grey processes (e.g., coagulation, sedimentation, flotation, adsorption, AOPs, photocatalysis, biological methods) and green processes (e.g., CWs, phytoremediation) (Table 4.5). The integrated technologies can be either grey-green processes or grey-grey processes

Table 4.5 Advantages and disadvantages of integrated technologies for decontamination of antibiotics in wastewater.

Treatment Technologies	Removal Efficiency or Adsorption Capacity	Advantages	Disadvantages	Reference
Multistage ozone and biological treatment	AMX: 99% (Ozonation) AMX: > 99%, COD: > 98%, > 99% color removal, > 90% toxicity (Ozonation + Biodegradation)	• Reducing operation costs • Enhancing removal efficiency	• More complicated configurations and higher costs than single treatment processes	Marcelino *et al.* (2017)
Staged anaerobic fluidized-bed membrane bioreactor (SAF-MBR) + granular activated carbon (GAC)	SMZ: 100%	• High effluent quality, less sludge generation, methane production thanks to AnMBR • Reduced membrane fouling and enhanced biofilm formation due to GAC fluidization		Lim *et al.* (2019)
Modified ultrafiltration membrane (PVDF–PVP–TiO_2–dopamine) filtration-photocatalysis system	SDZ: 91.40%	• AOPs increase the permeation flux by 56% and eliminate membrane fouling		Zhou *et al.* (2020)
CWs and microbial fuel cells (MFC)	TET: 99.45–100%, CMB: 97.84–100%	• Enhanced antibiotic degradation • Bioelectricity production		Maldonado *et al.* (2023)
MFC + CW	SMX: 99.4% TET: 97.8%			Wen *et al.* (2021)

**Note:* AMX: Amoxicillin. CMB: Chloramphenicol. COD: Chemical oxygen demand. SDZ: Sulfadiazine. SMX: Sulfamethoxazole. SMZ: Sulfamethazine. TET: Tetracycline. UV: Ultraviolet.

in combination. Both types of integrated processes result in better treatment performance, while mitigating drawbacks of single processes. However, integrated grey-green processes are usually better choices because of their more ecological friendliness.

The integrated UF filtration and photocatalysis is promising for sulfonamide abatement with high efficacy (91.40%). In addition, it led to the increment of the permeation flux (56%), and thus eliminating the membrane fouling (Zhou *et al.*, 2020). The coupling of a microbial fuel cell (MFC) and CWs resulted in high removal efficiencies of sulfamethoxazole (99.4%) and tetracycline (97.8%). This can be attributed to microorganisms as biocatalysts that promote antibiotic degradation. Furthermore, this integrated process generates bioelectricity with the maximum voltages of 386 (±20) mV, highest power density of 123.43 mW/m^3, and coulombic efficiency of 0.273% (Wen *et al.*, 2021). As a result of the combination of UV/ozone with NF in a WWTP, the antibiotic removal efficiency reached 87%, the biodegradability elevated 4.6 times, and ecotoxicity reduced by 58% (Huang *et al.*, 2021). A membrane bioreactor (MBR) is an integrated method for wastewater treatment that combines adsorption and biodegradation with membrane filtration. Consequently, the MBR shows a higher removal efficiency of antibiotics compared to AS and AD. The MBR is currently used for antibiotic removal from swine wastewater. The removal efficiency of sulfonamide and tetracycline in swine wastewater of a MBR was above 90% (Huang *et al.*, 2021). The MBR has main advantages of long SRT, flexible operation, and low sludge generation. However, a MBR also demonstrates limitations of high energy consumption and elevated expenses (Huang *et al.*, 2021).

4.2 ANTIBIOTIC REMOVAL BY CWS AND INFLUENTIAL FACTORS

The removal efficiencies of antibiotics by CWs are functions of multiple factors, such as antibiotic properties (e.g., adsorption coefficient – K_d, dissociation constant – pK_a, partition coefficient – K_{ow}, persistence), wastewaters characteristics (e.g., pH, co-existing pollutants), CWs design factors (e.g., substrates, plants, flow configuration), and CWs operation parameters (e.g., feeding mode, HLR, HRT, temperature) as shown in Figure 4.3 (Huang *et al.*, 2021).

4.2.1 Antibiotic categories and properties

Antibiotic categories and properties are dominant factors influencing the treatment performance of antibiotics by CWs. Avila *et al.* (2014), in a study on antibiotic removal using a HyF-CW, revealed that the removal efficiency was high (>90%) for enrofloxacin, moderate for doxycycline, and low for sulfamethoxazole, erythromycin, and lincomycin. Later, Dan *et al.* (2021), while employing a full-scale HyF-CW for decontamination of antibiotics from sewage treatment plant (STP) effluent, found that the antibiotic removal efficiencies were varied, heavily dependent on the antibiotic categories. Similarly, the average removal efficiency of antibiotic groups in CWs was ranked in the sequence: fluoroquinolones (90.9 ± 17.2%) > tetracyclines (86.1 ± 25.0%) > others (67.8 ± 32.1%) > sulfonamides (63.5 ± 33.2%) > macrolides (59.1 ± 38.5%) ~ polyether ionophores (59.9 ± 30.3%) (He *et al.*, 2021).

It was found that the removal of antibiotics can be predicted based on their physicochemical properties, such as pK_a, K_d, log K_{ow}, and molecular weight (Ilyas & van Hullebusch, 2020). A low pK_a shows a strong acidity of an antibiotic, suggesting its high adsorption on negatively charged soil via electrostatic interactions. This explains the high removal efficiencies of sulfamethoxazole (73%) and trimethoprim (90.5%) (Bayati *et al.*, 2021). The adsorption coefficient (K_d) is an important indicator for antibiotic adsorption on substrates. The higher K_d indicates better adsorption. This explains why FQs were easily adsorbed to substrates. The K_d values of FQs are the highest (119.2–10,594.73 L/kg), followed by TCs (325.04–8736.55 L/kg), MLs (119.45–1212.13 L/kg), and SAs (11.15–95.61 L/kg) (Tang *et al.*, 2019). Among the antibiotics investigated, ciprofloxacin was the only antibiotic that was removed in cold season thanks to its highest K_d value (Dan *et al.*, 2020). The *n*-octanol/water partition coefficient (K_{ow}) is another indicator for antibiotic adsorption. Antibiotics with lower log K_{ow} values are easier to dissolve in water and more difficult to be adsorbed by substrates (Ma *et al.*, 2022).

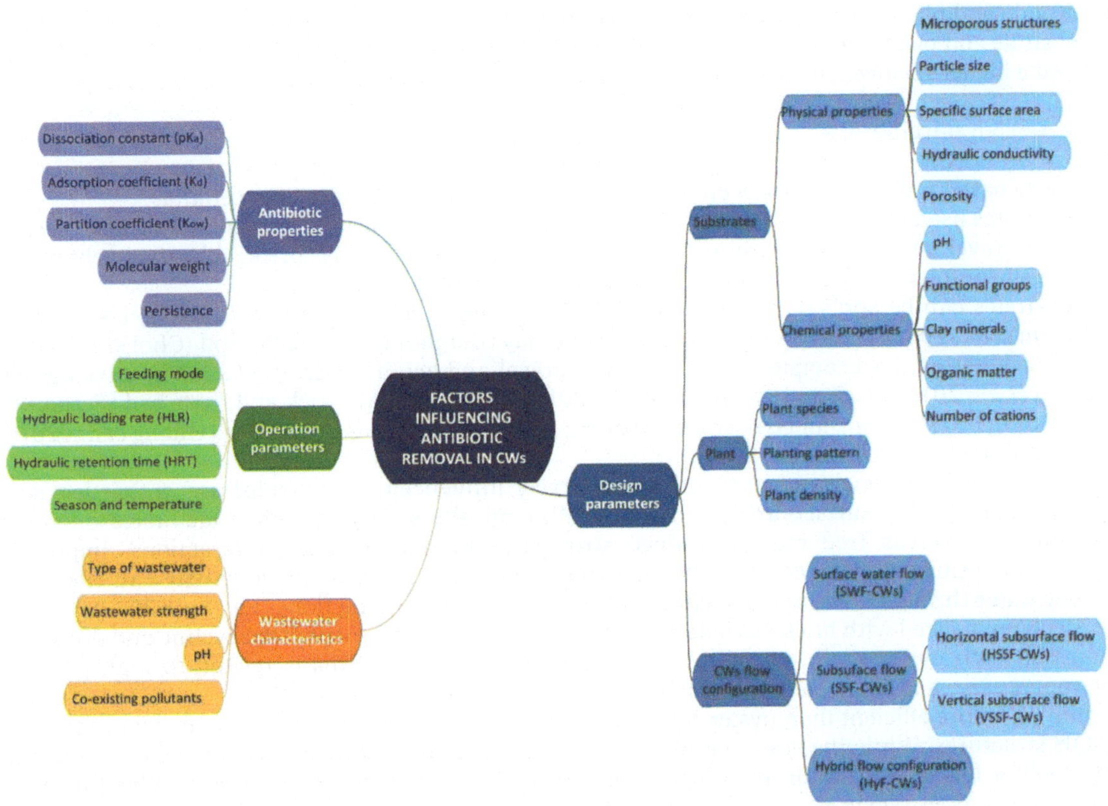

Figure 4.3 Factors influencing the antibiotic treatment performance by CWs.

Chen *et al.* (2016b) reported that among eight examined antibiotics, only erythromycin-H_2O and clarithromycin were detected in zeolite as the wetland substrate. These results agree well with their greater K_{ow} values, implying their lower solubility. According to Dan *et al.* (2020), antibiotics with low Henry coefficients ($< 8.7 \times 10^{-19}$) can be hardly removed via direct phytovolatilization. Conversely, antibiotics with medium hydrophobicity ($1.0 < \log K_{ow} < 3.5$) and/or low molecular weight (MW < 500) can be easily utilized by plants via numerous mechanisms (e.g., passive uptake via protein channels, active uptake via protein mediated and energy-dependent channels) (Lv *et al.*, 2022).

The addition of functional groups into the sulfonyl-phenylamine structure of antibiotics improved their adsorption in the descending order: pyridine > diazine plus methyl = diazine plus methoxy > diazine > amino. This suggested the adsorption of antibiotics could be promoted by their stronger adsorption sites (Thiele-Bruhn *et al.*, 2004).

Since FQs are chemically stable, they can hardly be degraded by hydrolysis or high temperature. As a result, they can be easily retained by soil. Similarly, with stable symmetric structure, trimethoprim is persistent to microbial degradation, resulting in its low removal efficiency (Polianciuc *et al.*, 2020). Also, enrofloxacin is characterized by low biodegradation, high adsorption and vulnerable to photolysis because it has two rings of quinolones with fluorine atoms (Huang *et al.*, 2019).

On the contrary, penicillin and cephalosporin are sensitive to hydrolysis, they are poorly adsorbed on the soil. However, they can be accumulated in sediments via complexation with cations (Polianciuc *et al.*, 2020). Likewise, microbial degradation is a major removal pathway of sulfonamides, whereas substrate adsorption and plant uptake play minor roles in their removal. This is evidenced by no detection of SAs in substrate and plant (Huang *et al.*, 2019). The integrated vertical CW (IVF-CW) with zeolite substrate showed a higher removal efficiency of sulfonamides than tetracyclines (Du *et al.*, 2020).

4.2.2 Design parameters

4.2.2.1 Substrates

Substrate-based CWs are more effective in removing antibiotics than floating wetlands.

Substrates used in CWs for removing antibiotics can be classified as (i) conventional substrates, for example, gravel, soil and (ii) emerging functional substrates, for example, brick, coke (fuel), chaff, lava rock, vesuvianite, zeolite, heavy clay, steel slag, and oyster shell.

Substrates can be applied in CWs separately, for example, gravel (Dan *et al.*, 2021; Hsieh *et al.*, 2015; Qu *et al.*, 2022), coke (Bai *et al.*, 2022), heavy clay (Berglund *et al.*, 2014), soil (Choi *et al.*, 2016), or in combination, for example, vesuvianite and gravel and zeolite (Dan *et al.*, 2020), mixed gravel (Button *et al.*, 2019), chaff and soil (Chen *et al.*, 2015), gravel and lava rock and roots bed (Fernandes *et al.*, 2015), brick granule and oyster shell (Huang *et al.*, 2015), zeolite and gravel (Huang *et al.*, 2019), gravel and steel slag (Tian *et al.*, 2023).

The removal efficiencies of antibiotics are largely influenced by the substrate properties, such as pore size, specific surface area, and chemical composition. High porosity facilitates microbial attachment, whereas free radicals induce stronger antibiotic–substrate interactions, improving removal of antibiotics (Dan *et al.*, 2021). Substrates with high hydraulic conductivity can effectively convey water through CWs, which enables antibiotic–substrate interactions (Fu *et al.*, 2022).

A column packed with brick exhibited a higher removal efficiency of oxytetracycline and difloxacin than that with oyster shell, regardless of the experimental scale. This is related to high porosity, micropore size, and high per cent of iron oxide in brick (Huang *et al.*, 2017). As a CW substrate, zeolite was more efficient than oyster shell, medical stone, and ceramic, thanks to its micropores and Si-OH structures. While the microporous structure results in high surface area for chemical sorption and biofilm formation, the bridging hydroxyls on the surface of zeolite may be involved in many chemical reactions (Du *et al.*, 2020). The removal efficiencies of sulfamethazine (73%), oxytetracycline (95%) and ciprofloxacin (85%) of the zeolite-based VSSF-CW were higher than those of the volcanic rock-based VSSF-CW by 5%, 4%, and 3%, respectively (Huang *et al.*, 2021). The integrated zeolite medium VSSF-CW showed high removal efficiencies of SAs and TCs (>95%). This is mainly due to the abundance of *Pseudomonas, Acinetobacter*, and *Bacillus* in zeolite, leading to the intensified microbial biodegradability and its adsorption capacity (Lv *et al.*, 2022). As compared to gravel, zeolite exhibited higher removal efficiencies of macrolides, quinolones, sulfonamides, and tetracyclines by 78%, 34%, 8%, and 25%, respectively (Cui *et al.*, 2023). Physicochemical properties of substrates are very important in antibiotic removal of CWs (Du *et al.*, 2020).

4.2.2.2 Plant

Plant species in CWs for removal of antibiotics are mainly emergent plants, in addition to the floating leave plants and submerged plants (Hu *et al.*, 2021). Among the studies included in this book chapter, 19 emergent plant species were employed in CWs for antibiotic abatement, namely *Acorus calamus, Acorus tatarinowii, Arundo donaz* var. *versicolor, Canna glauca, Canna indica, Cattail, Cyperus alternifolius, Desmodium styracifolium, Iris pseudacorus, Miscanthus sacchariflorus, Phalaris arundinacea, Pontederia cordata L., Phragmites australis, Phragmites karka, Thalia dealbata, Typha angustifolia, Typha latifolia*, and *Vetiveria*. Among these, *Phragmites* is used worldwide, in Europe, for example, Sweden (Berglund *et al.*, 2014), Netherlands (Sabri *et al.*, 2021), Spain (Avila *et al.*, 2014), Portugal (Fernandes *et al.*, 2015); in Asia, for example, in Korea, China, Taiwan; and in Africa, for example, Ethiopia (Mesenbet *et al.*, 2021).

The tolerance of CWs plants to antibiotics can vary. *P. australis* is tolerant with the normal levels of antibiotics in wastewater. *Brassica juncea* showed phytotoxicity at very high tetracycline concentrations. *P. subcapitata* can suffer from great toxicity of doxycycline, azithromycin and oxytetracycline, whereas *Alocasia macrorrhiza* and *Saururus chinensis* was sensitive to sulfadiazine pollution. As a result, CW plants show different capabilities of removing antibiotics from water. *Azolla* could remove sulfadimethoxine by 56%–86% (initial concentration, 50–450 mg/L). The removal efficiency of sulfadiazine, sulfamethazine, and sulfamethoxazole by ryegrass amounts to 89%–99% (initial concentration, 10–100 μg/L). The tetracycline removal rate by *Eichhornia crassipes* exceeded 70% (initial concentration, 3.0–15.0 mg/L) (Huang *et al.*, 2021). *P. australis* could remove sulfa drug antibiotics more effectively, whereas they were less effective in removing tetracycline antibiotics. The removal of antibiotics by *P. australis* was in the descending order: ciprofloxacin > oxytetracycline > sulfamethazine (Chen *et al.*, 2019a).

Antibiotic removal efficiencies in CWs can be affected by the presence of plants and plant categories. The planted CWs demonstrated better removal of ofloxacin (93–95%) compared to unplanted CWs in summer (Dan *et al.*, 2020). In another study, under the conditions of a HLR of 5 cm/d and operation duration of 60 d, the removal efficiencies of tetracycline, sulfonamide, and quinolone in the celery planted aquatic system (72%, 47%, and 22%) were substantially larger than those in the spinach planted aquatic system (33%, 20%, and 7%) in the summer (Huang *et al.*, 2021). *Phragmites australis* showed a higher enrofloxacin removal efficiency than *Iris pseudacorus*. That was because of the deeper and denser root system of *P. australis,* which promoted biofilm formation, reduced substrate clogging, and improved enrofloxacin retention. The antibiotic removal efficiencies in CWs can also be dependent on the planting method. The CW planted with the mixture of *I. pseudacorus* and *P. australis* had a lower removal efficiency of sulfamethazine than the CW planted with either *I. pseudacorus* or *P. australis*. This may be relevant to synergistic effects of stronger root exudates of *I. pseudacorus* and higher oxygen release of *P. australis* in the mixed planted CW, which led to the enhanced microbial diversity. Due to increased microbial competition, the quantity and activity of the microbial community for sulfamethazine degradation were reduced, inducing a decline in the sulfamethazine removal efficiency (Huang *et al.*, 2019).

On the contrary, unvegetated CWs resulted in a higher treatment efficiency of erythromycin, norfloxacin, and ofloxacin (36–92%) compared to vegetated CWs (0–88%) in winter. The poorer treatment performance of vegetated CWs can be explained by inhibition to anaerobic biodegradation of antibiotics, which was caused by O_2 exudated via plant roots. It may also result from the low Henry coefficients of these antibiotics ($<8.7 \times 10^{-19}$), which hindered their direct phytovolatilization (Dan *et al.*, 2020).

From all results mentioned above, it is clear that plants, seasons, and antibiotic categories may have synergistic effects on the antibiotic removal in CWs.

4.2.2.3 CWs flow configuration

CWs flow configuration can be classified as FWS-CWs, SSF-CWs (HSSF-CWs and VSSF-CWs), and hybrid CWs (HyF-CWs) (Fu *et al.*, 2022). It is reported that the flow configuration influences the antibiotic removal in CWs. That is because the flow configuration can affect the interaction between antibiotics and substrates, change in physicochemical conditions (e.g., dissolved oxygen, oxidation–reduction potential), and consequently influence the antibiotic conversion pathways (Dan *et al.*, 2020). The removal efficiencies of antibiotics by various CWs categories are given in Table 4.6.

Chen *et al.* (2016b) found that SWF-CWs were not as good as SSF-CWs in removing antibiotics. The antibiotic removal of SWF-CWs was comparable to those of conventional treatment methods (Chen *et al.*, 2019a).

Many studies found that VSSF-CWs performed better (>78%) than other single CWs in removing antibiotics (Lv *et al.*, 2022). This may be because of the better oxygen diffusion in VSSF-CWs, which can favour the oxidation of antibiotics (Fu *et al.*, 2022; Ma *et al.*, 2022). Furthermore, the upflow VSSF-CWs exhibited higher antibiotic removal efficiencies than the downflow CWs. This can be

Table 4.6 Potential of CWs for antibiotic removal from different types of wastewater.

Study Area	Wastewater			Design Parameters				Operation parameters				Removal Efficiency (%)	Reference
	Wastewater	Antibiotics	Initial Concentration of Antibiotics (ng/L)	Scale	Flow Configuration	Substrate	Plant	Feeding Mode	HRT (d)	HLR (cm/d)	T (°C)		
Chungnamdo, Korea	Real livestock wastewater	SMX	10,030–11,583,330	Full-scale (4492 m^2)	Hybrid	Soil	*Phragmites australis*, *Miscanthus sacchariflorus*	Continuous	2.02			49.43	Choi *et al.* (2016)
		STZ	1,263,330–57,833,330									81.86	
		SMZ	1,055,000–300,330,330									85	
		TRM	10,760–673,330									2.32	
		CTC	4,300,000–16,100,000									29.47	
		ENR	34,260–262,160									27.26	
Xiamen, South China	Real swine wastewater diluted with tap water	CTC	14,000	40 × 80 cm (D × H)	VSSF-CW (upflow)	Brick granule, oyster shells	*Phragmites australis*	Continuous	5	4		88.4–98.3	Huang *et al.* (2015)
		TET	5560									69–99.7	
		OTC	4320									92.7–99.9	
Portugal	Livestock wastewater	ENR	10,0000	Mesocosm (0.4 × 0.3 × 0.3 m)	VSSF-CW	Gravel, lava rock, root beds	*Phragmites australis*	Batch	7	0.118	16–28	98	Fernandes *et al.* (2015)
		TET	10,0000									94	
Changzhou, East China	Real aquaculture wastewater	SMZ	64–211.2	Mesocosm (15 × 40 × 80 cm) (L × W × H)	HSSF-CW	Zeolite, gravel	*Iris pseudacorus*, *Phragmites australis*	Continuous	3	8.4		54.3–68.7	Huang *et al.* (2019)
		ENR	2.57–6.74									75.6–81.1	
Taiwan	Real livestock aquaculture wastewater	CMP	59	Full-scale (6 sections, each, 340 × 140 × 150 cm, L × W × H)	FWS-CW	Gravel	Cattail, Cyperus, Phragmites, Vetiveria	Continuous	2.21			98.2	Hsieh *et al.* (2015)
		SMM	93.2									87	
		SDM	84.3									90.6	
		OTC	217.6									97	
		CIP	17,600									100	
		NP	1653									89.7	
		OP	1119									85.1	
Ethiopia	Real hospital wastewater	CIP	1230	2 × 0.68 × 0.45 m (L × W × H)	HSSF-CW		*Typha latifolia*, *Phragmites karka*	Continuous	4			2–8	Mesenbet *et al.* (2021)
		CTX	17,000									28–62	
Spain	Urban wastewater	SMX	2500 ± 580	Meso-scale VSSF-CWs (3 m^2) + HSSF-CW (2 m^2) + SWF-CW (2 m^2)	Hybrid	Sand, gravel, stone	*Phragmites australis*	Continuous		6, 13, 18		<40%	Avila *et al.* (2014)
		ERY	1900 ± 530									<40%	
		LIN	2900 ± 500									<40%	
		DXY	3700 ± 640									50–90%	
		ENR	1800 ± 350									>90%	
Guilin, South China	Domestic sewage	OTC	254,530	Mesocosm (0.3 × 1 m (D × H)	VSSF-CWs (upflow)	Coke	*Pontederia cordata* L.	Continuous	4		26–34	91%	Bai *et al.* (2022)
		TET	228,530									90%	
		DXY	160,490									85%	

Changsha, South China	Rural wastewater	OFL	193	Full-scale (981 m²)	Hybrid (regulating pool, 4 surface and subsurface flow CWs, a stabilization unit)	Chaff, soil	*Myriophyllum verticillatum*, *Pontederia cordata*	Continuous	1.5	0.7		100 ± 0.00	Chen *et al.* (2015)
		LIN	67									78.0 ± 0.39	
		LCM	120									100 ± 0.00	
		ERY-H_2O	44.5									85.1 ± 3.78	
		SFM	3.4									23.8 ± 3.27	
		SDZ	2.4									10.3 ± 1.81	
		CMZ	54.1									95.1 ± 0.71	
		SMM	50.8									100 ± 0.00	
		TRM	2									100 ± 0.00	
Guangzhou, South China	Domestic sewage	NFX	2034 (Summer) 128 (Winter)	Pilot-scale (12 tanks, each tank 80 × 60 × 80, W × L × H)	VSSF-CW, VSSF-CW (upflow), FWS-CW; HSSF-CW	Vesuvianite; gravel; zeolite	*Thalia dealbata*, *Arundo donaz* var. *versicolor*	Continuous		12.5; 25; 50	Summer (28.2–29.3°C) Winter (11.4–16.0 °C)	46–75% (Summer) 18–72% (Winter)	Dan *et al.* (2020)
		LOM	0 (Summer) 141 (Winter)									-54–94% (Winter)	
		OFL	536 (Summer) 265 (Winter)									38–96% (Summer) 77–98% (Winter)	
		CIP	58 (Summer) 2 (Winter)									0–46% (Summer) 0–80% (Winter)	
		TET	34 (Summer) 101 (Winter)									64–95% (Summer) 0–74% (Winter)	
		OTC	6.7 (Summer) 132 (Winter)									14–61% (Summer) 13–82% (Winter)	
		ROX	122 (Summer) 109 (Winter)									37–86% (Summer) 0–69% (Winter)	
Netherlands	WWTP effluent	TIA	200–20,000	Full-scale (7010 m³; 17.820 m³)	HSSF-CW		*Phragmites australis*	Continuous	1; 3		Winter and summer	28–100%	Sabri *et al.* (2021)
		OTC	201–20,000										
		SMX	202–20,000										
		TRM	203–20,000										

(*Continued*)

Table 4.6 Potential of CWs for antibiotic removal from different types of wastewater (*Continued*).

	Wastewater			Design Parameters				Operation parameters					
Study Area	**Wastewater**	**Antibiotics**	**Initial Concentration of Antibiotics (ng/L)**	**Scale**	**Flow Configuration**	**Substrate**	**Plant**	**Feeding Mode**	**HRT (d)**	**HLR (cm/d)**	***T* (°C)**	**Removal Efficiency (%)**	**Reference**
Huizhou, South China	WWTP effluent	SDZ	4.3 ± 3.2	Full-scale	Hybrid (VSSF-CW + HSSF-CW)	Gravel	*Typha angustifolia*, *Canna indica*, *Thalia dealbata*, and *Canna glauca* (VSSF-CW); *Desmodium styracifolium*, *Acorus tatarinowii*, *Arundo donax*, and *Cyperus alternifolius* (HSSF-CW)	Intermittent (12 h/day)	3.2–8.0	50–200		–70 ± 92	Dan *et al.* (2021)
		SFM	16 ± 8									–99 ± 158	
		SMX	47 ± 32									–145 ± 138	
		CMZ	54 ± 55									–118 ± 242	
		NFX	83 ± 61									95 ± 6	
		CIP	8.8 ± 6.5									70 ± 13	
		OFL	66 ± 71									86 ± 15	
		OTC	5.6 ± 4.1									67 ± 9	
		TET	3.1 ± 1.7									64 ± 5	
		ERY	933 ± 272									58 ± 13	
		ROX	130 ± 108									77 ± 6	
Canada	Synthetic wastewater	SMX	500,000	12 mesocosms, each 180 L	VSSF-CW (recirculated)	Mixed gravel	*Phalaris arundinacea*	Batch		2.2	4–10	99.7	Button *et al.* (2019)
		TCL	501,000									100	
Halmstad, Sweden	Ground water added antibiotics (methanol: ground water, v: v =2500:1)	TRM	1000	18 rectangular CWs, (4 × 10 × 1.2 m)	SWF-CWs	Heavy clay	*Phragmites australis*, *Typha latifolia*	Continuous	5.7	10		61	Berglund *et al.* (2014)
		OTC	400									97	
		NFX	1000									100	
		TET	1000									99	
		AZT	100									100	
		SMX	1000									77	
		ERY	2000									96	
Xian, Weat China	Synthetic wastewater	SMX	103,080	Lab-scale	Integrated VSSF-CW (upflow + downflow)	Gravel	*Acorus calamus*	Continuous	2	10	25	98	Qu *et al.* (2022)
Beijing, North China	Synthetic wastewater	OTC SMX	460,400	Lab-scale 0.4 × 0.2 × 0.45 m (L × W × H)	Integrated VSSF-CWs (downflow + upflow)	Gravel, steel slag	Cattail	Continuous	2		19–32	84.3 ± 3.5	Tian *et al.* (2023)

explained by the higher oxygen supply in the upflow CWs, which can facilitate aerobic biodegradation of antibiotics (Huang *et al.*, 2017).

Successful elimination of antibiotics can be achieved with HyF-CWs. Huang *et al.* (2015) found that HyF-CWs could remove tetracycline with a high efficiency of 69.0–99.9%. Chen *et al.* (2019a) reported that HyF-CWs successfully removed antibiotics by 78–100%. Sakurai *et al.* (2021) claimed that the hybrid CWs resulted in a successful removal of 12 out of 23 antibiotics with an efficacy of more than 92%. Ohore *et al.* (2022) noted that up to 99.4% of antibiotics could be successfully removed in a hybrid CW planted with *I. aquatica*.

HyF-CWs are superior to other CWs categories in removing antibiotics from water. Ilyas and van Hullebusch (2020) found that the HyF-CW resulted in a higher removal (96%) of trimethoprim as compared to other CWs flow configurations (65–70%). Similarly, Fu *et al.* (2022) found that the HyF-CW exhibited a higher antibiotic removal efficiency (87.94%) than other CWs flow configurations (50.39–80.44%). Likewise, Ohore *et al.* (2022) observed the enhancement in the antibiotic removal efficiency when using four hybrid systems of VSSF-CW and HSSF-CW in series.

The superior removal of antibiotics by HyF-CWs can be explained by simultaneous existence of aerobic and anaerobic conditions, as well as the higher HRT in HyF-CWs, which are beneficial for microbial degradation of antibiotics (Lv *et al.*, 2022). Moreover, the use of HyF-CW helps to address drawbacks of single types of CWs (e.g., low HLR) (Oberoi *et al.*, 2019). Since HyF-CWs integrate the merits of all CWs flow configurations, they are the best choice for antibiotic elimination (Fu *et al.*, 2022; Ohore *et al.*, 2022).

It is interesting to note that various CWs flow configurations may have their best applications to different antibiotics. Dan *et al.* (2020) reported that SWF-CWs were more effective for removing photodegradable antibiotics (e.g., FQs) thanks to their large, illuminated surface. VSSF-CWs were most suitable for the removal of biodegradable antibiotics (e.g., SAs) because of their better oxygen conveyance abilities, whereas HSSF-CWs could retain antibiotics with high sorption coefficients (e.g., FQs, MLs, TCs) more effectively.

4.2.3 Operation parameters

4.2.3.1 Feeding mode

CWs used for antibiotic removal from wastewater can be operated in different feeding modes. The prevalent feeding mode is continuous mode (Bai *et al.*, 2022; Chen *et al.*, 2015; Choi *et al.*, 2016; Dan *et al.*, 2020; Huang *et al.*, 2019; Qu *et al.*, 2022; Sabri *et al.*, 2021). Other feeding modes consist of batch mode (Button *et al.*, 2019; Fernandes *et al.*, 2015) and intermittent mode (Dan *et al.*, 2021). In batch mode, wastewater in CWs is entirely drained between batches. If CWs are rested after complete drainage, the batch mode is referred to as tide-flow mode. In intermittent mode, a new volume of wastewater is added to the CWs before the previous volume of wastewater is completely drained. In alternating mode, wastewater is filled and drained imperatively for predetermined periods of time. As a result of interchanging of wet/dry states in alternating mode, CWs are re-oxygenated. It was found that partially flooded VSSF-CWs demonstrated better treatment performance of antibiotics (trimethoprim 99%, sulfamethoxazole 64%) compared to completely flooded VSSF-CWs (trimethoprim 87%, sulfamethoxazole 4%) (He *et al.*, 2021). This can be ascribed to larger redox conditions in partially flooded VSSF-CWs compared to those in completely flooded VSSF-CWs, which enabled the aerobic and anaerobic degradation of antibiotics (Cui *et al.*, 2023).

Besides, the hydraulic loading frequency can influence the removal of antibiotics in alternating mode CWs. The sulfamethoxazole removal efficiency was increased by 16.2% when the number of pulses per day reduced from 4 to 1 (He *et al.*, 2021).

4.2.3.2 Hydraulic loading rate (HLR) and hydraulic retention time (HRT)

HLR is defined as the volume of wastewater that a CW can treat per unit area in a day. In CWs for antibiotic removal, HLR varies from 0.118 cm/d (Fernandes *et al.*, 2015) to 200 cm/d (Dan *et al.*,

2021). HRT is the retention time of wastewater in CWs. It fluctuates in the range of 1 d (Sabri *et al.*, 2021) to 8 d (Dan *et al.*, 2021).

Generally, higher HLR resulted in a better antibiotic removal efficiency (Avila *et al.*, 2014). That is because higher HLRs pose a CW at more risks of clogging and lower substrate porosity due to excessive growth of microorganisms and plants (He *et al.*, 2021). Contrarily, Salah *et al.* (2023) observed that the removal rate of sulfamethazine improved from 10.80% to 25.04% with an increment in HLR from 2 to 10 cm/d. Similarly, higher removal efficiencies of ofloxacin and ciprofloxacin (38–94%) were observed at low HLRs compared to those (47–90%) at high HLRs in summer. The inferior antibiotic removal at high HLRs can be attributed to lower porosity in CWs, which was caused by excessive growth of microorganisms and plants because of more nutrients and organic matter. In contrast, high HLRs resulted in better removal efficiencies of oxytetracycline and erythromycin (66–70%) compared to low HLRs (17–44%) in winter. This can be explained because low-strength inlet wastewater in winter may lack carbon and energy sources for microbial activity. Consequently, antibiotics as alternative carbon sources were biodegraded more vigorously at high HLRs in winter (Dan *et al.*, 2020).

The antibiotic removal efficiency is improved at longer HRT. This may be associated with enhancement in substrate adsorption and microbial degradation of antibiotics at longer HRT (He *et al.*, 2021). The removal rates of sulfamethoxazole by HSSF-CWs and integrated VSSF-CWs were increased by 55% and 52%, respectively, when the HRT was extended from 1 to 3 d. On the contrary, HRT showed negligible effects on the removal of enrofloxacin and florfenicol. This can be possibly related to various removal pathways of different antibiotics. While sulfamethoxazole was mainly removed via microbial degradation, adsorption was the main route for removal of enrofloxacin. The florfenicol was hardly removed by conventional CW (Huang *et al.*, 2021).

4.2.3.3 Influent quality

The removal of antibiotics in CWs is greatly influenced by the influent quality, especially the concentrations of COD, NH_4^+-N, total nitrogen (TN), and total phosphorus (TP). A review of 106 CWs showed that higher levels of COD and nutrients improved microbial activity, thereby enhancing removal of antibiotics. Generally, removal efficiencies of antibiotics, COD, and nutrients in CWs show positive correlations. Particularly, nitrifying bacteria could enhance antibiotic degradation by increasing the release of dioxygenases by both autotrophic and heterotrophic microbes (Dong *et al.*, 2016). However, influent concentrations of COD > 300 mg/L, NH_4^+-N > 48 mg/L, TN > 40 mg/L, and TP > 3.2 mg/L had adverse impacts on antibiotic removal due to competition of these contaminants with antibiotics for microbial degradation. Thus, the stable removal of COD and nutrients in CWs is needed to ensure successful elimination of antibiotics (He *et al.*, 2021).

According to Zhou *et al.* (2020), when a synthetic solution was replaced by the real wastewater, the removal efficiency of sulfadiazine by an integrated system of UF membrane and photocatalysis was reduced by 9.12%. The substitution also led to a reduction in the pseudo first-order kinetic reaction rate constant (k) from 0.216 to 0.0149 min^{-1}. The inhibitory effects on sulfadiazine removal from the real wastewater were due to the presence of Cl^- and SO_4^{2-}. In contrast, NO_3^- was found to favour sulfadiazine decomposition, whereas the effect of pH was negligible (Zhou *et al.*, 2020).

4.2.3.4 Season and temperature

The removal of antibiotics in CWs may vary according to the seasons owing to the changes in temperature and light intensity. In a study on removal of antibiotics from domestic wastewater using pilot-scale CWs, Dan *et al.* (2020) noted that the removal of tetracycline and roxithromycin was enhanced in summer (37–95%) compared to winter (−58–74%). This can be explained by the higher light intensity in summer, which promoted photodegradation of these antibiotics. Contrarily, better removal of ciprofloxacin was achieved in winter (0–80%) than in summer (−72–46%). This is possibly because ciprofloxacin, as a fluoroquinolone antibiotic, was easily adsorbed on substrates even in winter when photolysis played a less dominant role in antibiotic removal. There was no obvious seasonal difference in the removal of norfloxacin, ofloxacin, and oxytetracycline. It is obvious that the effects of seasons on antibiotic removal heavily rely on the types and properties of antibiotics.

Temperature influences reproduction of microorganisms, whereas light affects plant growth (Huang *et al.*, 2021). The variations in plant growth can lead to changes in the oxygen release and exudate secretion, which will affect the microbial quantity and microbial degradation of antibiotics (Du *et al.*, 2020). The reduction in temperature has a significant influence on the growth and activity of both plants and microorganisms. A dramatic decrease in microbial activity in the rhizosphere was observed at low soil and air temperatures, leading to inefficient elimination of antibiotics (Hu *et al.*, 2021).

Nevertheless, the effects of temperature on the removal of antibiotics can be different, depending on the CWs flow configuration. For example, in winter, HSSF-CWs were more efficient in removing antibiotics with high K_d values (e.g., FQs, MLs, and TCs). In summer, VSSF-CWs showed the worst removal of antibiotics. This can be attributed to high influent concentrations of organic matter and nutrients in summer, which induced excessive microbial proliferation and plant growth, thereby lowering the porosity of the filter bed (Fu *et al.*, 2022). SWF-CWs had a slightly higher removal efficiency of antibiotics (17–98%) than HSSF-CWs (−42–97%) in summer, whereas better removal of antibiotics was recorded for HSSF-CWs (37–95%) as compared to SWF-CWs (16–85%) in winter. The better removal of antibiotics in HSSF-CWs in winter may result from the low influent concentrations of organic matter and nutrients in winter, which generates an increasing demand for carbon and energy sources to ensure microbial activity. Results showed that antibiotic biodegradation in SWF-CWs was enhanced in summer, whereas antibiotic adsorption was accelerated in HSSF-CWs in winter (Dan *et al.*, 2020).

The effects of temperature on the removal of antibiotics are also dependent on the antibiotic categories. Though the planted CWs exhibited a better removal of ofloxacin than unplanted CWs in summer, the opposite trend was noted in winter (Dan *et al.*, 2020). While elimination of tetracycline and roxithromycin was more effective in summer (37–95%) than in winter (−58–74%), a higher removal of ciprofloxacin was observed in winter (0–80%) than in summer (−72–46%) (Dan *et al.*, 2020).

The effects of temperature on the removal of antibiotics are affected by the plants. Though the celery planted filter bed (72%, 47%, and 22%) exhibited substantially higher removal efficiencies of tetracycline, sulfonamide, and quinolone than the spinach planted filter bed (33%, 20%, and 7%) in summer, no substantial difference was observed in winter. This may be related to the synergetic effects of temperature and plant characteristics (Huang *et al.*, 2021).

4.2.3.5 Plant harvest

In a study on the effects of plants on the removal efficiency of tetracycline, Du *et al.* (2020) found that the tetracycline content in the substrate was reduced after the plant harvest. According to Wang *et al.* (2015), the plant radial oxygen loss, microbial population and activity in winter were adversely affected by biomass harvest in the late autumn. Due to improper management of CWs plant litter, antibiotics can be reintroduced into the CWs (Fu *et al.*, 2022). Hence, plant litter should be periodically harvested during CWs' operation.

4.3 MECHANISMS OF ANTIBIOTIC REMOVAL BY CWS

In CWs, antibiotics can be removed via a wide variety of physical, chemical, and biological processes, which are classified as substrate associated mechanisms, plant-related mechanisms, microorganisms-based mechanisms, and other mechanisms as shown in Figure 4.4.

4.3.1 Substrate associated mechanisms

The common substrates in CWs encompass soil, gravel, sand, and organic materials (e.g., plant litter). The mineral substrates affect the flow in CWs, provide sites for microbial attachment, retain, or

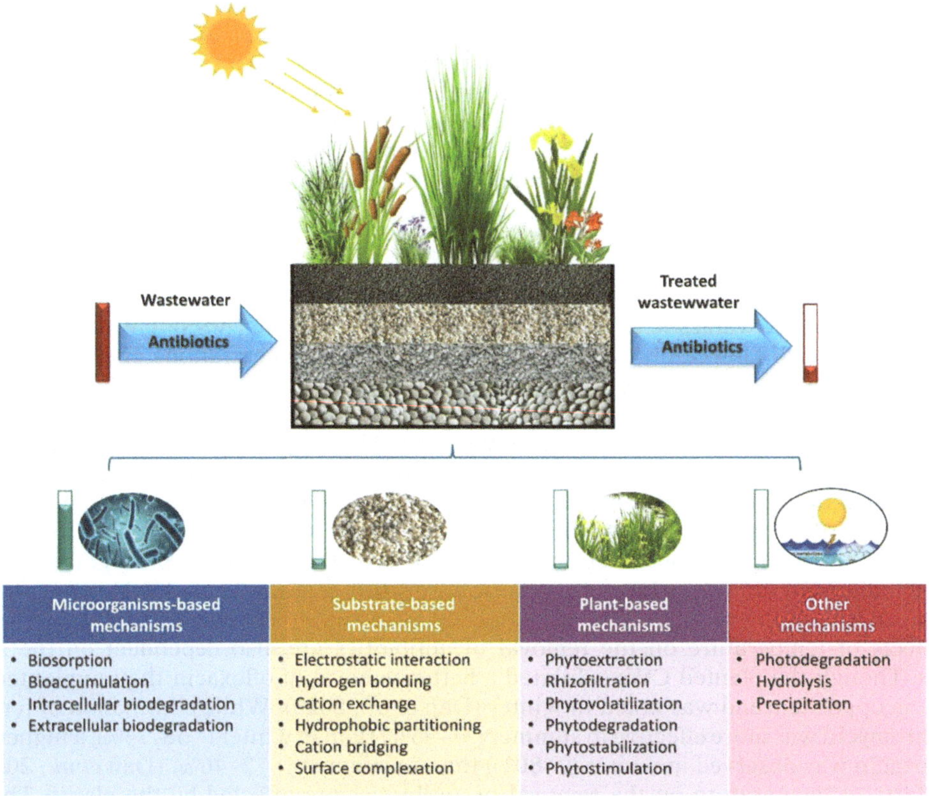

Figure 4.4 Mechanisms for antibiotic removal in CWs.

transform pollutants via various physical, chemical, and biological processes. Meanwhile, organic materials supply carbon and energy sources in addition to attachment sites for microorganisms (Omondi & Navalia, 2020).

Substrates play an important role in removing antibiotics (Bai *et al.*, 2022). In a CW based on siliceous gravel and sand, sulfamethoxazole was removed up to 98% and substrate adsorption was considered as the main removal pathway (Ohore *et al.*, 2022).

The removal of antibiotics by substrates can be done via substrate adsorption, chemical oxidation, filtration (direct pathways), and microbial degradation (indirect pathway) (Cui *et al.*, 2023). The antibiotic removal by substrates relies on their binding capacities (Ohore *et al.*, 2022). Many studies have demonstrated that chemisorption with sharing and exchange of electrons plays a more significant role than physisorption via electrostatic interactions in removing antibiotics (Cui *et al.*, 2023).

The efficient removal of enrofloxacin by zeolite and gravel in HSSF-CWs was attributed to their high content of Ca^{2+}, Mg^{2+}, and other metal ions, which enabled the formation of ligand complexes (Huang *et al.*, 2019). The replacement of gravel by Mn ore as the substrate in VSSF-CW led to an increment in the removal efficiency of sulfamethoxazole by 39.6%. The contribution of Mn oxidation and adsorption to sulfamethoxazole removal in Mn ore-based VSSF-CW was 33% and 15%, respectively. These were found to increase by 20% and 8% as compared to gravel-based VSSF-CW. The enhancement in sulfamethoxazole removal in Mn ore-based VSSF-CW can be associated with the high oxidation and adsorption capacity of biogenic Mn oxides (BioMnOx), which was produced by re-oxidation of Mn(II), being generated from reduction of Mn(III,IV) by sulfamethoxazole with microorganisms as biocatalysts (Xu *et al.*, 2022).

The removal of antibiotics in CWs depends on types and properties of substrates. Antibiotic adsorption is improved by substrates with higher porosity, more microporous structure, higher specific surface area, and smaller particle size (Cui *et al.*, 2023). A study on the removal of seven antibiotics from domestic sewage using HSSF-CWs with four substrates revealed that zeolite was the best substrate, evidenced by the highest removal efficiency and mass removal. This was presumably due to the microporous and Si-OH structures of zeolite (Chen *et al.*, 2016a). The antibiotic adsorption was affected by particle size as follows: fine silt > clay > sand > medium silt > whole soil > coarse silt. It also varied according to pH (Thiele-Bruhn *et al.*, 2004).

Substrate adsorption of antibiotics is affected by the antibiotic categories. Ionizable/polar and moderately water-soluble antibiotics are favourably captured via electrostatic interactions (e.g., hydrogen bonding, ion bridging, ion exchange, surface complexation) by charged substrates (e.g., soil, clay, zeolite, organic matter). The common substrates (e.g., soil, clay, zeolite, organic matter) are negatively charged. Ion exchange is associated with coulombic attraction of positively charged amine groups on antibiotics and negatively charged sites on the substrates surface. It was found that surface complexation was a main route for retention of fluoroquinolones by zeolite and gravel. Accordingly, the ligand complexes were formed between Ca^{2+} and Mg^{2+} cations of these substrates and carboxyl groups of fluoroquinolones. On the other hand, hydrophobic interactions (e.g., van der Waals force, electron donor and acceptor interaction) are the main removal pathways of neutral/nonpolar antibiotics by hydrophobic substrates (e.g., humic substances) (Cui *et al.*, 2023). Hydrophobic partitioning is playing a crucial role in adsorption of antibiotics with $\log K_{ow} > 4$, whereas those with $\log K_{ow} < 2.5$ cannot be adsorbed by substrates (Lv *et al.*, 2022).

On the other hand, antibiotic removal via substrate adsorption has some drawbacks. Antibiotics are not degraded after adsorption. In addition, the adsorption efficacy will be reduced by the competition of co-existing pollutants (Cui *et al.*, 2023).

4.3.2 Plant-related mechanisms

The removal of antibiotics in CWs was improved with the presence of plants (He *et al.*, 2021). The CW plants include nonvascular (e.g., algae) and vascular (e.g., macrophyte) plants (Omondi & Navalia, 2020). For removal of antibiotics in CWs, plants may have direct (e.g., phytoextraction, phytodegradation, phytovolatilization, phytostabilization) or indirect (e.g., enhanced microbial transformation) effects (Choi *et al.*, 2016).

Algae can increase the level of oxygen in water via photosynthesis, leading to an increment in antibiotic degradation (Omondi & Navalia, 2020).

Macrophytes can directly uptake, accumulate, and transform antibiotics via glycosylation and glutathione routes (Cui *et al.*, 2023; Ma *et al.*, 2022). They can also indirectly intensify the antibiotic removal in CWs via rhizosphere effects, namely offering a microbial habitat, providing oxygen and exudates (e.g., enzymes, carbon sources) to shape the microbial community and to strengthen the microbial metabolism, forming supramolecular structures based on antibiotics and humic acids (Choi *et al.*, 2016; Hu *et al.*, 2021; Lv *et al.*, 2022). Plants can protect CWs from cold weather (He *et al.*, 2021). These indirect effects were responsible for above 70% of the total antibiotic removal in CWs (Hu *et al.*, 2021). Furthermore, plants in CWs have additional usages, such as stabilizing substrate surface, enhancing filtration, preventing water clogging, insulating against frost in winter, and maintaining pH in CWs. While the pH in unvegetated CWs exceeded 8.05, the pH in vegetated CWs was around 7.5 (Carvalho *et al.*, 2013).

The plant uptake of antibiotics relies on plant categories, antibiotic properties (e.g., water solubility, hydrophobicity), and antibiotic concentrations (Hu *et al.*, 2021). While substrate-grown plants can uptake both hydrophilic and hydrophobic antibiotics, hydroponically grown plants are only efficient in removing dissolved antibiotics (Lv *et al.*, 2022). The plant uptake of antibiotics is proportional to the log K_{ow} of the antibiotic (Fu *et al.*, 2022). Many studies show that antibiotics with medium hydrophobicity ($1.0 < \log K_{ow} < 3.5$) and/or low molecular weight (MW < 500) can be taken up by plants via numerous mechanisms, for example passive uptake via protein channels as well as active uptake via protein-mediated and energy-dependent channels (Lv *et al.*, 2022).

Generally, planted CWs are better than unplanted CWs in terms of antibiotic removal, owing to the close relationship between plant rhizosphere and microbial transformation (Hu *et al.*, 2021). Likewise, Bai *et al.* (2022) observed that the removal of tetracyclines was boosted by the addition of plants. Each plant category has its own microbial population, which influences the antibiotic treatment performance of that plant (Hu *et al.*, 2021). The accumulation of antibiotics in different parts of the plant usually decreases in the following order: roots > stems > leaves (Fu *et al.*, 2022).

Many studies showed that plant uptake is an inefficient pathway for antibiotic removal from wastewater. For instance, in a floating CW, the contribution of plant uptake to the total removal of oxytetracycline, chlortetracycline, and tetracycline from wastewater was limited to 0.013–0.098%, 0.013–0.258%, and 0.013–0.085%, respectively (Lu *et al.*, 2014). Also, plant uptake is less important than substrate adsorption and biodegradation in removing antibiotics from wastewater (Chen *et al.*, 2016b; Xu *et al.*, 2022).

Although direct plant uptake contributes to only a small portion of the total antibiotic removal in CWs, it is significant in aspects of releasing oxygen and root exudates for microbial growth and biotransformation of antibiotics (Chen *et al.*, 2016b).

4.3.3 Microorganism-based mechanisms

Microbial degradation is a process that converts a large, complex antibiotic molecule into small, simpler molecules via the microbe's metabolism. During metabolism, each antibiotic is degraded by a different enzyme. On the contrary, in co-metabolism, the degradation of the second antibiotic relies on the enzyme that is catalysing the first antibiotic. Noteworthy, antibiotic concentrations need to be sufficiently high to induce enzymes for microbial degradation. So far, no information can be found on the essential antibiotic concentration that induces metabolism (Lv *et al.*, 2022).

Microbial degradation is the main route for the removal of antibiotics in CWs. There was a significant positive correlation between total antibiotics and total bacterial biomass, indicating that removal efficiencies of antibiotics were closely related to bacterial activities at the mesocosm-scale CWs (Huang *et al.*, 2019). Microbes contributed up to 94% to the removal of tetracycline and enrofloxacin (Ohore *et al.*, 2022). The aerobic degradation of antibiotics in VSSF-CWs is more effective than their anaerobic degradation in HSSF-CWs (Ilyas & van Hullebusch, 2020). Similarly, aerobic degradation was found to be more important than anaerobic degradation in the microbial transformation of erythromycin in CWs (Lv *et al.*, 2022).

Many studies showed that microbial degradation plays a more important role than substrate adsorption and plant uptake in removing antibiotics from water. For instance, the removal of sulfamethoxazole in Mn ore-based VSSF-CW was mainly via biodegradation (33%). The substrate adsorption contributed 15%, whereas the plant uptake was minor (Xu *et al.*, 2022). Similarly, the removal of sulfamethoxazole in HSSF-CW and VSSF-CW by biodegradation accounted for 56.9% and 64.85%, respectively, whereas the roles of substrate adsorption and plant uptake were insignificant (Chen *et al.*, 2016b). The contribution of biodegradation to the total removal of erythromycin was even higher, reaching 93.5% and 85.3%, respectively, for HSSF-CW and VSSF-CW. This is strongly supported by Cheng *et al.* (2021), stating that the contribution of different pathways to the antibiotic removal in CWs was ranked in the following order: biodegradation (85%) > substrate adsorption (14.3%) > plant uptake (0.44%).

On the other hand, biodegradation was shown to be inferior to substrate adsorption in the elimination of clarithromycin. The removal efficiencies of clarithromycin by substrate adsorption in HSSF-CW and VSSF-CW were 44.5% and 46.33%, whereas those by biodegradation were 29.6% and 20.86% (Chen *et al.*, 2016b).

It was noted that sulfamethazine could not be determined in the substrate and plant, suggesting the dominant role of microbial degradation in eliminating this antibiotic from water. Despite of their less crucial roles, substrate adsorption and plant uptake largely affect microbial biodegradation of antibiotics by providing attachment sites as well as essential conditions (e.g., oxygen, organic acids, amino acids, enzymes) for microbial growth and organic matter decomposition (Huang *et al.*, 2019).

The microbial degradation of antibiotics is associated with their properties. Antibiotics with higher water solubility can be degraded more easily. Besides, this process can be affected by many other factors, for example, substrate, plant, temperature, HLR, and nutrient concentration (Ohore *et al.*, 2022).

CWs have a complicated behaviour of ARGs in bacterial cells. Some studies indicated that planted CWs had a role in the elimination of ARGs (Berglund *et al.*, 2014; Fang *et al.*, 2017). Contrarily, others showed that plants could increase the abundance of ARGs in faecal bacterial populations (Christofilopoulos *et al.*, 2019). CWs can be a source of ARG pollution, from which ARGs can spread among various bacteria downstream. Consequently, ARGs should be monitored not only in the treated effluent but also in bacterial cells in CWs for risk control.

4.3.4 Other mechanisms

Although photolysis can play a certain role in removing antibiotics by SWF-CWs, their removal via this process is the lowest as compared to other removal pathways (Chen *et al.*, 2016b). In CWs, antibiotics can also be removed via precipitation on the plant root surfaces that are covered by ion plaque. Iron plaque, which is composed of ferric (Fe^{3+}) hydroxides and oxides, can be formed via oxidation of ferrous iron (Fe^{2+}) with radial oxygen released from plant roots. It was found that the removal of macrolides was enhanced with the increasing amount of iron plaque on the root surface (Hu *et al.*, 2021). That is because iron plaque has a high specific surface area, contains abundant functional groups, and induces a wide range of redox potentials (–250–230 mV) (Fu *et al.*, 2022).

4.4 STRATEGIES TO INTENSIFY ANTIBIOTIC REMOVAL BY CWS

To enhance the decontamination of antibiotic-rich wastewater in CWs, viable strategies include optimization of design and operation parameters, application of emerging substrates, microbial augmentation, and utilizing integrated processes (Lv *et al.*, 2022).

4.4.1 Aeration

Oxygen supply is one of the factors limiting the antibiotic treatment performance of CWs, especially for HSSF-CWs with permanently saturated substrates (Lv *et al.*, 2022). The removal of antibiotics in a shallower bed is normally more efficient. This may result from better passive aeration and a more positive redox potential, thereby strengthening microbial biodegradation of antibiotics (García *et al.*, 2020). The oxygen can be passively supplied into CWs via the influent, rhizosphere, and air. However, in cases of high-strength wastewater, winter season, and plant mortality, artificial aeration is a useful technical solution to ensure sufficient oxygen and efficient antibiotic removal (Cui *et al.*, 2023). Accordingly, oxygen can be intermittently or continuously introduced into CWs with pumps (He *et al.*, 2021).

A lab-scale study revealed that aerobic biodegradation was more efficient than anaerobic degradation in removing antibiotics (Avila *et al.*, 2014). Similarly, Wen *et al.* (2021) noted that because of 12 h aeration, the removal percentage of sulfamethoxazole and tetracycline in MFC-CWs increased by 4.98% and 4.34%, respectively. Moreover, aeration led to an improvement in the voltage and maximum power density, while reducing internal resistance. The enhancement in the antibiotic removal by aerated CWs is attributed to the formation of various micro-conditions, which promoted both aerobic and anaerobic transformations of antibiotics (Ilyas & van Hullebusch, 2020). Artificial aeration promoted microbial activities and antibiotic oxidation, thus leading to an increment in antibiotic removal (Ohore *et al.*, 2022). In addition to the increase in oxygen level, artificial aeration also helps to reduce substrate clogging (He *et al.*, 2021).

Nevertheless, the application of continuous aeration makes CW an expensive and unsustainable process. Therefore, intermittent aeration is recommended to achieve the double goal of high antibiotic removal efficiency and lower operational cost (Ohore *et al.*, 2022). Further studies should be conducted on impacts of aeration on plant growth, as well as optimization of the aeration position and duration (He *et al.*, 2021).

4.4.2 Innovative substrates

To improve the antibiotic removal and sustainability of CWs, there is a trend to use single functional substrates or combined substrates (Hu *et al.*, 2021).

As a result of the poor antibiotic removal by conventional substrates, there is a need to fabricate and apply innovative functional substrates. Examples of functional substrates that were used in CWs for antibiotic treatment include ceramsite, Mn ore, biochar, and light expanded clay aggregates (LECA) (Lv *et al.*, 2022). In the future, it is suggested to use nanomaterials as a substitute for conventional substrates or to combine nanomaterials with conventional substrates to eliminate antibiotics more effectively from water (Chen *et al.*, 2019a).

The combined substrates can be applied in CWs via two ways, as a mixture in one layer or as separate substrates in different layers. The examples for these two ways are gravel/sand/soil and lava rock/ceramic/zeolite, respectively (Cui *et al.*, 2023).

The addition of exogenous organic matter led to the enhanced retention of sulfonamides in anaerobic granular sludge (Oliveira *et al.*, 2016). Similarly, the supply of external carbon sources (e.g., glucose, sucrose, acetate, starch) intensified the adsorption of sulfamethoxazole and tetracycline onto the granular activated carbon as the substrate (Wen *et al.*, 2021).

The use of zeolite in an integrated vertical-flow constructed wetlands (IVCWs) resulted in high removal efficiencies of antibiotics of about 95% (Du *et al.*, 2020). Hamadeen and Elkhatib (2022) claimed that nanostructured activated biochar derived from pomegranate peels (PPAB) could effectively remove ciprofloxacin from real wastewater. The ciprofloxacin removal efficiencies by PPAB in batch and column tests reached 89.94 and 84.74%, respectively.

4.4.3 Bacterial augmentation

Bioaugmentation is an effective approach to enhance the biodegradation of antibiotics in CWs (Lv *et al.*, 2022).

A common method is the addition of certain bacterial strains with the ability to decompose specific antibiotics into CWs. As a result, the population, density, and activity of functional microbes will be enhanced. In contrast, the niche of indigenous microbes will be narrowed down, thus reducing their competitiveness with the functional microbes. This helps to accelerate the microbial degradation of target antibiotics (Zhao *et al.*, 2022). Shah *et al.* (2022) observed that the removal of ciprofloxacin from aqueous solution (100 mg/L) by a floating constructed wetland (FCW) with immobilized bacterial consortia (*Acinetobacter lwoffii* ACRH76, *Bacillus pumulis* C2A1, and *Acinetobacter* sp. HN3) reached up to 97%, which was considerably higher than that (48%) by the FCW without bacterial augmentation. Obviously, immobilized bacterial augmentation can be a promising strategy to strengthen FCW's antibiotic remediation in wastewater. Nevertheless, it should be noted that the augmentation of ARB can unbalance the ecosystem in CWs. Another greater risk is the possibility that ARGs escape from CWs to the downstream environment, inducing unforeseeable consequences.

Another method is to integrate microalgae into CWs. The living microalgae can generate oxygen via photosynthesis, whereas the dead algae can be used as organic substrates. Thanks to oxygen availability and organic substrate enrichment, the removal of antibiotics can be improved. Additionally, the integration of microalgae in CWs is useful to overcome the insufficient microbial population and treatment performance in winter (Lv *et al.*, 2022).

4.4.4 Integrated processes

The coupling of conventional treatment methods with CWs provides an effective solution for decontamination of antibiotics from wastewater (García *et al.*, 2020). The grey processes as the primary stage help to remove conventional pollutants (e.g., suspended solid, organic matters, nutrients) and antibiotics to some extent, whereas CWs as the polishing stage provide further treatment. Therefore, the integration of grey processes and CWs can improve antibiotic treatment, reduce the land investment, as well as simplify operation and maintenance of the hybrid process (Fu *et al.*, 2022).

Recently, activated sludge processes (ASP) have been combined with CWs in most large-scale WWTPs to enhance antibiotic treatment (Fu *et al.*, 2022). It was shown that it is feasible to use CWs for polishing treatment of the anaerobically treated blackwater. In this study, the combination of an UASB reactor and an IFCW system resulted in a high antibiotic removal efficiency of 93% (Sakurai *et al.*, 2021). Later, Wang and Chen (2022) claimed that a hybrid process of MBRs and CWs intensified the removal efficacy of antibiotics, demonstrating a great potential for antibiotic elimination. Lim *et al.* (2019) investigated the removal of sulfamethoxazole by a lab-scale staged anaerobic membrane bioreactor (SAF-MBR) coupled with a granular activated carbon (GAC) fluidized bed. Using a HRT of 8.7 h, initial sulfamethoxazole concentration of 0.1 mg/L, and a permeate flux of 5 $L/m^2/h$, the removal efficiency of sulfamethoxazole was 100% after 3 days of operation of the SAF-MBR. The excellent performance of the integrated process can be attributed to the synergetic effects of the MBR and GAC combination. While the MBR resulted in high effluent quality, low sludge production and methane generation, GAC fluidization reduced membrane fouling and enhanced biofilm formation on the GAC. Adsorption by GAC as well as adsorption and biodegradation by the biofilm on the GAC were responsible for complete removal of sulfamethoxazole in the SAF-MBR. Maldonado *et al.* (2023) employed CWs and MFC for the removal of tetracycline and chloramphenicol from lake water. Under the initial concentrations of tetracycline (4.25 μg/L) and chloramphenicol (1.87 μg/L), their minimal removal efficiencies were 99.45% and 97.84%, whereas their maximum removal efficiency reached up to 100%. This can be explained by the accelerated biodegradation in the anodic zone, phytoaccumulation, and substrate adsorption. The hybrid system also produced electricity. Zhou *et al.* (2020) applied a modified UF membrane filtration (PVDF-PVP-TiO_2-dopamine) and photocatalysis coupling system in abatement of sulfadiazine from water. This resulted in synergistic effects, such as increased antibiotic oxidation, improved permeation flux, and reduced membrane fouling. In addition, TiO_2 enhances the hydrophilic features of hydrophobic membranes. The removal of SR was as high as 91.4% in pure water. The coupled system produced $\cdot OH$, $\cdot O_2^-$, $HO_2\cdot$, H_2O_2 and O_2, indicating that $\cdot OH$ played a main role in sulfadiazine decontamination.

4.4.5 Other strategies

Button *et al.* (2019) intensified the removal of triclosan and sulfamethoxazole in VSSF-CW by recirculation. The removal efficiency of both triclosan and sulfamethoxazole was improved from > 80% to > 99.7% with the retention time increment from 1 to 168 h.

To strengthen the removal efficiencies of sulfamethoxazole and tetracycline in an MFC-CWs coupling system, the supply of external carbon sources was compared. Glucose was found superior to other carbon sources. The medium glucose concentration (200 mg/L) yielded the maximum removal efficiencies of sulfamethoxazole (99.4%) and tetracycline (97.8%). Besides, the increment in glucose levels in the range of 100–400 mg/L substantially improved the voltage and maximum power density, while lowering the internal resistance and coulombic efficiency (Wen *et al.*, 2021).

4.5 RESEARCH OBSTACLES AND FUTURE PERSPECTIVES

Although CWs are promising for antibiotic removal from wastewater, there remain several challenges to their antibiotic bioremediation.

The use of CWs packed with conventional substrates (e.g., soil, sand, and gravel) for antibiotic treatment in wastewater often results in unsatisfactory results. This may be related to their low contents of nutrients, poor abilities for antibiotic retention and biofilm formation (Lv *et al.*, 2022). Hence, it is recommended that future research should be carried out on fabrication and employment of novel functional substrates as a replacement for conventional ones. Especially, more attention should be paid to modified substrates with their enhanced physicochemical properties (e.g., hydraulic conductivity, mechanical strength, and electron supply), effects on plant growth and microbial proliferation (Lv *et al.*, 2022).

While many plants have the potential for antibiotic removal from water and low-strength wastewater, some plants suffer from phytotoxicity due to high levels of antibiotics in high-strength wastewater, for example, livestock or pharmaceutical wastewater. This highlights the significance of selecting potential plant species for antibiotic elimination in wastewater (Chen *et al.*, 2016b). Besides, the CWs design with proper planting patterns should be further explored (Huang *et al.*, 2019).

Despite the significant advantages of HyF-CWs, their full-scale applications with real wastewaters are still limited. Thus, additional work is needed to further develop HyF-CWs as an efficient, affordable, simple, and sustainable approach for antibiotic decontamination of real wastewaters (Huang *et al.*, 2021; Lv *et al.* 2022).

Further in-depth studies are necessary to shed light on antibiotic removal mechanisms in CWs, for example the fate and dynamics of antibiotics in CWs, competitive adsorption of co-existing pollutants in wastewater, formation and effects of antibiotic metabolites in plants, and response of the microbial community to antibiotics in the rhizosphere (Cui *et al.*, 2023; Ohore *et al.*, 2022).

Since CWs can be reservoirs for ARGs, more research is essential focusing on configuration and operational strategies to strengthen their ARGs treatment performance (Ohore *et al.*, 2022; Fu *et al.*, 2022). Also, it is essential to closely monitor levels of ARGs in the CWs effluent to minimize their risks to the environment and public health (Bai *et al.*, 2022; García *et al.*, 2020).

To detect antibiotics at concentrations in wastewaters, advanced analytical devices with high sensitivity and selectivity are needed. Consequently, in-depth studies on antibiotics in developing countries are still scare (Hu *et al.*, 2021).

4.6 CONCLUDING REMARKS

Livestock is a major source of antibiotic pollution, followed by pharmaceutical manufacturers, household consumption, hospitals, and WWTPs. Antibiotics are one of the emerging pollutants that can cause detrimental effects on ecosystems even at trace levels. Therefore, antibiotic pollution remediation is of great significance. Even though numerous technologies can be applied in removing antibiotics from wastewater, the integrated grey-green processes receive a growing interest because of their effective performance and environmental sustainability. Although CWs technology can offer a viable, affordable, and sustainable solution for decontamination of antibiotic-rich wastewaters, some bottlenecks still need to be overcome. Key strategies for intensifying the antibiotic abatement from wastewater by CWs include the utilization of novel functional substrates, microbial augmentation, optimization of process parameters, and coupling of grey-green technologies.

Antibiotic removal by CWs depends on the CWs design and operation parameters. Among these, the most influencing factors comprise wastewater composition, antibiotic properties, flow configuration, feeding mode, HLR, and HRT. The antibiotic removal efficiencies by CWs can range from negative values to 100%. Antibiotic removal pathways in CWs can be classified as four main groups based on substrates, plants, microorganisms, and other factors. Typically, the contribution of different mechanisms to the overall removal of antibiotics in CWs is ranked in the following order: biodegradation > substrate adsorption > plant uptake > other mechanisms.

Despite a high potential of CWs for antibiotic removal, there remain some challenges, such as unsatisfactory performance of conventional substrates, costly upscaling of CWs with emerging substrates, intolerance of some plants to antibiotic-rich wastewater, restricted application of full-scale CWs with real wastewaters, unelucidated antibiotic removal mechanisms, and potential risk of ARGs release in CWs effluent.

REFERENCES

Acosta-Rangel A., Sánchez-Polo M., Rozalen M., Rivera-Utrilla J., Polo A. M. S., Berber-Mendoza M. S. and López-Ramón M. V. (2020). Oxidation of sulfonamides by ferrate (VI): reaction kinetics, transformation byproducts and toxicity assessment. *Journal of Environmental Management*, **255**, 109927, https://doi.org/10.1016/j.jenvman.2019.109927

Alsager O. A., Alnajrani M. N., Abuelizz H. A. and Aldaghmani I. A. (2018). Removal of antibiotics from water and waste milk by ozonation: kinetics, byproducts, and antimicrobial activity. *Ecotoxicology and Environmental Safety*, **158**, 114–122, https://doi.org/10.1016/j.ecoenv.2018.04.024

Avila C., Matamoros V., Reyes-Contreras C., Piña B., Casado M., Mita L., Rivetti C., Barata C., García J. and Bayona J. M. (2014). Attenuation of emerging organic contaminants in a hybrid constructed wetland system under different hydraulic loading rates and their associated toxicological effects in wastewater. *Science of the Total Environment*, **470–471**, 1272–1280, https://doi.org/10.1016/j.scitotenv.2013.10.065

Bai S., Wang X., Zhang Y., Liu F., Shi L., Ding Y., Wang M. and Lyu T. (2022). Constructed wetlands as nature-based solutions for the removal of antibiotics: performance, microbial response, and emergence of antimicrobial resistance (AMR). *Sustainability* **14**(22), 14989, https://doi.org/10.3390/su142214989

Bayati M., Ho T. L., Vu D. C., Wang F., Rogers E., Cuvellier C., Huebotter S., Inniss E. C., Udawatta R., Jose S. and Lin C. H. (2021). Assessing the efficiency of constructed wetlands in removing PPCPs from treated wastewater and mitigating the ecotoxicological impacts. *International Journal of Hygiene and Environmental Health*, **231**, 113664, https://doi.org/10.1016/j.ijheh.2020.113664

Berglund B., Khan G. A., Weisner S. E., Ehde P. M., Fick J. and Lindgren P. E. (2014). Efficient removal of antibiotics in surface-flow constructed wetlands, with no observed impact on antibiotic resistance genes. *Science of the Total Environment*, **476–477**, 29–37, https://doi.org/10.1016/j.scitotenv.2013.12.128

Button M., Cosway K., Sui J. and Weber K. (2019). Impacts and fate of triclosan and sulfamethoxazole in intensified recirculating vertical flow constructed wetlands. *Science of the Total Environment*, **649**, 1017–1028, https://doi.org/10.1016/j.scitotenv.2018.08.395

Carvalho P. N., Araújo J. L., Mucha A. P., Basto M. C. P. and Almeida C. M. R. (2013). Potential of constructed wetlands microcosms for the removal of veterinary pharmaceuticals from livestock wastewater. *Bioresource Technology*, **134**, 412–416, https://doi.org/10.1016/j.biortech.2013.02.027

Carvalho P. N., Basto M. C. P., Almeida C. M. R. and Brix H. (2014). A review of plant-pharmaceutical interactions: from uptake and effects in crop plants to phytoremediation in constructed wetlands. *Environmental Science and Pollution Research*, **21**, 11729–11763, https://doi.org/10.1007/s11356-014-2550-3

Chen J., Liu Y. S., Su H. C., Ying G. G., Liu F., Liu S. S., He L. Y., Chen Z. F., Yang Y. Q. and Chen F. R. (2015). Removal of antibiotics and antibiotic resistance genes in rural wastewater by an integrated constructed wetland. *Environmental Science and Pollution Research*, **22**(3), 1794–1803, https://doi.org/10.1007/s11356-014-2800-4

Chen J., Wei X. D., Liu Y. S., Ying G. G., Liu S. S., He L. Y., Su H. C., Hu L. X., Chen F. R. and Yang Y. Q. (2016a). Removal of antibiotics and antibiotic resistance genes from domestic sewage by constructed wetlands: optimization of wetland substrates and hydraulic loading. *Science of the Total Environment*, **565**, 240–248, https://doi.org/10.1016/j.scitotenv.2016.04.176

Chen J., Ying G. G., Wei X. D., Liu Y. S., Liu S. S., Hu L. X., He L. Y., Chen Z. F., Chen F. R. and Yang Y. Q. (2016b). Removal of antibiotics and antibiotic resistance genes from domestic sewage by constructed wetlands: effect of flow configuration and plant species. *Science of the Total Environment*, **571**, 974–982, https://doi.org/10.1016/j.scitotenv.2016.07.085

Chen J., Liu Y. S., Zhang J. N., Yang Y. Q., Hu L. X., Yang Y. Y., Zhao J. L., Chen F. R. and Ying G. G. (2017). Removal of antibiotics from piggery wastewater by biological aerated filter system: treatment efficiency and biodegradation kinetics. *Bioresource Technology*, **238**, 70–77, https://doi.org/10.1016/j.biortech.2017.04.023

Chen M., Zhu M., Zhu Y., Wang D., Li Z., Zeng G., Zhang C., Huang J. and Xu P. (2019a). Collision of emerging and traditional methods for antibiotics removal: taking constructed wetlands and nanotechnology as an example. *NanoImpact*, **15**, 100175, https://doi.org/10.1016/j.impact.2019.100175

Chen J., Deng W. J., Liu Y. S., Hu L. X., He L. Y., Zhao J. L., Wang T. T. and Ying G. G. (2019b). Fate and removal of antibiotics and antibiotic resistance genes in hybrid constructed wetlands. *Environmental Pollution*, **249**, 894–903, https://doi.org/10.1016/j.envpol.2019.03.111

Cheng X. Q., Liu Y., Guo Z. and Shao L. (2015). Nanofiltration membrane achieving dual resistance to fouling and chlorine for 'green' separation of antibiotics. *Journal of Membrane Science*, **493**, 156–166, https://doi.org/10.1016/j.memsci.2015.06.048

Cheng Y. X., Chen J., Wu D., Liu Y. S., Yang Y. Q., He L. X., Ye P., Zhao J. L., Liu S. S., Yang B. and Ying G. G. (2021). Highly enhanced biodegradation of pharmaceutical and personal care products in a novel tidal flow constructed wetland with baffle and plants. *Water Research*, **193**, 116870, https://doi.org/10.1016/j.watres.2021.116870

Choi Y. J., Kim L. H. and Zoh K. D. (2016). Removal characteristics and mechanism of antibiotics using constructed wetlands. *Ecological Engineering*, **91**, 85–92, https://doi.org/10.1016/j.ecoleng.2016.01.058

Christofilopoulos S., Kaliakatsos A., Triantafyllou K., Gounaki I., Venieri D. and Kalogerakis N. (2019). Evaluation of a constructed wetland for wastewater treatment: addressing emerging organic contaminants and antibiotic resistant bacteria. *New Biotechnology*, **52**, 94–103, https://doi.org/10.1016/j.nbt.2019.05.006

Cui E., Zhou Z., Gao F., Chen H. and Li J. (2023). Roles of substrates in removing antibiotics and antibiotic resistance genes in constructed wetlands: a review. *Science of the Total Environment*, **859**, 160257, https://doi.org/10.1016/j.scitotenv.2022.160257

Dan A., Zhang X., Dai Y., Chen C. and Yang Y. (2020). Occurrence and removal of quinolone, tetracycline, and macrolide antibiotics from urban wastewater in constructed wetlands. *Journal of Cleaner Production*, **252**, 119677, https://doi.org/10.1016/j.jclepro.2019.119677

Dan A., Chen C. X., Zou M. Y., Deng Y. Y., Zhang X. M., Du J. J. and Yang Y. (2021). Removal efficiency, kinetic, and behaviour of antibiotics from sewage treatment plant effluent in a hybrid constructed wetland and a layered biological filter. *Journal of Environmental Management*, **288**, 112435.

Dong H., Yuan X., Wang W. and Qiang Z. (2016). Occurrence and removal of antibiotics in ecological and conventional wastewater treatment processes: a field study. *Journal of Environmental Management*, **178**, 11–19, https://doi.org/10.1016/j.jenvman.2016.04.037

Du L., Zhao Y., Wang C., Zhang H., Chen Q., Zhang X., Zhang L., Wu J., Wu Z. and Zhou Q. (2020). Removal performance of antibiotics and antibiotic resistance genes in swine wastewater by integrated vertical-flow constructed wetlands with zeolite substrate. *Science of the Total Environment*, **721**, 137765, https://doi.org/10.1016/j.scitotenv.2020.137765

Fang H., Zhang Q., Nie X., Chen B., Xiao Y. and Zhou Q. (2017). Occurrence and elimination of antibiotic resistance genes in a long-term operation integrated surface flow constructed wetland. *Chemosphere*, **173**, 99–106, https://doi.org/10.1016/j.chemosphere.2017.01.027

Fernandes J. P., Almeida C. M. R., Pereira A. C., Ribeiro I. L., Reis I., Carvalho P., Basto M. C. P. and Mucha A. P. (2015). Microbial community dynamics associated with veterinary antibiotics removal in constructed wetlands microcosms. *Bioresource Technology*, **182**, 26–33, https://doi.org/10.1016/j.biortech.2015.01.096

Fu J., Zhao Y., Yao Q., Addo-Bankas O., Ji B., Yuan Y., Wei T. and Esteve-Núñez A. (2022). A review on antibiotics removal: leveraging the combination of grey and green techniques. *Science of the Total Environment*, **838**, 156427, https://doi.org/10.1016/j.scitotenv.2022.156427

García J., García-Galán M. J., Day J. W., Boopathy R., White J. R., Wallace S. and Hunter R. G. (2020). A review of emerging organic contaminants (EOCs), antibiotic resistant bacteria (ARB), and antibiotic resistance genes (ARGs) in the environment: increasing removal with wetlands and reducing environmental impacts. *Bioresource Technology*, **307**, 123228, https://doi.org/10.1016/j.biortech.2020.123228

Gothwal R. and Shashidhar T. (2015). Antibiotic pollution in the environment: a review. *Clean-Soil, Air, Water*, **43**(4), 479–489, https://doi.org/10.1002/clen.201300989

Hamadeen H. M. and Elkhatib E. A. (2022). New nanostructured activated biochar for effective removal of antibiotic ciprofloxacin from wastewater: adsorption dynamics and mechanisms. *Environmental Research*, **210**, 112929, https://doi.org/10.1016/j.envres.2022.112929

Han C. H., Park H. D., Kim S. B., Yargeau V., Choi J. W., Lee S. H. and Park J. A. (2020). Oxidation of tetracycline and oxytetracycline for the photo-Fenton process: their transformation products and toxicity assessment. *Water Research*, **172**, 115514, https://doi.org/10.1016/j.watres.2020.115514

Hang J., Yi X. H., Wang C. C., Fu H., Wang P. and Zhao Y. (2022). Heterogeneous photo-Fenton degradation toward sulfonamide matrix over magnetic Fe_3S_4 derived from MIL-100 (Fe). *Journal of Hazardous Materials*, **424**, 127415, https://doi.org/10.1016/j.jhazmat.2021.127415

He Y., Zhang L., Jiang L., Wagner T., Sutton N. B., Ji R. and Langenhoff A. A. (2021). Improving removal of antibiotics in constructed wetland treatment systems based on key design and operational parameters: a review. *Journal of Hazardous Materials*, **407**, 124386, https://doi.org/10.1016/j.jhazmat.2020.124386

Hsieh C. Y., Liaw E. T. and Fan K. M. (2015). Removal of veterinary antibiotics, alkylphenolic compounds, and estrogens from the Wuluo constructed wetland in southern Taiwan. *Journal of Environmental Science and Health. Part A, Toxic/Hazardous Substances & Environmental Engineering*, **50**(2), 151–160.

Hu X., Xie H., Zhuang L., Zhang J., Hu Z., Liang S. and Feng K. (2021). A review on the role of plant in pharmaceuticals and personal care products (PPCPs) removal in constructed wetlands. *Science of the Total Environment*, **780**, 146637, https://doi.org/10.1016/j.scitotenv.2021.146637

Huang A., Yan M., Lin J., Xu L., Gong H. and Gong H. (2021). A review of processes for removing antibiotics from breeding wastewater. *International Journal of Environmental Research and Public Health*, **18**(9), 4909, https://doi.org/10.3390/ijerph18094909

Huang X., Liu C., Li K., Su J., Zhu G. and Liu L. (2015). Performance of vertical up-flow constructed wetlands on swine wastewater containing tetracyclines and tet genes. *Water Research*, **70**, 109–117, https://doi.org/10.1016/j.watres.2014.11.048

Huang X., Zheng J., Liu C., Liu L., Liu Y. and Fan H. (2017). Removal of antibiotics and resistance genes from swine wastewater using vertical flow constructed wetlands: effect of hydraulic flow direction and substrate type. *Chemical Engineering Journal*, **308**, 692–699, https://doi.org/10.1016/j.cej.2016.09.110

Huang X. F., Ye G. Y., Yi N. K., Lu L. J., Zhang L., Yang L. Y., Xiao L. and Liu J. (2019). Effect of plant physiological characteristics on the removal of conventional and emerging pollutants from aquaculture wastewater by constructed wetlands. *Ecological Engineering*, **135**, 45–53, https://doi.org/10.1016/j.ecoleng.2019.05.017

Huang S., Yu J., Li C., Zhu Q., Zhang Y., Lichtfouse E. and Marmier N. (2022). The effect review of various biological, physical and chemical methods on the removal of antibiotics. *Water*, **14**(19), 3138, https://doi.org/10.3390/w14193138

Ilyas H. and van Hullebusch E. D. (2020). Performance comparison of different types of constructed wetlands for the removal of pharmaceuticals and their transformation products: a review. *Environmental Science and Pollution Research*, **27**(13), 14342–14364, https://doi.org/10.1007/s11356-020-08165-w

Kovalakova P., Cizmas L., Feng M., McDonald T. J., Marsalek B. and Sharma V. K. (2021). Oxidation of antibiotics by ferrate (VI) in water: evaluation of their removal efficiency and toxicity changes. *Chemosphere*, **277**, 130365, https://doi.org/10.1016/j.chemosphere.2021.130365

Krall A. L., Elliott S. M., Erickson M. L. and Adams B. A. (2018). Detecting sulfamethoxazole and carbamazepine in groundwater: is ELISA a reliable screening tool? *Environmental Pollution*, **234**, 420–428, https://doi.org/10.1016/j.envpol.2017.11.065

Lim M., Ahmad R., Guo J., Tibi F., Kim M. and Kim J. (2019). Removals of micropollutants in staged anaerobic fluidized bed membrane bioreactor for low-strength wastewater treatment. *Process Safety and Environmental Protection*, **127**, 162–170, https://doi.org/10.1016/j.psep.2019.05.004

Lu X., Gao Y., Luo J., Yan S., Rengel Z. and Zhang Z. (2014). Interaction of veterinary antibiotic tetracyclines and copper on their fates in water and water hyacinth (*Eichhornia crassipes*). *Journal of Hazardous Materials*, **280**, 389–398, https://doi.org/10.1016/j.jhazmat.2014.08.022

Lv M., Zhang D., Niu X., Ma J., Lin Z. and Fu M. (2022). Insights into the fate of antibiotics in constructed wetland systems: removal performance and mechanisms. *Journal of Environmental Management*, **321**, 116028, https://doi.org/10.1016/j.jenvman.2022.116028

Ma J., Cui Y., Li A., Zou X., Ma C. and Chen Z. (2022). Antibiotics and antibiotic resistance genes from wastewater treated in constructed wetlands. *Ecological Engineering*, **177**, 106548, https://doi.org/10.1016/j.ecoleng.2022.106548

Maldonado I., Terrazas E. G. M., Mamani J. M. and Vilca F. Z. (2023). Removal of tetracycline and chloramphenicol through constructed wetlands: roles of plants, substrates, and microbial fuel cells. *Results in Engineering*, **17**, 100982, https://doi.org/10.1016/j.rineng.2023.100982

Marcelino R. B., Leão M. M., Lago R. M. and Amorim C. C. (2017). Multistage ozone and biological treatment system for real wastewater containing antibiotics. *Journal of Environmental Management*, **195**, 110–116, https://doi.org/10.1016/j.jenvman.2016.04.041

Mesenbet M., Birtukan G. and Adey F. (2021). Performance of a small-scale constructed wetland for the removal of antibiotics and antibiotic resistant bacteria from hospital wastewater. *Ethiopian Journal of Biological Sciences*, **20**(1), 69–86.

Oberoi A. S., Jia Y., Zhang H., Khanal S. K. and Lu H. (2019). Insights into the fate and removal of antibiotics in engineered biological treatment systems: a critical review. *Environmental Science & Technology* **53**(13), 7234–7264, https://doi.org/10.1021/acs.est.9b01131

Ohore O. E., Qin Z., Sanganyado E., Wang Y., Jiao X., Liu W. and Wang Z. (2022). Ecological impact of antibiotics on bioremediation performance of constructed wetlands: microbial and plant dynamics, and potential antibiotic resistance genes hotspots. *Journal of Hazardous Materials*, **424**, 127495, https://doi.org/10.1016/j.jhazmat.2021.127495

Oliveira G. H. D. D., Santos-Neto A. J. D. and Zaiat M. (2016). Evaluation of sulfamethazine sorption and biodegradation by anaerobic granular sludge using batch experiments. *Bioprocess and Biosystems Engineering*, **39**, 115–124, https://doi.org/10.1007/s00449-015-1495-3

Omondi D. O. and Navalia A. C. (2020). Constructed wetlands in wastewater treatment and challenges of emerging resistant genes filtration and reloading. In: Inland Waters: Dynamics and Ecology, M. M. Shah, J. Pan, A. A. Nuhu and A. Devlin (eds.), IntechOpen, London, 1–16, https://doi.org/10.5772/intechopen.93293

Östman M., Björlenius B., Fick J. and Tysklind M. (2019). Effect of full-scale ozonation and pilot-scale granular activated carbon on the removal of biocides, antimycotics and antibiotics in a sewage treatment plant. *Science of the Total Environment*, **649**, 1117–1123, https://doi.org/10.1016/j.scitotenv.2018.08.382

Oulebsir A., Chaabane T., Tounsi H., Omine K., Sivasankar V., Flilissa A. and Darchen A. (2020) Treatment of artificial pharmaceutical wastewater containing amoxicillin by a sequential electrocoagulation with calcium salt followed by nanofiltration. *Journal of Environmental Chemical Engineering*, **8**(6), 104597, https://doi.org/10.1016/j.jece.2020.104597

Polianciuc S. I., Gurzău A. E., Kiss B., Ştefan M. G. and Loghin F. (2020). Antibiotics in the environment causes and consequences. *Medicine and Pharmacy Reports*, **93**(3), 231.

Qu M., Liu Y., Hao M., Wang M., Chen R., Wang X. C., Zheng Y. and Dzakpasu M. (2022). Microbial community and carbon-nitrogen metabolism pathways in integrated vertical flow constructed wetlands treating wastewater containing antibiotics. *Bioresource Technology*, **354**, 127217, https://doi.org/10.1016/j.biortech.2022.127217

Sabri N. A., Schmitt H., van der Zaan B. M., Gerritsen H. W., Rijnaarts H. H. M. and Langenhoff A. A. M. (2021). Performance of full scale constructed wetlands in removing antibiotics and antibiotic resistance genes. *Science of the Total Environment*, **786**, 147368, https://doi.org/10.1016/j.scitotenv.2021.147368

Sakurai K. S. I., Pompei C. M. E., Tomita I. N., Santos-Neto ÁJ and Silva G. H. R. (2021). Hybrid constructed wetlands as post-treatment of blackwater: an assessment of the removal of antibiotics. *Journal of Environmental Management*, **278**, 111552, https://doi.org/10.1016/j.jenvman.2020.111552

Salah M., Zheng Y., Wang Q., Li C., Li Y. and Li F. (2023). Insight into pharmaceutical and personal care products removal using constructed wetlands: a comprehensive review. *Science of the Total Environment*, **885**, 163721, https://doi.org/10.1016/j.scitotenv.2023.163721

Shah S. W. A., Rehman M. U., Tauseef M., Islam E., Hayat A., Iqbal S., Arslan M. and Afzal M. (2022). Ciprofloxacin removal from aqueous media using floating treatment wetlands supported by immobilized bacteria. *Sustainability*, **14**(21), 14216, https://doi.org/10.3390/su142114216

Sosa-Hernández J. E., Rodas-Zuluaga L. I., López-Pacheco I. Y., Melchor-Martínez E. M., Aghalari Z., Limón D. S., Iqbal H. M. N. and Parra-Saldívar R. (2021). Sources of antibiotics pollutants in the aquatic environment under SARS-CoV-2 pandemic situation. *Case Studies in Chemical and Environmental Engineering*, **4**, 100127, https://doi.org/10.1016/j.cscee.2021.100127

Tang J., Wang S., Fan J., Long S., Wang L., Tang C., Tam N. F. and Yang Y. (2019). Predicting distribution coefficients for antibiotics in a river water-sediment using quantitative models based on their spatiotemporal variations. *Science of the Total Environment*, **655**, 1301–1310, https://doi.org/10.1016/j.scitotenv.2018.11.163

Thiele-Bruhn S., Seibicke T., Schulten H. R. and Leinweber P. (2004). Sorption of sulfonamide pharmaceutical antibiotics on whole soils and particle-size fractions. *Journal of Environmental Quality*, **33**(4), 1331–1342, https://doi.org/10.2134/jeq2004.1331

Tian Y., Wang X., Liang Y., Li B., Liu Y., Feng L., Li J. and Zhang L. (2023). Removal of oxytetracycline and sulfamethoxazole in vertical down-up flow constructed wetlands: efficiency, removal pathways, and occurrence of antibiotic resistance genes. *Journal of Water Process Engineering*, **55**, 104142, https://doi.org/10.1016/j.jwpe.2023.104142

Timm A., Borowska E., Majewsky M., Merel S., Zwiener C., Bräse S. and Horn H. (2019). Photolysis of four β-lactam antibiotics under simulated environmental conditions: degradation, transformation products and antibacterial activity. *Science of the Total Environment*, **651**, 1605–1612, https://doi.org/10.1016/j.scitotenv.2018.09.248

Tran N. H., Chen H., Reinhard M., Mao F. and Gin K. Y. H. (2016). Occurrence and removal of multiple classes of antibiotics and antimicrobial agents in biological wastewater treatment processes. *Water Research*, **104**, 461–472, https://doi.org/10.1016/j.watres.2016.08.040

Wang J. and Chen X. (2022). Removal of antibiotic resistance genes (ARGs) in various wastewater treatment processes: An overview. *Critical Reviews in Environmental Science and Technology*, **52**(4), 571–630, https://doi.org/10.1080/10643389.2020.1835124

Wang Q., Xie H., Zhang J., Liang S., Ngo H. H., Guo W., Liu C., Zhao C. and Li H. (2015). Effect of plant harvesting on the performance of constructed wetlands during winter: radial oxygen loss and microbial characteristics. *Environmental Science and Pollution Research*, **22**, 7476–7484, https://doi.org/10.1007/s11356-014-3966-5

Wang T., Pan X., Ben W., Wang J., Hou P. and Qiang Z. (2017). Adsorptive removal of antibiotics from water using magnetic ion exchange resin. *Journal of Environmental Sciences*, **52**, 111–117, https://doi.org/10.1016/j.jes.2016.03.017

Wang R., Feng F., Chai Y., Meng X., Sui Q., Chen M., Wei Y. and Qi K. (2019). Screening and quantitation of residual antibiotics in two different swine wastewater treatment systems during warm and cold seasons. *Science of the Total Environment*, **660**, 1542–1554, https://doi.org/10.1016/j.scitotenv.2019.01.127

Wang K., Zhuang T., Su Z., Chi M. and Wang H. (2021). Antibiotic residues in wastewaters from sewage treatment plants and pharmaceutical industries: occurrence, removal and environmental impacts. *Science of the Total Environment*, **788**, 147811, https://doi.org/10.1016/j.scitotenv.2021.147811

Wen H., Zhu H., Xu Y., Yan B., Shutes B., Bañuelos G. and Wang X. (2021). Removal of sulfamethoxazole and tetracycline in constructed wetlands integrated with microbial fuel cells influenced by influent and operational conditions. *Environmental Pollution*, **272**, 115988, https://doi.org/10.1016/j.envpol.2020.115988

Weng X., Cai W., Lan R., Sun Q. and Chen Z. (2018). Simultaneous removal of amoxicillin, ampicillin and penicillin by clay supported Fe/Ni bimetallic nanoparticles. *Environmental Pollution*, **236**, 562–569, https://doi.org/10.1016/j.envpol.2018.01.100

Xu D., Li B., Dou X., Feng L., Zhang L. and Liu Y. (2022). Enhanced performance and mechanisms of sulfamethoxazole removal in vertical subsurface flow constructed wetland by filling manganese ore as the substrate. *Science of the Total Environment*, **812**, 152554, https://doi.org/10.1016/j.scitotenv.2021.152554

Xue J., Lei D., Zhao X., Hu Y., Yao S., Lin K., Wang Z. and Cui C. (2022). Antibiotic residue and toxicity assessment of wastewater during the pharmaceutical production processes. *Chemosphere*, **291**, 132837, https://doi.org/10.1016/j.chemosphere.2021.132837

Ye Z. X., Shao K. L., Huang H. and Yang X. (2021). Tetracycline antibiotics as precursors of dichloroacetamide and other disinfection byproducts during chlorination and chloramination. *Chemosphere*, **270**, 128628, https://doi.org/10.1016/j.chemosphere.2020.128628

Zhao Q., Guo W., Luo H., Xing C., Wang H., Liu B., Si Q., Li D., Sun L. and Ren N. (2022). Insights into removal of sulfonamides in anaerobic activated sludge system: mechanisms, degradation pathways and stress responses. *Journal of Hazardous Materials*, **423**, 127248, https://doi.org/10.1016/j.jhazmat.2021.127248

Zhao W., Sui Q., Mei X. and Cheng X. (2018). Efficient elimination of sulfonamides by an anaerobic/anoxic/oxic-membrane bioreactor process: Performance and influence of redox condition. *Science of The Total Environment*, **633**, 668–676, https://doi.org/10.1016/j.scitotenv.2018.03.207

Zhou A., Jia R., Wang Y., Sun S., Xin X., Wang M., Zhao Q. and Zhu H. (2020). Abatement of sulfadiazine in water under a modified ultrafiltration membrane (PVDF-PVP-TiO_2-dopamine) filtration-photocatalysis system. *Separation and Purification Technology*, **234**, 116099, https://doi.org/10.1016/j.seppur.2019.116099

Zhou Y. and Wang J. (2023). Detection and removal technologies for ammonium and antibiotics in agricultural wastewater: Recent advances and prospective. *Chemosphere*, **334**, 139027, https://doi.org/10.1016/j.chemosphere.2023.139027

doi: 10.2166/9781789065015_0097

Chapter 5

Biological oxidation for mitigating methane emissions from solid wastes and wastewater

W. Chiemchaisri[1], C. Chiemchaisri[1], S. Towprayoon[2], Y. Ogata[3], Y. Ebie[3] and T. Ishigaki[3]

[1]Department of Environmental Engineering, Faculty of Engineering, Kasetsart University, Bangkok, Thailand
[2]Joint Graduate School of Energy and Environment (JGSEE), King Mongkut's University of Technology, Bangkok, Thailand
[3]National Institute for Environmental Studies (NIES), Tsukuba, Japan
Corresponding author: fengccc@ku.ac.th

ABSTRACT

Biological oxidation of emitting methane gas generated during solid waste disposal and wastewater treatment is a low-cost option for the mitigation of greenhouse gases from solid waste and wastewater disposal. Mitigation of methane emission results from the biological conversion of methane to carbon dioxide through methanotrophic activities. A variety of methanotrophs prevail in solid waste and wastewater treatment environments varying from aerobic to anaerobic types. Their existence is influenced by various environmental factors such as pH, temperature, and nutrient conditions. To promote methanotrophic activities for mitigating methane emissions from wastes and wastewater, process design, and operating conditions should be optimized to local conditions. This chapter overviews information and performance of methane oxidation in landfill and wastewater applications.

Keywords: Biofiltration, greenhouse gas, landfill cover, methane oxidation, methanotrophs, wastewater pond

5.1 INTRODUCTION

Solid waste landfills and wastewater stabilization ponds are the most common technologies for municipal solid waste disposal and wastewater treatment in developing countries (Agamuthu, 2013; Kadri *et al.*, 2021). Those technologies not only serve as pollution mitigation measures but also create adverse environmental impacts if not designed and operated properly. For example, the gas produced from the biodegradation of solid waste in landfills and wastewater in ponds contains various compounds including ammonia (NH_3), nitrogen (N_2), carbon dioxide (CO_2), hydrogen (H_2), hydrogen sulfide (H_2S), methane (CH_4) and volatile organic compounds (VOCs) that are normally released directly to the atmosphere (Campos *et al.*, 2016; Manfredi *et al.*, 2009). Methane and carbon dioxide are the principal gases generated which can cause the rising of the earth's temperature, the so-called greenhouse effect. Proper management of emitting gas from landfills and wastewater treatment ponds is therefore one of the important components in their operation. The gas produced from landfills or anaerobic ponds in large quantities could be utilized as a source of energy but it should be eliminated when the utilization is not feasible, especially in small facilities.

© 2025 IWAP. This is an Open Access book chapter distributed under the terms of the Creative Commons Attribution License (CC BY-NC-ND 4.0) which permits copying and redistribution for non-commercial purposes with no derivatives, provided the work is properly cited (https://creativecommons.org/licenses/by-nc-nd/4.0/). The chapter is from the book *Nature-Based Solutions for Urban Sustainability*, Piet N.L. Lens and Xuan-Thanh Bui (Editors).

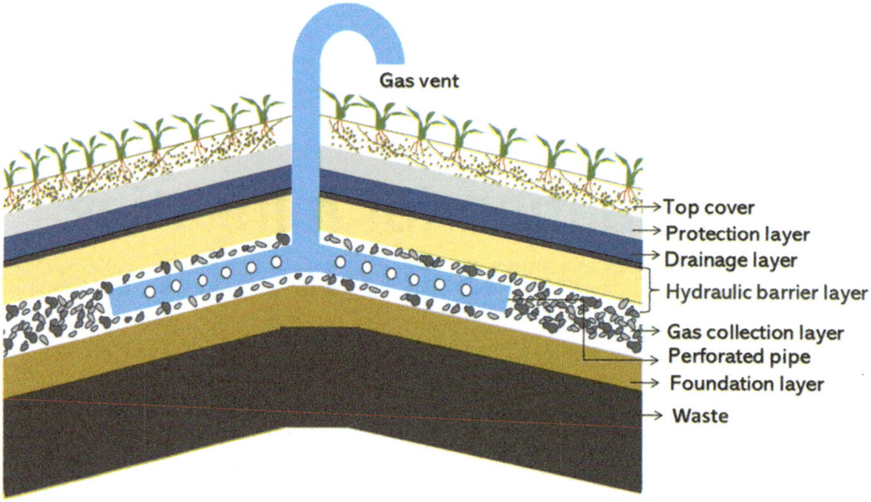

Figure 5.1 Multi-layered landfill final cover.

From the perspective of gas emission control from those sources, the biological layer covering landfill and pond surfaces could play an important role in regulating gas emissions (Bulter *et al.*, 2017; Scheutz *et al.*, 2009). At landfills, final cover layers consisting of multi-components including the top cover layer (vegetative support), a protection layer, drainage layer, hydraulic barrier layer, foundation layer, and gas collection (control) layer are available for protection of water migration into the landfills as well as controlling gas transport from the landfill to the external environment as illustrated in Figure 5.1 (USEPA, 1994). In wastewater stabilization ponds, aquatic plants and other floating matter, for example, foam, scum, or other attached media, could serve as a barrier between the gas production area inside the pond and the atmospheric environment (Syed *et al.*, 2017).

To minimize methane emissions from solid waste and wastewater disposal, methanotrophic bacteria attached to the media in landfills and pond surface layers can help utilize methane and convert it to carbon dioxide. Application of methane biodegradation via methanotrophic activity is preferable to control methane emissions as it is a natural process. Investigation of the proper design of the final cover layer of landfills and wastewater treatment ponds to promote their methanotrophic activities is essential for the reduction of methane emissions to the atmosphere. This technology will be an attractive mitigation measure for controlling emissions from solid waste disposal and wastewater treatment, especially in developing countries, as they require low investment, operation, and maintenance costs during their implementation.

5.2 METHANE OXIDATION MICROORGANISMS AND METABOLIC PATHWAYS

5.2.1 Microbial methane oxidation

Methane oxidation reactions are carried out by methanotrophic microorganisms or methanotrophs. Methanotrophs have been first classified by Whittenbury, Phillips, and Wilkinson as aerobic bacteria in a subset of a physiological group of bacteria known as methylotrophs which can consume methane (CH_4) as their carbon source and energy (Hanson & Hanson, 1996; Whittenbury *et al.*, 1970). Generally, they can survive under a wide range of environmental conditions where methane is available including aquatic systems (lakes, oceans, streams), soils, sediments, marshes, wetlands, rice paddies, landfills, and wastewater sludge. It is estimated that microbially mediated methane oxidation (methanotrophy) consumes >50% of the global methane flux each year (Houghton *et al.*, 2019). By taxonomy classification, methanotrophs are in the families *Methylococcaceae* and *Methylocystaceae* (Bowman, 2006). Currently, over 60 species of methanotrophs are classified by phylogenomic analysis of the whole genome which belongs to proteobacteria (Orata *et al.*, 2018), a phylum of Gram-negative bacteria including a total of 19 genera of γ-proteobacterial (Parte, 2014) and five genera of α-proteobacterial methanotrophs (Table 5.1).

Table 5.1 Classification of methanotrophs present in the environment.

Type	Classification	Assimilation Pathway	Methanotrophic Genera	Reference
Aerobic bacteria	γ-Proteobacteria: Type I	Ribulose monophosphate (RuMP) cycle	*Methylomonas, Methylobacter, Methylococcus, Methylomicrobium, Methylocaldum, Methylohalobius, Methylogaea, Methylosphaera, Methylomarinum, Methylosarcina, Methylosoma, Methylothermus, Methylo vulum,Methylotuvimicro bium, Methylococcace, Methyloglobulus, Methyloprofundus, Crenotrix, Methylocucumis*	Whittenbury *et al.* (1970), Bowman *et al.* (1995), Hanson and Hanson (1996), Bowman (2018) Orata *et al.* (2018)
	α-Proteobacteria: Type II	Serine pathway	*Methylocella, Methylocystis, Methylocapsa, Methyloferula, Methylosinus*	Whittenbury *et al.* (1970), Hanson and Hanson (1996), Bowman (2018)
	γ-Proteobacteria: Type X	RuMP cycle, sometimes serine pathway of formaldehyde assimilation	*Methylococcus capsulatus* (Bath), *Methylacidiphilum (verrucomicrobium), Methylomirabilis*	Hanson and Hanson (1996), Bowman, (2018)
Anaerobic bacteria	NC10 (unculture strains)	Use nitrite or nitrate as electron acceptors	*Candidatus methylomirabilis*	Ettwig *et al.* (2010)
Anaerobic Archaea (ANME)	Archaea	Use sulfate as electron acceptor	ANME-1, ANME-2a/b, ANME-2c, ANME-3	Bhattarai *et al.* (2017), Harrison *et al.* (2009), Raghoebarsing *et al.* (2006)
	Euryarchaeota	Use nitrite or nitrate as electron acceptors	ANME-2d *Candidatus Methanoperedens nitroreducens*	Vaksmaa *et al.*, (2017) Yao *et al.* (2023)
	Archaea	Use Mn, Fe as electron acceptors	ANME-2d *Candidatus Methanoperedens manganireducens/ manganicus, Candidatus Methanoperedens ferrireducens*	Ettwig *et al.* (2016), Cai *et al.* (2018), Leu *et al.* (2020)

5.2.2 Aerobic methane oxidation

In a methane oxidation reaction, there are three different metabolic pathways to assimilate the C1 intermediate formaldehyde (CH_2O, Figure 5.2), which divides the sub-categories of methanotrophs into type I, type II and type X. All types of methanotrophs normally oxidize methane to methanol (CH_3OH) via the enzyme methane monooxygenase (MMO), and to formaldehyde via the enzyme methanol dehydrogenase (MDH). Type I methanotrophs generate many intermediates for a central metabolic route through the ribulose monophosphate (RuMP) cycle in formaldehyde assimilation (equation 5.1), whereas the serine cycle is employed by type II methanotrophs in formaldehyde utilization (equation 5.2). Type X uses the RuMP cycle similarly to type I, but also generates low levels of enzymes of the serine pathway, ribulose-bisphosphate carboxylase, in the Calvin–Benson cycle Bassham (CBB cycle) (Hanson & Hanson, 1996). Additionally, type X methanotrophs can grow in thermophilic conditions (45°C) while both types I and II cannot survive at thermophilic conditions.

$$\text{Type I, RuMP Pathway: } 3CH_2O + ATP \rightarrow \text{Glyceraldehyde-3-phosphate} + ADP \tag{5.1}$$

$$\text{Type II, Serine Pathway: } 2CH_2O + CO_2 + 3\,ATP + 2\,NADH \rightarrow \text{2-Phosphoglycerate} + 2\,ADP + Pi + NAD^+ \tag{5.2}$$

The methane oxidation capability of each species of methanotrophs can vary widely depending on environmental conditions. MMO has two distinct forms: a particulate membrane-associated form (pMMO) and a soluble cytoplasmic form (sMMO). All known methanotrophs are capable of forming a particulate or membrane-bound MMO (pMMO) (Hanson & Hanson, 1996), whereas the ability to form sMMO has been observed only in some type II methanotrophs, the type X methanotroph (*Methylococcus capsulatus*), and one type I methanotroph (*Methylomonas methanica*). In terms of substrate affinity, sMMO has a broader substrate specificity than pMMO including more complex

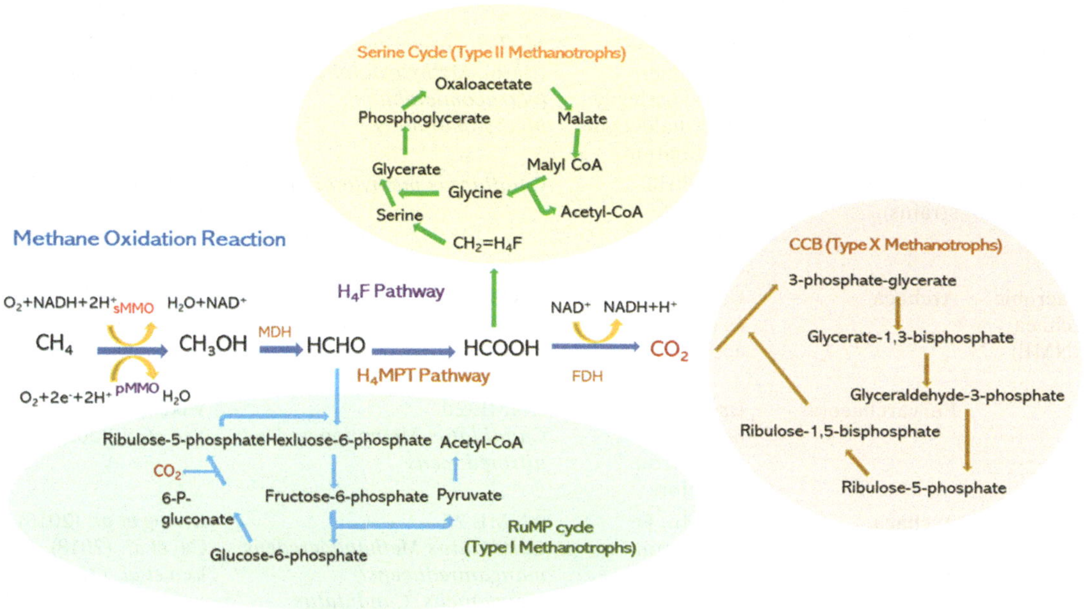

Figure 5.2 Microbial methane oxidation metabolic pathways.

organic compounds such as long-chain, cyclic, and aromatic organic compounds. Therefore, methanotrophs have a great potential for application in the biodegradation of recalcitrant organic pollutants contained in wastes and wastewater. Meanwhile, methanotrophs having pMMO have higher growth yields and a greater affinity for methane oxidation than those with sMMO. The sMMO functioning catalytic reaction requires [NADH + H^+] as an electron donor, meanwhile, the pMMO employs a higher-potential electron donor. The presence of reduced NAD^+ is often a growth-limiting factor when methane is used as a substrate.

5.2.3 Anaerobic methane oxidation

In addition to aerobic methanotrophic bacteria, several studies have reported methane oxidation potential via other electron acceptors such as sulfate, nitrate, nitrite, manganese (IV), and iron (III) in the absence of oxygen by other microorganisms, both archaea and bacteria (Raghoebarsing *et al.*, 2006). This gave another category of methanotrophs, anaerobic methanotrophs (AMNE). ANME coupled with the reduction of different electron acceptors are shown in equations (5.3–5.7) (Caldwell *et al.*, 2008; Harrison *et al.*, 2009). To date, five different clusters of ANME archaea have been found: ANME-1, ANME-2a/b, ANME-2c, ANME-2d, and ANME-3 (Baker *et al.*, 2020). ANME-1, ANME-2a/b, ANME-2c, and ANME-3 archaea use sulfate as an electron acceptor in methane oxidation, namely sulfate-dependent anaerobic methane oxidation or SAMO (Barker & Fritz, 1981) functioning with their partner sulfate-reducing bacteria (SRB) giving sulfide and carbonate as the products (equation 5.3). Denitrifying anaerobic methane oxidation (DAMO) uses nitrite or nitrate as the electron acceptors. They include NC10 phylum bacteria (nonculture), Candidatus *Methylomirabilis* species and ANME-2d archaea (equation 5.5), *Candidatus Methanoperedens nitroreducens* (Arshad *et al.*, 2015; Graf *et al.*, 2018; Haroon *et al.*, 2013; He *et al.*, 2016). The ANME oxidizing methane using iron- and manganese as electron acceptors (equations 5.6 and 5.7) are *Candidatus Methanoperedens ferrireducens*, and *Candidatus Methanoperedens manganireducens/manganicus*, respectively affiliated with the ANME-2d cluster (Cai *et al.*, 2018; Ettwig *et al.*, 2016; Leu *et al.*, 2020).

$$CH_4 + SO_4^{2-} \rightarrow HS^- + HCO_3^- + H_2O \tag{5.3}$$

$$CH_4 + 2.67\,NO_2^- + 1.67\,H^+ \rightarrow HCO_3^- + 1.33\,N_2 + 2.33\,H_2O \tag{5.4}$$

$$CH_4 + 4\,NO_3^- \rightarrow 4\,NO_2^- + HCO_3^- + H^+ + H_2O \tag{5.5}$$

$$CH_4 + 8\,Fe^{3+} + 2\,H_2O \rightarrow CO_2 + 8\,Fe^{2+} + 8\,H^+ \tag{5.6}$$

$$CH_4 + 4\,MnO_2 + 7\,H^+ \rightarrow HCO_3^- + 4\,Mn^{2+} + 5\,H_2O \tag{5.7}$$

5.3 ENVIRONMENTAL FACTORS AFFECTING BIOLOGICAL METHANE OXIDATION

There are several limiting factors attributed to microbial activities such as their oxygen availability, by-products, and inappropriate environments. Recently, more exploration of novel species of methanotrophs detected in extreme environmental conditions has been reported. For instance, thermo-acidophilic methanotrophs isolated from geothermal areas showed a distinct lineage within the bacterial phylum *Verrucomicrobia* (Op den Camp *et al.*, 2009). They can be isolated in environmental samples covering a broad temperature range (22.5–81.6°C), but only in acidic conditions (pH 1.8–5.0) and only in geothermal environments, but not in acidic bogs or fens (Sharp *et al.*, 2014). Methane oxidation has been found under wide environmental conditions and even under some extreme conditions. Several factors affect their survival and growth such as pH, temperature, and nutrient availability.

5.3.1 pH

Among several environmental factors, pH is known as an important factor in the natural selection of species and activities of environmental microorganisms. Earth soils and water generally vary in pH on a global scale due to multiple geological and ecological processes, but additional changes in terrestrial pH are also influenced by anthropogenic activities such as land farming, livestock, and waste landfilling. Moreover, climate change can reconstruct the spatial distribution pattern of soil pH to more acidification or alkalinization (Sun *et al.*, 2023). Microbes can adapt to live in fluctuated pH in habitats by synthesizing some metabolites with specific structures or compositions to control the proton exchange flux and maintain the intracellular pH close to neutral (Semrau *et al.*, 2008). Likewise, most methanotrophs can grow normally in neutral, slightly acidic, and alkaline environments (Table 5.2). Nevertheless, recent studies have shown that many of them can survive in acid (pH 3–6) and alkaline (pH 8–11) environments (Nguyen *et al.*, 2018; Schmitz *et al.*, 2021; Semrau *et al.*, 2008; Trotsenko & Khmelenina, 2002; Zhao *et al.*, 2020).

The methanotrophic acidophiles are tolerant to strong acidity (pH 0.5–3). They are in the *Verrucomicrobia* phylum: *Methylacidiphilum/Methylacidimicrobium* and NC10 phylum: *Candidatus Methylomirabilis*. The acidophilic verrucomicrobial methanotrophs have saturated fatty acids inside the cell membrane to minimize proton permeability in an extremely acidic environment, while those of proteobacterial methanotrophs contain unsaturated fatty acids (Op den Camp *et al.*, 2009; Erikstad *et al.*, 2019; Siliakus *et al.*, 2017). In contrast, under extremely alkaline environments, alkaliphilic species such as *Methylomicrobium alcaliphilum* and *Methylomicrobium buryatense* generate an outside monolayer of hydrophobic glycoproteins (Kaluzhnaya *et al.*, 2001; Khmelenina *et al.*, 2010) which exhibits acidity to neutralize the surrounding negative charges (Trotsenko & Khmelenina, 2002). Besides, methanotrophs also produce some cellular membrane compounds to regulate the proton flux (Kaluzhnaya *et al.*, 2001). Many alkaliphilic methanotrophs are proteobacteria and some are in NC10 or Euryarchaea (Table 5.2).

5.3.2 Temperature

Besides pH, temperature is one of the factors controlling microbial metabolic processes. Several studies have reported various optimum temperatures for the growth of isolated methanotrophs from different habitats. Temperature highly affects methanotrophic growth and methane oxidation rates

Table 5.2 Methanotrophic genera reported under different pH ranges.

Phylum	0.5 < pH <3	3 < pH <6	6 > pH < 8	8 > pH > 11
Verrucomicrobia	*Methylacidiphilum* *Methylacidimicrobium*	*Methylacidiphilum*	–	–
Proteobacteria	–	*Methylocapsa* *Methylocella* *Methylocystis* *Methyloferula* *Methylobacter* *Methylomonas* *Methylovulum* *Candidatus Methylospira*	*Most of methanotrophs*	*Methylobacter* *Methylomicrobium* *Methylomicrobium*
NC10	*Candidatus Methylomirabilis*	–	–	*Candidatus Methylomirabilis*
Euryarchaeota	–	*Candidatus Methanoperedens*	–	*Candidatus Methanoperedens*

Source: Yao *et al.* (2023).

(Houghton & Stewart, 2020; Li *et al.*, 2021). The optimum temperature of aerobic methanotrophs in normal soils is between 20 and 45°C (Shukla *et al.*, 2013), whereas those of acid soils in the Arctic show optimum growth temperatures <10°C (Hanson & Hanson, 1996). The temperature sensitivity of methanotrophs is vulnerable to other environmental factors, such as pH, water content, and sediment properties (Duc *et al.*, 2010; Einola *et al.*, 2007; Xing *et al.*, 2018), consequently resulting in various responses of the methane oxidation rates.

In marine sediments with different water depths and temperature conditions, the aerobic methane oxidation rates of the shallow sediment increased by an order of magnitude from 4 to 28°C, whereas the deep sediment reached the maximum methane oxidation at 15°C. Those methanotrophs in sediments are found more sensitive to low temperatures than high temperatures (Li *et al.*, 2021). Likewise, the thermotolerant methanotrophs residing in heated Roman baths, such as *M. capsulatus* Bath, display a maximum growth temperature between 42 and 50°C (Houghton *et al.*, 2019). True thermophilic methanotrophs are found in geothermal soils or hot springs which exhibit an optimum growth temperature >40°C such as *Methylothermus*, *Methylocaldum*, and *Methylacidiphilum* (Houghton *et al.*, 2019). At high temperatures, the increased activity of cytochrome c peroxidase during methanotrophic growth at elevated temperatures is adapted to cope with the increased concentration of reactive oxygen species produced (Medvedkova *et al.*, 2009). Additionally, thermophilic methanotrophs accumulate compatible solutes or thermolytes, such as polyamines, sucrose, α-glutamate, di-myo-inositol phosphate or hydroxyectoine, in response to both osmotic and heat stress (Houghton *et al.*, 2019).

5.3.3 Nutrients

Methanotrophs require nitrogen sources for their metabolism like other microorganisms. However, most methanotrophs favor to use nitrate as a nitrogen source for synthesizing cellular biomolecules (Whittenbury *et al.*, 1970). Meanwhile, high ammonium concentrations completely inhibited methane oxidation in various habitats such as lake water, marine sediment, paddy fields, croplands, forest soils, and landfill cover soils (Hanson & Hanson, 1996). This is because competition of ammonium with methane for MMO activity. Moreover, nitrite production during ammonium oxidation by methanotrophs is toxic to them (King & Schnell, 1994). High nitrate concentrations can also adversely inhibit methane oxidation rates indirectly due to possible accumulated nitrite from incomplete nitrification reactions in the landfill cover soils (Chiemchaisri *et al.*, 2012).

5.4 APPLICATIONS OF METHANE OXIDATION IN LANDFILLS AND WASTEWATER SYSTEMS

The development of scientific knowledge on methanotrophic functions and performance in the mitigation of methane emissions from various anthropogenic sources and activities has been accumulating in the past several decades. Two major sources of methane emission in the waste sector are solid waste landfills and anaerobic wastewater treatment. Therefore, those sources are potential sites for adopting biological methane oxidation technologies to mitigate methane emissions (Figure 5.3). Among them, the studies and investigations on methane oxidation at the cover layer or biological filter treating landfill gas have been extensively carried out at different application scales, ranging from laboratory-scale investigations to full-scale implementation. Meanwhile, the studies in natural water bodies or wastewater treatment systems are still relatively limited despite being reported in wide environmental conditions. This section reviews methane oxidation performance in landfills and natural water/wastewater sources reported in the literature.

5.4.1 Methane oxidation in landfills

Mitigation of methane emissions from landfills using biological methane oxidation has been widely investigated ranging from laboratory-scale to full-scale applications. Two major applications of methane oxidation processes in landfills are either located at the landfill cover layer using passive

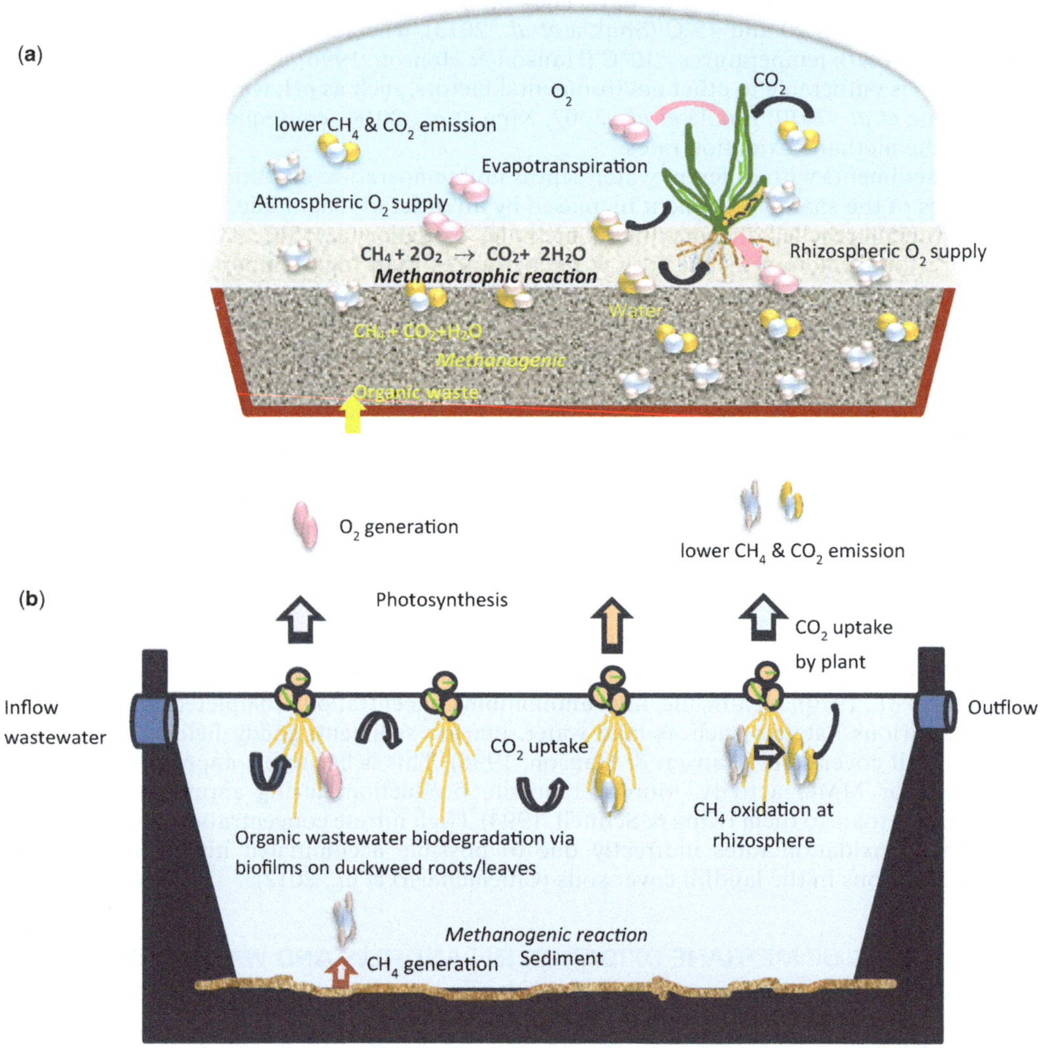

Figure 5.3 Biological methane oxidation applications in (a) solid waste landfills and (b) wastewater treatment ponds.

aeration or in methane biofilters using active aeration. The passive aeration utilizes natural diffusion of atmospheric air into the cover bed in which biological methane oxidation generally occurs under limited oxygen supply. The active aeration is performed in closed biofilters which is provided through force aeration to accelerate biological activities.

5.4.1.1 Methane oxidation at landfill cover layer (passive aeration)

Biological methane oxidation in a final landfill cover layer has been studied under different environments varying from laboratory-controlled conditions to real applications at landfill sites located in different climatic zones. This naturally occurring process utilizes the natural aeration supply (passive aeration) for microbial activities. This mitigation technique has been named differently as bio-cover, bio-window, or passive biofilter. Several influencing factors on the performance of methane oxidation capacities of the final cover were identified such as type of material, material depth, temperature, moisture content or landfill age. The optimum operating and environmental conditions intended to maximize the methane oxidation rate applied for a landfill final cover have been investigated.

The selection of cover materials is one of the most important factors in developing methane oxidation in landfill covers because oxygen availability for aerobic methanotrophic activities relies on its porosity. Different types of materials found suitable for promoting methanotrophic activities are summarized in Table 5.3. Generally, sandy loam soil and compost were effective in promoting methane oxidation. Moreover, vegetation also provides a positive effect on methane oxidation through the promotion of oxygen transfer through plant roots, whereas leachate irrigation on landfill covers also helped to maintain a proper moisture content and nutrient supply for methanotrophic microorganisms (Chiemchaisri *et al.*, 2010; Tanthachoon *et al.*, 2008). Alternatively, stabilized waste has been used as landfill cover material showing moderate-to-high methane removal efficiencies (Chiemchaisri *et al.*, 2013), whereas incorporation of biochar (Reddy *et al.*, 2021) and scum from wastewater treatment ponds (Schirmer *et al.*, 2022) into landfill cover soil gave a profound impact in terms of methane removal efficiencies.

When operated in natural landfill environments, the methane oxidation performance can fluctuate due to the variation in environmental factors governing methanotrophic activities. For instance, variations in the moisture content of landfill covers along different operating seasons in landfills significantly adversely impacted methane emissions and oxidation rates. In this regard, the influence of rainfall during wet seasons on reduction of methane removal rates in tropical landfills was observed (Chan *et al.*, 2023). Water logging in the cover material during wet seasons reduced fugitive methane emission, but yielded poor methane oxidation from the limitation of oxygen supply into the cover materials. Temperature is another key factor affecting methane oxidation as it regulates microbial activities and growth. Generally, the optimum temperatures for methane oxidation in a bio-cover range from 25 to 35°C (Sadasivam & Reddy, 2014). Nevertheless, moderate-to-high methane oxidation activities have been detected at a lower temperature (close to 10°C) even though those were higher at moderate (21°C) and higher (35°C) temperatures (Safari *et al.*, 2017). The effect of temperature may not necessarily affect the microbial activity alone but also other processes such as methane adsorption and diffusion in the cover materials as well. Additionally, the pH can influence the selection of microbial species. While most methanotrophs prefer neutral pH conditions (6–8), higher methane oxidation rates have also been achieved at a lower pH of 5.5 (Schirmer *et al.*, 2022).

Besides environmental conditions, maintenance of the final cover such as irrigation with landfill leachate can help promote methane oxidation activities. The supplement of moisture and appropriate nutrients could explain the beneficial impact of leachate irrigation. Nevertheless, the inhibitory effect of ammonium nitrogen at high concentrations on methanotrophic activities have also been reported (Lü *et al.*, 2012) as the activities of methane-consuming bacteria (type I) are suppressed by ammonium. Moreover, nitrate accumulation from nitrification strengthened the ammonium inhibition on types I and II methanotrophs. Therefore, ammonium accumulation resulting from excessive and repetitive irrigation should be avoided to sustain methane oxidation.

5.4.1.2 Methane oxidation in biofilters (active aeration)

To promote higher methane oxidation capacities in treating landfill gas, biofilters with active aeration and operation in optimized conditions have been employed. Most previous investigations in the development of effective biofilters focused on determining appropriate aeration as well as biofilter bed design to maximize its methane oxidation capacities to specific characteristics of the feeding gas. They have been performed at different scales of operation ranging from laboratory-scale to full-scale operation (Table 5.4). The most common materials used in biofilters are compost or compost amended with other organic (sludge, woodchip, leaf, biochar) or inorganic (sand, perlite, silica) materials. Much higher methane oxidation capacities are generally achieved in biofilters compared to biocovers

Table 5.3 Methane oxidation or removal performance in open (passively aerated) biocovers.

System (Scale)	Media	Operating Conditions	CH_4 Flux/Loading Rate	Environmental Conditions	Methane Oxidation Rate or Removal Efficiencies	Reference
Open column (lab-scale, 0.15 m diameter)	Sandy loam (0.6 m depth) with tropical grass (*S. virginicus*)	Synthetic gas (60% CH_4) feed, leachate irrigation	14 mol/m^3/d	28–30°C, 10–15% moisture	8.5–10.05 mol/m^3/d	Chiemchaisri *et al.* (2010)
Open column (lab-scale, 0.15 m diameter)	Compost (0.6 m depth) with tropical grass (*P. repens*)	Synthetic gas (60% CH_4) feed, leachate irrigation	14 mol/m^3/d	28–30°C, 50–55% moisture	12 mol/m^3/d	Tanthachoon *et al.* (2008)
Open column (lab-scale, 0.15 m diameter)	Stabilized wastes (0.8 m depth)	Synthetic gas (60% CH_4) feed, leachate irrigation	8.32–51.28 g/m^3/d	28–30°C, 10–15% moisture	7.58–34.06 g/m^3/d, 66–91%	Chiemchaisri *et al.* (2013)
Biocover (lab-scale, 0.15 m diameter)	Soil (30%): compost (70%) mixture	CH_4 gas (99%) feed	100–180 g/m^2/d	Temperature control at room (21.3°C), fridge (11.2°C) and heated (35°C) conditions, 30–35% moisture	78.8–95.48% (fridge), 99.1–99.6% (room), 90.1–92.3% (heated)	Safari *et al.* (2017)
Biochar amended landfill cover (lab-scale, 0.2 m diameter)	Silty clay amended with biochar at 10% w/w (0.6 m depth)	Synthesis gas (60% CH_4) feed, atmosphere air flushing headspace	50–200 g/m^2/d	22–25°C, pH 7.3–7.5, 8–17% moisture (v/v)	>90%	Reddy *et al.* (2021)
Biofilter (0.6 m diameter)	Sandy clayey silt mixed with scum (0.6 m depth)	Fugitive landfill gas	6–9 g/m^2/d	25°C, pH 5.5, C/N = 9.4	89.5% (maximum)	Schirmer *et al.* (2022)
Biocover (field, 2.45 × 9.45 m)	Sandy: compost mixture (0.8 m depth)	Landfill gas (62.8% CH_4) feed	20–580 g/m^2/d	Air temperature 3–30°C (18°C avg)	576 g/m^2/d (max), 99%	Roncato and Cabral (2012)
Open lysimeter (field, 1.50 m diameter)	Compost (0.6 m depth)	Landfill gas feed	10.5 mol/m^3/d (wet period), 15.3 mol/m^3/d (dry period)	28–30°C, 58.6–60.3% moisture	8.3 mol/m^3/d (wet period),10.1 mol/m^3/d (dry period)	Chan *et al.* (2023)

Table 5.4 Methane oxidation performance of forced aeration/active biofilters.

System (Scale)	Media	Operating Conditions	CH_4 Flux/ Loading Rate	Environmental Conditions	Methane Oxidation Rates or Removal Efficiencies	Reference
Biofilter (lab-scale, 0.15 m dia.)	Sludge/woodchip/ leaf compost, sand, perlite (0.50 m depth)	CH_4 gas (99%) feed, sweep air 200 ml/min	530 g/m²/d	50–80% moisture	82% (for leaf compost)	Mancebo and Hettiaratchi (2015)
Biofilter (lab-scale, 0.1 m dia.)	Composted leaves/ expanded vermiculite (1.0 m depth)	CH_4 gas: air mixture (CH_4 0.17–3.63% v/v), gas residence time 29.5–42.8 min	3.0–148.8 gCH_4/ m³/h	19–36°C, 58.7% moisture	95.0%	Brandt *et al.* (2016)
Biofilter (lab-scale, 0.14 m dia.)	Leaf compost (0.90 m depth)	CH_4 gas (99%) fed with 10 times air flow rate injected at multi-levels	529–1589 g/ m³/d	60% moisture	420–1217 g/m³/d	Farrokhzadeh *et al.* (2017)
Biofilter (lab-scale, 0.14 m dia.)	Compost, biochar, lava rock mixture (0.70 m depth)	CH_4 gas (99%) fed at 10 ml/min with 2.5 times air flow rate, gas residence time 82.8 min	877 g/m³/d	22°C, 40–50% moisture	100%	La *et al.* (2018)
Biofilter (lab-scale, 0.05 m dia.)	Compost (70%): porous silica pellet (30%) mixture (1.80 m depth)	Synthetic gas (59.94%CH_4, 39.96%CO_2, 0.10%N_2O) fed with air zero 0.5, 1.5 ml/min	110 g/m³/d	28–30°C, 10–15% moisture, pH 7.5	70.1 g/m³/d	Tung *et al.* (2021)
Biofilter (full-scale, 4 m³ vol.)	Leaf compost	Waste gas (CH_4 80.26%), 25 m³/d flow rate, air:CH_4 ratio 3–11, gas retention time 11–61 min	529–2647 g/ m³/d	18–40°C, 25–66% moisture	197–1919 g/m³/d	Gunasekera *et al.* (2017)
Biofilter (full-scale, 270 m²)	Compost: sand mixture of 1:5 (v/v) (1.50 m depth)	Diluted landfill gas (20%CH_4, 13.5%CO_2, 10.5%O_2) fed at 20 m³/h	10.9 g/m²/h	−4.3–30.6°C (ambient), 30% moisture	5.7 ± 2.3 g/m²/h (58.1±46.5%)	Pecorini *et al.* (2020)
Biofilter (full-scale, 30 m³ vol.)	Mature compost (18 m³, 1.1–1.3 m depth)	Diluted landfill gas with air (CH_4 4.4–9.2%)	106–794 g/m²/d	46–60°C, 45–55% moisture	460 g/m²/d (58%)	Fjelsted *et al.* (2020)
Biofilter (full-scale, 14 × 84 m)	Mature compost (0.8 m depth)	Landfill gas mixed with air (CH_4 20%, O_2 7.9%$_2$, CO_2 6.7%), 78–95 m³/h feed rate, gradual vegetation	61 g/m²/d	32–60°C	>99%	Duan *et al.* (2021)

due to a sufficient supply of oxygen through forced aeration, but they could gradually decline over long-term operation attributed to excessive methanotrophic biofilm growth and decay processes, and the production of extracellular polysaccharides (Mancebo & Hettiaratchi, 2015). Maintaining uniform performance of biofilters can be achieved through supplied aeration such as multiple-level air distribution (Farrokhzadeh *et al.*, 2017). The design of biofilters with a mixture of organic and inorganic media also helps maintain the mechanical properties and structure of the media bed to facilitate gas flow during long-term operation. For instance, compost-based materials amended with vermiculite (Brandt *et al.*, 2016) and porous silica pellets (Tung *et al.*, 2021) have shown promising results and superior performance compared to other media types (Table 5.4). Moreover, inorganic media incorporated into the mixed media could also serve as nutrient sources for methanotrophs as well (La *et al.*, 2018).

Full-scale implementations of biofilters treating real waste gas have been successfully demonstrated (Table 5.4). Actively aerated high-rate methane biofilter (HMBF) systems were designed to overcome the shortcomings of passively aerated systems by ensuring the entire filter bed is utilized for methane oxidation (Gunasekera *et al.*, 2017). A high methane oxidation rate could be achieved by maintaining higher temperatures in the HMBFs when operated under low ambient temperature conditions. Moderate-to-complete removal of methane contained in landfill gas mixed with air can be removed in full-scale biofilters (Duan *et al.*, 2021; Fjelsted *et al.*, 2020; Pecorini *et al.*, 2020). Even when operated under extremely wide temperature conditions, the biofilter performance could be maintained while keeping relatively high and stable temperatures in the biofilters. Together with methane gas, other trace compounds in landfill gas such as aromatics and aliphatic hydrocarbons were also completely removed (Duan *et al.*, 2021), whereas odor was highly eliminated as well (Pecorini *et al.*, 2020).

5.4.2 Methane oxidation in water environments

Methane oxidation prevails under a wide range of water environments varying from polluted natural water bodies to natural and mechanical wastewater treatment systems. In polluted water bodies, methane is produced under anaerobic conditions especially in the underlying sediment layer, whereas the oxidation can take place in the aerobic zone of the water column or surface of the sediment layer. In the presence of vegetation, methane oxidation could also take place at plant roots which are usually found in natural or constructed wetlands. In wastewater treatment, methane produced during preceding anaerobic conditions can be treated in biofilters for the emitted biogas as well as subsequent aerobic wastewater treatment units for the methane that remained in dissolved form. Figure 5.4 presents some potential habitats of methanotrophs in natural water and wastewater treatment systems.

5.4.2.1 Methane oxidation in natural waters

Polluted aquatic environments such as wetlands and lakes are major methane emission sites. Therefore, methanotrophs have been detected in polluted lake water and sediments in which the surface of sediments shows high methane oxidation activity. Moreover, methanotrophs inhabit hydrophytes (Iguchi *et al.*, 2019). As presented in Table 5.5, methane oxidation prevails in an eutrophic freshwater lake in Japan where methanotrophs were found associated with duckweed and several aquatic plant species (Figure 5.4). It was estimated that up to 160 tons/year of CH_4 was potentially consumed by 2500 tons of aquatic plants in the lake basin which corresponds to 17% of the total CH_4 emitted from the lake surface (Yoshida *et al.*, 2014). A previous study also revealed that the majority (up to 70%) of methane produced may become oxidized as a result of a strong decrease in gas exchange combined with high methane oxidation activity in shallow aquatic systems with free-floating plants (Kosten *et al.*, 2016). The CH_4 consumed by microbial communities associated with aquatic plants took place in the forms of dissolved CH_4 in the water column as well as uptake in the sediment. Aquatic plants such as duckweed can also be found stimulating methanotrophic growth. In turn, methane emissions through the plant (water hyacinth) were found to increase with plant density, and higher emissions were observed under elevated temperature conditions (Struik *et al.*, 2022).

Figure 5.4 Possible habitats of methanotrophs in natural waters and wastewater treatment systems: (a) scum in anaerobic pond, (b) duckweed mat on pond, (c) aquatic plants, for example, water hyacinth, and (d) synthetic media used in Thailand.

5.4.2.2 Methane oxidation in wastewater treatment systems

During wastewater treatment, methane oxidation has been reported in natural treatment systems such as treatment ponds and constructed wetlands, especially those operated with floating plants. In free water surface constructed wetlands, an increase of plant biomass under high nutrient conditions was promoted, leading to a reduction of the methane flux at the water–atmosphere interface whereas the methanotrophs were found predominant in the rhizospheric biofilm of plants (He *et al.*, 2023). Methane emission fluxes in constructed wetlands with different plant species vary due to plant physiological effects, especially aerenchyma proportions of roots, stems, and leaves whereas the methane oxidation potential varies between the plant species with type I methanotrophs becoming predominant (Wang *et al.*, 2024).

In a vertical subsurface constructed wetland utilizing submerged media and emergent plants, the majority of methanotrophs inhabit its media (Schalk *et al.*, 2019). The effect of plants on methane oxidation could thus be less pronounced. The study has also revealed that although the methane in this wetland type was mainly oxidized by oxygen, there could be some degree of denitrifying anaerobic methane oxidation developed in the media. A major part of the removed methane was attributed to biomass built-up leading to organic matter accumulation and production of extracellular polymeric substances.

Table 5.5 Methane oxidation capacities/activities in natural waters and wastewater treatment systems.

Site or System (Scale)	Methanotroph Habitat	Operating Conditions	CH_4 Flux/ Loading Rate	Environmental Conditions	Methane Oxidation Rate or Removal Efficiencies	Reference
Freshwater lake	Duckweed	Direct sampling and measurement	–	Summer season, 28°C incubation	0.0067–0.89 μmol/h/g wet wt.	Iguchi *et al.* (2019)
Shallow eutrophic lake	Aquatic plants (10 species)	Direct sampling and measurement	–	28°C incubation	3.7–37 μmol/h/g dry wt.	Yoshida *et al.* (2014)
Wastewater stabilization ponds (lab-scale)	Floating plants (*Eichhornia crassipes*)	Synthetic sewage, batch operation, 0.6 m water and 0.25 m sediment depths, 31 d experiment	20–130 mg/m^2/h	17.6–24.1°C, pH 6.02–7.63, DO 0.07–3.93 mg/l	0.83 mgCH_4/kg wet wt./d, 52.3–83.2%	He *et al.* (2023)
Simulated wetlands (lab-scale)	Floating plants (*Eichhornia crassipes, Hydrocharis dubia, Trapa natans*)	Batch operation, 80% plant occupancy on water surface, 7 d experiment	265.53–456.08 g/m^2/d	Outdoor experiment, 28°C	84.3–97.3% (*E. crassipies*) 4.98–88.9% (*H. dubia*) 43.6–92.5% (*T. natans*)	Wang *et al.* (2024)
Vertical flow constructed wetland (small-scale trial, 0.66 m dia.)	Sand media (0–2 mm grain size, 0.6 m depth)	Municipal wastewater fed at 34 L/m^2/d, CH_4 gas (100%) fed at half depth of sand bed	1.0–27 g/m^2/d	17–21°C, pH 4.2–5.9	100% for <19 g/m^2/d loading	Schalk *et al.* (2019)
Biofilter (lab-scale, 0.05 m dia.)	Compost bed (0.54 m depth, 1.1 L bed vol.)	Simulated anaerobic reactor off-gas (CH_4 4%) gas residence time 19 min	75 g/m^3/h	16–23°C, pH 6.3, 55% moisture	0.07 gCH_4/g dry wt./h, 60–80%	Gómez-Borraz *et al.* (2017)
Down-flow hanging sponge reactor (lab-scale, 6.8 cm dia.)	Sponge cube of 2 × 2 × 2 cm size (31 pieces, total depth 1.0 m)	Continuous operation, artificial wastewater with dissolved CH_4, HRT 2 h, Air flow rate 3.21 L/d	Influent dissolved CH_4 60.8 mgCOD/l	20°C, pH 8.1	79%	Hatamoto *et al.* (2011)

Biological filter media generally serve as inhabitants of methanotrophs treating methane gas released from wastewater treatment. It can be generally applied in the form of an exhausted gas treatment unit as a biofilter (Gómez-Borraz *et al.*, 2017). Due to the presence of a lower methane content in off-gas from wastewater treatment, the biofilter is operated at a lower methane loading rate than those in active biofilters treating landfill gas (Table 5.4) so high methane removal efficiencies could be expected. The biofilter performance was also found less sensitive to temperature conditions as it fluctuated less in the wastewater treatment environment.

The methane oxidation process was also applied in aerobic wastewater treatment with attached media for methanotrophs such as downflow hanging sponges (Hatamoto *et al.*, 2011) or other similar media (Figure 5.4d) for removing dissolved methane. Under the oxygen-abundant conditions of a short hydraulic retention time of 2 h, more than 90% oxidation of dissolved methane, ammonium, sulfide, and organic matter was achieved. Moreover, methanotrophs were also detected in the aerobic zone of suspended (activated sludge) and attached (rotating biological contactor) growth wastewater treatment systems (Ho *et al.*, 2013) even though their methane uptake capacities varied due to the presence of different environmental conditions, for example pH, temperature, dissolved oxygen, and nutrient contents.

So far, most of the practical applications of methane oxidation in wastewater treatment have been focused solely on its aerobic pathway while the use of denitrifying anaerobic methane oxidation processes has not yet been applied at full scale although many approaches have been made to achieve a better comprehension of its potential wastewater treatment applications (Monica-Macías *et al.*, 2023).

5.5 OTHER BENEFICIAL EFFECTS OF METHANE OXIDATION

Apart from its utilization to mitigate greenhouse gas emissions, methane oxidation also provided other benefits when applied to landfills or as off-gas treatment units. One of the beneficial effects of promoting methane oxidation in landfills such as in the case of semi-aerobic landfill conditions is the concurrent enhanced biodegradation of plastic wastes (Muenmee *et al.*, 2016). Under the presence of methane and oxygen in the waste layer, methanotrophic activities are promoted under which heterotrophs, autotrophs, and methanotrophs colonized simultaneously on plastic wastes (Muenmee *et al.*, 2015). Under such conditions, methanotrophs types I and II (*Methylobacter* sp./*Methylocella* sp.) could grow and function as principal decomposers of plastics via co-metabolism. The plastic degradation was also found to correlate with methane oxidation rate (Muenmee *et al.*, 2015).

Bioconversion of methane by aerobic methanotrophs is also considered an efficient and cost-effective methanol production method (Park & Kim, 2019). Along the metabolic pathways of aerobic methanotrophs, methane is oxidized to methanol as an intermediate which can be recovered as a chemical or fuel by inhibiting the methanol degrading enzyme (methanol dehydrogenase) while maintaining MMO activity. Other high-value products which can be obtained from methane oxidation include biofuel, biopolymers, and single-cell proteins (Wang *et al.*, 2021). The use of anaerobic methanotrophic archaea and '*Candidatus Methylomirabilis*' (NC10) bacteria are also promising candidates for a methane biorefinery through anaerobic or intra-aerobic oxidation to save energy and costs (Ni, 2023). Nevertheless, the gas supply strategy and bioreactor design need to be optimized to realize its application at a larger scale.

5.6 CONCLUSIONS

Biological methane oxidation has been successfully applied to remediate greenhouse gas emissions from solid waste disposal and wastewater treatment. The variety of methanotrophic microorganisms and their metabolic pathways made them versatile and adapted to vast environmental conditions such as oxygen level, pH, temperature, availability of nutrients, and electron acceptors. From an extensive literature survey, methane oxidation has been demonstrated to reduce methane emissions from landfills effectively in different implementation scales varying from laboratory to full-scale applications. The applied technologies also varied from passive or naturally operated systems to high-rate or actively operated in optimized reactor systems. Nevertheless, there are still challenges in sustaining methanotrophic microorganisms while maintaining their activities under fluctuating environmental conditions anticipated in the long-term operation of pollution control systems. To develop an effective biological methane oxidation process, optimization of methane and oxygen supply as well as system design to enhance gas transfer to methanotrophic microorganisms should be considered.

ACKNOWLEDGEMENTS

This work was carried out as activities under research projects titled 'Enabling waste management scheme in response to disaster and climate change' supported by the National Science, Research and Innovation Fund via the Program Management Unit for Human Resources and Institutional Development, Research and Innovation (Grant number B16F640136) and 'Development of the Duckweed Holobiont Resource Values towards BCG Economy' supported by the Japan International Cooperation Agency (JICA) and Japan Science and Technology Agency (JST).

REFERENCES

Agamuthu P. (2013). Landfilling in developing countries. *Waste Management & Research*, **31**(1), 1–2, https://doi.org/10.1177/0734242X12469169

Arshad A., Speth D. R., de Graaf R. M., Op den Camp H. J. M., Jetten M. S. M. and Welte C. U. (2015). A metagenomics-based metabolic model of nitrate dependent anaerobic oxidation of methane by *Methanoperedens*-like archaea. *Frontiers in Microbiology*, **6**, 1423, https://doi.org/10.3389/fmicb.2015.01423

Baker B. J., De Anda V., Seitz K. W., Dombrowski N., Santoro A. E. and Lloyd K. G. (2020). Diversity, ecology and evolution of archaea. *Nature Microbiology*, **5**, 1–14.

Barker J. F. and Fritz P. (1981). Carbon isotope fractionation during microbial methane oxidation. *Nature*, **293**, 289–291, https://doi.org/10.1038/293289a0

Brandt E. M. F., Duarte F. V., Vieira J. P. R., Melo V. M., Souza C. L., Araújo J. C. and Chernicharo C. A. L. (2016). The use of novel packing material for improving methane oxidation in biofilters. *Journal of Environmental Management*, **182**, 412–420, https://doi.org/10.1016/j.jenvman.2016.07.075

Bhattarai S., Cassarini C., Gonzalez-Gil G., Egger M., Slomp C. P., Zhang Y., Esposito G. and Lens P. N. L. (2017). Anaerobic methane-oxidizing microbial community in a coastal marine sediment: anaerobic methanotrophy dominated by ANME-3. *Microbial Ecology*, **74**, 608–622, https://doi.org/10.1007/s00248-017-0978-y

Bowman J. P. (2006). The methanotrophs – the families *Methylococcaceae* and *Methylocystaceae*. In: The Prokaryotes, M. Dworkin, S. Falkow, E. Rosenberg, K. H., Schleifer and E. Stackebrandt (eds), Springer, New York, pp. 266–289.

Bowman J. P. (2018). 'Methylococcales,' in Bergey's Manual of Systematics of Archaea and Bacteria. John Wiley & Sons, New Jersey.

Bowman J. P., Sly L. I. and Stackebrandt E. (1995). The phylogenetic position of the family Methylococcaceae. *International Journal of Systematic Bacteriology*, **45**, 182–185, https://doi.org/10.1099/00207713-45-1-182

Bulter E., Hung Y. T., Al Ahmad M. S., Yeh R. Y. L., Liu R. L. H. and Fu Y. P. (2017). Oxidation pond for municipal wastewater treatment. *Applied Water Science*, **7**, 31–51, https://doi.org/10.1007/s13201-015-0285-z

Cai C., Leu A. O., Xie G. J., Guo J., Feng Y., Zhao J. X., Tyson G. W., Yuan Z. and Hu S. (2018). A methanotrophic archaeon couples anaerobic oxidation of methane to Fe (III) reduction. *The ISME Journal*, **12**, 1929–1939, https://doi.org/10.1038/s41396-018-0109-x

Caldwell S. L., Laidler J. R., Brewer E. A., Eberly J. O., Sandborgh S. C. and Colwell F. S. (2008). Anaerobic oxidation of methane: mechanisms, bioenergetics, and the ecology of associated microorganisms. *Environmental Science & Technology*, **42**, 6791–6799, https://doi.org/10.1021/es800120b

Campos J. L., Valenzuela-Heredia D., Pedrouso A., Val del Río Belmonte M. and Mosquera-Corral A. (2016). Greenhouse gases emissions from wastewater treatment plants: minimization, treatment, and prevention. Journal of Chemistry, **2016**, 3796352.

Chan R., Chiemchaisri W. and Chiemchaisri C. (2023). Exploring effective bio-cover materials for mitigating methane emission at a tropical landfill. *Applied Science*, **13**, 1990, https://doi.org/10.3390/app13031990

Chiemchaisri C., Chiemchaisri W., Chittanukul K., Soontornlerdwanich W. and Tanthachoon N. (2010). Effect of leachate irrigation on methane oxidation in tropical landfill cover soil. *Journal of Material Cycles and Waste Management*, **12**, 161–168, https://doi.org/10.1007/s10163-009-0284-y

Chiemchaisri C., Chiemchaisri W., Kumar S. and Wicramarachchi P. N. (2012). Reduction of methane emission from landfill through microbial activities in cover soil: a brief review. *Critical Reviews in Environmental Science and Technology*, **42**, 412–434, https://doi.org/10.1080/10643389.2010.520233

Chiemchaisri W., Chiemchaisri C. and Boonchaiyuttasak J. (2013). Utilization of stabilized wastes for reducing methane emission from municipal solid waste disposal. *Bioresource Technology*, **141**, 199–204, https://doi.org/10.1016/j.biortech.2013.03.035

Duan Z., Hansen P. O. R., Scheutz C. and Kjeldsen P. (2021). Mitigation of methane and trace gas emissions through a large-scale active biofilter system at Glatved landfill, Denmark. *Waste Management*, **126**, 367–376, https://doi.org/10.1016/j.wasman.2021.03.023

Duc N. T., Crill P. and Bastviken D. (2010) Implications of temperature and sediment characteristics on methane formation and oxidation in lake sediments. *Biogeochemistry*, **100**, 185–196, https://doi.org/10.1007/s10533-010-9415-8

Einola J. K. M., Kettunen R. H. and Rintala J. A. (2007). Responses of methane oxidation to temperature and water content in cover soil of a boreal landfill. *Soil Biology and Biochemistry*, **39**, 1156–1164, https://doi.org/10.1016/j.soilbio.2006.12.022

Erikstad H. A., Ceballos R. M., Smestad N. B. and Birkeland N. K. (2019). Global biogeographic distribution patterns of thermoacidophilic *Verrucomicrobia methanotrophs* suggest allopatric evolution. *Frontiers in Microbiology*, **10**, 1129, https://doi.org/10.3389/fmicb.2019.01129

Ettwig K. F., Butler M. K., Le Paslier D., Pelletier E., Mangenot S., Kuypers M. M. M., Schreiber F., Dutilh B. E., Zedelius J., de Beer D. and Gloerich J. (2010). Nitrite-driven anaerobic methane oxidation by oxygenic bacteria. *Nature*, **464**, 543–548, https://doi.org/10.1038/nature08883

Ettwig K. F., Zhu B., Speth D., Keltjens J. T., Jetten M. S. M. and Kartal B. (2016). Archaea catalyze iron-dependent anaerobic oxidation of methane. *Proceedings of the National Academy of Sciences of the United States of America*, **113**(45), 12792–12796, https://doi.org/10.1073/pnas.1609534113

Farrokhzadeh H., Hettiaratchi J. P. A., Jayasinghe P. and Kumar S. (2017). Aerated biofilters with multiple-level air injection configurations to enhance biological treatment of methane emissions. *Bioresource Technology*, **239**, 219–225, https://doi.org/10.1016/j.biortech.2017.05.009

Fjelsted L., Scheutz C., Christensen A. G., Larsen J. E. and Kjeldsen P. (2020). Biofiltration of diluted landfill gas in an active loaded open-bed compost filter. *Waste Management*, **103**, 1–11, https://doi.org/10.1016/j.wasman.2019.12.005

Gunasekera S. S., Hettiaratchi J. P. A., Bartholameuz E. M., Farrokhzadeh H. and Irvine E. (2017). A comparative evaluation of the performance of full-scale high-rate methane biofilter (HMBF) systems and flow-through laboratory columns. *Environmental Science and Pollution Research*, **25**, 35845–35854, https://doi.org/10.1007/s11356-018-3100-1

Gómez-Borraz T. L., González-Sánchez A., Bonilla-Blancas W., Revah S. and Noyola A. (2017). Characterization of the biofiltration of methane emissions from municipal anaerobic effluents. *Process Biochemistry*, **63**, 204–213, https://doi.org/10.1016/j.procbio.2017.08.011

Graf J. S., Mayr M. J., Marchant H. K., Tienken D., Hach P. F., Brand A., Schubert C. J., Kuypers M. M. M. and Milucka J. (2018). Bloom of a denitrifying methanotroph, 'candidatus methylomirabilis limnetica', in a deep stratified lake. *Environmental Microbiology*, **20**, 2598–2614, https://doi.org/10.1111/1462-2920.14285

Hanson R. S. and Hanson T. E. (1996). Methanotrophic bacteria. *Microbiological Reviews*, **60**, 439–471, https://doi.org/10.1128/mr.60.2.439-471.1996

Haroon M. F., Hu S., Shi Y., Imelfort M., Keller J., Hugenholtz P., Yuan Z. and Tyson G. W. (2013). Anaerobic oxidation of methane coupled to nitrate reduction in a novel archaeal lineage. *Nature*, **500**, 567–570, https://doi.org/10.1038/nature12375

Harrison B. K., Zhang H., Berelson W. and Orphan V. J. (2009). Variations in archaeal and bacterial diversity associated with the sulfate-methane transition zone in continental margin sediments (Santa Barbara Basin, California). *Applied and Environmental Microbiology*, **75**, 1487–1499, https://doi.org/10.1128/AEM.01812-08

Hatamoto M., Miyauchi T., Kindaichi T., Ozaki N. and Ohashi A. (2011). Dissolved methane oxidation and competition for oxygen in down-flow hanging sponge reactor for post-treatment of anaerobic wastewater treatment. *Bioresource Technology*, **102**, 10299–10304, https://doi.org/10.1016/j.biortech.2011.08.099

He Z., Cai C., Wang J., Xu X., Zheng P., Jetten M. S. M. and Hu B. (2016). A novel denitrifying methanotroph of the NC10 phylum and its microcolony. *Scientific Reports*, **2016**, 33241.

He X., Zhang S., Lv X., Liu M., Ma Y. and Guo S. (2023). *Eichhornia crassipes*-rhizosphere biofilms contribute to nutrients removal and methane oxidization in wastewater stabilization ponds receiving simulative sewage treatment plants effluents. *Chemosphere*, **322**, 138100, https://doi.org/10.1016/j.chemosphere.2023.138100

Ho A., Vlaeminck S. E., Ettwig K. F., Schneider B., Frenzel P. and Boon N. (2013). Revisiting methanotrophic communities in sewage treatment plants. *Applied and Environmental Microbiology*, **79**(8), 2841–2846, https://doi.org/10.1128/AEM.03426-12

Houghton K. M. and Stewart L. C. (2020). Temperature-gradient incubation isolates multiple competitive species from a single environmental sample. *Access Microbiology*, **2**, acmi000081, https://doi.org/10.1099/acmi.0.000081

Houghton K. M., Carere C. R., Stott M. B. and McDonald I. R. (2019). Thermophilic methanotrophs: in hot pursuit. *FEMS Microbiology Ecology*, **95**, 125, https://doi.org/10.1093/femsec/fiz125

Iguchi H., Umeda R., Taga H., Oyama T., Yurimoto H. and Sakai Y. (2019). Community composition and methane oxidation activity of methanotrophs associated with duckweeds in a fresh water lake. *Journal of Bioscience Bioengineering*, **128**, 450–455.

Kadri S. U. T., Tavanappanavar A. N., Babu R. N., Bilal M., Sigh B., Gupta S. K., Bharagava R. N., Muthusamy G., Savanur M. A. and Mulla S. I. (2021). Overview of waste stabilization ponds in developing countries. In: Cost-efficient Wastewater Treatment Technologies: Natural Systems, M. Nasr and A. M. Negm (eds), Springer Nature, Switzerland, pp. 153–175.

Kaluzhnaya M., Khmelenina V., Eshinimaev B., Suzina N., Nikitin D., Solonin A., Lin L. J., McDonald I., Murrell C. and Trotsenko Y. (2001). Taxonomic characterization of new alkaliphilic and alkalitolerant methanotrophs from soda lakes of the southeastern transbaikal region and description of *Methylomicrobium buryatense* sp.nov. *Systematic and Applied Microbiology*, **24**, 166–176, https://doi.org/10.1078/0723-2020-00028

Khmelenina V. N., Shchukin V. N., Reshetnikov A. S., Mustakhimov I. I., Suzina N. E., Eshinimaev B. T. and Trotsenko Y. A. (2010). Structural and functional features of methanotrophs from hypersaline and alkaline lakes. *Microbiology*, **79**, 472–482, https://doi.org/10.1134/S0026261710040090

King G. M. and Schnell S. (1994). Ammonium and nitrite inhibition of methane oxidation by *Methylobacter albus* BG8 and *Methylosinus trichosporium* OB3b at low methane concentrations. *Applied and Environmental Microbiology*, **60**, 3508–3513 https://doi.org/10.1128/aem.60.10.3508-3513.1994

Kosten S., Piñeiro M., de Goede E., de Klein J., Lamers L. P. M. and Ettwig K. (2016). Fate of methane in aquatic systems dominated by free-floating plants. *Water Research*, **104**, 200–207, https://doi.org/10.1016/j.watres.2016.07.054

La H., Hettiaratchi J. P. A., Achari G., Verbeke T. J. and Dunfield P. F. (2018). Biofiltration of methane using hybrid mixtures of biochar, lava rock and compost. *Environmental Pollution*, **241**, 45–54, https://doi.org/10.1016/j.envpol.2018.05.039

Leu A. O., Cai C., McIlroy S. J., Southam G., Orphan V. J., Yuan Z., Hu S. and Tyson G. W. (2020). Anaerobic methane oxidation coupled to manganese reduction by members of the *Methanoperedenaceae*. *The ISME Journal*, **14**, 1030–1041, https://doi.org/10.1038/s41396-020-0590-x

Li J., Xu X., Liu C., Wu N., Sun Z., He X. and Chen Y. (2021) Active methanotrophs and their response to temperature in marine environments: an experimental study. *Journal of Marine Science and Engineering*, **9**, 1261, https://doi.org/10.3390/jmse9111261

Lü F., He P., Guo M., Yang N. and Shao L. (2012). Ammonium-dependent regulation of aerobic methane-consuming bacteria in landfill cover soil by leachate irrigation. *Journal of Environmental Sciences*, **24**(4), 711–719, https://doi.org/10.1016/S1001-0742(11)60813-9

Mancebo J. U. and Hettiaratchi J. P. A. (2015). Rapid assessment of methanotrophic capacity of compost-based materials considering the effects of air-filled porosity, water content and dissolved organic carbon. *Bioresource Technology*, **177**, 125–133, https://doi.org/10.1016/j.biortech.2014.11.058

Manfredi S., Tonini D. and Christensen T. H. (2009). Landfilling of waste: accounting of greenhouse gases and global warming contributions. *Waste Management & Research*, **27**, 825–836, https://doi.org/10.1177/0734242X09348529

Medvedkova K. A., Khmelenina V. N., Suzina N. E. and Trotsenko Y. A. (2009). Antioxidant systems of moderately thermophilic methanotrophs *Methylocaldum szegediense* and *Methylococcus capsulatus*. *Microbiology*, **78**, 670–677, https://doi.org/10.1134/S0026261709060022

Monica-Macías A. K., Londono Y. A. and Penuela G. A. (2023). Denitrifying anaerobic methane oxidation and its application for wastewater treatment. *International Journal of Environmental Science and Technology*, **20**, 2209–2228, https://doi.org/10.1007/s13762-022-04155-2

Muenmee S., Chiemchaisri W. and Chiemchaisri C. (2015). Microbial consortium involving biological methane oxidation in relation to the biodegradation of waste plastics in a solid waste disposal open dump site. *International Biodeterioration & Biodegradation*, **102**, 172–181 https://doi.org/10.1016/j.ibiod.2015.03.015

Muenmee S., Chiemchaisri W. and Chiemchaisri C. (2016). Enhancement of biodegradation of plastic wastes via methane oxidation in semi-aerobic landfill. *International Biodeterioration & Biodegradation*, **113**, 244–255, https://doi.org/10.1016/j.ibiod.2016.03.016

Nguyen N. L., Yu W. J., Gwak J. H., Kim S. J., Park S. J., Herbold C. W., Kim J. G., Jung M. Y. and Rhee S. K. (2018). Genomic insights into the acid adaptation of novel methanotrophs enriched from acidic forest soils. *Frontiers in Microbiology*, **9**, 1982, https://doi.org/10.3389/fmicb.2018.01982

Ni G. (2023). Mitigating greenhouse gas emissions from wastewater treatment through microbiological innovation. *Microbiology Australia*, **44**(1), 22–26, https://doi.org/10.1071/MA23006

Op den Camp H. J. M., Islam T., Stott M. B., Harhangi H. R., Hynes A., Schouten S., Jetten M. S. M., Birkeland N. K., Pol A. and Dunfield P. F. (2009). Environmental, genomic and taxonomic perspectives on methanotrophic *Verrucomicrobia*. *Environmental Microbiology Reports*, **1**, 293–306, https://doi.org/10.1111/j.1758-2229.2009.00022.x

Orata F. D., Meier-Kolthoff J. P., Sauvageau D. and Stein L. Y. (2018). Phylogenomic analysis of the gammaproteobacterial methanotrophs (order *Methylococcales*) calls for the reclassification of members at the genus and species levels. *Frontiers in Microbiology*, **9**, 3162, https://doi.org/10.3389/fmicb.2018.03162

Park S. Y. and Kim C. G. (2019.) Application and development of methanotrophs in environmental engineering. *Journal of Material Cycles and Waste Management*, **21**, 415–422, https://doi.org/10.1007/s10163-018-00826-w

Parte A. C. (2014). LPSN – list of prokaryotic names with standing in nomenclature. *Nucleic Acids Research*, **42**, D613–D616, https://doi.org/10.1093/nar/gkt1111

Pecorini I., Rossi E. and Iannelli R. (2020). Mitigation of methane, NMVOCs and odor emissions in active and passive biofiltration systems at municipal solid waste landfills. *Sustainability*, **12**, 3203, https://doi.org/10.3390/su12083203

Raghoebarsing A. A., Pol A., van de Pas-Schoonen K. T., Smolders A. J. P., Ettwig K. F., Rijpstra W. I. C., Schouten S., Sinninghe Damsté J. S., Op den Camp H. J. M., Jetten M. S. M. and Strous M. (2006). A microbial consortium couples anaerobic methane oxidation to denitrification. *Nature*, **440**, 918–921, https://doi.org/10.1038/nature04617

Reddy K. R., Yargicoglu E. N. and Chetri J. K. (2021). Effects of biochar on methane oxidation and properties of landfill cover soil: long-term column incubation tests. *Journal of Environmental Engineering, ASCE*, **147**(1), 04020144, https://doi.org/10.1061/(ASCE)EE.1943-7870.0001829

Roncato C. D. L. and Cabral A. R. (2012). Evaluation of methane oxidation efficiency of two biocovers: field and laboratory results. *Journal of Environmental Engineering, ASCE*, **138**(2), 164–173, https://doi.org/10.1061/(ASCE)EE.1943-7870.0000475

Sadasivam B. Y. and Reddy K. R. (2014). Landfill methane oxidation in soil and bio-based cover systems: a review. *Reviews in Environmental Science and Bio/Technology*, **13**, 79–107, https://doi.org/10.1007/s11157-013-9325-z

Safari E., Al-Suwaidi G. and Rayhani M. T. (2017). Methane emissions from municipal solid waste landfills in cold climates. *Journal of Environmental Engineering, ASCE*, **143**(5), 06017003, https://doi.org/10.1061/(ASCE)EE.1943-7870.0001191

Schalk T., Effenberger J., Jehmlich A., Nowak J., Rustige H., Krebs P. and Kühn V. (2019). Methane oxidation in vertical flow constructed wetlands and its effect on denitrification and COD removal. *Ecological Engineering*, **128**, 77–88, https://doi.org/10.1016/j.ecoleng.2018.12.029

Scheutz C., Kjeldsen P., Bogner J. E., De Visscher A., Gebert J., Hilger H. A., Huber-Humer M. and Spokas K. (2009). Microbial methane oxidation processes and technologies for mitigation of landfill gas emissions. *Waste Management & Research*, **27**(5), 409–455, https://doi.org/10.1177/0734242X09339325

Schirmer W. N., Stroparo E. C., Gueri M. V. D., Capanema M. A., Mazur D. L., Jucá J. F. T. and Martins K. G. (2022). Biofiltration of fugitive methane emissions from landfills using scum from municipal wastewater treatment plants as alternative substrate. *Journal of Material Cycles and Waste Management*, **24**, 2041–2053 https://doi.org/10.1007/s10163-022-01468-9

Schmitz R. A., Peeters S. H., Versantvoort W., Picone N., Pol A., Jetten M. S. M. and Op den Camp H. J. M. (2021). Verrucomicrobial methanotrophs: ecophysiology of metabolically versatile acidophiles. *FEMS Microbiology Reviews*, **45**(5), fuab007, https://doi.org/10.1093/femsre/fuab007

Semrau J. D., Dispirito A. A. and Murrell J. C. (2008). Life in the extreme: thermoacidophilic methanotrophy. *Trends in Microbiology*, **16**, 190–193, https://doi.org/10.1016/j.tim.2008.02.004

Sharp C. E., Smirnova A. V., Graham J. M., Stott M. B., Khadka R., Moore T. R., Grasby S. E., Strack M. and Dunfield P. F. (2014). Distribution and diversity of Verrucomicrobia methanotrophs in geothermal and acidic environments. *Environmental Microbiology Reports*, **16**, 293–306.

Shukla P. N., Pandey K. and Mishra V. K. (2013). Environmental determinants of soil methane oxidation and methanotrophs. *Critical Reviews in Environmental Science and Technology*, **43**, 1945–2011, https://doi.org/10.1080/10643389.2012.672053

Siliakus M. F., van der Oost J. and Kengen S. W. M. (2017). Adaptations of archaeal and bacterial membranes to variations in temperature, pH and pressure. *Extremophiles*, **21**, 651–670, https://doi.org/10.1007/s00792-017-0939-x

Struik Q., Oliveira Junior E. S., Veraart A. J. and Kosten S. (2022). Methane emissions through water hyacinth are controlled by plant traits and environmental conditions. *Aquatic Botany*, **183**, 103574, https://doi.org/10.1016/j.aquabot.2022.103574

Sun W., Li S., Zhang G., Fu G., Qi H. and Li T. (2023). Effects of climate change and anthropogenic activities on soil pH in grassland regions on the Tibetan Plateau. *Global Ecology and Conservation*, **45**, e02532.

Syed R., Sagger S., Tate K., Rehm B. H. A. and Berben P. (2017). Assessing the performance of floating biofilters for oxidation of methane from dairy effluent ponds. *Journal of Environmental Quality*, **46**, 272–280, https://doi.org/10.2134/jeq2016.08.0310

Tanthachoon N., Chiemchaisri C., Chiemchaisri W., Tudsri S. and Kumar S. (2008). Methane oxidation in compost-based landfill cover with vegetation during wet and dry conditions in the tropics. *Journal of the Air & Waste Management Association*, **58**, 603–612, https://doi.org/10.3155/1047-3289.58.5.603

Trotsenko Y. A. and Khmelenina V. N. (2002). Biology of extremophilic and extremotolerant methanotrophs. *Archives of Microbiology*, **177**, 123–131, https://doi.org/10.1007/s00203-001-0368-0

Tung N. T., Chiemchaisri W., Chiemchaisri C. and Sanguanpak S. (2021). Removal of greenhouse gas in biofilter using organic and inorganic media. *Vietnam Journal of Science and Technology*, **63**(1), 83–89.

USEPA. (1994). Design, Operation, and Closure of Municipal Solid Waste Landfills. Seminar Publication, United States Environmental Protection Agency, Washington, DC.

Vaksmaa A., Guerrero-Cruz S., van Alen T. A., Ettwig K. F., Lüke C. and Jetten M. S. M. (2017). Enrichment of anaerobic nitrate-dependent methanotrophic '*Candidatus Methanoperedens nitroreducens*' archaea from an Italian paddy field soil. *Applied Microbiology and Biotechnology*, **101**, 7075–7084, https://doi.org/10.1007/s00253-017-8416-0

Wang S., Liu Q., Li J. and Wang Z. (2021). Methane in wastewater treatment plants: status, characteristics, and bioconversion feasibility by methane oxidizing bacteria for high value-added chemicals production and wastewater treatment. *Water Research*, **198**, 117122, https://doi.org/10.1016/j.watres.2021.117122

Wang F., Zhang S., Hu X., Lv X., Liu M., Ma Y. and Manirakiza B. (2024). Floating plants reduced methane fluxes from wetlands by creating a habitat conducive to methane oxidation. *Journal of Environmental Sciences*, **135**, 149–160, https://doi.org/10.1016/j.jes.2023.01.013

Whittenbury R., Phillips K. C. and Wilkinson J. F. (1970). Enrichment, isolation and some properties of methane-utilizing bacteria. *The Journal of General Microbiology*, **61**, 205–218, https://doi.org/10.1099/00221287-61-2-205

Xing Z., Zhao T., Zhang L., Gao Y., Liu S. and Yang X. (2018). Effects of copper on expression of methane monooxygenases, trichloroethylene degradation, and community structure in methanotrophic consortia. *Engineering in Life Sciences*, **18**, 236–243, https://doi.org/10.1002/elsc.201700153

Yao X., Wang J. and Hu B. (2023) How methanotrophs respond to pH: a review of ecophysiology. *Frontiers in Microbiology*, **13**, 1034164, https://doi.org/10.3389/fmicb.2022.1034164

Yoshida N., Iguchi H., Yurimoto H., Murakami A. and Sakai Y. (2014). Aquatic plant surface as a niche for methanotrophs. *Frontiers in Microbiology*, **5**, 30.

Zhao J., Cai Y. and Jia Z. (2020). The pH-based ecological coherence of active canonical methanotrophs in paddy soils. *Biogeoscience*, **17**, 1451–1462, https://doi.org/10.5194/bg-17-1451-2020

Part 3

Re-naturing cities through water-centric development

doi: 10.2166/9781789065015_0119

Chapter 6

Regenerative living walls for urban wastewater management and climate adaptation

Thammarat Koottatep[1], Chawalit Chaiwong[1], Tatchai Pussayanavin[2] and Chongrak Polprasert[3]*

[1]*Asian Institute of Technology, Thailand*
[2]*Ramkhamheang University, Thailand*
[3]*Thammasat University, Thailand*
**Correspondence: poktatchai@gmail.com*

ABSTRACT

Regenerative living walls offer an innovative solution for urban wastewater management and climate adaptation, particularly for the challenges posed by the inadequate sanitation infrastructure prevalent in developing countries such as Thailand, India and Cambodia. The absence of proper sanitation systems results in uncontrolled release of human wastes, including feces, urine, washing water, and cleansing fluids, into the environment. This not only causes environmental pollution, but also contributes to significant public health issues. In many developing countries, on-site wastewater treatment systems, such as septic tanks or modified septic tanks with aerated systems, are commonly employed as the primary treatment for urban household or building wastewater. However, primarily due to short hydraulic retention times (HRTs) and unpredictable organic loading rates (OLRs), effluents of these on-site systems usually contain high concentrations of organic compounds (e.g. chemical oxygen demand (COD)) and nutrients (e.g. nitrogen), resulting in detrimental effects on the environment. To address these challenges, a paradigm shift is essential in the design concept of sanitation systems, especially in urban contexts. Moving away from the traditional 'end-of-pipe' approach, there is a growing need to embrace strategies centered around 'reuse, recycle, and resource recovery.' In this context, novel constructed wetland systems, such as living garden walls and employing novel media, emerge as promising post-treatment solutions, demonstrating potential improvements in treatment efficiencies, cost-effectiveness, ease of maintenance, and installation footprint for densely populated urban areas.

Keywords: Living walls, urban wastewater, treatment performance, nature-based solution, resource recovery

6.1 CONCEPTS OF NATURE-BASED SOLUTIONS FOR URBAN WASTEWATER MANAGEMENT

Nature-based solutions (NBS) as constructed wetlands (CWs) emulate natural treatment conditions, leveraging the interplay of microorganisms, soil, and plants to facilitate waste degradation. Incorporating ecological elements like plants, water, and various media (soil, sand, and gravel), CWs naturally enhance biological, photosynthetic, and physical processes, effectively eliminating organic matter, nutrients, and pathogens from wastewater (Koottatep *et al.*, 2021). These systems demonstrate

© 2025 IWAP. This is an Open Access book chapter distributed under the terms of the Creative Commons Attribution License (CC BY-NC-ND 4.0) which permits copying and redistribution for non-commercial purposes with no derivatives, provided the work is properly cited (https://creativecommons.org/licenses/by-nc-nd/4.0/). The chapter is from the book *Nature-Based Solutions for Urban Sustainability*, Piet N.L. Lens and Xuan-Thanh Bui (Editors).

substantial removal capabilities for chemical oxygen demand (COD), biochemical oxygen demand (BOD), suspended solids (SS), nitrogen (N), phosphorus (P), inorganics, trace organics (pharmaceuticals and personal care products (PPCPs)), and pathogens present in wastewater. Particularly in tropical climates, CWs are gaining attraction as an alternative technology for wastewater treatment in polishing effluent from on-site wastewater treatment systems.

In urban settings, green walls, with different plant varieties and growth patterns, offer a unique approach (Prodanovic *et al.*, 2019). Façade-based greenery relies on plants supported by media, while green walls integrate plants into the structure itself, utilizing soil, stone, and water. For the living wall system normally applied in the vertical systems, wastewater undergoes infiltration and percolation vertically through planted filter media, categorized as up-flow vertical subsurface flow constructed wetlands. This configuration facilitates efficient water drainage and promotes efficient air diffusion. The vertical sub-surfaces play a crucial role in fostering the nitrification process, resulting in a high efficacy in nitrogen removal. This concept contributes to urban ecosystem establishment through strategic plant application and integrated monitoring of climatic, energy, and technical aspects. Green buildings, including both vertical and roof greening, create environmentally friendly spaces with numerous benefits for the people, buildings, infrastructures, biodiversity, and micro-climates. Green walls serve as versatile buffers, purifying air, mitigating heat, reducing noise, conserving energy and water, all while enhancing aesthetic values. The utilization of vertical-flow constructed wetlands further complements these efforts, boasting low investment and operation costs alongside promising treatment performance (Koottatep *et al.*, 2021). Together, these innovative approaches in constructed wetlands and green wall technologies represent significant strides in sustainable urban wastewater management and climate-responsive adaptation (Pradhan *et al.*, 2019).

Recent discussions have centered on advancing notions that advocate for reuse in decentralized contexts, promoting resource recovery through source separation and management rather than relying on centralized treatment. The aim is to enhance the performance of existing urban wastewater management technology in terms of treatment efficiencies, cost-effectiveness, and operational maintenance ease (Addo-Bankas *et al.*, 2021; Vymazal *et al.*, 2021). There is a call for a shift in the design concept from traditional 'end-of-pipe' approaches to a more holistic approach encompassing 'reuse, recycle, and resource recovery,' especially in urban contexts. To address on-site wastewater treatment in decentralized settings, Koottatep *et al.* (2021) have introduced innovative small-scale constructed wetlands achieving over 80% removal efficiency for both organic and nitrogen compounds. Recent research emphasizes the transition towards a more 'circular economy', advocating for the reuse of treated wastewater in response to the increasing demand for natural resources, including freshwater. Despite reduced fecal contamination, urban wastewater or graywater must still undergo cleansing to meet water reuse standards (Boano *et al.*, 2021; Jackson, 2017). Various techniques and technologies are proposed to transform current water management systems into circular systems. NBS, emerging as a promising alternative, are derived from natural processes, integrating nature into urban environments for resource recovery with lower energy consumption, increased biodiversity, and the restoration of ecosystem services, all while addressing specific water-related challenges. As a NBS, green walls are highlighted as energy-efficient and low-maintenance decentralized treatment systems for graywater (Sami *et al.*, 2023). Factors such as hydraulic loading rates, media composition, flow rates, and hydraulic retention times (HRT) are crucial in designing and operating NBS effectively. However, the impact of HRTs on different media or bed layers in NBS remains underexplored, and understanding the void volume of the media is essential for ecological engineering design. Media composition, including actual porosity, influences the NBS treatment efficiencies (Boano *et al.*, 2021; Prodanovic *et al.*, 2020).

6.2 TREATMENT MECHANISMS OF LIVING WALLS

Living walls, also known as green walls or vertical gardens, can be employed as innovative and sustainable solutions for wastewater treatment. The treatment mechanisms of living walls for wastewater involve a combination of physical, chemical, and biological processes. The mechanisms responsible for pollutant removal in these living walls are described below.

6.2.1 Organic matter removal

Microorganisms play a pivotal role in wastewater treatment by converting organic matter (COD and BOD) into up to 50% CO_2. The remaining 50% is sludge biomass. A diverse array of microorganisms, adapted to both aerobic and anaerobic conditions, engage in the removal of easily degradable organic compounds through a combination of chemical, physical, and biological processes (Tietz *et al.*, 2007). The primary phyla involved in organic matter removal are Proteobacteria, Actinobacteria, and Bacteroidetes. Proteobacteria, a metabolically diverse group, exhibits a strong correlation with the presence of oxygen. Firmicutes, which include Gram-positive bacteria capable of forming endospores, such as Clostridia (anaerobes) and Bacilli (obligate or facultative aerobes), are prevalent in the MSL and CW systems, offering resistance to desiccation and survivability in extreme conditions (Wakatsuki *et al.*, 1993). Clostridia, predominantly found in the soil layer, utilize various anaerobic energy-yielding mechanisms. Bacilli, on the other hand, produce extracellular hydrolytic enzymes that break down complex polymers in MSL and CW systems.

Essentially, oxygen produced in the root zone and dissolved in the top layer is utilized by aerobic microorganisms during organic matter decomposition, as observed in the respiration reaction (equation 6.1). The role of emergent plants in oxygen transfer within constructed wetlands (CWs) becomes crucial. The symbiotic relationship between phototrophic micro-algae and microorganisms is often portrayed as mutual reactions between aerobic bacteria and algae cells. Koottatep *et al.* (2021) identified biofilm bacteria on surfaces of the media and root zones in CW beds as major contributors to BOD removal (equations 6.1–6.3). Furthermore, fermentation and denitrification reactions within the multi-soil layers (MSL) occur, particularly in the middle and bottom sections of a CW. This underscores the intricate interplay of microorganisms in driving the treatment processes within constructed wetlands.

$$\text{Respiration}\,(\textit{aerobic zone}) \qquad C_6H_{12}O_6 + 6O_2 \rightarrow 6CO_2 + 6H_2O \tag{6.1}$$

$$\text{Fermentation}\,(\textit{anoxic and anaerobic zone}) \qquad \begin{aligned} &C_6H_{12}O_6 \rightarrow 2CH_3CHOHCOOH \\ &C_6H_{12}O_6 \rightarrow 2CH_3CH_2OH + 2CO_2 \end{aligned} \tag{6.2}$$

$$\text{Denitrification}\,(\textit{anoxic and anaerobic zone}) \qquad C_6H_{12}O_6 + 4NO_3^- \rightarrow 6H_2O + 6CO_2 + 2N_2 + 4e^- \tag{6.3}$$

6.2.2 Nitrogen removal

The nitrogen cycling processes include ammonification, a biologically mediated transformation of organic nitrogen into ammonia under both aerobic and anaerobic conditions, influenced by factors such as pH, soil redox potential, temperature, nutrient availability, carbon/nitrogen ratio, soil texture, microbial biomass, and soil structure (Coppini *et al.*, 2019). The nitrification process (equation 6.6) converts ammonium into nitrate under aerobic conditions, with subsequent denitrification being essential for ammonia removal in CWs (Grismer *et al.*, 2003; Udert & Wächter, 2012).

Denitrification, depicted as $2NO_3^- \rightarrow 2NO_2^- \rightarrow 2NO \rightarrow N_2O \rightarrow N_2$, occurs under anaerobic conditions and can be enhanced by adding an external carbon source, although organic carbon substrates may not be optimal for heterotrophic denitrification. Nutrient uptake by emergent plants, particularly nitrogen (N) and phosphorus (P), supports their growth and contributes to the removal of these elements from influent wastewater, with reported removal rates ranging from 10% to 50%. Nitrification and denitrification processes play also a significant role in nitrogen removal in MSL systems, requiring conducive environmental conditions for the growth of nitrifying (high dissolved oxygen) and denitrifying (low dissolved oxygen) bacteria. Koottatep *et al.* (2018) observed a higher abundance of ammonia-oxidizing archaea (AOA) and ammonia-oxidizing bacteria (AOB) in MSL beds compared to CW beds. This abundance contributes to effective ammonia (NH_4^+-N) removal, as AOA and AOB oxidize NH_3-N to NO_2^--N and NO_3^--N, which is further denitrified to N_2 gas by denitrifying bacteria. Notably, the *Nitrososphaera* cluster, classified under the *Thaumarchaeota* family, represents AOA in MSL. Most AOB sequences are affiliated with *Nitrosospira* and *Nitrosomonas* clusters. The prevalence of AOA and AOB in the MSL units highlights the positive impact of novel MSL media in supporting microbial growth, crucial for NH_4^+-N, COD, and BOD removal.

Ammonium sorption $$RH^+ + NH_4^+ + OH^- \leftrightarrow RNH_4^+ + H_2O \tag{6.4}$$

Nitration (*Nitrosomonas*) $$2\,NH_4^+ + 3\,O_2 \rightarrow 2\,NO_2^- + 2\,H_2O + 4\,H^+ \tag{6.5}$$

Nitrification (*Nitrobacter*) $$2\,NO_2^- + 2\,O_2 \rightarrow 2\,NO_3^- \tag{6.6}$$

Denitrification

Based on methanol (CH_3OH): $$NO_3^- + 0.833\,CH_3OH \rightarrow 0.5\,N_2 + 0.833\,CO_2 + 1.167\,H_2O + OH^- \tag{6.7}$$

Based on glucose $(C_6H_{12}O_6)$: $$NO_3^- + 0.208\,C_6H_{12}O_6 \rightarrow 0.5\,N_2 + 1.25\,CO_2 + 0.75\,H_2O + OH^- \tag{6.8}$$

$$\begin{aligned} &84.8\,NO_3^- + (CH_2O)_{106}\,(NH_3)_{16}\,(H_3PO_4) \\ &\rightarrow 42.4\,N_2 + 106\,CO_2 + 148.8\,H_2O + 16\,NH_3 + H_3PO_4 \end{aligned} \tag{6.9}$$

Biological nitrogen removal could also involve the anaerobic ammonium oxidation (ANAMMOX) pathway, transforming NO_2^- and NH_4^+ into N_2, depending on factors like pH, free ammonia, temperature, dissolved oxygen (DO), HRT, hydroxylamine, free nitrous acids, organic compounds, and salts. Plant uptake and assimilation contribute significantly to wastewater treatment by converting inorganic nitrogen forms (ammonium, nitrate and amino acids) into organic forms, with effective removal of 66–71% of total nitrogen through plant uptake in CWs.

Ammonia adsorption in CWs involves ionized ammonia being adsorbed onto various substrates like soil sediments, inorganic sediments, or detritus through cation-exchange reactions, and the absorbed ammonia be released back into wastewater. The process depends on the substrate materials such as biochar, vermiculite, blast furnace slag, and zeolite (Vymazal & Kröpfelová, 2011).

6.2.3 Phosphorus removal

Phosphorus uptake in plants primarily occurs through their root systems, with submerged plant species additionally absorbing phosphorus from water bodies through leaves and shoots (Grismer *et al.*, 2003; Naja & Volesky, 2011; Shi *et al.*, 2017). The assimilation of nutrients, including nitrogen (N) and phosphorus (P), is positively correlated with plant biomass yield. Moreover, the phosphorus removal capacity varies based on plant species, specific plant parts, and seasonal factors (Koottatep *et al.*, 2021). Microorganisms, encompassing bacteria, fungi, algae, and invertebrates, play a dual role as both a source and sink for phosphorus. These organisms actively transform phosphorus compounds, although microbial uptake remains relatively low in quantity due to their slow growth rates.

Phosphorus adsorption is characterized by the movement of soluble inorganic phosphorus from soil pore water to the mineral surface of the soil. The soil adsorption capacity is influenced by the mineral and clay content, with media substrates, containing iron (Fe) and a carbon source, impacting the rate of phosphorus adsorption. Additionally, phosphorus adsorption is subject to various interactions involving the redox potential, aluminum (Al), pH, calcium (Ca) minerals, iron (Fe), and native soil phosphorus (Vymazal & Kröpfelová, 2011). Besides, the CWs could be integrated with some novel adsorptive media, resulting in increased water pH level, which facilitated phosphorus precipitation.

At high pH values (>9), the PO_4^{3-} in the wastewater was adsorbed and reacted then with the Mg^{3+}, Ca^{2+}, Al^{3+} and Fe^{3+} present in a ceramsite adsorbents to form phosphorus precipitates, according to equation (6.10) (Chaiwong *et al.*, 2023; Wang *et al.*, 2021):

$$\times \mathrm{PO}_4^{3-} + 3M^{\times +} = M_3(\mathrm{PO}_4)\times \downarrow \tag{6.10}$$

where M represents metal elements Mg, Ca, Al, Fe and and so on.

6.2.4 Pathogen removal

The effectiveness of sedimentation and adsorption in pathogen removal by constructed wetlands relies on media particle size and density, with additional influence from interactions between media grains, biofilm, and plant roots (Kadam *et al.*, 2008). Plants, crucial for surface stabilization, facilitate physical filtration through roots, providing extensive surface areas for microbial attachment and growth in constructed wetlands. The natural die-off of pathogens is influenced by the HRT, predation activities, and microbial starvation, exhibiting a higher rate in water bodies compared to sediments (Ansari *et al.*, 2016; Chaiwong *et al.*, 2020; Naja & Volesky, 2011). The presence of oxygen in water bodies or the rhizosphere and plant exudates with antimicrobial properties result in reduced populations of *fecal coliform*, *enterococcus*, *Salmonella*, *Shigella*, *Yersinia*, and coliphages. Predation processes involve bacteriophages, *Bdellovibrio*-like organisms (BLOs), and protozoa. These contribute to pathogen reduction, with protozoa predation activities enhancing the reduction of bacterial populations.

Elevated wastewater temperatures in summer enhance pathogen removal, potentially influencing annual variation patterns and impacting the reduction of enteric bacteria and viruses, with increased non-pathogenic activities like grazing protozoa potentially boosting pathogen ingestion rates. Enteric bacteria, being facultative and obligate anaerobes, face unfavorable conditions in aerobic environments. Plant roots contribute to oxygenation in the rhizosphere, leading to higher dissolved oxygen (DO) levels and increased removal of bacteria, like *E. coli*. Biofilms play a pivotal role in contaminant removal, influencing pathogen removal through attachment and protozoa grazing, with rhizosphere biofilms supporting the proliferation of bacterial populations with antibiotic activity, such as *Pseudomonas*.

Living wall systems can be integrated with novel adsorptive media, made of low-cost materials such as combined sewage sludge with zeolite, bentonite, and magnesium oxide (MgO). These can inactivate the pathogens by raised pH levels (>8.5) (Chaiwong *et al.*, 2023). When the added MgO powder contacts wastewater, it releases hydroxide ions (OH^-) that increase the pH of the solution, interfering the enzymatic activity of pathogens and disrupting their cellular functions, and ultimately leading to their inactivation (Chaiwong *et al.*, 2023).

HRT and hydraulic loading rate significantly influence pathogen removal in constructed wetlands, with increased HRT leading to higher microbial inactivation ratios. During wet periods, the pathogen removal efficiency is lower due to decreased HRT and loading rates, and the microbial inactivation ratio at saturation depends on the granular medium in the bed.

6.3 IMPLICATIONS FOR DESIGN AND OPERATION OF LIVING WALLS

6.3.1 Plant selection

The types of plants used in a living wall system significantly affect its wastewater treatment capabilities. Some plants are more effective at nutrient uptake and water filtration than others. Plants with a high tolerance for water and nutrient fluctuations are usually preferred. About 90% of BOD, COD, TSS, TN, and TP removal were achieved through the utilization of plant roots in constructed wetlands. This suggests the plant selection is a significant factor that can influence the treatment performance of the LW systems of treatment-vegetative building systems (Picard *et al.*, 2005). Figure 6.1 illustrates examples of the removal of various types of pollutants by diverse plant roots (Pradhan *et al.*, 2019).

Types of Plant		Removal of Pollutants
• *Ipomea alpina* and *Arabidopsis thaliana*	⟹	Cu
• *Thlaspi caerulescens, Corydalis pterygopetala, Sedum alfredii, Arabidopsis thaliana* and *Arabidopsis halleri*	⟹	Zn
• *Corydalis pterygopetala, Sedum alfredii* and *Viola calaminaria*	⟹	Pb
• *Thlaspi caerulescens, Brassica oleracea, Corydalis pterygopetala, Salsola kali, Helianthus annus* and *Viola calaminaria*	⟹	Cd
• *Helianthus annus, Thlaspi goesingens, Alyssum bertolonii* and *Stackhousia tryonii*	⟹	Co
• *Sebertia acuminata, Sebertia acuminata* and *Helianthus annus*	⟹	Ni and Cr
• *Astragalus racemosus, Genus astragalus*, and *Astragalus bisulcatus*	⟹	Se
• *Arabidopsis thaliana, Hemidesmus indicus* and *Sedum alfredii*	⟹	Mn
• *Pteris vittata*	⟹	As
• *Phragmites australis, Iris pseudacorus, Schoenoplectus tabernaemontani, Typha augustifolia, Lemna, Elodea canadensis, Nuphar lutea, Carex sp., Acorus sp., Sparganium sp.* and *Juncus sp.*	⟹	SS, BOD, phosphates, nitrates, Ammonia, organophosphates and heavy metals

Figure 6.1 Removal of different types of pollutants by various plants (Pradhan *et al.*, 2019).

The plants with high and moderate growth rates, namely *C. appressa, N. obliterata, D. tasmanica, A. praecox, L. muscari, P. tenax,* and *M. parvifolium* are notably effective in removing nitrogen and phosphorus from graywater (Prodanovic *et al.*, 2019). Plants with well-established root systems and robust growth patterns, such as *C. appressa, N. obliterata,* and *L. muscari,* exhibit resilience to changes in operational conditions, making them suitable plants for nutrient treatment in green walls (Prodanovic *et al.*, 2019).

Because the local climate and environmental conditions such as extreme temperatures, drought, or excessive rainfall affect the overall growth of the employed plants in the living wall systems, plant selection should be based on those various mentioned conditions (Prodanovic *et al.*, 2019).

6.3.2 Substrate composition

The substrate or growth media in the living wall is crucial for supporting plant growth and providing a habitat for beneficial microorganisms. The composition of the substrates and media influences nutrient adsorption, water drainage, and the overall growth of the plants (Pradhan *et al.*, 2019). Traditionally, natural porous substrates such as soil and sand have been utilized in green roofs and living walls (Pradhan *et al.*, 2019). However, there is a growing preference for modern lightweight materials due to their lower structural support requirements, resulting in significant cost savings. Although the pollutant removal capabilities of traditional natural media are well-documented, there is limited research on the newer lightweight materials. Various lightweight media such as Coco coir, Fyto-foam, Growstone, expanded clay, vermiculite, perlite, river sand, LECA (lightweight expanded clay aggregate)-coconut fibers, LECA-sand, and biochar have been utilized in living walls for wastewater treatment (Dalahmeh *et al.*, 2018; Farhan *et al.*, 2018; Lakho *et al.*, 2021; Masi *et al.*, 2016; Prodanovic *et al.*, 2017). The selection of materials was based on criteria such as local availability, weight, sustainability, water distribution, nutrient and water-retention capacity, porosity, and ability to support plant growth (Prodanovic *et al.*, 2017).

6.3.3 Hydraulic loading rate

The rate at which wastewater flows through the living walls, known as the hydraulic loading rate, is an important factor. If the flow rate is too high, the plants may not have enough time to absorb nutrients and pollutants effectively as well as insufficiently inactivate pathogens. One key parameter affected by the HLR is the retention time (HRT), which represents the duration wastewater remains in contact with the employed media and plant roots (Akratos & Tsihrintzis, 2007). A lower HLR corresponds to a longer retention time, facilitating enhanced physical, chemical, and biological processes and, consequently improving the treatment efficiency. Conversely, a higher HLR results in a reduced HRT, potentially limiting the system performance to achieve effective wastewater treatment. The nutrient uptake by plants is also impacted by the HLR. Higher HLR values may diminish nutrient uptake due to the shorter contact time or HRT between wastewater and plant roots, potentially compromising the treatment system effectiveness in removing nutrients (Akratos & Tsihrintzis, 2007; Prodanovic *et al.*, 2019). Additionally, clogging and saturation of the utilized media are concerns associated with elevated HLR (Pradhan *et al.*, 2019). A high HLR could lead to reduced permeability, potentially allowing water to bypass the media or substrates without sufficient contact time for effective treatment processes. Furthermore, the plant health is affected by the HLR, as excessive rates may induce water stress, compromising the plant ability to contribute effectively to the treatment process (Prodanovic *et al.*, 2017).

6.3.4 Inflow water quality

The quality of the incoming wastewater can impact the living wall performance. High levels of contaminants or pollutants may overwhelm the plants and reduce their ability to purify water. Pre-treatment of wastewater may be necessary in some cases. The concentrations and types of contaminants, nutrient content, pH, alkalinity, toxic substances, suspended solids, biodegradability, temperature, and flow variability in the influent wastewater collectively impact the efficiency of these treatment systems. High pollutant concentrations could challenge the green walls' ability to effectively remove contaminants, while imbalances in nutrients or the presence of toxic substances can compromise plant health and microbial activity (Pradhan *et al.*, 2019). Factors such as suspended solids and biodegradability can influence physical and biological treatment processes, while temperature and flow variability affect the stability of the system. A comprehensive understanding of these influent characteristics is crucial for designing and operating green wall systems to achieve optimal treatment performance under diverse wastewater conditions.

6.4 APPLICATION OF LIVING WALLS WITH POROUS MEDIA FOR URBAN WASTEWATER MANAGEMENT AND CLIMATE ADAPTATION

6.4.1 Living walls with porous media for wastewater treatment and recycling

Due to lack of proper sanitation systems, human wastes such as feces, urine, washing water, and cleansing fluids, are released into the environment without being adequately treated, causing environmental pollution and public health issues. Most developing countries, including Thailand, India, and Cambodia, generally use on-site wastewater treatment systems or sanitation systems, including septic tanks or modified septic tanks with aerated systems, as the primary treatment for the treatment of black and gray water from households or buildings in urban areas. The short HRTs and unpredictable organic loading rates (OLRs) usually cause the effluent of septic tanks to typically contain high concentrations of organic matter and nutrients such as COD and nitrogen, which have a negative impact on the environment. Therefore, the existing design concept of sanitation systems, especially for the urban context, has to shift from 'end-of-pipe' to 'reuse, recycle, and resource recovery'.

To improve the performance of the existing urban (dense area) wastewater management technologies in terms of treatment efficiencies, costs, and ease of maintenance, installation footprint, novel constructed wetland systems have been proposed as effective post-treatment systems. The novel constructed wetlands with specific uses for treating septic tank effluents can remove more than 80% of both organic matter and nitrogen. To overcome the limitation of the large horizontal area requirement of constructed wetlands, the 'Living Garden Wall' (LGW) is being proposed in several areas (Figure 6.2). The influent is pumped to the first reactor and passed through to the next ones and the LGW utilized soil as the medium and was planted with *Spider Ivy, Coleus, and Selaginella frosty*, as shown in Figure 6.2.

The vertically constructed living garden walls (LGW) have been deliberately adorned with ornamental plants such as *Chlorophyta Como sum Vittate, Hydrocotyle umbellate* L. and *Pandanus sanderi* Sander ex M.T. Mast. These plants are strategically planted in various media types, including soil, expanded clay, and innovative media produced from coffee grounds, eggshells, and sewage sludge

(a) (b)

Figure 6.2 LGW at (a) Asian Institute of Technology and (b) Ramkhamheang University, Thailand.

(Chaiwong *et al.*, 2023). Wastewater is evenly distributed across the LGW surface, allowing it to flow vertically downward through the media and the roots and rhizomes of the plants. The incorporation of these novel media serves multiple purposes. First, they aim to enhance the treatment performance by providing the conducive environment for the growth of heterogeneous bacteria, crucial for organic matter removal, and nitrifying–denitrifying bacteria, essential for nitrogen removal. Additionally, these media contribute to the adsorption capacity, aiding in the removal of phosphorus and heavy metals, while also mitigating clogging issues.

The LGW are positioned not only as an efficient wastewater treatment system, but also as a multifaceted solution offering green landscapes and recreational spaces in urban areas. This dual-purpose approach aligns with the broader vision of creating a more livable city. By integrating wastewater treatment functionality with aesthetic and recreational elements, these living structures contribute to the overall well-being of urban environments. In essence, the LGWs transcend their utilitarian role by not only addressing wastewater treatment needs, but also enhancing the quality of urban life through the creation of visually appealing and functional green spaces.

The novel multifunctional media used in the LGW were also effectively removing TP (97%) and *fecal coliform* (3.4 log) form the wastewater (Chaiwong *et al.*, 2023), addressing the pressing need for more sustainable wastewater treatment methods (Chaiwong *et al.*, 2023). In addition, many types of living wall systems have also been previously implemented for graywater treatment and recycling. For examples, Fowdar *et al.* (2017) applied plants in filter columns for graywater treatment, affirming the potential of these systems for on-site treatment. Similarly, a small-scale green wall system for treating graywater could effectively treat urban gray wastewater, and provide recovered treated graywater for purposes of toilet flushing and irrigation activities (Masi *et al.*, 2016). The employed vertical ecosystem for treatment of graywater to meet regulatory standards for garden and golf course irrigation and toilet flushing has been implemented by Gattringer *et al.* (2016). Likewise, the full-scale demonstration of a green wall for graywater treatment and recycling has been successfully conducted (Lakho *et al.*, 2021). Those indicate the high potential of the application of the living walls with porous media for urban wastewater management.

6.4.2 Living walls for climate adaptation

Living wall systems can offer a multifaceted approach to climate adaptation by addressing temperature regulation, air quality improvement, stormwater management, biodiversity support, heat island mitigation, and providing aesthetic and psychological benefits (Pradhan *et al.*, 2019).

6.4.2.1 Temperature regulation

The living walls provide additional insulation to buildings, reducing the need for heating in winter and cooling in summer (Wong *et al.*, 2010). This helps in maintaining a more stable indoor temperature, which can contribute to energy efficiency and climate adaptation. The cost-effective method of using evapotranspiration to naturally cool urban areas is a valuable strategy for mitigating the urban heat island effect. For instance, the cooling effect extended up to 0.6 m away from a living wall, resulting in a temperature reduction of up to 3.3°C at a distance of 15 cm (Wong *et al.*, 2010). The plant leaves function as insulation and generate cool and moist air that can decrease the ambient temperature of the surrounding building (Fioretti *et al.*, 2010). The green roofs and living walls with vegetation have even more substantial impacts on structural surfaces, reducing roof temperatures by over 20°C during the summer (Liu & Minor, 2005). The investigations by Sternberg *et al.* (2011) indicate that the presence of the plants led to between 13°C and 16°C reduction in building wall temperature (Sternberg *et al.*, 2011). Hence, integrating these green technologies into urban planning and design can contribute to more sustainable and resilient cities in the face of climate change.

6.4.2.2 Air quality improvement

Plants can absorb carbon dioxide during photosynthesis, helping to mitigate climate change. Living walls, by increasing the plant biomass in urban areas, contribute to carbon sequestration, acting as a natural carbon sink. Additionally, the plants can filter pollutants from the air, including volatile organic compounds (VOCs) and particulate matter (Pradhan *et al.*, 2019; Ysebaert *et al.*, 2021). Over 100 plant species have been investigated for their efficiency to purify air by removing pollutants (Dela Cruz *et al.*, 2014; Soreanu *et al.*, 2013). A study revealed that the plants on the roof could remove particulate matter from the air (Ysebaert *et al.*, 2021). Plant species are crucial in the planning of green walls, as they significantly influence the accumulation of particulate matter (PM), since they highly determine PM accumulation values ranging from 10^{10} to 10^{11} particles per m^2 leaf area along a high intensity traffic road (Ysebaert *et al.*, 2021). The filtration process involving plants, the media and the rhizosphere microorganisms had a remarkable capacity to remove 90% of volatile organic carbons (VOCs), CO_2, SO_2, CO, and NO_2 (Wood *et al.*, 2006). Improved air quality is beneficial for human health and can mitigate the impact of climate change-related events, such as heatwaves and air pollution (Ysebaert *et al.*, 2021).

6.4.3 Aesthetic values of living walls for landscape and decorative purposes

Living walls, also known as green walls or vertical gardens, offer a plethora of aesthetic values for landscape and decorative purposes. The key benefit of this application is that living walls bring nature into urban environments, adding lush greenery and vibrant colors to otherwise dull spaces, softening the harsh lines of buildings. Additionally, the use of various plant species, such as *Spider Ivy, Coleus,* and *Selaginella frosty,* introduces texture and depth to a space, creating visual interest (Figure 6.2). Different leaf shapes, sizes, and colors create a dynamic and layered appearance, enhancing the overall aesthetic appeal. Living walls can also display seasonal changes, such as blooming flowers in spring or changing leaf colors in autumn. Designing and arranging plants on a vertical surface allows for endless creative possibilities. Living walls can be customized to reflect different themes, patterns, or even incorporate artistic elements such as sculptures or decorative features. Moreover, the living walls can help reduce the noise pollution in the buildings which usually originates from human activities and is a prevalent issue for urban residents (Ghazalli *et al.*, 2019). Since the living walls are covered with vegetation that is typically installed vertically on the exterior or interior walls of buildings, they can act as natural sound absorbers, dampening and absorbing sound waves. The dense vegetation on the living walls create physical barriers that assist to block and deflect sound waves, resulting in further reductions of the transmissions of noise from outside to inside the buildings (Ghazalli *et al.*, 2019).

6.5 CONCLUSION

Regenerative living walls provide an innovative solution for urban wastewater management, especially in regions like Thailand, India, and Cambodia where inadequate sanitation infrastructure poses significant challenges. Traditional on-site treatment systems often fall short, leading to environmental pollution and public health risks. Embracing novel strategies centered around reuse and resource recovery, such as living garden walls with innovative media, offers promising improvements in treatment efficiency, cost-effectiveness, and ease of maintenance, making them ideal for densely populated urban areas. These approaches pave the way for sustainable and resilient urban environments, addressing critical sanitation issues while promoting environmental health.

REFERENCES

Addo-Bankas O., Zhao Y., Vymazal J., Yuan Y., Fu J. and Wei T. (2021). Green walls: a form of constructed wetland in green buildings. *Ecological Engineering*, **169**, 106321, https://doi.org/10.1016/j.ecoleng.2021.106321

Akratos C. S. and Tsihrintzis V. A. (2007). Effect of temperature, HRT, vegetation and porous media on removal efficiency of pilot-scale horizontal subsurface flow constructed wetlands. *Ecological Engineering*, **29**(2), 173–191, https://doi.org/10.1016/j.ecoleng.2006.06.013

Ansari A. A., Gill S. S., Gill R., Lanza G. R. and Newman L. (2016). Phytoremediation: management of environmental contaminants. Springer Cham, pp 1–409, https://doi.org/10.1007/978-3-319-41811-7

Boano F., Costamagna E., Caruso A., Fiore S., Chiappero M., Galvão A., Pisoeiro J., Rizzo A. and Masi F. (2021). Evaluation of the influence of filter medium composition on treatment performances in an open-air green wall fed with greywater. *Journal of Environmental Management*, **300**, 1136–1146, https://doi.org/10.1016/j.jenvman.2021.113646

Chaiwong C., Koottatep T. and Polprasert C. (2020). Comparative study on attached-growth photobioreactors under blue and red lights for treatment of septic tank effluent. *Journal of Environmental Management*, **260**, 110134, https://doi.org/10.1016/j.jenvman.2020.110134

Chaiwong C. C., Koottatep T., Chirasuwannaphot Y., Thanasrilungkul C., Panchai P., Chanamarn W., Noophan P. L., Kasahara T., Wongkiew S. and Polprasert C. (2023). Novel multifunctional sewage sludge-based adsorbents for treatment of municipal wastewater. *Journal of Applied Science and Engineering*, **27**(9), 3127–3246.

Coppini E., Palli L., Antal A., Del Bubba M., Miceli E., Fani R. and Fibbi D. (2019). Design and start-up of a constructed wetland as tertiary treatment for landfill leachates. *Water Science and Technology*, **79**(1), 145–155, https://doi.org/10.2166/wst.2019.030

Dalahmeh S., Ahrens L., Gros M., Wiberg K. and Pell M. J. S. (2018). Potential of biochar filters for onsite sewage treatment: adsorption and biological degradation of pharmaceuticals in laboratory filters with active, inactive and no biofilm. *Science of the Total Environment*, **612**, 192–201, https://doi.org/10.1016/j.scitotenv.2017.08.178

Dela Cruz M., Christensen J. H., Thomsen J. D. and Müller R. (2014). Can ornamental potted plants remove volatile organic compounds from indoor air? – A review. *Environmental Science and Pollution Research*, **21**(22), 13909–13928, https://doi.org/10.1007/s11356-014-3240-x

Farhan A., Zakaria A., Mat N. and Mohd K. S. (2018). Soilless media culture – a propitious auxiliary for crop production. *Asian Journal of Crop Sciecne*, **10**, 1–9.

Fioretti R., Palla A., Lanza L. G. and Principi P. J. B. (2010). Green roof energy and water related performance in the Mediterranean climate. *Building and Environment*, **45**(8), 1890–1904, https://doi.org/10.1016/j.buildenv.2010.03.001

Fowdar H. S., Hatt B. E., Breen P., Cook P. L. and Deletic A. (2017). Designing living walls for greywater treatment. *Water Research*, **110**, 218–232, https://doi.org/10.1016/j.watres.2016.12.018

Gattringer H., Claret A., Radtke M., Kisser J., Zraunig A., Rodriguez-Roda I. and Buttiglieri G. (2016). Novel vertical ecosystem for sustainable water treatment and reuse in tourist resorts. *International Journal of Sustainable Development*, **11**(3), 263–274.

Ghazalli A. J., Brack C., Bai X. and Said I. (2019). Physical and non-physical benefits of vertical greenery systems: a review. *Journal of Urban Technology*, **26**(4), 53–78, https://doi.org/10.1080/10630732.2019.1637694

Grismer M. E., Carr M. A. and Shepherd H. L. (2003). Evaluation of constructed wetland treatment performance for winery wastewater. *Water Environment Research*, **75**(5), 412–421, https://doi.org/10.2175/106143003×141213

Jackson W. (2017). Regenerative sanitation foundations. In: Regenerative Sanitation: A New Paradigm for Wastewater Management, T. Koottatep, P. E. Cookey, and C. Polprasert (eds), IWA Publishing, pp. 20–45.

Kadam A., Oza G., Nemade P., Dutta S. and Shankar H. (2008). Municipal wastewater treatment using novel constructed soil filter system. *Chemosphere*, **71**(5), 975–981, https://doi.org/10.1016/j.chemosphere.2007.11.048

Koottatep T., Suksiri P., Pussayanavin T. and Polprasert C. (2018). Development of a novel multi-soil layer constructed wetland treating septic tank effluent with emphasis on organic and ammonia removals. *Water, Air, and Soil Pollution*, **229**(8), 258, https://doi.org/10.1007/s11270-018-3907-x

Koottatep T., Pussayanavin T., Khamyai S. and Polprasert C. (2021). Performance of novel constructed wetlands for treating solar septic tank effluent. *Science of the Total Environment*, **754**, 142447, https://doi.org/10.1016/j.scitotenv.2020.142447

Lakho F. H., Vergote J., Khan H. I., Depuydt V., Depreeuw T., Van Hulle S. W. and Rousseau D. P. J. (2021). Total value wall: full-scale demonstration of a green wall for greywater treatment and recycling. *Journal of Environmental Management*, **298**, 113489, https://doi.org/10.1016/j.jenvman.2021.113489

Liu K. and Minor J. J. P. (2005). Performance evaluation of an extensive green roof. NRCC-48204, Washington DC, 1–11.

Masi F., Bresciani R., Rizzo A., Edathoot A., Patwardhan N., Panse D. and Langergraber G. (2016). Green walls for greywater treatment and recycling in dense urban areas: a case study in Pune. *Journal of Water, Sanitation and Hygiene for Development*, **6**(2), 342–347, https://doi.org/10.2166/washdev.2016.019

Naja G. M. and Volesky B. (2011). Constructed wetlands for water treatment. In: Comprehensive Biotechnology (3rd edn.), M. Moo-Young (ed.), Pergamon, Vol. **6**, pp. 307–322, https://doi.org/10.1016/B978-0-444-64046-8.00362-1

Picard C. R., Fraser L. H. and Steer D. (2005). The interacting effects of temperature and plant community type on nutrient removal in wetland microcosms. *Bioresource Technology*, **96**(9), 1039–1047, https://doi.org/10.1016/j.biortech.2004.09.007

Pradhan S., Al-Ghamdi S. G. and Mackey H. R. (2019). Greywater recycling in buildings using living walls and green roofs: A review of the applicability and challenges. *Science of The Total Environment*, **652**, 330–344, https://doi.org/10.1016/j.scitotenv.2018.10.226

Prodanovic V., Hatt B., McCarthy D., Zhang K. and Deletic A. (2017). Green walls for greywater reuse: Understanding the role of media on pollutant removal. *Ecological Engineering*, **102**, 625–635, https://doi.org/10.1016/j.ecoleng.2017.02.045

Prodanovic V., McCarthy D., Hatt B. and Deletic A. (2019). Designing green walls for greywater treatment: the role of plants and operational factors on nutrient removal. *Ecological Engineering*, **130**, 184–195, https://doi.org/10.1016/j.ecoleng.2019.02.019

Prodanovic V., Hatt B., McCarthy D. and Deletic A. (2020). Green wall height and design optimisation for effective greywater pollution treatment and reuse. *Journal of Environmental Management*, **261**(2019), 110173, https://doi.org/10.1016/j.jenvman.2020.110173

Sami M., Hedström A., Kvarnström E., McCarthy D. T. and Herrmann I. (2023). Greywater treatment in a green wall using different filter materials and hydraulic loading rates. *Journal of Environmental Management*, **340**, 117998, https://doi.org/10.1016/j.jenvman.2023.117998

Shi J., Yang Z., Dai H., Lu X., Peng L., Tan X., Shi L. and Fahim R. (2017). Preparation and application of modified zeolites as adsorbents in wastewater treatment. *Water Science and Technology*, **2017**(3), 621–635, https://doi.org/10.2166/wst.2018.249

Soreanu G., Dixon M. and Darlington A. (2013). Botanical biofiltration of indoor gaseous pollutants – A mini-review. *Chemical Engineering Journal*, **229**, 585–594, https://doi.org/10.1016/j.cej.2013.06.074

Sternberg T., Viles H. and Cathersides A. J. B. (2011). Evaluating the role of ivy (*Hedera helix*) in moderating wall surface microclimates and contributing to the bioprotection of historic buildings. *Building and Environment*, **46**(2), 293–297, https://doi.org/10.1016/j.buildenv.2010.07.017

Tietz A., Kirschner A., Langergraber G., Sleytr K. and Haberl R. (2007). Characterisation of microbial biocoenosis in vertical subsurface flow constructed wetlands. *Science of the Total Environment*, **380**(1–3), 163–172, https://doi.org/10.1016/j.scitotenv.2006.11.034

Udert K. M. and Wächter M. (2012). Complete nutrient recovery from source-separated urine by nitrification and distillation. *Water Research*, **46**(2), 453–464, https://doi.org/10.1016/j.watres.2011.11.020

Vymazal J. and Kröpfelová L. (2011). A three-stage experimental constructed wetland for treatment of domestic sewage: first 2 years of operation. *Ecological Engineering*, **37**(1), 90–98, https://doi.org/10.1016/j.ecoleng.2010.03.004

Vymazal J., Zhao Y. and Mander Ü. (2021). Recent research challenges in constructed wetlands for wastewater treatment: a review. *Ecological Engineering*, **169**, 106318, https://doi.org/10.1016/j.ecoleng.2021.106318

Wakatsuki T., Esumi H. and Omura S. (1993). High performance and N and P-removable on-site domestic waste water treatment system by multi-soil-layering method. *Water Science and Technology*, **27**(1), 31–40, https://doi.org/10.2166/wst.1993.0010

Wang H., Xu J., Liu Y. and Sheng L. J. J. (2021). Preparation of ceramsite from municipal sludge and its application in water treatment: a review. *Journal of Environmental Management*, **287**, 112374, https://doi.org/10.1016/j.jenvman.2021.112374

Wong N. H., Tan A. Y. K., Chen Y., Sekar K., Tan P. Y., Chan D., Chiang K. and Wong N. C. J. B. (2010). Thermal evaluation of vertical greenery systems for building walls. *Building and Environment*, **45**(3), 663–672, https://doi.org/10.1016/j.buildenv.2009.08.005

Wood R. A., Burchett M. D., Alquezar R., Orwell R. L., Tarran J. and Torpy F. J. W. (2006). The potted-plant microcosm substantially reduces indoor air VOC pollution: I. Office field-study. *Water, Air, and Soil Pollution*, **175**, 163–180, https://doi.org/10.1007/s11270-006-9124-z

Ysebaert T., Koch K., Samson R. and Denys S. (2021). Green walls for mitigating urban particulate matter pollution – a review. *Urban Forestry & Urban Greening*, **59**, 127014, https://doi.org/10.1016/j.ufug.2021.127014

doi: 10.2166/9781789065015_0131

Chapter 7

Application of green roof systems for domestic wastewater treatment

*Thi-Kim-Ngan Nguyen[1,2], Thi-Kim-Quyen Vo[3], Cong-Sac Tran[1,2], Thi-Viet-Huong Dao[1,2], Mai-Duy-Thong Pham[1,2], Thi-Yen-Phuong Nguyen[1,2], Veeriah Jegatheesan[4], Hau-Vuong Tran[5], Phuoc-Dan Nguyen[1,2], Thanh-Dai Tran[6], Piet N. L. Lens[7] and Xuan-Thanh Bui[1,2]**

[1]Key Laboratory of Advanced Waste Treatment Technology, Faculty of Environment and Natural Resources, Ho Chi Minh City University of Technology (HCMUT), 268 Ly Thuong Kiet Street, District 10, Ho Chi Minh City, Vietnam
[2]Vietnam National University Ho Chi Minh (VNU-HCM), Linh Trung Ward, Ho Chi Minh City 700000, Vietnam
[3]Faculty of Biology and Environment, Ho Chi Minh City University of Industry and Trade (HUIT), 140 Le Trong Tan Street, Tay Thanh Ward, Tan Phu District, Ho Chi Minh City 700000, Vietnam
[4]School of Engineering and Water: Effective Technologies and Tools (WETT) Research Centre, RMIT University, Melbourne, Victoria, Australia
[5]Ho Chi Minh City University of Natural Resources and Environment (HCMUNRE), 236B Le Van Sy Street, Tan Binh District, Ho Chi Minh City, Vietnam
[6]Faculty of Technology, Dong Nai Technology University, Bien Hoa City, Vietnam
[7]National University of Ireland Galway, University Road, Galway H91 TK33, Ireland
**Corresponding author: bxthanh@hcmut.edu.vn*

ABSTRACT

Green roofs are vegetated spaces located atop buildings, also known as 'contained' green spaces, considered a nature-based solution aligned with sustainable development goals (SDGs) such as SDG6 (Clean water and sanitation) and SDG13 (Climate action). These roofs utilize organic compounds and nutrients, including nitrogen and phosphorus, from domestic wastewater for plant growth. Therefore, irrigating green roofs with domestic wastewater can not only alleviate a significant burden on urban wastewater treatment plants but also reduce costs for irrigation and wastewater treatment. The application of green roofs has become widespread worldwide, benefiting environmental, economic, and social sustainability. In Vietnam, where rapid building development has led to a scarcity of green areas and sanitary facilities, green roofs have garnered significant attention. They are broadly categorized into extensive, semi-intensive, and intensive types. This chapter provides details of the fundamental components, pollutant removal mechanisms, factors affecting performance, and benefits associated with green roofs. It also offers technical parameters for implementing green roof systems.

Keywords: Green roof, nature-based solution, phytoremediation, plant uptake, sustainable development, wastewater treatment.

© 2025 IWAP. This is an Open Access book chapter distributed under the terms of the Creative Commons Attribution License (CC BY-NC-ND 4.0) which permits copying and redistribution for non-commercial purposes with no derivatives, provided the work is properly cited (https://creativecommons.org/licenses/by-nc-nd/4.0/). The chapter is from the book *Nature-Based Solutions for Urban Sustainability*, Piet N.L. Lens and Xuan-Thanh Bui (Editors).

7.1 INTRODUCTION

Currently, more than half of the world's population resides in urban areas, a figure that continues to grow rapidly due to the swift pace of urbanization and economic expansion. By 2050, ~68% of the projected 9.7 billion global population will inhabit large cities (World Bank, 2022). This burgeoning population brings with it an increasing demand for freshwater resources. Consequently, larger volumes of wastewater are being generated and discharged into the environment, placing significant strain on wastewater treatment plants, and posing a serious threat to sustainable development. Moreover, rapid urban development often results in a scarcity of green spaces and inadequate sanitary facilities due to spatial constraints. Green infrastructure, recognized as a natural ecosystem, plays a crucial role in mitigating the negative impacts of urbanization, enhancing climate resilience, and reducing urban pollution (Mihalakakou *et al.*, 2023). Although the concept of vegetated roofs has existed for centuries, modern green roof technology was first pioneered in Germany in the 20th century. Since then, green roofs have gained widespread acceptance and implementation globally (Zapater-Pereyra *et al.*, 2016). Particularly in densely populated cities, green roof systems are heralded as a natural solution for improving environmental quality, owing to their myriad benefits.

Green roofs offer numerous ecological and social advantages (Figure 7.1), including mitigating climate change, enhancing air quality, reducing building energy consumption, managing stormwater runoff, and enhancing aesthetic value for buildings (Liu *et al.*, 2021a). Additionally, green roofs can serve as effective wastewater treatment units, utilizing organic compounds and nutrients such as nitrogen and phosphorus from domestic wastewater to foster plant growth (Vo *et al.*, 2018). Consequently, irrigating green roofs with domestic wastewater not only alleviates the burden on urban wastewater treatment facilities, but also yields cost savings for irrigation purposes.

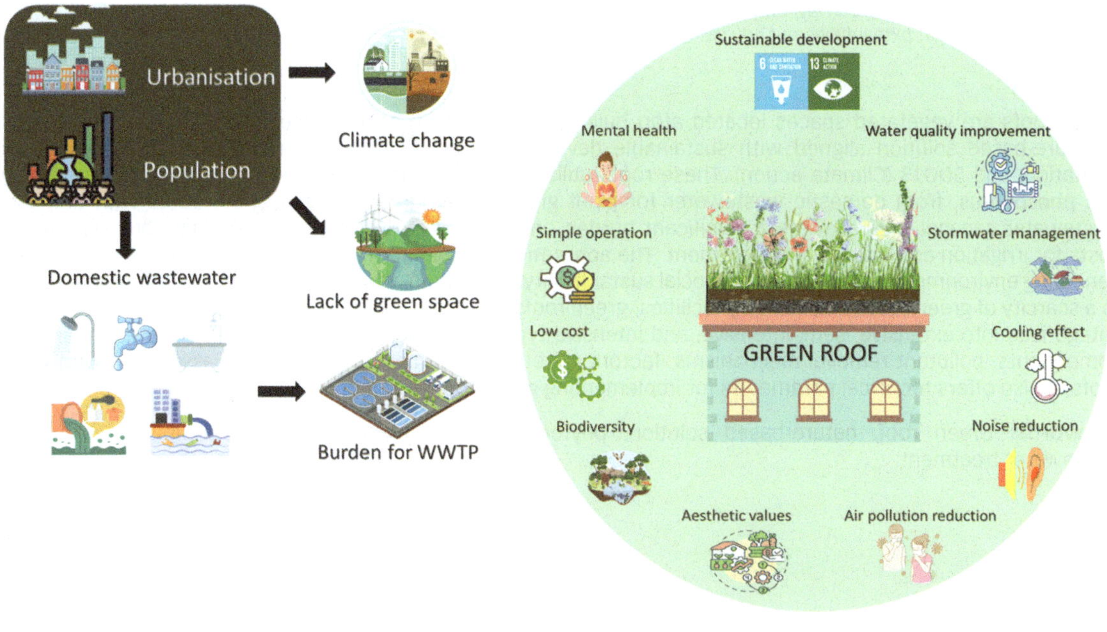

Figure 7.1 Strengths of green roofs and their potential to address the challenges of urbanization and population growth.

7.2 CHARACTERISTICS OF GREEN ROOFS

7.2.1 Classification of green roofs

Figure 7.2 illustrates three types of green roofs: extensive, semi-intensive, and intensive. The categorization depends on vegetation and substrate depth. Extensive green roofs consist of shallow substrate depths (5–15 cm), lightweight materials (60–150 kg/m^2), and low capital and maintenance costs. A substrate depth greater than 10 cm is more effective in removing pollutants from wastewater compared to shallower depths (Shahmohammad *et al.*, 2022). Due to the shallow substrate layer, only low-lying plants such as sedum, succulents, and grasses, capable of withstanding dry conditions and low-nutrient substrates, can be planted on extensive green roofs. These roofs are suitable for large areas, flat or slightly sloped roofs, and can often be installed without major structural alterations. In contrast, intensive green roofs consist of deeper substrate layers (>25 cm), heavier weight (200–500 kg/m^2), and higher capital and maintenance costs, allowing for a wider variety of plants, shrubs, and other herbaceous species. Semi-intensive green roofs combine features of both extensive and intensive types. Particularly, semi-intensive green roofs are characterized by substrate depths (15–25 cm), weights (120–200 kg/m^2), and capital and maintenance costs that are lower than intensive green roofs but higher than extensive green roofs. Due to moderate substrate depths, semi-intensive green roofs can support grasses, ground covers, herbs, and small shrubs (Shahmohammad *et al.*, 2022). Among the three types of green roofs, extensive green roofs are the most popular worldwide due to their lightweight construction and lower costs (Vijayaraghavan, 2016).

7.2.2 Structure of green roofs

Green roof systems comprise various components, including vegetation, substrate, filter fabric, drainage elements, and waterproofing membranes (Figure 7.3). Each component fulfills a specific role in green roof systems, and the selection process for these components varies based on geographic location (Vijayaraghavan, 2016). Vegetation, situated atop the layers, is the essential component. The benefits of green roofs primarily hinge on plant species, as plants directly contribute to the improvement of air and water quality and the reduction of building energy demand (Cascone, 2019). Plant selection must consider numerous factors such as geographic location, climatic conditions, types of green roofs, and types of substrates (Vijayaraghavan, 2016).

Due to their impact on building load and structure, the selection of appropriate substrates is crucial for green roofs. Substrates not only provide physical support for plant growth but also influence the treatment performance of green roofs (Mihalakakou *et al.*, 2023). The weight and thickness of substrates depend on the types of green roofs and selected plant species. Substrate characteristics encompass physical parameters (particle size, density, and water permeability) and chemical parameters (pH and organic content) (Cascone, 2019).

The primary function of a filter fabric is to separate the substrate from the drainage element and prevent small particles such as fine soil or plant debris from clogging the drainage system.

Figure 7.2 Different types of green roofs.

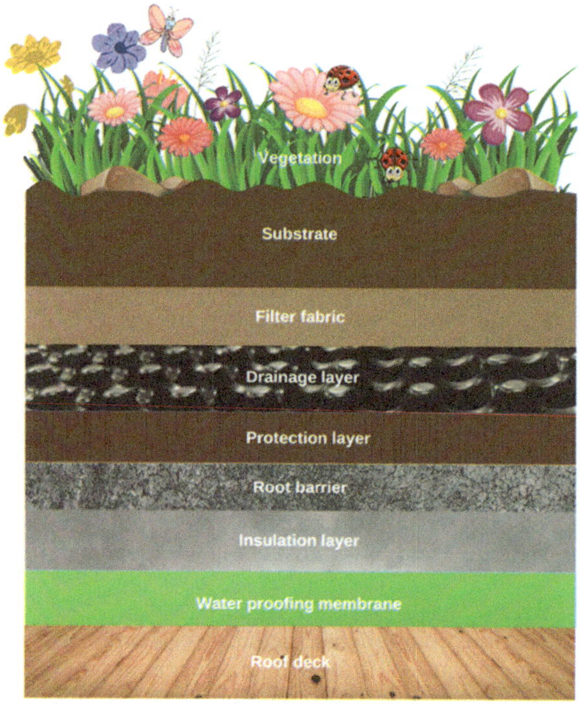

Figure 7.3 Main components of a typical green roof system.

Additionally, a filter fabric layer serves as a barrier membrane for plants with short and soft roots. The characteristics of a filter fabric must include tensile strength to sustain high loads and narrow holes for water permeability. Geotextile fabrics, renowned for their high water-holding capacity, are commonly used as a filter fabric in green roofs (Vijayaraghavan, 2016).

A drainage layer's function is to remove excess water from the substrate, protect the waterproofing membrane, and reduce building temperature and energy consumption. Two primary types of drainage layers are commonly applied in green roofs: modular panels for drainage composed of polyethylene or polystyrene, which feature many perforations to facilitate water drainage, and granular materials for drainage, such as stone, broken brick, coarse gravel, expanded shale, and lightweight expanded clay aggregates, known for their large pore spaces that allow them to hold more water (Shafique *et al.*, 2018).

A waterproofing membrane is a prerequisite in green roofs as it prevents building leakage caused by high moisture levels from upper layers. Therefore, water-tightness is the main requirement for this layer. Due to the complexity of maintenance in case of leaks, the selection of a waterproofing membrane should be thoroughly considered during the design phase. Several available options for waterproofing membranes include thermoplastic membranes, liquid-applied membranes, modified bitumen sheets, polymer cement systems, and single-ply sheet membranes (Vijayaraghavan, 2016).

7.2.3 Pollutant removal mechanisms

Pollutant removal processes from wastewater using green roof systems consist of physical processes (sedimentation and filtration), chemical processes (precipitation and adsorption), and biological processes (microbial degradation and plant uptake) (Arden and Ma, 2018). A summary of the different mechanisms involved in the removal of pollutants by green roofs is illustrated in Figure 7.4.

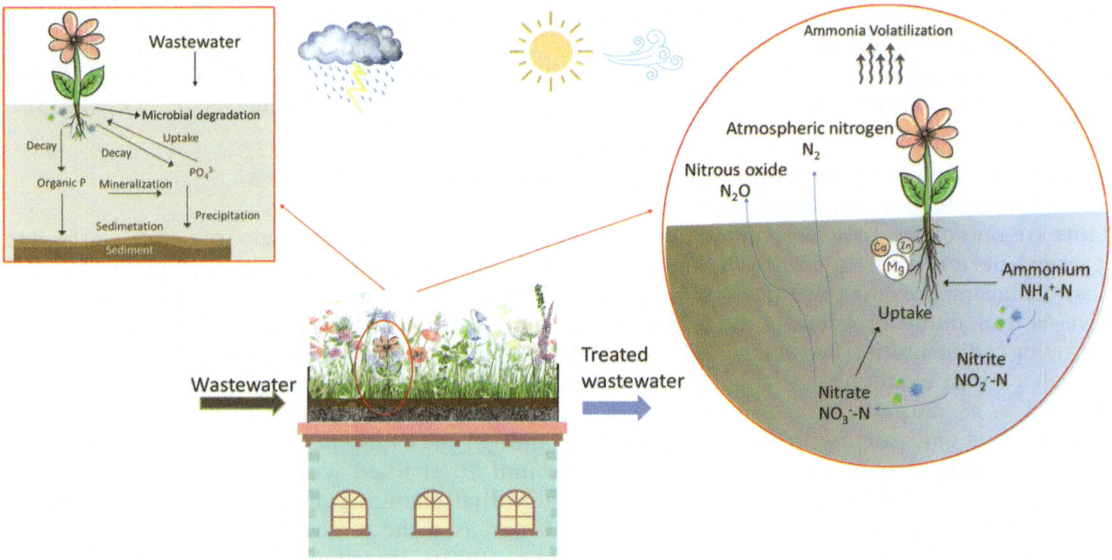

Figure 7.4 Mechanisms of the removal of pollutants in green roofs.

Physical processes primarily account for the removal of total suspended solids (TSS) in wastewater. Plant roots contribute to the filtration and solids retention process (Arden and Ma, 2018). The main source of carbon for microbial activity is organic matter. Organic matter is removed under aerobic and anaerobic conditions through biodegradation, accumulation, and filtration processes, with organic matter mainly biodegraded by the biofilm attached to submerged parts of plants, root systems, and substrate layers (Van *et al.*, 2015). Aerobic decomposition occurs in the presence of oxygen generated from the transfer and convection of plant root systems, as well as air diffusion, whereas anaerobic degradation occurs in the absence of oxygen by heterotrophic bacteria (Nguyen *et al.*, 2021).

Nitrogen is one of the main pollutants in domestic wastewater that requires treatment before discharge. Additionally, nitrogen is a major contributor to eutrophication, which limits oxygen diffusion into water, thus impacting aquatic ecosystems. In green roofs, the mechanism of nitrogen removal consists of ammonia volatilization, ammonia adsorption, denitrification, plant uptake, and bacterial bioconversions (Vo *et al.*, 2017). Ammonia volatilization is a physical process that occurs depending on the pH values of wastewater, with significant volatilization of ammonia observed at pH levels above 9.3 (Lyu *et al.*, 2018). For the nitrification process, ammonium (NH_4^+-N) is oxidized to nitrite (NO_2^--N) by ammonia-oxidizing bacteria or ammonium-oxidizing archaea. Nitrite is then converted to nitrate (NO_3^--N) by nitrite-oxidizing bacteria. Denitrification, on the contrary, occurs through various bacteria or archaea, where nitrate is converted to nitrous oxide (N_2O), nitric oxide (NO), or nitrogen (N_2) (Vo *et al.*, 2018). Nitrogen is essential for plant growth and is primarily taken up through root systems. NO_3^--N and NH_4^+-N are the main forms of nitrogen taken up by plants (Vo *et al.*, 2017).

Plants not only provide carriers for microbial attachment but also deliver oxygen from stems to roots, creating favorable conditions for the growth of different bacteria. These rhizospheric bacteria degrade various pollutants present in wastewater (Mahmoudi *et al.*, 2021). A denitrification process removes 60–95% of nitrogen (Tang *et al.*, 2020), whereas plants uptake less than 20% of nitrogen (Zhuang *et al.*, 2019). Therefore, denitrification is the dominant pathway for nitrogen removal.

An adsorption process supports the removal of pollutants such as ammonium, phosphorus, and heavy metals. Pollutants are adsorbed on the surface of substrates, with molecules flowing across the liquid film covering substrate particles and diffusing into substrate pores (Ji *et al.*, 2022). Phosphorus is also necessary for the development and metabolism of microorganisms and plants. It is removed through various mechanisms including physical retention, adsorption, precipitation, and plant uptake (Gong *et al.*, 2020). Furthermore, green roofs have the ability to retain heavy metals, with above-ground plant tissues and roots of green roof plants absorbing significant amounts of heavy metals (Ye *et al.*, 2013).

7.2.4 Factors affecting the treatment performance of green roof systems

7.2.4.1 Effects of plant species

Plants are one of the critical components of green roofs because they strongly influence wastewater treatment performance and green space enhancement. The roots and rhizomes of plants stabilize substrate layers, prevent clogging, and uptake nutrients and heavy metals. Additionally, root systems serve as a medium for bacterial attachment and contribute to reducing water flow rates, thereby improving sedimentation (Vo *et al.*, 2017). Moreover, plant roots can enhance denitrification processes through carbon excretion and improve nitrification by releasing oxygen (Omotade *et al.*, 2019). Thomaidi *et al.* (2022) observed that the average chemical oxygen demand (COD) removal efficiency of green roofs cultivated with *Polygala myrtifolia* and *Atriplex halimus* (78 and 82%, respectively) was higher than that of an unplanted system (around 70%) filled with 10-cm vermiculite under similar operating conditions. Another study indicated that changing plant species from *Axonopus compressus* to *Wedelia trilobata* in a wetland roof increased the removal efficiency by 11–13% for total nitrogen (TN) and 6–12% for COD (Nguyen *et al.*, 2021). *W. trilobata* is known as an invasive plant species with high-nutrient utilization efficiency, tolerance under stressful conditions, and rapid growth rates (Huang *et al.*, 2022). Additionally, a study demonstrated that green roofs using C4 plant species (*Cynodon dactylon* and *Eremochloa ophiuroides*) and C3 plant species (*Festuca arundinacea* and *Poa pratensis*) exhibited higher runoff reduction and pollutant removal compared to crassulacean acid metabolism plant species (*Callisia repens* and *Sedum lineare*) (Liu *et al.*, 2021a). In general, plants have a significant impact on the treatment performance of green roofs, which vary depending on the characteristics of the plant species.

7.2.4.2 Effects of substrate layers

The selection of materials for the substrate layer is crucial to achieve the long-term benefits of green roofs, including peak flow reduction and water quality improvement (Cascone, 2019). Thomaidi *et al.* (2022) demonstrated that a green roof filled with a 20-cm depth of vermiculite had a higher TSS and COD removal efficiency (over 90%) compared to the one filled with perlite (63–75% of TSS, 54–70% of COD) under similar operational conditions. These results were likely due to the smaller particle size of vermiculite (0.5–3.0 mm) compared to perlite (1.0–5.0 mm). In general, finer particles improve filtration, provide a higher surface area for microbial activity, and increase contact time between porous materials and wastewater (Vo *et al.*, 2019).

According to findings investigated by Prodanovic *et al.* (2017), a variety of materials with various properties were used to identify suitable materials for pollutant treatment. Among eight potential materials (river sand, expanded clay, growstone, vermiculite, perlite, fyto-foam, rockwool, and coco coir), perlite and coco coir showed the highest potential for pollutant removal using green walls for gray wastewater treatment. Perlite has a high porosity ($\sim$30%) but a low cation exchange capacity, whereas vermiculite has a low porosity ($<$10%), restricted drainage, and poor aeration (Pradhan *et al.*, 2019). Another study demonstrated that charcoal exhibited a higher pollutant removal rate than sand, with removal rates increasing by 8% for COD, 45–64% for TN, and 125% for total phosphorus (TP) (Nguyen *et al.*, 2021). Additionally, charcoal is considered an absorbent for pollutant removal due to its high porosity (Omotade *et al.*, 2019). These studies indicate that the selection of materials for the substrate layer is based on several factors, including weight, particle size, sustainability, water distribution, porosity, water and nutrient retention, and local availability.

The selection of substrate depth depends on the type of green roof. Deeper substrates promote better plant growth due to their higher water-retention capacity (Pradhan *et al.*, 2019). Additionally, substrate depth positively impacts stormwater management. A study investigated that a green roof filled with a 12.5-cm substrate had a higher runoff retention capacity than the one filled with a 7.5-cm substrate (32.9 and 23.2% of total precipitation volume, respectively) (Schultz *et al.*, 2018). From the results of the study conducted by Schultz *et al.* (2018), it was recognized that an increase of 50 mm in substrate depth could lead to a 10% rise in annual runoff retention.

There is also evidence that increasing substrate depth enhances the pollutant removal efficiency in gray wastewater treatment. Thomaidi *et al.* (2022) indicated that 20-cm deep substrates of perlite (63–75% of TSS, 54–67% of COD, 34–48% of TN, 29–38% of TP) resulted in a higher pollutant removal efficiency compared to 10-cm deep substrates (39–55% of TSS, 41–45% of COD, 23–30% of TN, 28–31% of TP) in green roofs. Gong *et al.* (2020) demonstrated that TN removal in green roofs increased by 15% with a 50 mm increase in substrate depth. Overall, substrate depth has a significant impact on plant survival and the treatment performance of green roofs.

7.2.4.3 Effects of hydraulic loading rate

Hydraulic loading rate (HLR) is a key design parameter that impacts the treatment performance of both wetland roofs and green roofs. Rahman *et al.* (2023) indicated that the highest removal efficiency (over 90% for nutrients and organic compounds) was achieved at an HLR of 67–80 L/m^2/day in pilot wetland roofs treating gray wastewater from households, which operated within a range of 10–109 L/m^2/day. An increase in HLR within the range of 80–109 L/m^2/day resulted in a decrease in the biochemical oxygen demand (BOD_5) removal rate from 40 to 20 g BOD_5/m^2/L. From Rahman *et al.*'s study (2023), it can be inferred that a higher HLR contributes to a reduction in hydraulic retention time (HRT), consequently slowing down the biodegradation of organic pollutants.

Furthermore, the treatment of nitrogen compounds also requires lower HLR values to achieve a better efficiency (Rahman *et al.*, 2023). Therefore, determining an appropriate HLR value for green roofs is essential.

7.2.4.4 Effects of feeding patterns

Wetland roofs or green roofs have two feeding modes: intermittent or continuous. A feeding mode has been shown to significantly impact the denitrification process and COD removal efficiency. Nguyen *et al.* (2021) found that the intermittent feeding pattern (96%) had a significantly higher COD removal efficiency than the continuous feeding (92%). Moreover, TN removal with intermittent feeding (80–87%) was higher than that with continuous feeding (73–80%) under the same plants and substrate conditions in wetland roofs (Nguyen *et al.*, 2021). From the results of the study, it is clear that an intermittent feeding pattern enhances oxygen transfer and diffusion, leading to the improvement of ammonium removal by nitrification processes. Additionally, an intermittent feed of wastewater to systems provides a larger volume of the green roof being exposed to the atmosphere, increasing wastewater–biomass contact (Vo *et al.*, 2019).

7.2.4.5 Effects of climatic conditions and seasons

Dry climates have been shown to exhibit higher stormwater retention performance in green roofs compared to wet climates. Sims *et al.* (2016) demonstrated that Calgary (semi-arid, continental) had significantly better stormwater retention percent than both London (humid, continental) and Halifax (humid, maritime) (67, 48, and 34%, respectively). Zhang *et al.* (2014) indicated that the concentrations of TN and TP in green roof runoff were significantly lower in the summer and higher in the spring. Another study conducted by Gong *et al.* (2020) showed that seasons affected the effluent quality of green roofs. The effluent concentrations of nitrate-nitrogen (NH_3^+-N), TN, and TP were higher in the spring and lower in the summer and fall, whereas the ammonium-nitrogen (NH_4^+-N) concentration was higher in the fall. These results could be explained by the different evapotranspiration rates as well as rainfall intervals and intensities in different seasons. The summer had a higher rainfall depth and shorter rainfall intervals than the spring, leading to NO_3^--N leaching and subsequently higher effluent concentrations in summer. The effluent NH_4^+-N concentration was higher in fall as plants release NH_4^+-N through a decay process (from September to October) (Gong *et al.*, 2020; Zhang *et al.*, 2014).

7.3 DESIGN AND OPERATING PARAMETERS

7.3.1 Selection of plants

Plants or vegetation are selected based on the following criteria: ease of growth, good coverage (green area), shallow or fibrous roots, low growth, thriving or active throughout the years/seasons, light weight at maturity, drought and heat tolerance, wind and heavy rain resistance, aesthetic values, preferably locally available, low cost, and ability to treat wastewater (phytoremediation) (Shahmohammad *et al.*, 2022). Many plant species are suitable for green roofs, including species belonging to the Crassulaceae, Asteraceae, Poaceae, Apocynaceae, Scrophulariaceae, Plantaginaceae, Caprifoliaceae, Lamiaceae, Fabaceae, and Portulacaceae families (Shahmohammad *et al.*, 2022). However, when considering the geographical scope and purposes of green roofs (e.g., landscaping, carbon fixation, water retention, and wastewater treatment), priority is given to native plant species due to their adaptation to local climate conditions and the potential for reducing irrigation and maintenance costs. Figure 7.5 shows suggested vegetation for the application of suitable green roofs.

(a) *Sedum spurium* (Crassulaceae)
(b) *Kalanchoe blossfeldiana* (Crassulaceae)
(c) *Centratherum punctatum* (Asteraceae)
(d) *Melampodium paludosum* (Asteraceae)
(e) *Vernonia elliptica* (Asteraceae)
(f) (f) *Axonopus compressus* (Poaceae)
(g) *Catharanthus roseus (L.) G. Don* (Apocynaceae)
(h) *Myoporum parvifolium* (Scrophulariaceae)
(i) *Plantago lanceolata* (Plantaginaceae)
(j) *Lonicera pileata* (Caprifoliaceae)
(k) *Satureja repandens* (Lamiaceae)
(l) *Origanum vulgare* (Lamiaceae)
(m) *Arachis pintoi* (Fabaceae)
(n) *Portulaca grandiflora* (Portulacaceae)
(o) *Portulaca quadrifida L.* (Portulacaceae)

Figure 7.5 Suitable plants for green roofs.

7.3.2 Selection of substrates

The selection of an appropriate substrate is crucial for green roofs due to its impact on the building structure. Additionally, a green roof substrate also affects stormwater retention and plant survival (Vijayaraghavan, 2016). A substrate should be chosen based on weight, hydraulic conductivity, moisture retention, longevity, plant support capacity, and adsorption capacity (Vijayaraghavan and Raja, 2014). Substrates provide physical support for plant growth, serve as a carrier for biofilm growth, and function as an adsorbent for pollutant treatment, sedimentation, and filtration through systems (Mihalakakou *et al.*, 2023). Therefore, the medium for a substrate layer is often a combination of organic and inorganic materials. The mass ratio between inorganic and organic materials is usually around 4:1, where inorganic materials create long-term stability and organic materials reduce substrate weight while increasing water retention of green roof substrates. Most substrates are divided into three types based on their origin and category, including natural, industrial, and artificial substrates (Vijayaraghavan and Raja, 2014). The ideal substrate characteristics have the following properties (Vijayaraghavan, 2016): (i) good anchorage to plants and ability to support a wide range of plants; (ii) locally available, low cost, and light weight; (iii) high water-holding capacity (WHC), hydraulic conductivity, high porosity, sorption capacity, and stability; (iv) effective adsorbent; and (v) low organic content, low bulk density, and minimal leaching.

Some commonly used media for green roofs are presented in Table 7.1 and Figure 7.6. Recently, biochar has been used as a substrate for green roofs because it has the ability to increase both WHC and plant-available water (Vijayaraghavan, 2016). In addition, organic wastes (e.g., fine coir, coarse coir, and composted green waste) have been applied to create lightweight green roof substrates (Vijayaraghavan and Raja, 2014).

7.3.3 Design and operating parameters

When designing green roof systems, HLR is considered as the basis for determining the required area of green roof systems. Additionally, HLR is related to treatment performance. A higher HLR is preferred because it has the ability to collect a large volume of wastewater. However, higher HLRs increase the flow velocity of wastewater, limiting the contact time between microorganisms and pollutants in wastewater, which reduces the HRT and the removal efficiency (Rahman *et al.*, 2023). Overall, an appropriate range of HLR must be carefully selected. To achieve good wastewater treatment performance, the design parameters of a green roof are proposed as an HLR in the range of 1000–4170 L/m^2/day, an organic loading rate (OLR) in the range of 120–490 g COD/m^2/day, and an HRT in the range of 14–96 h (Table 7.2).

Table 7.1 Substrates and their characteristics.

No.	Substrates	Characteristics			Reference
		Diameter (mm)	Density (kg/m^3)	Porosity (%)	
1	Cobble	80–100	270	49	Vo *et al.* (2024)
2	Oyster shell	200–500	250	88	Vo *et al.* (2024)
3	Coconut shell	10–20	600–800	15–25	Ntoulas *et al.* (2015)
4	Ceramsite	30–40	330	>36	Cheng *et al.* (2014)
5	Pumice	0.05–8	N/A	N/A	Ntoulas *et al.* (2015)
6	Zeolite	0.8–2.5	N/A	N/A	Ntoulas *et al.* (2015)
7	Soil	N/A	1000–1500	N/A	Vo *et al.* (2024)
8	Charcoal	9–10	180–220	38	Vo *et al.* (2018)
9	Mussel shell	90–140	2620	87–89	Vo *et al.* (2024)

N/A, not available.

Figure 7.6 Substrates used in green roofs for wastewater treatment.

Table 7.2 Design parameters of green roofs for wastewater treatment.

Design Parameter	Criteria	Reference
HLR (L/m^2/day)	1000–4170	Bui *et al.* (2012); Masi *et al.* (2016); Nguyen *et al.* (2021); Van *et al.* (2015); Vo *et al.* (2018); Zapater-Pereyra *et al.* (2016)
OLR (g COD/m^2/day)	120–490	Bui *et al.* (2012); Nguyen *et al.* (2021); Van *et al.* (2015); Vo *et al.* (2018); Zapater-Pereyra *et al.* (2016)
HRT (h)	14–96	Bui *et al.* (2012); Fowdar *et al.* (2017); Zapater-Pereyra *et al.* (2016)
Materials for substrate	Cobble, oyster shell, coconut shell, ceramsite, pumice, zeolite, soil, charcoal, mussel shell	Cheng *et al.* (2014); Ntoulas *et al.* (2015); Vo *et al.* (2018, 2024)
Plant families	Crassulaceae, Asteraceae, Poaceae, Apocynaceae, Scrophulariaceae, Plantaginaceae, Caprifoliaceae, Lamiaceae, Fabaceae, Portulacaceae	Shahmohammad *et al.* (2022)

7.4 BENEFITS AND CHALLENGES OF GREEN ROOFS

7.4.1 Benefits of green roofs

7.4.1.1 Water quality improvement

Domestic wastewater, as a polluted source, needs to be treated before being discharged into the environment. Urban residents release a large amount of domestic wastewater, placing a burden on wastewater treatment plants. Rainwater also contains pollutants due to atmospheric deposition, vehicular transportation, and industrial activities (Liu *et al.*, 2021b). During heavy precipitation events, stormwater can exceed the capacity of wastewater treatment plants, resulting in untreated stormwater flowing directly into rivers, lakes, and oceans. This problem is one of the causes of environmental pollution. Using wastewater for green roof irrigation is an environmentally sustainable solution, especially in areas with water shortages (Shahmohammad *et al.*, 2022). Green roofs have been demonstrated as nature-based solutions to not only take advantage of wastewater resources for irrigation, but also reduce costs for wastewater treatment (Liu *et al.*, 2021b; Mihalakakou *et al.*, 2023). Green roofs irrigated with domestic wastewater showed a high pollutant removal efficiency for domestic wastewater treatment (79–97% of TSS, 80–99% of COD, 65–91% of TN, 83–96% of NH_4^+-N, 78–94% of NO_3^--N, and 94–99% of TP) (Liu *et al.*, 2021b). Another study of green roofs planted with *A. halimus* for gray wastewater treatment also showed a high COD (91%), BOD (91%), TSS (93%), and turbidity (93%) removal efficiency (Thomaidi *et al.*, 2022). Plants and soil substrates of green roofs can enhance rainwater quality by absorbing and filtering pollutants (Mihalakakou *et al.*, 2023). A green roof using *Ipomoea imperati* and *Sphagneticola trilobata* decreased the concentration of COD up to 99% and BOD to 17% in effluent stormwater (Romali *et al.*, 2021).

7.4.1.2 Retention of rainwater runoff

Rapid urbanization has led to an increase in impervious surfaces such as asphalt and concrete. Due to the impermeable nature of these surfaces, rainwater flows quickly and accumulates into drainage systems instead of percolating through the ground, causing floods and erosion. Green roofs can mitigate this issue by retaining stormwater and attenuating the peak flow. Water retained by substrates and plants will ultimately evaporate or return to the atmosphere through transpiration (Shafique *et al.*, 2018). The extent of runoff reduction depends on various factors, including plant species, substrate thickness, drainage materials, the slope of green roofs, and rainfall intensity (Shafique *et al.*, 2018). Another study on green roofs in Poland indicated that the water-retention performance was up to 65% on the surface (Pęczkowski *et al.*, 2020). In green roofs, plant species significantly reduce rainwater runoff through canopy interception. A study conducted in China investigated the water retention and hydrological performance of four turfgrass species (*Agrostis stolonifera*, *F. arundinacea*, *P. pratensis*, and *Lolium perenne*) and two succulent plant species (*C. repens* and *S. lineare*). The results showed that *A. stolonifera* had the best water-retention capacity, contributing from 20% (in the spring) to 23% (in the winter) of increased rainwater runoff retention capacity compared to unvegetated roofs. *F. arundinacea* exhibited the highest evapotranspiration in all four seasons, contributing from 47% (in the summer) to 116.7% (in the winter) of increased water loss compared to the non-vegetated control (Li *et al.*, 2018).

7.4.1.3 Reduction of heat island effects

The heat island phenomenon refers to urbanized areas having higher temperatures than the surrounding rural areas. This increase in air temperature results from the absorption of solar and infrared radiation, which dissipates into the atmosphere. The heat island exacerbates the climate and compromises human health (Filho *et al.*, 2017). Green roofs contribute to reducing heat island effects through vegetation and substrate layers, which facilitate heat and water transfer processes. In green roofs, plants can absorb 60% of the total solar radiation, whereas only 13% can penetrate the substrate (Ziaee *et al.*, 2022). Through radiation and convection, a portion of the absorbed heat returns to the atmosphere. Loss of water by the substrate through evaporation and by plants through transpiration increases air humidity, exerting a cooling effect and reducing urban heat island effects. The heat island effect has caused many cities in America to experience higher air temperatures, up to 5.6°C compared to the surrounding natural land cover (Filho *et al.*, 2017). A study in Duhok City demonstrated that green roofs can reduce temperatures by 2–3°C between urban centers and the surrounding areas (AbdulBaqi, 2022). Additionally, the average temperature on buildings covered with green roofs was recorded at 35°C, compared to 38°C on traditional roofs (AbdulBaqi, 2022).

7.4.1.4 Mitigation of air pollution

Overpopulation and unplanned urbanization worsen air quality, negatively impacting the physical and mental health of hundreds of millions of people worldwide. In recent years, the number of deaths caused by lung cancer and respiratory diseases has significantly increased due to air pollution. Specifically, there were 275,000 premature deaths related to exposure to NO_2 and particulate matter (PM2.5), and 28,000 to ozone (O_3) across 41 countries (Soares *et al.*, 2022). Green roofs can reduce air pollution by absorbing gaseous pollutants through the stomata of plants and intercepting particulate matter with leaves (Mihalakakou *et al.*, 2023). Additionally, the evapotranspiration of plants and the shade created by the plants on green roofs also lower air temperatures, thereby reducing air pollutants (Kostadinović *et al.*, 2023).

Carbon dioxide (CO_2) contributes to the warming of the Earth's surface, leading to climate change. Green roofs can reduce CO_2 emissions in two ways. First, CO_2 is removed through the photosynthesis process; plants absorb CO_2 while releasing half of it back into the atmosphere through respiration. The total yearly carbon sequestration of three grass species (*F. arundinacea*, *C. dactylon*, and *Zoysia matrella*) for irrigation treatment was ~2.5 kg CO_2/m^2/year, higher than that of *Sedum aizoon* (1.7 kg CO_2/m^2/year) (Kuronuma *et al.*, 2018). Second, green roofs act as excellent insulators for houses, thereby reducing the energy needed for cooling and heating. This helps reduce CO_2 emissions, which are by-products of combustion with fossil fuels used to produce energy for humans in thermal power plants and individual furnaces (Kostadinović *et al.*, 2023; Mihalakakou *et al.*, 2023).

Comparative studies of air quality between buildings with integrated green roofs and those without in Barangaroo (Sydney) showed that the average concentration of O_3 on green roofs was lower. Green roofs can remove 2.3, 6.9, and 0.5 kg/year of NO_2, O_3, and PM2.5, respectively (Irga *et al.*, 2022). A 1-year-old green roof in Belgrade showed lower concentrations of PM10, PM2.5, and PM1.0 than conventional roofs (Kostadinović *et al.*, 2023). Another study in Ontario (Canada) revealed that green roofs reduced the average concentration of O_3, NO_2, and CO_2 by 31, 65,/ and 6%, respectively (Anderson and Gough, 2020).

7.4.1.5 Biodiversity

Urbanization has posed a threat to biodiversity due to human-driven activities such as construction, deforestation, and the exploitation of natural resources. These activities lead to the destruction of natural habitats for plants and animals. Additionally, population growth limits the availability of green spaces in cities. Green roofs partially mitigate the loss of green spaces and provide substitute habitats for flora and fauna (Wooster *et al.*, 2022). Green roofs have demonstrated the potential for developing biologically diverse cities on a widespread scale. Figure 7.7 shows some insects, birds, and mollusk that have been observed on green roofs. They support more than four times the bird diversity and seven times the arthropod diversity compared to traditional roofs. Furthermore, green roofs provide habitat for gastropods, which are not typically found on traditional roofs. These roofs also attract rare and unexpected arthropods, such as metallic shield bugs (*Scutiphora pedicellata*) and blue banded bees (*Amegilla cingulata*) (Wooster *et al.*, 2022).

A study by Partridge and Clark (2018) at four sites throughout New York City evaluated the biodiversity of green roofs compared to traditional roofs. During spring migration, more bird species were recorded on green roofs (~20 species per roof) than on traditional roofs (~8 species per roof). A total of 41 bird species was recorded on green roofs, compared to 14 species on traditional roofs. Additionally, over 2 consecutive years, eight roofs (four green roofs and four traditional roofs) hosted a total of 13,313 arthropods from 17 taxa. Green roofs collected around 9939 individuals of arthropods and 17 taxa, whereas traditional roofs recorded 3374 individuals of arthropods and 9 taxa.

7.4.1.6 Reduction of noise

Noise pollution has emerged as one of the leading environmental causes of health issues after air pollution (Irga *et al.*, 2022). It originates from various sources such as transportation, industry, and

Figure 7.7 Biodiversity on two pilot-scale green roofs in Ho Chi Minh City (Vietnam).

community activities, including traffic noise, catering and nightlife, animals, and construction sites. In terms of its impact on human health, noise pollution not only leads to hearing loss but also causes psychological trauma (Mihalakakou *et al.*, 2023).

Green roofs have been studied for their potential to reduce noise levels through transmission, absorption, and diffraction mechanisms. The plants and substrate layers of green roofs can absorb sound waves and prevent them from being redirected back into the environment (Mihalakakou *et al.*, 2023). Green roofs are effective in reducing the diffraction and transmission of sound waves. Research has shown that the soil and plants of green roofs can absorb and trap sound wave frequencies. Green roofs can reduce noise heard near or within a building by 20–30 dB, compared to 10–20 dB by non-vegetated roofs (Van Renterghem, 2018).

7.4.1.7 Economic benefits

Besides environmental and social benefits, green roofs also have the potential for economic benefits. First, green roofs help to save energy costs. Acting as a thermal insulating barrier, green roofs reduce heat transfer between the building and the external environment. Therefore, the energy used for cooling and heating in buildings can be reduced through cooling effects facilitated by soil reactions, photosynthetic processes, and evapotranspiration during summer. In winter, the vegetation and substrate of green roofs protect building and reduce wind speed, saving energy required for heating rooms (Ziaee *et al.*, 2022). A green roof can reduce heat flow by 70–90% in summer and by 10–30% in winter (Besir and Cuce, 2018), saving 45% of total energy compared to conventional roofs (Bevilacqua, 2021). A case study in Italy showed an annual economic benefit of using green roofs, saving heating energy costs of €0.19–0.28/m^2/year and cooling costs of €1.62–2.69/m^2/year (Rosasco and Perini, 2019).

In addition, green roofs also provide intangible economic value by improving the aesthetic aspects of buildings and the surrounding area, which in turn lead to economic benefits. It has been reported that a green roof can contribute to an increase in real-estate value ranging from 4–20% (Rosasco and Perini, 2019). Studies conducted in Czechia and Portugal indicated that buildings with green roofs saw a 5% increase in property value and a 1% increase in rental value. In addition, surrounding buildings that combined installation of green roofs with a view of green roofs experienced a 5% increase in property value and a 5% increase in rental value (Hekrle *et al.*, 2023). Moreover, some plant species can be used to produce bioenergy which is an alternative to fossil fuels (Rozina *et al.*, 2024).

Treatment of gaseous pollutants by green roofs also brings economic value. It was estimated that the absorption of NO_2, sulfur dioxide (SO_2), O_3, and PM10 by green roofs could save €12/m^2 in the first year and €0.11/m^2/year thereafter (Hekrle *et al.*, 2023).

7.4.2 Challenges of green roofs

In addition to their advantages, green roofs also present some difficulties that should be considered. Initially, the costs of green roofs are quite high. The installation of green roofs requires a more significant investment and operation cost compared to conventional roofs. A study conducted in Washington, D.C. showed that installing a 1795 m^2 green roof costs 27% more than conventional roof installation in the same area (Shafique *et al.*, 2018). In Canada, specifically British Columbia, the initial construction cost of a typical large green roof ranged from \$130 to \$165/m^2 (\$12 to \$15/ft^2) (Bianchini and Hewage, 2012). Besides, green roofs may also incur additional costs such as operation, maintenance, replacement, and disposal costs. An estimation of operation and maintenance cost of green roofs ranged from \$0.7 to \$13.5/m^2 (\$0.06 to \$1.25/ft^2) per year (Bianchini and Hewage, 2012). The thickness of the soil and heavy plants could overload the roofs of old buildings. When green roofs are not properly designed, this can cause damage to the structural integrity of buildings (Shafique *et al.*, 2018). It is essential to ensure that the roof has sufficient strength to support the weight of a total green roof system before installation.

This could lead to paying more to reinforce the home. Green roofs might become sources of pollution (Liu *et al.*, 2021a). The use of fertilizer for cultivation of plants on green roofs will influence the quality of water from the system. A study indicated that nutrient contents in the leachate of green roofs are higher than those in non-vegetated roofs and rainwater (Marín *et al.*, 2023). A comparison of TP concentrations in drainage water from green roofs and traditional roofs in a study in Porto Alegre (southern Brazil) showed that the TP concentration (1.10 mg/L) in leachate from green roofs was higher than that from traditional roofs (0.10 mg/L) (Castro *et al.*, 2020). Another study in Taiwan showed that the TP concentration in the effluent of green roofs (0.11 mg/L) was much higher than that in rainwater (0.01 mg/L). The TN concentration in runoff from green roofs was 1.71 mg/L, compared to a concentration of 0.59 mg/L TN in conventional roofs (Marín *et al.*, 2023). Therefore, it is necessary to consider the use of fertilizer on green roofs to avoid negative impacts on the effluent from the system.

The presence of pollutants in leachate might be due to the type of material used in green roofs (Marín *et al.*, 2023). Polymer materials are commonly used for green roofs, but these might cause pollution. Most components in green roofs are mainly made of polymer materials, which can cause pollution. This material is easily damaged over time and can release pollutants when incinerated. Finding a sustainable substitute for these polymer materials is necessary to enhance the performance of green roofs over a long-term period without causing pollution problems (Shafique *et al.*, 2018). Furthermore, consumer psychology is one of the challenges that affect how individuals make decisions about using green roofs. Green roofs act as wastewater treatment systems on roofs. A common concern is the worry of having a wastewater roof over their heads.

7.5 CONCLUSION AND PERSPECTIVES

In developing countries, improperly controlled urbanization is significantly exacerbating the air and water environment, causing deterioration in life quality and urban landscape. Green roofs are one of the nature-based solutions available that can help to achieve sustainable urban development and enhance human health and well-being. Applying green roofs brings various benefits, including water quality improvement, rainwater runoff retention, reduction of heat island effects, mitigation of air pollution, biodiversity, noise reduction, and economic benefits. Future studies on the long-term and large-scale green roof systems should be implemented to gain a comprehensive insight of the holistic potential of green roofs under different environmental conditions. Incorporation of a wide range of climate conditions, geographic locations, building structure, vegetations, and substrates provides global users more specific knowledge to apply green roof systems successfully. Furthermore, future works should also focus on energy-saving benefits of green roof systems for urban sustainability and climate-adaptation solutions.

ACKNOWLEDGEMENTS

We acknowledge Ho Chi Minh City University of Technology (HCMUT) and VNU-HCM for supporting this study. Professor Jegatheesan would like to appreciate funding from APN (Project number: CRRP2021-06MY-Jegatheesan).

REFERENCES

AbdulBaqi F. K. (2022). The effect of afforestation and green roofs techniques on thermal reduction in Duhok City. *Trees, Forests and People*, **8**, 100267, https://doi.org/10.1016/j.tfp.2022.100267

Anderson V. and Gough W. A. (2020). Evaluating the potential of nature-based solutions to reduce ozone, nitrogen dioxide, and carbon dioxide through a multi-type green infrastructure study in Ontario, Canada. *City and Environment Interactions*, **6**, 100043, https://doi.org/10.1016/j.cacint.2020.100043

Arden S. and Ma X. (2018). Constructed wetlands for greywater recycle and reuse: a review. *Science of the Total Environment*, **630**, 587–599, https://doi.org/10.1016/j.scitotenv.2018.02.218

Besir A. B. and Cuce E. (2018). Green roofs and facades: a comprehensive review. *Renewable and Sustainable Energy Reviews*, **82**, 915–939, https://doi.org/10.1016/j.rser.2017.09.106

Bevilacqua P. (2021). The effectiveness of green roofs in reducing building energy consumptions across different climates. A summary of literature results. *Renewable and Sustainable Energy Reviews*, **151**, 111523, https://doi.org/10.1016/j.rser.2021.111523

Bianchini F. and Hewage K. (2012). How 'green' are the green roofs? Lifecycle analysis of green roof materials. *Building and Environment*, **48**, 57–65, https://doi.org/10.1016/j.buildenv.2011.08.019

Bui X.-T., Hien V. T. D., Dan N. P., Van P. T. H. and Tin N. T. (2012). Performance of wetland roof treating domestic wastewater in the tropic urban area. *Journal of Water Sustainability*, **2**(1), 79–86.

Cascone S. (2019). Green roof design: state of the art on technology and materials. *Sustainability*, **11**(11), 3020, https://doi.org/10.3390/su11113020

Castro A. S., Goldenfum J. A., da Silveira A. L., DallAgnol A. L. B., Loebens L., Demarco C. F., Leandro D., Nadaleti W. C. and Quadro M. S. (2020). The analysis of green roof's runoff volumes and its water quality in an experimental study in Porto Alegre, southern Brazil. *Environmental Science and Pollution Research*, **27**(9), 9520–9534, https://doi.org/10.1007/s11356-019-06777-5

Cheng Y., Fan W. and Guo L. (2014). Coking wastewater treatment using a magnetic porous ceramsite carrier. *Separation and Purification Technology*, **130**, 167–172, https://doi.org/10.1016/j.seppur.2014.04.030

Filho W. L., Icaza L. E., Emanche V. O. and Al-Amin A. Q. (2017). An evidence-based review of impacts, strategies and tools to mitigate urban heat islands. *International Journal of Environmental Research and Public Health*, **14**(12), 1600, https://doi.org/10.3390/ijerph14121600

Fowdar H. S., Hatt B. E., Breen P., Cook P. L. M. and Deletic A. (2017). Designing living walls for greywater treatment. *Water Research*, **110**, 218–232, https://doi.org/10.1016/j.watres.2016.12.018

Gong Y., Zhang X., Li J., Fang X., Yin D., Xie P. and Nie L. (2020). Factors affecting the ability of extensive green roofs to reduce nutrient pollutants in rainfall runoff. *Science of The Total Environment*, **732**, 139248, https://doi.org/10.1016/j.scitotenv.2020.139248

Hekrle M., Liberalesso T., Macháč J. and Matos Silva C. (2023). The economic value of green roofs: A case study using different cost–benefit analysis approaches. *Journal of Cleaner Production*, **413**, 137531, https://doi.org/10.1016/j.jclepro.2023.137531

Huang P., Shen F., Abbas A., Wang H., Du Y., Du D., Hussain S., Javed T. and Alamri S. (2022). Effects of different nitrogen forms and competitive treatments on the growth and antioxidant system of *Wedelia trilobata* and *Wedelia chinensis* under high nitrogen concentrations. *Frontiers in Plant Science*, **13**, 851099, https://doi.org/10.3389/fpls.2022.851099

Irga P. J., Fleck R., Arsenteva E. and Torpy F. R. (2022). Biosolar green roofs and ambient air pollution in city centres: mixed results. *Building and Environment*, **226**, 109712, https://doi.org/10.1016/j.buildenv.2022.109712

Ji Z., Tang W. and Pei Y. (2022). Constructed wetland substrates: a review on development, function mechanisms, and application in contaminants removal. *Chemosphere*, **286**, 131564, https://doi.org/10.1016/j.chemosphere.2021.131564

Kostadinović D., Jovanović M., Bakić V. and Stepanić N. (2023). Mitigation of urban particulate pollution using lightweight green roof system. *Energy and Buildings*, **293**, 113203, https://doi.org/10.1016/j.enbuild.2023.113203

Kuronuma T., Watanabe H., Ishihara T., Kou D., Toushima K., Ando M. and Shindo S. (2018). CO_2 payoff of extensive green roofs with different vegetation species. *Sustainability*, **10**(7), 2256, https://doi.org/10.3390/su10072256

Li X. X., Cao J. J., Xu P. X., Fei L., Dong Q. and Wang Z. L. (2018). Green roofs: effects of plant species used on runoff. *Land Degradation and Development*, **29**(10), 3628–3638, https://doi.org/10.1002/ldr.3102

Liu H., Kong F., Yin H., Middel A., Zheng X., Huang J., Xu H., Wang D. and Wen Z. (2021a). Impacts of green roofs on water, temperature, and air quality: a bibliometric review. *Building and Environment*, **196**, 107794, https://doi.org/10.1016/j.buildenv.2021.107794

Liu L., Cao J., Ali M., Zhang J. and Wang Z. (2021b). Impact of green roof plant species on domestic wastewater treatment. *Environmental Advances*, **4**, 100059, https://doi.org/10.1016/j.envadv.2021.100059

Lyu T., He K., Dong R. and Wu S. (2018). The intensified constructed wetlands are promising for treatment of ammonia stripped effluent: nitrogen transformations and removal pathways. *Environmental Pollution*, **236**, 273–282, https://doi.org/10.1016/j.envpol.2018.01.056

Mahmoudi A., Mousavi S. A. and Darvishi P. (2021). Greywater as a sustainable source for development of green roofs: characteristics, treatment technologies, reuse, case studies and future developments. *Journal of Environmental Management*, **295**, 112991, https://doi.org/10.1016/j.jenvman.2021.112991

Marín C., El Bachawati M. and Pérez G. (2023). The impact of green roofs on urban runoff quality: A review. *Urban Forestry and Urban Greening*, **90**, 128138, https://doi.org/10.1016/j.ufug.2023.128138

Masi F., Bresciani R., Rizzo A., Edathoot A., Patwardhan N., Panse D. and Langergraber G. (2016). Green walls for greywater treatment and recycling in dense urban areas: a case-study in Pune. *Journal of Water, Sanitation and Hygiene for Development*, **6**(2), 342–347, https://doi.org/10.2166/washdev.2016.019

Mihalakakou G., Souliotis M., Papadaki M., Menounou P., Dimopoulos P., Kolokotsa D., Paravantis J. A., Tsangrassoulis A., Panaras G., Giannakopoulos E. and Papaefthimiou S. (2023). Green roofs as a nature-based solution for improving urban sustainability: progress and perspectives. *Renewable and Sustainable Energy Reviews*, **180**, 113306, https://doi.org/10.1016/j.rser.2023.113306

Nguyen T.-B., Bui M.-H., Nguyen D.-T., Nguyen D.-D. and Chang S. W. (2021). Influence of plant types, bed media and feeding patterns on wastewater treatment performance of wetland roofs. *Journal of Water Process Engineering*, **40**, 101972, https://doi.org/10.1016/j.jwpe.2021.101972

Ntoulas N., Nektarios P. A., Kapsali T.-E., Kaltsidi M.-P., Han L. and Yin S. (2015). Determination of the physical, chemical, and hydraulic characteristics of locally available materials for formulating extensive green roof substrates. *HortTechnology*, **25**(6), 774–784, https://doi.org/10.21273/HORTTECH.25.6.774

Omotade I. F., Alatise M. O. and Olanrewaju O. O. (2019). Recycling of aquaculture wastewater using charcoal based constructed wetlands. *International Journal of Phytoremediation*, **21**(5), 399–404, https://doi.org/10.1080/15226514.2018.1537247

Partridge D. R. and Clark J. A. (2018). Urban green roofs provide habitat for migrating and breeding birds and their arthropod prey. *PLoS ONE*, **13**(8), e0202298, https://doi.org/10.1371/journal.pone.0202298

Pęczkowski G., Szawernoga K., Kowalczyk T., Orzepowski W. and Pokładek R. (2020). Runoff and water quality in the aspect of environmental impact assessment of experimental area of green roofs in lower silesia. *Sustainability*, **12**(11), 4793, https://doi.org/10.3390/su12114793

Pradhan S., Al-Ghamdi S. G. and Mackey H. R. (2019). Greywater recycling in buildings using living walls and green roofs: a review of the applicability and challenges. *Science of The Total Environment*, **652**, 330–344, https://doi.org/10.1016/j.scitotenv.2018.10.226

Prodanovic V., Hatt B., McCarthy D., Zhang K. and Deletic A. (2017). Green walls for greywater reuse: understanding the role of media on pollutant removal. *Ecological Engineering*, **102**, 625–635, https://doi.org/10.1016/j.ecoleng.2017.02.045

Rahman K. Z., Chen X., Blumberg M., Bernhard K., Müller R. A., Mackenzie K., Trabitzsch R. and Moeller L. (2023). Effect of hydraulic loading rate on treatment performance of a pilot wetland roof treating greywater from a household. *Water*, **15**(19), 3375, https://doi.org/10.3390/w15193375

Romali N. S., Othman N. S. and Mhd Ramli N. N. (2021). The application of green roof for stormwater quantity and quality improvement. *IOP Conference Series: Earth and Environmental Science*, **682**(1), 012029.

Rosasco P. and Perini K. (2019). Selection of (green) roof systems: a sustainability-based multi-criteria analysis. *Buildings*, **9**(5), 134, https://doi.org/10.3390/buildings9050134

Rozina Emmanuel O., Ahmad M., Ezeji T. C., Qureshi N. and Abidin S. Z. U. (2024). Bioenergy production from waste seed oil biomass of *Cupressus sempervirens*: A strategy for reducing environmental pollution. *Sustainable Chemistry One World*, **2**, 100008, https://doi.org/10.1016/j.scowo.2024.100008

Schultz I., Sailor D. J. and Starry O. (2018). Effects of substrate depth and precipitation characteristics on stormwater retention by two green roofs in Portland, OR. *Journal of Hydrology: Regional Studies*, **18**, 110–118, https://doi.org/10.1016/j.ejrh.2018.06.008

Shafique M., Kim R. and Rafiq M. (2018). Green roof benefits, opportunities and challenges – a review. *Renewable and Sustainable Energy Reviews*, **90**, 757–773, https://doi.org/10.1016/j.rser.2018.04.006

Shahmohammad M., Hosseinzadeh M., Dvorak B., Bordbar F., Shahmohammadmirab H. and Aghamohammadi N. (2022). Sustainable green roofs: a comprehensive review of influential factors. *Environmental Science and Pollution Research*, **29**(52), 78228–78254, https://doi.org/10.1007/s11356-022-23405-x

Sims A. W., Robinson C. E., Smart C. C., Voogt J. A., Hay G. J., Lundholm J. T., Powers B. and O'Carroll D. M. (2016). Retention performance of green roofs in three different climate regions. *Journal of Hydrology*, **542**, 115–124, https://doi.org/10.1016/j.jhydrol.2016.08.055

Soares J., González Ortiz A., Gsella A., Horálek J., Plass D. and Kienzler S. (2022). Health risk assessment of air pollution and the impact of the new WHO guidelines. *Eionet Report – ETC HE, 10.*

Tang S., Liao Y., Xu Y., Dang Z., Zhu X. and Ji G. (2020). Microbial coupling mechanisms of nitrogen removal in constructed wetlands: a review. *Bioresource Technology*, **314**, 123759, https://doi.org/10.1016/j.biortech.2020.123759

Thomaidi V., Petousi I., Kotsia D., Kalogerakis N. and Fountoulakis M. S. (2022). Use of green roofs for greywater treatment: role of substrate, depth, plants, and recirculation. *Science of The Total Environment*, **807**, 151004, https://doi.org/10.1016/j.scitotenv.2021.151004

Van P. T. H., Tin N. T., Hien V. T. D., Quan T. M., Bui X.-T., Hang V. T., Tuc D. Q., Dan N. P., Van Khoa L., Le Phu V., Son N. T., Luong N. D., Kwon E., Park C., Jung J., Yoon I. and Lee S. (2015). Nutrient removal by different plants in wetland roof systems treating domestic wastewater. *Desalination and Water Treatment*, **54**(4–5), 1344–1352, https://doi.org/10.1080/19443994.2014.915767

Van Renterghem T. (2018). Green roofs for acoustic insulation and noise reduction. In: Nature Based Strategies for Urban and Building Sustainability, Butterworth-Heinemann, pp. 167–179.

Vijayaraghavan K. (2016). Green roofs: a critical review on the role of components, benefits, limitations and trends. *Renewable and Sustainable Energy Reviews*, **57**, 740–752, https://doi.org/10.1016/j.rser.2015.12.119

Vijayaraghavan K. and Raja F. D. (2014). Design and development of green roof substrate to improve runoff water quality: plant growth experiments and adsorption. *Water Research*, **63**, 94–101, https://doi.org/10.1016/j.watres.2014.06.012

Vo T. D. H., Do T. B. N., Bui X.-T., Nguyen V. T., Nguyen D. D., Sthiannopkao S. and Lin C. (2017). Improvement of septic tank effluent and green coverage by shallow bed wetland roof system. *International Biodeterioration and Biodegradation*, **124**, 138–145.

Vo T.-D.-H., Bui X.-T., Nguyen D.-D., Nguyen V.-T., Ngo H.-H., Guo W., Nguyen P.-D., Nguyen C.-N. and Lin C. (2018). Wastewater treatment and biomass growth of eight plants for shallow bed wetland roofs. *Bioresource Technology*, **247**, 992–998, https://doi.org/10.1016/j.biortech.2017.09.194

Vo T.-D.-H., Bui X.-T., Lin C., Nguyen V.-T., Hoang T.-K.-D., Nguyen H.-H., Nguyen P.-D., Ngo H. H. and Guo W. (2019). A mini-review on shallow-bed constructed wetlands: a promising innovative green roof. *Current Opinion in Environmental Science & Health*, **12**, 38–47, https://doi.org/10.1016/j.coesh.2019.09.004

Vo T. K. Q., Tran C. S., Ha T. L., Hoang Q. H., Dao T. V. H., Du M. L. and … Bui X.-T. (2024). Guide to green roofs for wastewater treatment: a Vietnam perspective. In: Water Treatment in Urban Environments: A Guide for the Implementation and Scaling of Nature-Based Solutions: Examples from South/Southeast Asia, Springer Nature, Switzerland, pp. 115–131.

Wooster E. I. F., Fleck R., Torpy F., Ramp D. and Irga P. J. (2022). Urban green roofs promote metropolitan biodiversity: a comparative case study. *Building and Environment*, **207**, 108458, https://doi.org/10.1016/j.buildenv.2021.108458

World Bank (2022). World development indicators, 2022. Data Bank. Last updated: 12/22/2022.

Ye J., Liu C., Zhao Z., Li Y. and Yu S. (2013). Heavy metals in plants and substrate from simulated extensive green roofs. *Ecological Engineering*, **55**, 29–34, https://doi.org/10.1016/j.ecoleng.2013.02.012

Zapater-Pereyra M., Lavrnić S., van Dien F., van Bruggen J. J. A. and Lens P. N. L. (2016). Constructed wetroofs: a novel approach for the treatment and reuse of domestic wastewater. *Ecological Engineering*, **94**, 545–554, https://doi.org/10.1016/j.ecoleng.2016.05.052

Zhang Q.-L., Hu P., Zhao Y., Feng G.-W., Zhang Y.-Q., Zhu B.-X. and Tao Z. (2014). Anion-controlled assembly of silver-di(aminophenyl)sulfone coordination polymers: syntheses, crystal structures, and solid state luminescence. *Journal of Solid State Chemistry*, **210**(1), 178–187, https://doi.org/10.1016/j.jssc.2013.11.027

Zhuang L.-L., Yang T., Zhang J. and Li X. (2019). The configuration, purification effect and mechanism of intensified constructed wetland for wastewater treatment from the aspect of nitrogen removal: a review. *Bioresource Technology*, **293**, 122086, https://doi.org/10.1016/j.biortech.2019.122086

Ziaee S., Gholampour Z., Soleymani M., Doraj P., Eskandani O. H. and Kadaei S. (2022). Optimization of energy in sustainable architecture and green roofs in construction: a review of challenges and advantages. *Complexity*, **2022**(1), 8534810, https://doi.org/10.1155/2022/8534810

doi: 10.2166/9781789065015_0149

Chapter 8

Plant–microbial fuel cells in green roofs of tropical urban regions for sustainable water treatment and energy generation

Que Vo Nguyen Xuan[1,2], Hoang Dung Nguyen[1,2], Nguyen Xuan Phuong Vo[3] and Tae Kwon Lee[4]*

[1]Faculty of Environment and Natural Resources, Ho Chi Minh City University of Technology (HCMUT), 268 Ly Thuong Kiet Street, District 10, Ho Chi Minh City, Vietnam

[2]Vietnam National University Ho Chi Minh City, Linh Trung Ward, Thu Duc City, Ho Chi Minh City, Vietnam

[3]Faculty of Applied Sciences, Ton Duc Thang University, No. 19 Nguyen Huu Tho Street, Tan Phong Ward, District 7, Ho Chi Minh City, Vietnam

[4]Department of Environmental and Energy Engineering, Yonsei University, Wonju 26493, Republic of Korea

**Corresponding author: vnxque@hcmut.edu.vn*

ABSTRACT

A plant–microbial fuel cell (PMFC) is a derived technology of microbial fuel cell technology that integrates plant photosynthesis and rhizosphere microbial metabolism for bioelectricity generation and organic pollutant treatment. With living plants as a structural component, PMFCs can also function as green roofs. Integration of PMFCs in a green roof is a new and innovative approach that has huge scope for sustainable development in urban areas, particularly with tropical climates. Solving of bottlenecks in system configuration and operational optimization are both essential for successful implementation and performance of PMFC–green roofs. Adaptive features of PMFC–green roofs in tropical urban regions include intensive and natural growth media, regular watering with supplementary wastewater sources, and diverse plant adaptation with superior resistance to strong irradiation and toxic pollutants in wastewaters. These critical issues determine the plant vitality and plant–microbe interactions. Plant species adopting C4 carbon fixation and crassulacean acid metabolism are prioritized to achieve high day and night transpiration rates, photosynthetic efficiency, and tolerance capacity to harsh tropical rooftop conditions. Runoff contamination and management of end-of-life disposals are specific concerns to ensure public safety and the full assessment of wastewater amended PMFC–green roofs in tropical urban regions. This chapter describes the working principles underlying PMFC technology and its possibility to integrate into tropical rooftops for a simultaneous wastewater treatment. This chapter also highlights challenges that need to be overcome for large-scale applications of PMFC–green roofs in tropical urban regions.

Keywords: Bioelectricity generation, integration configuration, plant–microbial fuel cell, sustainable development, tropical urban green roof, wastewater treatment, water–energy nexus.

8.1 INTRODUCTION

The water–energy nexus in tropical urban areas is changing in response to population growth, climate change, and technological development (Wu *et al.*, 2020). Population growth results in an increase in water and energy demand. The optimization of water and energy to support sustainable

© 2025 IWAP. This is an Open Access book chapter distributed under the terms of the Creative Commons Attribution License (CC BY-NC-ND 4.0) which permits copying and redistribution for non-commercial purposes with no derivatives, provided the work is properly cited (https://creativecommons.org/licenses/by-nc-nd/4.0/). The chapter is from the book *Nature-Based Solutions for Urban Sustainability*, Piet N.L. Lens and Xuan-Thanh Bui (Editors).

urban development is necessarily entwined: water is an important resource for energy production and fulfilling the urban life, whereas energy is consumed in every stage of the urban water cycle (Cano-Rodríguez *et al.*, 2022; Krarti & Aldubyan 2021; Liu & Mauter 2021; Qiu *et al.*, 2022; Singh & Tayal 2021; Wakeel & Chen 2016). Global challenges such as climate change additionally make the sustainable use of energy and water resources more imperative, specifically in tropical regions. In addition, dominant buildings, impermeable pavement, and artificial surfaces in cities reinforce microclimate change causing urban heat island (UHI) effects as well as the large loss of surface runoff. These effects reduce the climate regulation efficiency of rainfalls and increase the energy cost for air conditioning (Mohajerani *et al.*, 2017).

It is well known that urban green spaces have important functions, which define the quality of life in a city. Particularly, green spaces provide important ecological and environmental benefits such as climate amelioration and water remediation in tropical urban areas. The UHI effect and the huge discharge of wastewater are two issues highlighting the water–energy provision within urban planning (Wagner *et al.*, 2022). Wastewater treatment plants (WWTPs) require a high-energy demand for the operation while the increased temperature under UHI effects raises the power demand for cooling, thus the water demand for electricity production. Energy consumption of WWTPs ranges between 0.5 and 2.0 kWh/m^3, depending on the type of technology, the age of the facility, and the plant scheme (Maziotis *et al.*, 2023). One-year older WWTPs could lead to an average reduction of 0.101% of energy efficiency, probably due to the largest electricity consumption for pumping and aeration in biological processes (Longo *et al.*, 2016). The overall water demand could increase by 1.44% for each additional increase of 1°C in the city of Phoenix (Arizona, USA) during the summer peak (Guhathakurta & Gober, 2007). The simulation of summer peak water demand for a complete mitigation of the UHI effect in Rhode Island (USA) demonstrated a reduction of ~8% in total (Wagner *et al.*, 2022). A sufficient network of green spaces is indirectly beneficial to the urban water–energy nexus by solving the related issue of urban biodiversity reduction (Martens *et al.*, 2022). The expanding areas for infrastructure and facilities have resulted in a cut-down of vegetated lands. Unpaved soils in vegetated lands receiving urban combined surface runoff are hotspots of microbial communities that are critical for filtering pollutants entering local streams (Delgado-Baquerizo *et al.*, 2021; Eldridge *et al.*, 2024). The vegetation, on the contrary, functions to capture carbon dioxide (Moon *et al.*, 2024; Zhang *et al.*, 2024). Vegetated lands are ecological niches for many species in cities (Gómez-Baggethun *et al.*, 2013). Traditional green spaces (i.e., urban parks, public gardens, and street afforestation), however, constitute an issue of high maintenance cost with simultaneously low biodiversity (Castelli *et al.*, 2021). Comprehensively, nature-based alternatives integrating nature into the urban fabric are preferred to implement for short-term benefits and long-term sustainability, serving both to increase urban biodiversity and to reduce urban ecological footprints.

Among the nature-based alternatives, green roofs are potentially promising to support life in cities. Green roofs are primarily considered as a mitigation strategy for high UHI effects that intensify the energy crisis in cities (Oberndorfer *et al.*, 2007). Their vegetation is expected to adapt to cliff-like habitats, characterized by extreme drought conditions, high wind speeds, high light intensity and temperature, and shallow layers of nutrient-poor substrates. Consequently, only a few specialists can survive and persist under harsh roof conditions over long term. Non-native succulent stonecrop species (*Sedum*, *Phedimus*, or *Allium*) are widely utilized in roof gardens (Doyle, 2021; Dunnett & Kingsbury, 2004). These living plants require only a minimal intervention and maintenance. Lately, studies suggest incorporating green roofs as a conservation tool in supporting urban biodiversity (Ksiazek-Mikenas, 2014). Accordingly, the design and maintenance scheme of green roofs should be modified with special concerns to provide a myriad of living spaces for diverse native plant species and other organisms (Köhler & Ksiazek-Mikenas, 2018). As so, a variety of microhabitats with different moisture levels, slope degrees, and material thicknesses should be established within green roofs to enhance beta-diversity (Braaker *et al.*, 2017; Knapp *et al.*, 2019).

A plant–microbial fuel cell (PMFC) is a derived technology of microbial fuel cell (MFC) technology that integrates plant photosynthetic and rhizosphere microbial metabolism for bioelectricity generation and organic pollutant degradation. For better electricity generation, PMFCs are constructed under submerged soils (Kabutey *et al.*, 2019). Thus, wetland plants with high photosynthetic efficiencies usually comprise the major plant group integrated in PMFCs (Guan & Yu, 2021). With living plants as a structural component, PMFCs can also function as an urban green roof in which the household wastewater feed is converted into additional electricity (Abuseif, 2023; Guan & Yu, 2021; Helder *et al.*, 2013a, 2013b; Liu *et al.*, 2021; Tapia *et al.*, 2017, 2018). Plants require water for their normal growth. Soil water content significantly affects the PMFC performance (Nguyen & Nitisoravut, 2019). Among the plant species with distinctive photosynthetic and water-use strategies, plant species adapting C4 carbon fixation perform high transpiration rates and superior canopies during daytime, thus providing the greatest cooling effect. On the contrary, plant species with crassulacean acid metabolism (CAM) including many succulent species contribute to the significant night cooling (Cao *et al.*, 2019). The attractive application of PMFCs as green roofs for sustainable water–energy nexus in urban areas should focus on system configuration to improve plant species richness, *in-situ* performance (i.e., cooling effect, wastewater treatment, and bioelectricity generation), and full-scale feasibility. Additionally, plants and rhizosphere microbes are in positive synergistic relationship. The association allows the plants to overcome abiotic stresses such as nutrient deficiency and drought stress across a wide range of soils (Pantigoso *et al.*, 2022). By selectively recruiting microbes using root exudates, plants achieve the positive effects on nutrition sources, health, and overall productivity. Plant rhizodeposits serve as food for microbial communities including electrochemically active bacteria (EABs) via oxidation reactions, which produce electrons for bioelectricity generations of PMFCs.

The integration of PMFCs into urban green roofs in tropical regions is promising but also challenging. PMFCs are solar powered but not directly dependent on sunlight with continuous plant excretion. On the contrary, long-term operation of PMFCs under tropical rooftop conditions can be difficult due to the constraint on water resources for plant vitality during dry seasons. Regardless of the challenge mentioned, rooftop plantation for electricity production is a novel approach for urban development in tropical regions. To bring PMFC–green roof installations to the masses with value-added benefits, bottlenecks must be identified and solved (Figure 8.1). High efficiency, least maintenance, and sustainability in nature are all key aspects toward the development and practical applicability of PMFC–green roofs in tropical urban areas. Optimization of system design and operating conditions is essential for improving PMFC performance while maintaining the original functions of green roofs. In terms of system design, PMFC configurations can be simply designed to fit the state-of-the-art structural designs of green roofs. Moreover, it is necessary to configure PMFC–green roofs as a hybrid energy system that converts organic substances, in the forms of either internal plant exudates or external organic sources such as wastewater, to electricity (Lepikash *et al.*, 2024). Ideally, the designs that improve specific rhizosphere plant–microbe interactions are preferred (Rusyn, 2021). Plant ecophysiology and rhizosphere activity are essentially extended even under dry and hot conditions. Electrode placement should be carefully considered to avoid oxygen intrusion to anode (Helder *et al.*, 2013b). Electrode materials should be conductive, inexpensive, and long-lasting under the harsh rooftop conditions (Maddalwar *et al.*, 2021). Operating factors such as water level, wastewater loading rate, pH, and conductivity are highly important for PMFC performance and plant vitality, regulating the soil–root consortium relationship (Lepikash *et al.*, 2024).

If the major challenges in system design, configuration, and operation of PMFC–green roofs are solved, the integration of PMFC technology into green roofs can become a competitive strategy in the coming decades, providing outperformed advantages in the new era of water revolution and energy transition. The particular limitations of individual technologies, either PMFC or green roof, are compensated. This hybrid strategy can help to mitigate the significance of water shortages and fossil-fuel phase-out. Despite the embedded challenges, the potential of wastewater treatment and water

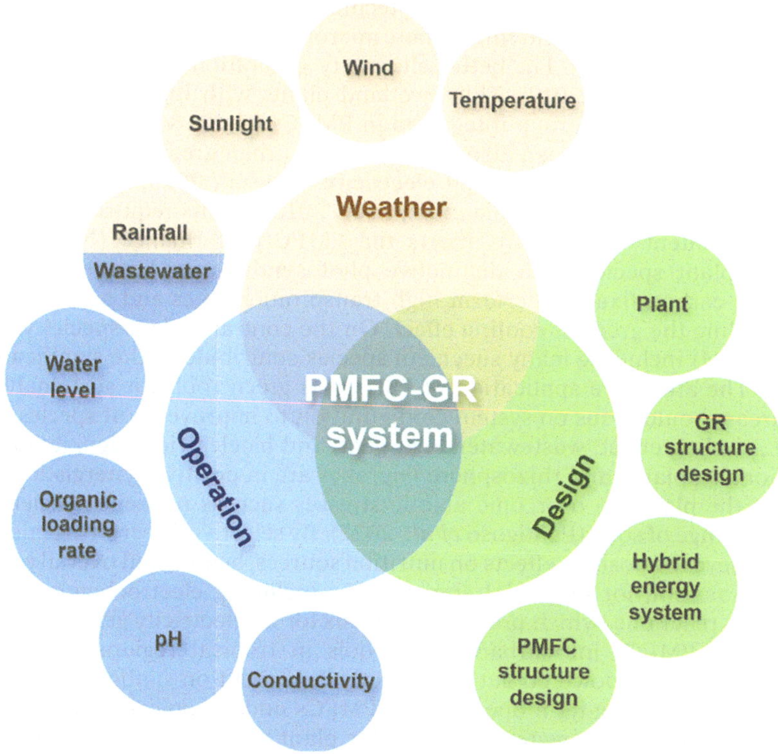

Figure 8.1 Identifying bottlenecks in PMFC–green roof development.

reuse for household purposes in coupling with the power output for local demands and the reduction of greenhouse gases prospectively distinguish the PMFC–green roof technology from conventional treatment methods and energy sources despite the low power output.

8.2 THEORETICAL BACKGROUND

8.2.1 Fundamental operational principles of PMFCs

PMFCs are innovative renewable energy systems that harness the symbiotic relationship between plants and microorganisms to generate electricity. These systems integrate the natural process of photosynthesis in plants with MFC technology to capture energy efficiently (Nitisoravut & Regmi, 2017). PMFCs operate based on the establishment of two main zones, that is, the anode and the cathode. The anode zone is typically anaerobic, present in the deeper soil layer near the plant roots. This zone is crucial for microbial colonization and electron transfer to the anode. The cathode zone is ideally aerobic and exposed to the air, supplying oxygen for reduction reactions on the cathode. This material setup allows for continuous electrochemical reactions sustained by the ongoing release of root exudates and microbial degradation processes. In PMFCs, plant roots are grown in an anode chamber where they excrete organic matter for cohabiting diverse microbial communities. The loss of soluble and insoluble organic matter by plant roots into the soil is known as rhizodeposition. Rhizodeposits comprise of exudates (sugars and organic acids), secretions (polymeric carbohydrates and enzymes), lysates (dead plant cell materials) as well as ethylene and CO_2, which are oxidized by EABs and other rhizosphere microbes. This oxidation process releases electrons, protons, and

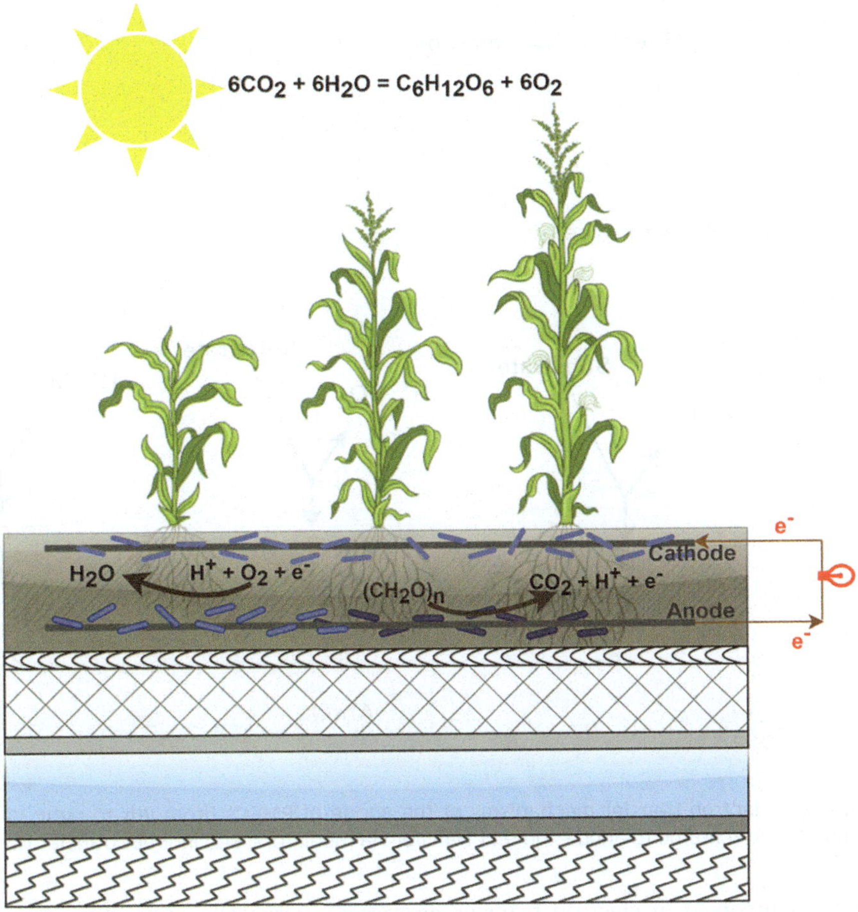

Figure 8.2 Structural integration of the PMFC in green roofs and its operational mechanism.

carbon dioxide (Figure 8.2). Electrons are transferred to the anode as an electron acceptor in PMFCs. Electrons travel continuously through an external circuit to the cathode, generating electric currents, whereas protons migrate through the soil to the cathode, where they combine with oxygen to form water (Rusyn, 2021).

Electricity generation by PMFCs depends on the electron flow harvested from microbial metabolism and their transfer to electrode surfaces. Three mechanisms of extracellular electron transfer in MFCs have been widely proposed, including (i) the direct contact between the outer-membrane cytochromes of microbes (e.g., *Shewanella* and *Geobacter* species) and the electrode surface, known as direct electron transfer mechanisms, (ii) the indirect electron transfer with the presence of natural mediators (i.e., thionine, flavin, quinine, phenoxazine, phenazine, and phenothiazine) released from microbial respiration (e.g., *Shewanella* and *Pseudomonas*), and (iii) the indirect electron transfer via cellular nanowires in the bacterial electron transport chain (e.g., *Geobacter* and *Shewanella*) (Figure 8.3) (Kabutey *et al.*, 2019). The efficiency of an extracellular electron transfer process is thus critical to the electricity generation capacity of PMFCs. Multiple factors are known to affect extracellular electron transfer processes. Temperature, pH, and redox potential are critical environmental parameters regulating the enzymatic

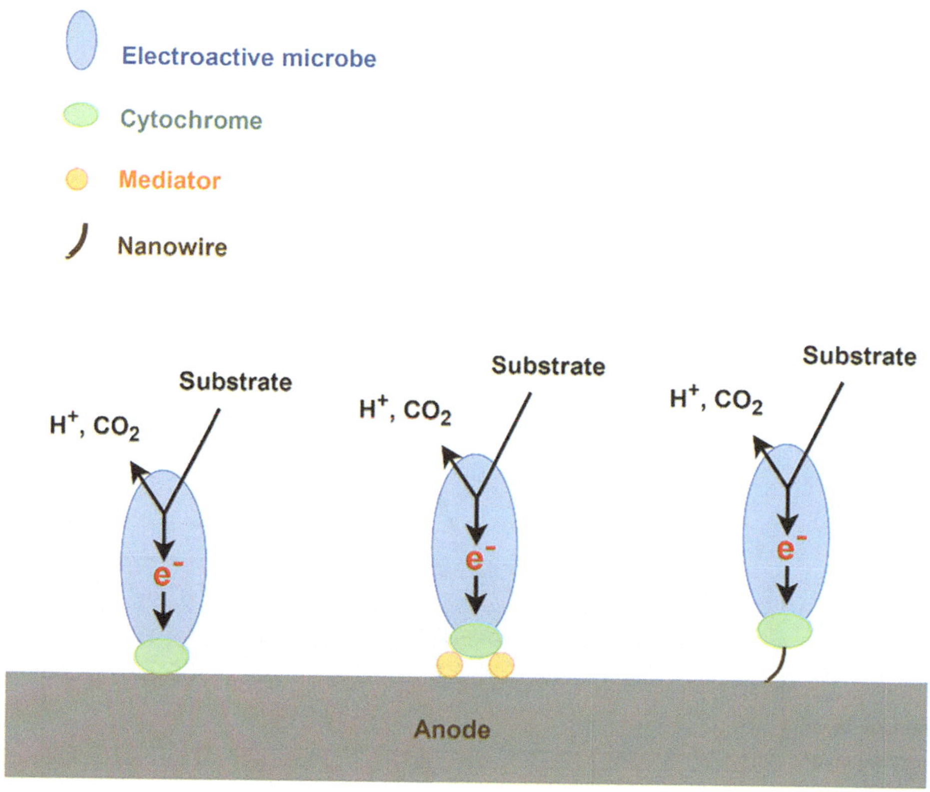

Figure 8.3 Electron transfer mechanisms at the anode of PMFCs (from left to right: direct transfer from outer-membrane cytochrome to anode; indirect transfer via mediators; and indirect transfer via cellular nanowires).

reactions involved in electron transfer processes. Increased temperature is well known to improve bioreaction rates. Enzyme activities and proton gradients across cellular membranes are pH dependent. Specifically, an increase in pH can disrupt proton gradients across microbial membranes and hinder enzyme activities. Redox potential represents the availability of electron acceptors and donors. The existence of mediators and electron acceptors and donors can significantly drive chemical reactions and electron transport to the electrode's surface (Hazzan *et al.*, 2023).

The operation of PMFCs with wastewater, for feasible integrating in tropical green roofs, requires further consideration of factors such as hydraulic retention time, organic loading rate, electrode material, and substrate toxicity. Obviously, organic loading rate and hydraulic retention determine the amount of organic matter present in a system at a given period, thus influencing the performance of PMFCs (Lepikash *et al.*, 2024; Nitisoravut & Regmi, 2017). Deterioration of electrode material is a major challenge in the long-term operation of PMFCs in salt-enriched wastewater. Ideal electrode materials must be supportive for electron transfer, chemically inert and bio-compatible toward microbes (Gupta *et al.*, 2023). Toxic constituents such as heavy metals and ammonium in wastewater may lead to acute and chronic toxicity on plants and microbes, thus impeding the performance of PMFCs (Ergettie & Dagbasi, 2021; Roy *et al.*, 2023; Sorgato *et al.*, 2023).

Electrodes play a fundamental role in electrochemical reactions, the formation of electrochemical active biofilms, and electron transfer process of PMFCs. Practically, the same material is used for both the anode and cathode. The common material used to fabricate electrodes is carbon-based materials

Table 8.1 Advantages and disadvantages of carbon-based electrode materials.

Carbon-based Materials	Advantages	Disadvantages
Carbon cloth	High surface area Relatively high porosity High conductivity Flexibility Mechanical strength	Relatively high cost
Carbon brush	Relatively high surface area High conductivity	High cost due to its titanium core
Carbon rods	Affordable cost	Low surface area, so primarily utilized as current collectors rather than electrodes
Carbon mesh	Relatively low cost	Low conductivity Low mechanical strength Low porosity
Carbon veil	Very cheap Relatively high conductivity High porosity	Quite fragile
Carbon paper	Relatively porous	Expensive Fragile
GAC	Biocompatibility Low cost Very porous High surface area	Relatively low conductivity
Granular graphite	Similar to GAC, except for lower surface area	Higher conductivity than GAC
Graphite plate	Relatively low cost High conductivity High mechanical strength	Low surface area and surface/ volume ratio
Reticulated vitreous carbon	Very conductive Great porosity	Quite fragile Very expensive

with specific advantages and disadvantages listed in Table 8.1 (Azri *et al.*, 2023). Accordingly, carbon-based electrodes can be categorized by their configuration into brush-like, fully packed, and plane structures. Carbon-based electrodes with the plane structure are made from carbon paper, graphite plate, and carbon cloth. Packed electrodes encompass graphite granules, granular rods, and granular activated carbon (GAC) with porosities. Graphite brush electrodes are created by twisting carbon fibers of specified lengths around non-corrosive titanium wires. To enhance electrode efficiency, chemical and morphological modification of the electrode surface is recommended. Current generation by PMFCs is determined by electrode surface area and pore sizes. The geometric surface area of electrodes is fundamental in accommodating microbial colonization and electron transfer from microbes. The fabrication of electrodes for surface area increase is achieved with the application of a thin layer of metals (e.g., Cu^{2+}, Cd^{2+}, and Ca^{2+}) or carbon conductive materials (e.g., graphene, carbon nanotubes, and composites between carbon nanomaterials with non-carbon materials). Large-scale PMFC–green roofs can use carbon veil electrodes with excellent porosity and relatively low resistance (Hazzan *et al.*, 2023). The distance between anodes and cathodes depends on the thickness of growing materials on green roofs and the plant root morphology. In standard designs, the anode is buried deep underground near the rhizosphere zone, and the cathode is ideally exposed to the air on the soil surface. The distance between anodes and cathodes is determined by the density of plant distribution on the green roof, which also determines the power density of PMFCs. A moderate increase in planting density could enhance the power density of PMFCs, whereas too high planting density reduces bioelectricity production due to an increased internal resistance from root occupancy and soil cracking (Chen *et al.*, 2024). To maximize the power output of PMFCs in green roofs, the stacking of PMFCs is needed. Particularly, stacking in parallel could conserve voltage between cells (Pamintuan *et al.*, 2020).

8.2.2 Taxonomy of PMFC configurations

There are several PMFC configurations with the same operation principle. The interaction between plants and microorganisms is critical for the electricity generation of PMFCs. Optimal contact between plant roots, soil, and electrodes is necessary, with carbon-based materials commonly used for anodes due to their high conductivity and ability to support microbial colonization. For cathodes, metal-based materials facilitating oxygen reduction reactions are preferred. Different configurations of PMFCs are categorized based on chamber design, structure, and plant species, with each having unique advantages and challenges. The selection of PMFC configurations depends on the specific application and the desired performance characteristics (i.e., operational simplicity, cost-effectiveness, sustainability, environment friendly, and long-term management).

PMFCs are classified as single- and double-chamber units. Single-chamber PMFCs are featured with a single-membrane-less-compartment housing both the anode and cathode. This simple design is cost-effective but may suffer from higher internal resistance due to the proximity of the electrodes. Practically, different models of the single-chamber category are used for dual benefits, that is, wastewater treatment and electricity recovery from wastewater. These models integrate PMFC units into water columns and submerged soils of constructed wetlands or green roofs, adapting plants and microbial activities for energy recovery (Shaikh *et al.*, 2021). PMFC-constructed wetlands (PMFC-CWs) combine the functions of PMFCs with constructed wetlands, providing a habitat for diverse microbial populations. Plants play a role in aeration, nutrient uptake, and extra supply of organic rhizodeposits for microbial activity, including EABs, in PMFC-CWs (Vo *et al.*, 2023). PMFC-CWs focus on wastewater treatment while producing electricity, differing from typical PMFCs, striving for green power production. A PMFC-floating wetland is a specific type of PMFC-CWs that uses floating macrophytes for pollutant removal from effluent streams, including heavy metals (Greenman *et al.*, 2024; Kabutey *et al.*, 2019; Shaikh *et al.*, 2021). The other integrated model of PMFC technology is PMFC–green roofs, which basically generates electricity based on organic compounds released by plant roots present on green roofs. This setup is originally operated under non-waterlogged conditions, offering benefits such as stormwater management and cooling effect improvement (Liao & He, 2023). However, hot and dry seasons in tropical regions necessitate the additional wastewater input in PMFC–green roofs for plant vitality and wastewater treatment, simultaneously (Abuseif, 2023; Liu *et al.*, 2021). The power output of PMFC–green roofs is relatively low (Guan & Yu, 2021; Helder *et al.*, 2013a). Despite the potential, PMFC-CWs and PMFC–green roofs remain challenges hindering their large-scale application. Some of these challenges include high internal resistance, microbial competition with diverse substrates in wastewater, and electrode biofouling (Lepikash *et al.*, 2024).

On the contrary, double-chamber PMFCs include two distinct chambers for the anode and cathode, which are separated by a proton exchange membrane. The membranes allow protons to transfer from the anode to the cathode while restricting the diffusion of oxygen from the cathode to the anode (Kabutey *et al.*, 2019). This design reduces internal resistance and improves efficiency but is more costly to install and replace membranes. Applicable models belonging to the two-chamber category are tubular and flat-plate PMFCs. Tubular PMFCs use cylindrical anodes and cathodes separated with a membrane in a rolling structure. The anode, with external contact with the rhizosphere, accepts electrons produced by EAB and transfers them to the cathode in a closed circuit. The protons generated in the anode move to the cathode and get reduced. There is a flexibility in choosing materials for the construction of tubular PMFCs, in which polyvinyl chloride is the most common material. Electrodes were made of graphite felt connected to a golden wire as a current (Wetser *et al.*, 2017). Flat-plate PMFCs involve flat anodes and cathodes placed horizontally, next to each other, in the soil and separated by the membrane. Optimization of the electrode distance and size is important for the performance improvement. However, the elongation of microbiome adaptation duration and the cathode deterioration can limit system efficiencies over long-term actions (Shaikh *et al.*, 2021).

Regardless of the potential of PMFCs for wastewater treatment and energy recovery, specific issues challenging the technology application under harsh conditions such as saline waters and drought conditions have been less discussed. Plants and microbes demonstrate different tolerances to salinity and water deficiency. Thus, the configuration of PMFCs requires careful selection of plant species. Beyond the stress on plant growth and microbial activities, drought conditions lead to a decline in bioelectricity generation by disrupting the liquid phase where the ions are transported (Chen *et al.*, 2023). Plants with drought resistance are the most selective to adapt in tropical PMFC–green roofs where a high rate of irradiance and low soil moisture characteristics are inherent. At the same time, even though high salinity positively affects PMFC performance by increasing electrical conductivity, which is beneficial for efficient electron and ion transport, PMFCs loading saline wastewaters require plants to adapt to salinity tolerance capacity (Khudzari *et al.*, 2018). The selection of suitable plant species impacts the PMFC's efficiency significantly. By refining these factors, the potential of PMFC technology can be maximized, offering sustainable solutions for energy generation and wastewater treatment.

Each configuration has unique benefits and drawbacks which necessitate careful considerations based on specific application and commercial viability. The ongoing development of PMFC configurations will provide environmentally friendly alternatives in energy and waste management. Especially, hybrid models of the PMFC technology innovate urban life with green services such as greenspace modification, sustainable energy generation, and wastewater reduction.

8.2.3 PMFCs in green roofs: a nature-based solution for tropical urban regions

The integration of PMFC technology in green roofs offers a benefit-added strategy for sustainable urban development. PMFC–green roofs are particularly well-suited for urban areas where space is limited for conventional ground-based system installations. This innovative technology generates renewable electricity by harnessing the natural interactions between plant root exudates and microbial activity, providing a complementary energy supply year-round for buildings (Guan & Yu, 2021; Maddalwar *et al.*, 2021). Additionally, PMFCs can enhance urban biodiversity, improve air quality, and provide thermal insulation, thus reducing building energy consumption for heating and cooling (Guan & Yu, 2021; Knapp *et al.*, 2019; Rusyn, 2021). In tropical urban regions prone to heavy rainfall, PMFC–green roofs can significantly aid stormwater management by enhancing water retention and reducing runoff, thereby preventing flooding and managing urban water resources more effectively (Knapp *et al.*, 2019). Even with relatively low power output compared to other renewable energy sources, PMFC–green roofs are attractive for additional benefits without compensation to the original advantages of standalone green roofs and PMFC units.

Research on PMFC–green roofs is still in the early stage, and many concerns about out-door design, long-term performance, and cost-effectiveness remain unanswered (Guan & Yu, 2021; Guan *et al.*, 2019). The performance of PMFC–green roofs can fluctuate depending on local seasonal conditions. Extreme temperature and limited water availability are important factors that reduce plant growth and microbial activities for urban greenspace provision and other life-support functions (Vaz Monteiro *et al.*, 2017). The actual electrical output of PMFCs often falls below the theoretical estimates and depends heavily on weather, making consistent energy production and production cost minimizing significantly challenging (Maddalwar *et al.*, 2021). The high initial costs associated with the installation of PMFC–green roof systems, including expenses for materials, labor for construction and design, make economic feasibility a critical consideration (Maddalwar *et al.*, 2021). The substantial initial investment and uncertain payback period pose significant barriers to widespread adoption. Additionally, similar to conventional green roofs, PMFC–green roofs require regular maintenance, including watering, fertilization, and weed control. Regular irrigation, pruning, and monitoring to ensure optimal plant vitality and growth as well as microbial activities are labor-intensive and necessitates specialized expertise (Rayner *et al.*, 2016; Tapia *et al.*, 2018; Thomaidi *et al.*, 2022; Vijayaraghavan, 2016; Zhang *et al.*, 2024).

Despite these challenges, the theoretical background and practical applications of PMFCs highlight their potential as innovative and sustainable energy systems, particularly when integrated with green roofs. With continued research and development, PMFCs in green-roof configurations could become a key component in the sustainable urban energy landscape, offering significant environmental and economic benefits.

8.3 DESIGN AND OPERATION OF PMFC–GREEN ROOFS

The combination of PMFCs in green roofs in tropical urban areas involves a multidisciplinary approach encompassing design, construction, operation, and long-term maintenance. This section delves into the intricacies of implementing PMFC technology on green roofs, highlighting the criteria for design and construction, the significance of operational optimization involving plant species and rhizosphere microorganisms, and the imperative of robust maintenance strategies to ensure long-term viability.

8.3.1 Criteria for design and construction: a multidisciplinary approach

Basically, the integration of PMFC technology into green roofs is just the involvement of electrodes in a submerged matrix of green roofs along with an external circuit to collect bioelectricity (Strik *et al.*, 2008). However, designing and constructing PMFC–green roofs necessitates a multifaceted strategy that combines elements of architecture and civil engineering, urban planning, and environmental and ecological engineering. This multidisciplinary approach aims to maximize the performance of PMFC–green roof systems while seamlessly integrating them with the urban environment for multiple benefit ecosystem services to human health and well-being (Ode Sang *et al.*, 2022).

Green roofing is currently considered as a strategic tool in urban planning for its significant role in sustainable development (Francis & Jensen, 2017). In fact, it has become an increasingly popular choice in urban planning, particularly in dense areas, over the past two decades (Dunnett & Kingsbury, 2004). In some countries such as Canada, USA, Switzerland, Hong Kong, and Japan, green roofing is law enforced for all new building constructions (Chow *et al.*, 2018). The roof is a main structural part of a building that demonstrates its relationship with the surrounding nature. Therefore, green roofs must harmonize with the aesthetic and structural aspects of the buildings they top. Achieving the integration of a PMFC unit in green roofs is thus limited by the state of the art in the structural design of green roofs (Figure 8.4). Generally, major components of green roofs including waterproofing membrane,

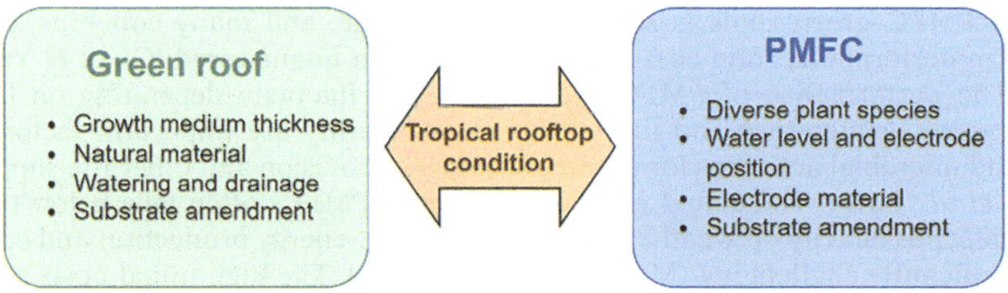

Figure 8.4 Criteria for design and construction of PMFC–green roofs under tropical rooftop conditions.

drainage layer, filtration membrane, growth substrate medium, and vegetation are arranged in the order from the structural supporting panel (Vijayaraghavan, 2016). The detailed features of green roofs are measured according to purpose of use, vegetation type, substrate thickness, supporting structural requirements, and irrigation and maintenance requirement. Particularly, for the PMFC integration into green roofs of tropical urban areas, growth substrate type and thickness, vegetation type, and seasonal irrigation scheme are critical to consider (Alqahtany, 2022).

Green roofs are classified as intensive and extensive green roofs based on the thickness of the growing substrate layer. The thickness of an extensive green roof is less than 15–20 cm, whereas the thickness of an intensive green roof can be up to 50 cm or even more (Alqahtany, 2022; Kaewpraek *et al.*, 2021). Thanks to the thicker growing layer, the intensive green roofs can support a wider variety of plant species than extensive green roofs, thus supporting biodiversity and better landscapes (Nagase *et al.*, 2018). With this advance, intensive green roofs can be superior for the integration of PMFC units. Plants in PMFCs play an important role in converting solar energy into chemical energy, that is, rhizospheric carbon sources that support the activity of electrochemically active microbes (Kabutey *et al.*, 2019). A deeper underwater matrix in intensive green roofs comprises anaerobic zones that can be coupled with the aerobic air–water interface by the PMFC electrode reactions to harvest electricity (Gupta *et al.*, 2023). In addition, intensive green roofs are especially beneficial in tropical urban regions, helping to reduce building temperature and save energy on air conditioners (Guan & Yu, 2021). Facing with heavy rains in tropical urban regions, intensive green roofs are also superior in retaining water, delaying the rainwater drainage, supporting the urban drainage system, and minimizing the risk of flooding. With excellent water retention ability, intensive green roofs can cover up to 60% of a year's rainfall (Knapp *et al.*, 2019; Liao & He, 2023).

Growth substrates in green roofs tend to be light weight, high porosity, and commonly have a low organic matter content (10% by weight) (Oberndorfer *et al.*, 2007). On the contrary, a substrate matrix supporting PMFC performance is constituted with a wide range of soil types, vermiculite, and graphite granules (Kabutey *et al.*, 2019). The type of soils inoculated in green roofs determines the living plants and microbial communities, in which natural soils are preferred with rich microbial communities, hence greatly affecting the performance of PMFC units. Natural soils such as agricultural and forest soils, rice field or paddy soil, flooded soil, red soil, peat soil, and salt marsh soil have been employed in PMFC units. The power generation of PMFCs depends on the soil physicochemical and biological properties which are determined by soil structure and texture (Nitisoravut & Regmi, 2017). Well-structured soils, not compacted or too dense, are recommended for the better growth of rhizobacterial microorganisms in PMFCs. Soil texture that includes too many sand or clay particles should be restricted because it limits access to nutrients and rhizobacterial activities. A substrate matrix can also be amended with a nitrate-less, ammonium-rich plant growth medium, composts, domestic sewage, or fermented distillery wastewater from biohydrogen production to enhance the pollutant removal and bioelectricity generation of PMFCs (Kabutey *et al.*, 2019). However, a careful amendment must be concerned to avoid the competition between the electrogenic communities and others. The organic matter compounds and nutrients in the enrichment medium will shape the microbial communities to utilize these readily available substrates rather than the organic matter derived from rhizodeposits. Denitrifiers compete directly with exoelectrogens by accepting electrons at the anode region, thereby reducing the performance of PMFCs fed with a nitrate-rich modified Hoagland solution (Lu *et al.*, 2015; Timmers *et al.*, 2012). Soil C/N ratios and soil microbial diversity were emphasized as drivers for the performance discrepancy between forest soil-based PMFCs and agricultural soil-based PMFCs (Nitisoravut & Regmi, 2017). Soil pH is also a pivotal factor affecting specific rhizobacterial activities and plant ecophysiology (Neina, 2019). Soil inorganic matter can affect the redox potential (Nanzyo & Kanno, 2018). The amount and pattern of carbon release in the form of root exudates are affected by soil pH, anoxia, and defoliation of plants (Nitisoravut & Regmi, 2017).

Although there are obvious environmental advantages, the application of intensive green roofs to buildings is associated with an additional economic aspect, which requires more expensive construction costs (Razzaghmanesh *et al.*, 2014; Silva *et al.*, 2016). The structural integrity of buildings must be evaluated to ensure they can support the added weight of PMFC–green roof systems. Engineering expertise is essential to guarantee safety. Columns, beams, and roof floors require a lot of reinforcement compared to normal designs (Bianchini & Hewage, 2012). Proper waterproofing and drainage systems are imperative to prevent water infiltration and manage excess moisture effectively. These components must be designed to coexist with PMFC technology. Waterproofing solutions such as waterproof paint, waterproofing additives for concrete, and self-adhesive waterproof membranes must also be given special attention. A drainage system on the roof (hollows, trenches) needs to be ensured effective to prevent water from seeping into the structure. In addition, transporting a large amount of soil to the rooftop also requires a lot of labor or energy from the crane (Kaewpraek *et al.*, 2021). Intensive green roofs also need a large amount of steady daily irrigation. Although this can be supported during the rainy season in tropical urban regions, feeding with household wastewater during the dry season is essential to mitigate the drought conditions while reusing wastewater for electricity recovery in integrated PMFC–green roofs (Knapp *et al.*, 2019; Liu *et al.*, 2021). With many difficulties in construction and maintenance of intensive green roofs, combining it with PMFC technology requires consideration of the payback time (Maddalwar *et al.*, 2021). In this case, it is the period of time for the bioelectricity value to be generated, the energy value saved from the cooling effect of green roofs for buildings, and the landscape benefit value equal to construction and maintenance cost during the life cycle of the building.

8.3.2 Operational optimization

8.3.2.1 The role of plant species

The success of PMFC–green roofs hinges on optimizing the biological interactions between plants and rhizosphere microorganisms. Plant species selection and the fostering of a conducive rhizosphere environment are fundamental for effective cooling effects and bioelectricity generation. Besides, a PMFC integrating with green roofs is expected to add new advantages without seriously interfering the advantages of green roofs (Helder *et al.*, 2013a). Therefore, plant species selection for PMFC–green roofs should consider the primary determining factors such as photosynthesis rate, transpiration capacity, root morphology, salt tolerance, and adaptability to specific local conditions within green roofs (i.e., waterlogged and drought conditions). In addition, the rhizosphere microbial community and its interaction with plant roots are equally significant. Plant species with the outstanding capacity in promoting rhizodeposition, which effectively fuels the exoelectrogen activity, are preferred.

Plants are categorized into three groups with different photosynthesis pathways, that is, C3, C4, and CAM (Maddalwar *et al.*, 2021). The classical C3 plants can fix carbon in the form of CO_2 through a series of dark reactions, the so-called Calvin cycle. This cycle is initiated with the binding of CO_2 molecules thanks to the ribulose-1,5-bisphosphate carboxylase enzyme (RuBisCO enzyme). The two three-carbon molecules created then undergo a series of further chemical reactions to ultimately produce a net molecule of glucose. These glucose production reactions need the activation energy from high-energy compounds such as adenosine triphosphate (ATP) and nicotinamide adenine dinucleotide phosphate (NADPH). These compounds are products of light reactions taken place at the chloroplasts of plant cells (Lopez & Barclay, 2017). C3 plants typically grow well under cool and moist conditions, thus hardly adapt to comparatively harsh rooftop conditions in tropical areas.

On the contrary, C4 plants evolve a different photosynthesis pathway with an add-on C4 carbon fixation in mesophyll cells by the phosphoenolpyruvate carboxylase enzyme in the very first step. Plant species with the C4 photosynthesis pathway develop a specific mechanism to deliver the pre-fixed CO_2 to the RuBisCO enzyme in the next step and achieve high photosynthetic efficiency. The C4 photosynthesis requires more energy (ATP and NADPH) than the C3 photosynthesis. C4 plants are able to grow under high light intensity (Lopez & Barclay, 2017). Theoretically, C4 plants never saturate with light intensity and have a maximum photosynthetic efficiency of 6.0%, higher than 4.6% of C3 plants (Deng *et al.*, 2012). C4 plants can also tolerate drought conditions by closing the stomata to limit water loss. By growing well even in hot and dry weather, C4 plants are suitable to adopt in tropical PMFC–green roofs (Wang *et al.*, 2012). The C4 photosynthesis is found only in angiosperms or vascular plants, particularly common in the *Cyperaceae* (sedges) and *Poaceae* (grasses), together making up ~79% of the C4 plant group. Some C4 plant species in green roofs such as *Cynodon dactylon* and *Eremochloa ophiuroides* achieve better cooling effects than succulent plants (Blanusa *et al.*, 2013; MacIvor *et al.*, 2018; Vaz Monteiro *et al.*, 2017). *C. dactylon* and *E. ophiuroides* provide superior canopy cooling effects than the representative C3 plants such as *Poa pratensis* and *Festuca arundinacea* during daytime, whereas CAM plant species such as *Sedum lineare* and *Callisia repens* perform the cooling effect effectively at night (Cao *et al.*, 2019). C4 plants, for example monocots and grass plants, also operate more reliably over the long term in PMFC systems than C3 and CAM plants, thanks to their longevity and vigor under harsh environmental circumstances while sustaining essential amounts of rhizodeposits (Chiranjeevi *et al.*, 2019; Helder *et al.*, 2012).

Compared to C4 plants, CAM plants can adapt better under xeric conditions by opening stomata only at night to absorb CO_2 and closing stomata during the daytime while performing photosynthesis to reduce water loss (Farrell *et al.*, 2012). CAM photosynthesis is found in a number of terrestrial and aquatic plant groups. Particularly, desert succulents (e.g., *Aizoaceae, Cactaceae, Portulacaceae, Crassulaceae,* and *Euphorbiaceae*), some epiphytic plants (e.g., *Bromeliaceae* and *Orchidaceae*), and even aquatic groups (e.g., *Crassulaceae, Isoetaceae,* and *Plantaginaceae*) are commonly adopted to grow in green roofs as they can survive under drought conditions of the shallow green-roof material (Lopez & Barclay, 2017). However, with the typical slow growth rates, CAM plants release limited amounts of rhizodeposits which are less supportive for the PMFC performance.

Overall, C4 plants are the most potential species to adapt in PMFC–green roofs in tropical areas. Some C4 plant species that have been adapted in PMFCs are listed in Table 8.2. By applying household wastewater irrigation for the PMFC–green roof operation in dry and hot seasons, the modification of rooftop plant communities with C3 and CAM plants can additionally extend the system's performance during extreme periods and potentially improve the whole ecosystem's services (Rayner *et al.*, 2016). In addition, more and more recent studies have shown that the cooling effects can be improved with diverse green roof plants (Cao *et al.*, 2019). In a recent study, wetland plants, either C3 plant species (i.e., *Rotala rotundifolia* and *Typha angustifolia*) or C4 plant species (i.e., *Pennisetum alopecuroides*) with different stem heights and seasonal growth rates, were investigated for their effects on the performance of PMFC–green roofs in Taipei City, a subtropical metropolis in Taiwan (Guan &

Table 8.2 C4 plant species adapted in PMFCs for bioelectricity generation.

Scientific Name	Common Name	Reference
Arundinella anomala	Tropical common grass	Helder *et al.* (2010)
Vetiveria zizanioides Nash	Vetiver grass	Regmi *et al.* (2018a)
Pennisetum setaceum, *P. alopecuroides*	Fountain grass	Guan and Yu (2021), Chiranjeevi *et al.* (2012)
Spartina anglica	Common cordgrass	Helder *et al.* (2010), Wetser *et al.* (2015)
Sporobolus arabicus	Arabian dropseed	Gilani *et al.* (2016)
Cyperus papyrus nanus	Dwarf papyrus	Gulamhussein and Randall (2020)
P. alopecuroides	Fountain grass	Guan and Yu (2021)
Amaranthus viridis	Slender amaranth	Arulmani *et al.* (2021)
A. calamus	Sweet flag	Yan *et al.* (2015)
C. indica	Canna lily	Lu *et al.* (2015)
Eichhornia crassipes	Water hyacinth	Venkata Mohan *et al.* (2011)
Echinochloa glabrescens	Barnyard grass	Bombelli *et al.* (2013)

Yu, 2021). The results suggest the most effective electricity generation from the *P. alopecuroides*-based PMFC–green roof from March to June while the output voltage of the *R. rotundifolia*-based PMFC–green roof approached to zero under different weather conditions. Faster growth rate, massive biomass, vigorous photosynthesis, wide and deep root morphology, and strong root exudation all together determine the outperformance of *P. alopecuroides*-based PMFC–green roofs. PMFC–green roofs also provided a comparable effect on floor slab temperature in spring and summer. The floor slab temperature was lower with the installation of PMFC–green roofs in comparison to conventional extensive green roofs. Among the plant species, *R. rotundifolia* grew horizontally and covered the ground surface, and thus provided the highest vegetation coverage and the lowest change of soil heat flux. The household wastewater irrigation of PMFC–green roofs can not only solve the water source limitation in dry seasons but also the wastewater treatment. This approach is particularly critical to reduce the pressure on water consumption and water footprint of large cities. The green roofs with C3, C4, and CAM plants could effectively treat household kitchen wastewater at chemical oxygen demand (COD) of 483.5 mg/L, total nitrogen (TN) of 14.47 mg/L, and total phosphorus of 7.32 mg/L to comply with the discharge standard limits. Among the plant species, C3 and C4 plants contributed to better runoff reduction, COD, and nutrient removal than CAM plants (Liu *et al.*, 2021). In another study of graywater recirculation in green roofs with different plant species, higher removal efficiencies of COD, biochemical oxygen demand, and total suspended solids were recorded in green roofs with C4 plant species (i.e., *Atriplex halimus*) in comparison to green roofs with C3 plants (i.e., *Geranium zonale* and *Polygala myrtifolia*) under the typical Mediterranean climatic conditions in Greece. Green-roof vegetation also improved TN removal significantly (Thomaidi *et al.*, 2022). Laundry graywater-fed green-roof systems in the arid region (Al Ain City, United Arab Emirates) with *Phalaris arundinacea* L. (C3 plant species) performed better with a deeper layer of growing material (Chowdhury & Sulaiman Abaya, 2018).

Besides the photosynthetic efficiency, the plant selection for PMFCs integrated in green roofs essentially considers plant hardiness, plant root morphology, canopy coverage, native plant availability, decorativeness, biomass yield, and applicability in agriculture or industry. With frequent high winds at roof height, strong tree trunks are necessary to avoid breakage and further dangerous accidents. Extensively developed root morphology assists the wider spread of plant exudates, thus minimizing polarization concentration and decreasing internal resistance (Chiranjeevi *et al.*, 2012). In addition, extensive root zones also make electrode installation convenient. On the contrary, densely adapting plants with wide canopies are not recommended because their shade can block light to reach to other plants in PMFC–green roofs (Knapp *et al.*, 2019). Native plant varieties are given priority because imported plant species can invade, causing an ecological imbalance (Regmi *et al.*, 2018b; Xu *et al.*, 2015). Plant species in PMFC–green roofs should be decorative such as ornamental plants, ensuring the aesthetics of buildings. Challenges arising from operating PMFCs can be attributed to using wild grass species with limited commercial value. High-yield plants with value-added benefits in agriculture or industry, such as Napier grass (*Pennisetum purpureum*), can provide inexpensive heating fuel (García *et al.*, 2024) or pollutant phytoremediation (Ujang *et al.*, 2021), which makes a PMFC–green roof an appealing strategy for sustainable development of tropical urban areas.

8.3.2.2 Role of rhizosphere microorganisms

Rhizosphere microorganisms are different in PMFCs with different plant communities, and thus play a pivotal role in determining the performance of PMFC–green roofs (Maddalwar *et al.*, 2021; Rusyn, 2021). The beneficial interaction between plants and rhizosphere microbes was revealed in a *Canna indica*-planted PMFC operated with the complex bacterial and archaeal communities using rhizodeposits as the sole electron donor (Lu *et al.*, 2015). In particular, to generate electricity, fermentative bacteria belonging to the *Anaerolineaceae* family in symbiosis with electroactive bacteria such as *Geobacter* that metabolizes short-chain fatty acids, which are fermentation products of specific small molecular sugars excreted by *C. indica* is observed (Lu *et al.*, 2015). In PMFC–green roofs of subtropical climates,

Firmicutes and *Proteobacteria* were the most abundant phyla with slightly different relative abundances among different plant species (i.e., *P. alopecuroides*, *R. rotundifolia*, and *T. angustifolia*). In addition, *Geobacter* was also among the most dominant genera with higher relative abundance in anode soils than in cathode soils of *P. alopecuroides* and *T. angustifolia*-planted PMFCs, thus achieving higher output voltage compared to *R. rotundifolia*-planted PMFCs. *Geobacter*, an anaerobic bacterium, was more abundant in cathode soils than anode soils in *R. rotundifolia*-planted PMFCs. The anaerobic conditions established in cathode soils due to the high ground coverage by the creeping growth of *R. rotundifolia*, coupled with the high surface substrate availability, could explain for the specific localization of *Geobacter* in *R. rotundifolia*-planted PMFCs (Guan & Yu, 2021). *Proteobacteria* including *Anaeromyxobacter*, *Geobacter*, *Phenylobacterium*, and *Azospirillum* genera were dominant in the anode region of *Typha orientalis* and *Oryza rufipogon*-planted PMFCs with Cd(II)-contaminated soil (Tongphanpharn *et al.*, 2021). In unsaturated PMFCs, comparing among different *Sedum* species, the power generation was highest with *Sedum hybridum*. *Micrococcaceae* members were recorded as the most abundant group in the anodic biofilms of *S. hybridum*-planted PMFCs. However, the most commonly reported electrogens such as *Geobacter* were not detected in all *Sedum*-based PMFCs. In addition, *Proteobacteria* were among the most abundant anodic communities and possibly competed with electrogens for the availability of oxygen and nitrate. Due to the limited CAM photosynthesis with the shallow root system of *Sedum* species, soil water content, instead of total organic carbon of exudates, was the main driver for power generation (Tapia *et al.*, 2017). Water saturation of soils in PMFCs shifts the system microenvironment from aerobic to anaerobic conditions and regulates the plant exudation as well as the subsequent fermentation (Chiranjeevi *et al.*, 2012).

In PMFC–green roofs supplemented with substrates and wastewaters, microbial communities can be reshaped in response to additional organic matter, mostly toward more competitive structures. Substrate types are well known to affect the adaptation time to achieve a stable current in MFCs. Compared to the studies of conventional MFCs treating a variety of wastewater sources, less research has been conducted on wastewater-fed PMFCs (Nitisoravut & Regmi, 2017). *Geobacter*, in a symbiosis with *Clostridium sporosphaeroides* species of the *Ruminococcaceae* family, was predominant in acetate-fed PMFCs (Timmers *et al.*, 2012). *Proteobacteria* were also among the most abundant phylotypes on either anode or cathode surfaces of PMFCs fed with a nutrient solution containing Cr(VI) and acetate (Habibul *et al.*, 2016). Typical microbial communities were enriched on the anode biofilm of *Acorus calamus*-planted PMFCs with polycyclic aromatic hydrocarbon (PAH)-contaminated sediments. *Denitratisoma*, in the interaction with *Geobacter*, *Desulfuromonas*, *Longilinea*, and *Bellilinea*, appeared as the distinctive genus dominated in the anode biofilm (4.8%) in the presence of macrophytes. *A. calamus* root exudates including secondary plant metabolites (i.e., aromatic organic acids) possibly stimulated PAH biodegradation by inducing the enzymatic expression or acting as the intermediates for microbial co-metabolism (Yan *et al.*, 2015).

In PMFCs, the anode is necessarily located within the growth areas of plant roots where the electrogens can use rhizodeposits, the most abundant electron donors, for their metabolism and generate electricity simultaneously. Within the intricate anodic zone of PMFCs, different microorganisms can compete for electron donors and contribute to a reduction of power generation. The composition of these microbial communities is shaped by various biological and abiotic factors, leading to the diverse PMFC performance. The supply of oxygen into the anodic anaerobic condition through plant roots may facilitate facultative anaerobes in PMFCs, which results in substrate unavailability for electrogens for power generation (Shaikh *et al.*, 2021). Different soil C/N ratios, for example forest and agricultural soils, shape the bacterial diversity and the abundance of electrogens, thus the power generation of PMFCs (Roesch *et al.*, 2007). Introducing PMFCs into rice paddy fields with lower C/N ratios significantly reduced methane emissions by the domination of exoelectrogens at the anode surface compared to acetate-utilizing methanogens (Arends *et al.*, 2014).

Geobacter, an anodic electroactive genus, possesses a structure that facilitates the impressive long-range transport of electrons (>10 μm) within soil bacteria (Wang *et al.*, 2019). Beyond exoelectrogens, cellulolytic and fermentative bacteria emerge as crucial contributors to electricity generation (Rusyn, 2021). Their collaborative efforts extend to the conversion of complex organic compounds into low-molecular organic compounds within the interelectrode space of a PMFC, providing essential nourishment to electroactive microorganisms (Nitisoravut & Regmi, 2017). Cellulolytic bacteria, primarily from the *Ruminococcaceae* family, dominated by *C. sporosphaeroides*, along with a few cellulolytic bacteria from the *Rhodocyclaceae* family, play a key role in breaking down cellulose in the rhizosphere (Timmers *et al.*, 2012). In addition, these fermentative and cellulolytic bacteria act as mediators, facilitating the transfer of electrons from plants to electroactive microorganisms in PMFCs (Rusyn, 2021). Some species recommended for effective electricity generation in PMFCs include *Geobacter*, *Ruminococcaceae*, *Desulfobulbus*, *Bacillus*, *Geothrix*, *Pseudomonas*, *Shewanella*, and *Acidoba*, all contributing to achieving higher magnitudes of voltage (Apollon *et al.*, 2021).

Microorganisms on the cathode surface include both aerobic microorganisms utilizing oxygen as the terminal electron acceptor and contributing to metal oxidation (manganese, iron, and chromium) and anaerobic microorganisms utilizing various compounds for electron reception (nitrite, nitrate, and sulfate) (Kabutey *et al.*, 2019). The reduction of Cr(VI) in *Lolium perenne*-planted PMFCs was due to the biocatalyst of the cathodic microbes. *Cyanobacteria* (32.1%) followed by *Proteobacteria* (23.2%), *Bacteroidetes* (10.2%), and *Firmicutes* were the most abundant members of microbial communities on the cathode surface (Habibul *et al.*, 2016). Depending on the concentrations of Cr(VI), significant shifts of microbial communities on the cathode surface could occur in PMFCs with two species of wetland plants (*P. alopecuroides* and *Phragmites communis*). By increasing the amounts of Cr(VI) spiked in soils, *Bacteroidetes*, together with *Alphaproteobacteria*, significantly outcompeted *Firmicutes* and *Cyanobacteria* on the cathodes (Guan *et al.*, 2019).

These findings highlight the intricate interplay of rhizosphere microorganisms in PMFCs, paving the way for sustainable bioelectricity generation. Understanding and enhancing the effective interactions between plant species and the microbial population within the rhizosphere can increase the electrical output of PMFCs and further enhance the feasibility to implement green roofs of tropical urban areas. Bioaugmentation and rhizodeposition stimulation are both necessary to adopt for the increased performance of PMFCs integrated in green roofs.

8.3.3 Maintenance strategies and long-term viability

Maintenance strategies and long-term viability are critical for the sustainable operation of PMFC–green roofs in tropical regions. By integrating PMFCs in green roofs, irrigation is indispensable because soil water content emerges as a key factor for increasing cation mobility and decreasing internal resistance (Chiranjeevi *et al.*, 2012). Studies revealed that the exudation of carbohydrates from plants rises with increased soil water content, thus enhances power generation (Nelson & Mele, 2007). However, in terms of economic effectiveness, irrigation strategies on PMFC–green roofs must be selective to promote water saving.

Dripping or discontinuous irrigation was reported to create distinct current generation patterns, highlighting the impact of irrigation strategies on PMFC performance (Tapia *et al.*, 2017). Soil moisture monitoring is thus needed for fine-tune irrigation scheduling. The direct correlation between current density and soil moisture content implied the potential application of PMFCs as moisture biosensors (Tapia *et al.*, 2018). This is particularly valuable for the efficient water management in tropical urban areas. Efforts to harness bioelectrochemical technologies for developing water-content biosensors mark a pioneering step, aiming to create affordable indicators for the effective use of water in green roofs. Salinity augmentation could increase the performance of PMFCs (Guo *et al.*, 2021). Therefore, the concentrated stream from a reverse osmosis system supplying water to a building can be a potential feed component for PMFCs on green roofs. However, it is necessary to monitor soil pH because too high salt concentrations lead to a decreased anolyte pH, negatively affecting PMFC performance. The introduction of growth media (coconut-based biochar, compost) in substrates can serve the dual goal of helping plant growth to harvest products (fruits, flowers) and amplifying the electricity output of

PMFCs (Kabutey *et al.*, 2019). Strategic use of plant by-products, such as crushed materials and straw, also enhances bioenergy harvest in PMFCs (Nitisoravut & Regmi, 2017).

Regular inspections are essential to monitor the health of green roofs, plant growth, and the condition of PMFC components. The frequency of inspections may vary based on local conditions and plant species. In addition, weeding and pruning of plant species are necessary to ensure continued root contact with the anode. Pruning also aids in maintaining a tidy and visually appealing green roof. Biofilm formation on the electrode surface contributes to an increased internal resistance during operation (Munjal *et al.*, 2020). Moreover, the buildup of debris and redox reaction products adhered to the electrode can impede oxygen diffusion. Therefore, electrode cleaning or periodic replacement is essential for sustained power production. Rotational farming should also be taken into account. Aging plants and changing microbial communities in the rhizosphere over time constrain the prolonged use of PMFCs (Gomathy *et al.*, 2021). Finally, design considerations for PMFC–green roofs should include protection against extreme weather events such as heavy rainfall or storms, ensuring the long-term viability of the system in tropical climates.

8.4 FUTURE PROSPECTS OF PMFC–GREEN ROOFS IN TROPICAL URBAN REGIONS

Green roofs producing electricity with integrated PMFCs are interesting to improve ecological and environmental quality of urban areas. The implementation of PMFC–green roof technology fits well in the developing concept of decentralized reuse of resources for sustainable development. Electricity production and wastewater treatment on roofs with plants can improve urban water–energy management, promoting on-site water retention and treatment. PMFC–green roofs can be integrated into various settings, both in public or private buildings and in urban or rural areas, due to their extended services in energy generation as well as environmental and aesthetic improvement. Unfortunately, standalone PMFCs and green roofs are still in the technological development stage, requiring in-depth research toward the upscaling process. One of the main limitations of PMFCs is the low electrical output. Factors that affect the electrical output of PMFCs include organic sources, redox conditions, plants, rhizosphere microorganisms, growth substrates, and electrode materials. Many of these are associated with structural and maintenance aspects of green roofs installed in tropical regions, such as the type and amount of rhizodeposits, water sources for irrigation, the thickness of the growth substrates, and wastewater salinity. Thus, future research should focus on enhancing the power output of PMFC–green roofs to make them a more competitive energy source. Identifying low-cost materials for PMFC components is crucial for wider adoption and viability. Developing optimal designs for PMFC–green roofs to maximize benefits, considering factors such as plant selection, substrate composition, and electrode configuration, is also necessary.

Irrigation of PMFC–green roofs with wastewater, specifically during dry seasons, requires frequent monitoring and detailed operational planning. The variable composition of wastewater and its potential impact on plant health are among the prioritized issues to consider. Runoff quality from PMFC–green roofs depends on the remediation by plants, the substrate material, and the aging of the green roofs (Shahmohammad *et al.*, 2022). Different substrate materials possess varying water retention capacity, sorption, and leaching characteristics, impacting the retention and release of pollutants. The possibility of runoff contamination becomes more concerned in PMFC–green roofs irrigated with wastewater. Green roofs are designed to retain a certain amount of rainwater, less than traditional roofs. Pollutants retained in the green-roof medium can be released into harvested rainwater, turning rainwater brownish. Therefore, it is recommended that the collected water should be used for activities outside the building for non-potable purposes (Agra *et al.*, 2018). Runoff quality can also vary from a newly developed green roof to a maturely established green roof. Rainfall intensity, historical maintenance activity, plant uptake capacity, and other biological activities are expected to flush pollutants out of mature green roofs over time. The use of fertilizers and composts for plant growth is a significant source of nutrients discharged from green roofs (Lim, 2023; Nektarios *et al.*, 2022; Sophia & Sreeja, 2017).

Public awareness on safety relating to PMFC–green roof irrigation with wastewaters can be a barrier for broader application. Although using wastewater for PMFC–green roof irrigation offers ecological and economic benefits, it is essential to acknowledge the potential health risks associated with pathogens and toxic contaminants (Petousi *et al.*, 2023). Vitally, bacteria can proliferate during water storage in the presence of organics. Thus, wastewater feeding into green roofs is essentially restricted to non-edible plants, requiring a lower microbiological quality standard compared to edible plants. Implementation and regulation re-enforcement for wastewater irrigation in PMFC–green roofs should highlight not only the importance of considering substrate type, depth, and vegetation for effective pathogen removal in green roofs but also the need for disinfection of effluents for indoor reuse. Post-treatment disinfection using chlorination is often necessary to achieve microbiological quality for safe reuse of green-roof effluents. However, high organic matter content in green-roof effluents may lead to increased chlorine demand. Higher chlorine doses for extended storage are also recommended to prevent pathogen regrowth (Petousi *et al.*, 2023). PAHs and heavy metals can be absorbed by plant tissues and may eventually leach into runoff (Arellano-Leyva *et al.*, 2017; Speak *et al.*, 2014). In green roofs amended with biochar, plants are beneficial with enhanced water supply and rhizosphere microbial growth. However, biochar can also be a source of PAHs causing plant toxicity. The contamination of PAHs on the biochar surface depends on the pyrolysis conditions and biomass composition (Haeldermans *et al.*, 2023).

To fully evaluate the sustainability of PMFC–green roofs, life-cycle assessments (LCAs) must be conducted to quantify both the positive and negative environmental impacts from system production to end-of-life disposal (Brachet *et al.*, 2019; Rasul & Arutla, 2020). Particularly, the end-of-life stage of component materials in PMFC–green roofs can significantly influence the environmental impacts. A comprehensive assessment should also integrate economic and social aspects (Morau *et al.*, 2017). This includes considering factors such as the costs of materials, installation, maintenance, and disposal, as well as social benefits such as improved aesthetics and well-being. Landfilling, incineration, reuse, and recycling are options for reducing the impacts of waste disposal. The application of standardized and consistent methods for conducting an LCA of PMFC–green roofs contributes to precise results (Fiorentin *et al.*, 2024). Specific guidelines for the end-of-life management of PMFC–green roof materials should be codified.

8.5 CONCLUSION

The integration of PMFC technology in green roofs is a new and innovative approach that has huge scope for sustainable development in urban areas, particularly with tropical climates. Its practical applications are still far to reach particularly with the unsolved technical issues associated with standalone PMFC units and green roofs. However, the operation of PMFC–green roofs at pilot scale is shifting toward a specific range of practices by considering the potential to implement the water–energy nexus in urban areas, where demands on water and energy are critically entwined. This is necessary to a certain extent to particular weather conditions in tropical regions, which currently are more exaggerated by global climate change. Mitigation of the rising heat in buildings in combination with sustainable wastewater treatment and electricity recovery using PMFC–green roofs is competitive but also challenging, requiring a multidisciplinary strategy that combines elements of architecture and civil engineering, urban planning, as well as environmental and ecological engineering. If the major challenges in system design, configuration, and operation of PMFC–green roofs are solved, the integration of PMFC technology into green roofs successfully turns this green technology more attractive and robust, distinguishing it from conventional treatment methods and energy sources.

Solving bottlenecks is required to bring PMFC–green roof applications to the masses with value-added benefits. High efficiency, least maintenance, and sustainability in nature are the key aspects toward the development and practical applicability of PMFC–green roofs in tropical urban areas. System design development and operational optimization are essential for successful implementation and performance of PMFC–green roofs, achieving comprehensive environmental, economic, and social services. Solutions for critical factors affecting PMFC performance and green-roof maintenance, particularly for sustaining plant vitality and regulating the plant–microbe relationships, are proposed. Intensive and natural growth media, regular watering with supplementary wastewater sources, and diverse plant species with competitive resistance to strong irradiation and toxic pollutants in wastewaters are among the adaptive features for PMFC–green roofs in tropical urban regions. Runoff contamination and management of end-of-life disposals are specific concerns to ensure public safety of PMFC–green roofs in tropical urban regions.

REFERENCES

Abuseif M. (2023). Exploring influencing factors and innovative solutions for sustainable water management on green roofs: a systematic quantitative review. *Architecture*, **3**(2), 294–327, https://doi.org/10.3390/architecture3020017

Agra H., Solodar A., Bawab O., Levy S., Kadas G. J., Blaustein L. and Greenbaum N. (2018). Comparing grey water versus tap water and coal ash versus perlite on growth of two plant species on green roofs. *Science of The Total Environment*, **633**, 1272–1279, https://doi.org/10.1016/j.scitotenv.2018.03.291

Alqahtany A. (2022). Green roofs as an approach to enhance urban sustainability: a study of public perception in Riyadh, Saudi Arabia. *Buildings*, **12**(12), 2202, https://doi.org/10.3390/buildings12122202

Apollon W., Luna-Maldonado A. I., Kamaraj S.-K., Vidales-Contreras J. A., Rodríguez-Fuentes H., Gómez-Leyva J. F. and Aranda-Ruíz J. (2021). Progress and recent trends in photosynthetic assisted microbial fuel cells: a review. *Biomass and Bioenergy*, **148**, 106028, https://doi.org/10.1016/j.biombioe.2021.106028

Arellano-Leyva E., Hernández-Quiróz M., Huerta-Guzmán R. G. and Collazo-Ortega M. (2017). Extensive green roofs as a means to capture polycyclic aromatic hydrocarbons. *Polycyclic Aromatic Compounds*, **37**(4), 280–291, https://doi.org/10.1080/10406638.2015.1105827

Arends J. B. A., Speeckaert J., Blondeel E., De Vrieze J., Boeckx P., Verstraete W., Rabaey K. and Boon N. (2014). Greenhouse gas emissions from rice microcosms amended with a plant microbial fuel cell. *Applied Microbiology and Biotechnology*, **98**(7), 3205–3217, https://doi.org/10.1007/s00253-013-5328-5

Arulmani S. R. B., Gnanamuthu H. L., Kandasamy S., Govindarajan G., Alsehli M., Elfasakhany A., Pugazhendhi A. and Zhang H. (2021). Sustainable bioelectricity production from *Amaranthus viridis* and *Triticum aestivum* mediated plant microbial fuel cells with efficient electrogenic bacteria selections. *Process Biochemistry*, **107**, 27–37, https://doi.org/10.1016/j.procbio.2021.04.015

Azri Y. M., Tou I. and Sadi M. (2023). Electrodes materials evaluation in plant microbial fuel cells: a comparison of graphite and stainless steels. *Biofuels*, **14**(10), 1077–1086, https://doi.org/10.1080/17597269.2023.2212987

Bianchini F. and Hewage K. (2012). How 'green' are the green roofs? Lifecycle analysis of green roof materials. *Building and Environment*, **48**, 57–65, https://doi.org/10.1016/j.buildenv.2011.08.019

Blanusa T., Vaz Monteiro M. M., Fantozzi F., Vysini E., Li Y. and Cameron R. W. F. (2013). Alternatives to *Sedum* on green roofs: can broadleaf perennial plants offer better 'cooling service'? *Building and Environment*, **59**, 99–106, https://doi.org/10.1016/j.buildenv.2012.08.011

Bombelli P., Iyer D. M. R., Covshoff S., McCormick A. J., Yunus K., Hibberd J. M., Fisher A. C. and Howe C. J. (2013). Comparison of power output by rice (*Oryza sativa*) and an associated weed (*Echinochloa glabrescens*) in vascular plant bio-photovoltaic (VP-BPV) systems. *Applied Microbiology and Biotechnology*, **97**(1), 429–438, https://doi.org/10.1007/s00253-012-4473-6

Braaker S., Obrist M. K., Ghazoul J. and Moretti M. (2017). Habitat connectivity and local conditions shape taxonomic and functional diversity of arthropods on green roofs. *Journal of Animal Ecology*, **86**(3), 521–531, https://doi.org/10.1111/1365-2656.12648

Brachet A., Schiopu N. and Clergeau P. (2019). Biodiversity impact assessment of building's roofs based on life cycle assessment methods. *Building and Environment*, **158**, 133–144, https://doi.org/10.1016/j.buildenv.2019.04.014

Cano-Rodríguez S., Rubio-Varas M. and Sesma-Martín D. (2022). At the crossroad between green and thirsty: carbon emissions and water consumption of Spanish thermoelectricity generation, 1969–2019. *Ecological Economics*, **195**, 107363, https://doi.org/10.1016/j.ecolecon.2022.107363

Cao J., Hu S., Dong Q., Liu L. and Wang Z. (2019). Green roof cooling contributed by plant species with different photosynthetic strategies. *Energy and Buildings*, **195**, 45–50, https://doi.org/10.1016/j.enbuild.2019.04.046

Castelli K. R., Silva A. M. and Dunning J. B. (2021). Improving the biodiversity in urban green spaces: a nature based approach. *Ecological Engineering*, **173**, 106398, https://doi.org/10.1016/j.ecoleng.2021.106398

Chen B., Cai W. and Garg A. (2023). Relationship between bioelectricity and soil–water characteristics of biochar-aided plant microbial fuel cell. *Acta Geotechnica*, **18**(7), 3529–3542, https://doi.org/10.1007/s11440-022-01787-z

Chen B., Li L., Cai W. and Garg A. (2024). Bioelectricity generation in plant microbial fuel cells: influence of vegetation density and unsaturated soil properties. *Biomass and Bioenergy*, **181**, 107053, https://doi.org/10.1016/j.biombioe.2024.107053

Chiranjeevi P., Mohanakrishna G. and Venkata Mohan S. (2012). Rhizosphere mediated electrogenesis with the function of anode placement for harnessing bioenergy through CO_2 sequestration. *Bioresource Technology*, **124**, 364–370, https://doi.org/10.1016/j.biortech.2012.08.020

Chiranjeevi P., Yeruva D. K., Kumar A. K., Mohan S. V. and Varjani S. (2019). Chapter 3.8 - Plant-Microbial Fuel Cell Technology. In: Microbial Electrochemical Technology. S.V. Mohan, S. Varjani and A. Pandey (eds), Elsevier, Amsterdam, Netherlands, pp. 549–564, https://doi.org/10.1016/B978-0-444-64052-9.00022-4

Chow M. F., Bakar M. F. A. and Wong J. K. (2018). An overview of plant species and substrate materials or green roof system in tropical climate urban environment. *AIP Conference Proceedings*, **2030**(1), 020004, https://doi.org/10.1063/1.5066645

Chowdhury R. and Sulaiman Abaya J. (2018). An experimental study of greywater irrigated green roof systems in an arid climate. *Journal of Water Management Modeling*, **26**(C437), 1–10.

Delgado-Baquerizo M., Eldridge D. J., Liu Y.-R., Sokoya B., Wang J.-T., Hu H.-W., He J.-Z., Bastida F., Moreno J. L., Bamigboye A. R., Blanco-Pastor J. L., Cano-Díaz C., Illán J. G., Makhalanyane T. P., Siebe C., Trivedi P., Zaady E., Verma J. P., Wang L., Wang J., Grebenc T., Peñaloza-Bojacá G. F., Nahberger T. U., Teixido A. L., Zhou X.-Q., Berdugo M., Duran J., Rodríguez A., Zhou X., Alfaro F., Abades S., Plaza C., Rey A., Singh B. K., Tedersoo L. and Fierer N. (2021). Global homogenization of the structure and function in the soil microbiome of urban greenspaces. *Science Advances*, **7**(28), eabg5809, https://doi.org/10.1126/sciadv.abg5809

Deng H., Chen Z. and Zhao F. (2012). Energy from plants and microorganisms: progress in plant–microbial fuel cells. *ChemSusChem*, **5**(6), 1006–1011, https://doi.org/10.1002/cssc.201100257

Doyle L. (2021). Essential Green Roof Construction: The Complete Step-by-Step Guide. New Society Publishers, Grabriola, Canada.

Dunnett N. and Kingsbury N. (2004). Planting Green Roofs and Living Walls. Timber Press, Oregon, United States.

Eldridge D. J., Cui H., Ding J., Berdugo M., Sáez-Sandino T., Duran J., Gaitan J., Blanco-Pastor J. L., Rodríguez A., Plaza C., Alfaro F., Teixido A. L., Abades S., Bamigboye A. R., Peñaloza-Bojacá G. F., Grebenc T., Nahberger T. U., Illán J. G., Liu Y.-R., Makhalanyane T. P., Rey A., Siebe C., Sun W., Trivedi P., Verma J. P., Wang L., Wang J., Wang T., Zaady E., Zhou X., Zhou X.-Q. and Delgado-Baquerizo M. (2024). Urban greenspaces and nearby natural areas support similar levels of soil ecosystem services. *NPJ Urban Sustainability*, **4**(1), 15, https://doi.org/10.1038/s42949-024-00154-z

Ergettie A. B. and Dagbasi M. (2021). Impact of wastewater concentration and feed frequency on ammonia inhibition in microbial fuel cells. *Biofuels*, **12**(6), 655–661, https://doi.org/10.1080/17597269.2018.1519760

Farrell C., Mitchell R. E., Szota C., Rayner J. P. and Williams N. S. G. (2012). Green roofs for hot and dry climates: interacting effects of plant water use, succulence and substrate. *Ecological Engineering*, **49**, 270–276, https://doi.org/10.1016/j.ecoleng.2012.08.036

Fiorentin D. P., Martín-Gamboa M., Rafael S. and Quinteiro P. (2024). Life cycle assessment of green roofs: a comprehensive review of methodological approaches and climate change impacts. *Sustainable Production and Consumption*, **45**, 598–611, https://doi.org/10.1016/j.spc.2024.02.004

Francis L. F. M. and Jensen M. B. (2017). Benefits of green roofs: a systematic review of the evidence for three ecosystem services. *Urban Forestry & Urban Greening*, **28**, 167–176, https://doi.org/10.1016/j.ufug.2017.10.015

García J. C., Alfaro A., Loaiza J. M., Lozano-Calvo S. and López F. (2024). Cold alkaline extraction of elephant grass for optimal subsequent extraction of hemicelluloses and energy production. *Biomass Conversion and Biorefinery*, **14**(7), 8307–8320, https://doi.org/10.1007/s13399-022-03054-3

Gilani S. R., Yaseen A., Zaidi S. A. R., Zahra M. and Mahmood Z. J. (2016). Photocurrent generation through plant microbial fuel cell by varying electrode materials. *Journal of the Chemical Society of Pakistan*, **38**, 17.

Gomathy M., Sabarinathan K. G., Subramanian K. S., Ananthi K., Kalaiyarasi V., Jeyshri M. and Dutta P. (2021). Rhizosphere: Niche for Microbial Rejuvenation and Biodegradation of Pollutants. In: Microbial Rejuvenation of Polluted Environment: Volume 1. D. G. Panpatte and Y. K. Jhala (eds), Springer, Singapore, pp. 1–22, https://doi.org/10.1007/978-981-15-7447-4_1

Gómez-Baggethun E., Gren Å., Barton D. N., Langemeyer J., McPhearson T., O'Farrell P., Andersson E., Hamstead Z. and Kremer P. (2013). Urban ecosystem services. In: Urbanization, Biodiversity and Ecosystem Services: Challenges and Opportunities: A Global Assessment, T. Elmqvist, M. Fragkias, J. Goodness, B. Güneralp, P. J. Marcotullio, R. I. McDonald, S. Parnell, M. Schewenius, M. Sendstad, K. C. Seto and C. Wilkinson (eds), Springer, Dordrecht, Netherlands, pp. 175–251, https://doi.org/10.1007/978-94-007-7088-1_11

Greenman J., Thorn R., Willey N. and Ieropoulos I. (2024). Energy harvesting from plants using hybrid microbial fuel cells; potential applications and future exploitation. *Frontiers in Bioengineering and Biotechnology*, **12**, 1276176, https://doi.org/10.3389/fbioe.2024.1276176

Guan C.-Y. and Yu C.-P. (2021). Evaluation of plant microbial fuel cells for urban green roofs in a subtropical metropolis. *Science of The Total Environment*, **765**, 142786, https://doi.org/10.1016/j.scitotenv.2020.142786

Guan C.-Y., Hu A. and Yu C.-P. (2019). Stratified chemical and microbial characteristics between anode and cathode after long-term operation of plant microbial fuel cells for remediation of metal contaminated soils. *Science of The Total Environment*, **670**, 585–594, https://doi.org/10.1016/j.scitotenv.2019.03.096

Guhathakurta S. and Gober P. (2007). The impact of the Phoenix urban heat island on residential water use. *Journal of the American Planning Association*, **73**(3), 317–329, https://doi.org/10.1080/01944360708977980

Gulamhussein M. and Randall D. G. (2020). Design and operation of plant microbial fuel cells using municipal sludge. *Journal of Water Process Engineering*, **38**, 101653, https://doi.org/10.1016/j.jwpe.2020.101653

Guo F., Luo H., Shi Z., Wu Y. and Liu H. (2021). Substrate salinity: a critical factor regulating the performance of microbial fuel cells, a review. *Science of The Total Environment*, **763**, 143021, https://doi.org/10.1016/j.scitotenv.2020.143021

Gupta S., Patro A., Mittal Y., Dwivedi S., Saket P., Panja R., Saeed T., Martínez F. and Yadav A. K. (2023). The race between classical microbial fuel cells, sediment-microbial fuel cells, plant–microbial fuel cells, and constructed wetlands-microbial fuel cells: applications and technology readiness level. *Science of The Total Environment*, **879**, 162757, https://doi.org/10.1016/j.scitotenv.2023.162757

Habibul N., Hu Y., Wang Y.-K., Chen W., Yu H.-Q. and Sheng G.-P. (2016). Bioelectrochemical chromium(VI) removal in plant–microbial fuel cells. *Environmental Science & Technology*, **50**(7), 3882–3889, https://doi.org/10.1021/acs.est.5b06376

Haeldermans T., Puente Torres J., Vercruysse W., Carleer R., Samyn P., Vandamme D., Yperman J., Cuypers A., Vanreppelen K. and Schreurs S. (2023). An experimentally validated selection protocol for biochar as a sustainable component in green roofs. *Waste*, **1**(1), 176–194, https://doi.org/10.3390/waste1010013

Hazzan O. O., Zhao B. and Xiao Y. (2023). Strategies for enhancing extracellular electron transfer in environmental biotechnology: a review. *Applied Sciences*, **13**(23), 12760, https://doi.org/10.3390/app132312760

Helder M., Strik D. P. B. T. B., Hamelers H. V. M., Kuhn A. J., Blok C. and Buisman C. J. N. (2010). Concurrent bioelectricity and biomass production in three plant–microbial fuel cells using *Spartina anglica*, *Arundinella anomala* and *Arundo donax*. *Bioresource Technology*, **101**(10), 3541–3547, https://doi.org/10.1016/j.biortech.2009.12.124

Helder M., Strik D. P., Hamelers H. V. M. and Buisman C. J. N. (2012). The flat-plate plant–microbial fuel cell: the effect of a new design on internal resistances. *Biotechnology for Biofuels*, **5**(1), 70, https://doi.org/10.1186/1754-6834-5-70

Helder M., Chen W.-S., van der Harst E. J. M. and Strik D. P. B. T. B. (2013a). Electricity production with living plants on a green roof: environmental performance of the plant–microbial fuel cell. *Biofuels, Bioproducts and Biorefining*, **7**(1), 52–64, https://doi.org/10.1002/bbb.1373

Helder M., Strik D. P. B. T. B., Timmers R. A., Raes S. M. T., Hamelers H. V. M. and Buisman C. J. N. (2013b). Resilience of roof-top plant–microbial fuel cells during Dutch winter. *Biomass and Bioenergy*, **51**, 1–7, https://doi.org/10.1016/j.biombioe.2012.10.011

Kabutey F. T., Zhao Q., Wei L., Ding J., Antwi P., Quashie F. K. and Wang W. (2019). An overview of plant microbial fuel cells (PMFCs): configurations and applications. *Renewable and Sustainable Energy Reviews*, **110**, 402–414, https://doi.org/10.1016/j.rser.2019.05.016

Kaewpraek C., Ali L., Rahman M. A., Shakeri M., Chowdhury M. S., Jamal M. S., Mia M. S., Pasupuleti J., Dong L. K. and Techato K. (2021). The effect of plants on the energy output of green roof photovoltaic systems in tropical climates. *Sustainability*, **13**(8), 4505, https://doi.org/10.3390/su13084505

Khudzari J. Md., Kurian J., Gariépy Y., Tartakovsky B. and Raghavan G. S. V. (2018) Effects of salinity, growing media, and photoperiod on bioelectricity production in plant microbial fuel cells with weeping alkaligrass. *Biomass and Bioenergy*, **109**, 1–9, https://doi.org/10.1016/j.biombioe.2017.12.013

Knapp S., Schmauck S. and Zehnsdorf A. (2019). Biodiversity impact of green roofs and constructed wetlands as progressive eco-technologies in urban areas. *Sustainability*, **11**(20), 5846, https://doi.org/10.3390/su11205846

Köhler M. and Ksiazek-Mikenas K. (2018). Chapter 3.14 - Green Roofs as Habitats for Biodiversity. In: Nature Based Strategies for Urban and Building Sustainability, G. Pérez and K. Perini (eds), Butterworth-Heinemann, Oxford, United Kingdom, pp. 239–249, https://doi.org/10.1016/B978-0-12-812150-4.00022-7

Krarti M. and Aldubyan M. (2021). Mitigation analysis of water consumption for power generation and air conditioning of residential buildings: case study of Saudi Arabia. *Applied Energy*, **290**, 116767, https://doi.org/10.1016/j.apenergy.2021.116767

Ksiazek-Mikenas K. (2014). The potential of green roofs to support urban biodiversity. In: Green cities in the world, J. Briz, M. Köhler and I. de Felipe (eds), Pronatur, Madrid, Spain, pp. 101–124.

Lepikash R., Lavrova D., Stom D., Meshalkin V., Ponamoreva O. and Alferov S. (2024). State of the art and environmental aspects of plant microbial fuel cells' application. *Energies*, **17**(3), 752, https://doi.org/10.3390/en17030752

Liao J. and He Z. (2023). Integrating microbial fuel cell with sedum green roof for stormwater retention and renewable energy generation. *Journal of Emerging Investigators*, **6**, 1–7.

Lim H. S. (2023). What happens to nitrogen and phosphorus nutrient contributions from green roofs as they age? A review. *Environmental Advances*, **12**, 100366, https://doi.org/10.1016/j.envadv.2023.100366

Liu Y. and Mauter M. S. (2021). Marginal energy intensity of water supply. *Energy & Environmental Science*, **14**(8), 4533–4540, https://doi.org/10.1039/D1EE00925G

Liu L., Cao J., Ali M., Zhang J. and Wang Z. (2021). Impact of green roof plant species on domestic wastewater treatment. *Environmental Advances*, **4**, 100059, https://doi.org/10.1016/j.envadv.2021.100059

Longo S., d'Antoni B. M., Bongards M., Chaparro A., Cronrath A., Fatone F., Lema J. M., Mauricio-Iglesias M., Soares A. and Hospido A. (2016). Monitoring and diagnosis of energy consumption in wastewater treatment plants. A state of the art and proposals for improvement. *Applied Energy*, **179**, 1251–1268, https://doi.org/10.1016/j.apenergy.2016.07.043

Lopez F. B. and Barclay G. F. (2017). Chapter 4 - Plant Anatomy and Physiology. In: Pharmacognosy: Fundamentals, Applications and Strategy, S. Badal and R. Delgoda (eds), Academic Press, Massachusetts, United States, pp. 45–60, https://doi.org/10.1016/B978-0-12-802104-0.00004-4

Lu L., Xing D. and Ren Z. J. (2015). Microbial community structure accompanied with electricity production in a constructed wetland plant microbial fuel cell. *Bioresource Technology*, **195**, 115–121, https://doi.org/10.1016/j.biortech.2015.05.098

MacIvor J. S., Sookhan N., Arnillas C. A., Bhatt A., Das S., Yasui S.-L. E., Xie G. and Cadotte M. W. (2018). Manipulating plant phylogenetic diversity for green roof ecosystem service delivery. *Evolutionary Applications*, **11**(10), 2014–2024, https://doi.org/10.1111/eva.12703

Maddalwar S., Kumar Nayak K., Kumar M. and Singh L. (2021). Plant microbial fuel cell: opportunities, challenges, and prospects. *Bioresource Technology*, **341**, 125772, https://doi.org/10.1016/j.biortech.2021.125772

Martens D., Öztürk Ö., Rindt L., Twarok J., Steinhardt U. and Molitor H. (2022). Supporting biodiversity: structures of participatory actions in urban green spaces. *Frontiers in Sustainable Cities*, **4**, 952790, https://doi.org/10.3389/frsc.2022.952790

Maziotis A., Sala-Garrido R., Mocholi-Arce M. and Molinos-Senante M. (2023). A comprehensive assessment of energy efficiency of wastewater treatment plants: an efficiency analysis tree approach. *Science of The Total Environment*, **885**, 163539, https://doi.org/10.1016/j.scitotenv.2023.163539

Mohajerani A., Bakaric J. and Jeffrey-Bailey T. (2017). The urban heat island effect, its causes, and mitigation, with reference to the thermal properties of asphalt concrete. *Journal of Environmental Management*, **197**, 522–538, https://doi.org/10.1016/j.jenvman.2017.03.095

Moon T., Kim M. and Chon J. (2024). Adaptive green space management strategies for sustainable carbon sink parks. *Urban Forestry & Urban Greening*, **94**, 128236, https://doi.org/10.1016/j.ufug.2024.128236

Morau D., Rabarison T. N. and Rakotondramiarana H. T. (2017). Life cycle analysis of green roof implemented in a global south low-income country. *International Journal of Environment and Climate Change*, **7**(1), 43–55, https://doi.org/10.9734/BJECC/2017/30796

Munjal M., Tiwari B., Lalwani S., Sharma M., Singh G. and Sharma R. K. (2020). An insight of bioelectricity production in mediator less microbial fuel cell using mesoporous cobalt ferrite anode. *International Journal of Hydrogen Energy*, **45**(22), 12525–12534, https://doi.org/10.1016/j.ijhydene.2020.02.184

Nagase A., Yamada Y., Aoki T. and Nomura M. (2018). Developing biodiverse green roofs for Japan: arthropod and colonizer plant diversity on Harappa and biotope roofs. *Urban Naturalist*, **1**, 16–38.

Nanzyo M. and Kanno H. (2018). Inorganic Soil Constituents Sensitive to Varying Redox Conditions. In: Inorganic Constituents in Soil: Basics and Visuals, M. Nanzyo and H. Kanno (eds), Springer, Singapore, pp. 97–131, https://doi.org/10.1016/B978-0-12-802104-0.00004-4

Neina D. (2019). The role of soil pH in plant nutrition and soil remediation. *Applied and Environmental Soil Science*, **2019**(1), 5794869.

Nektarios P. A., Ischyropoulos D., Kalozoumis P., Savvas D., Yfantopoulos D., Ntoulas N., Tsaniklidis G. and Goumenaki E. (2022). Impact of substrate depth and fertilizer type on growth, production, quality characteristics and heavy metal contamination of tomato and lettuce grown on urban green roofs. *Scientia Horticulturae*, **305**, 111318, https://doi.org/10.1016/j.scienta.2022.111318

Nelson D. R. and Mele P. M. (2007). Subtle changes in rhizosphere microbial community structure in response to increased boron and sodium chloride concentrations. *Soil Biology and Biochemistry*, **39**(1), 340–351, https://doi.org/10.1016/j.soilbio.2006.08.004

Nguyen V. and Nitisoravut R. (2019). Bioelectricity Generation in Plant Microbial Fuel Cell Using Forage Grass under Variations of Circadian Rhythm, Ambient Temperature, and Soil Water Contents. IEEE Asia Power and Energy Engineering Conference (APEEC), pp. 240–244.

Nitisoravut R. and Regmi R. (2017). Plant microbial fuel cells: a promising biosystems engineering. *Renewable and Sustainable Energy Reviews*, **76**, 81–89, https://doi.org/10.1016/j.rser.2017.03.064

Oberndorfer E., Lundholm J., Bass B., Coffman R. R., Doshi H., Dunnett N., Gaffin S., Kohler M. and Rowe B. (2007). Green roofs as urban ecosystems: ecological structures, functions, and services. *BioScience*, **57**(10), 823–833, https://doi.org/10.1641/B571005

Ode Sang Å., Thorpert P. and Fransson A.-M. (2022). Planning, designing, and managing green roofs and green walls for public health – an ecosystem services approach. *Frontiers in Ecology and Evolution*, **10**, 804500, https://doi.org/10.3389/fevo.2022.804500

Pamintuan K. R. S., Katipunan A. M. C., Palaganas P. A. O. and Caparanga A. R. (2020). An analysis of the stacking potential and efficiency of plant–microbial fuel cells growing green beans (*Vigna ungiculata* ssp. *sesquipedalis*). *International Journal of Renewable Energy Development*, **9**, 439–447, https://doi.org/10.14710/ijred.2020.29898

Pantigoso H. A., Newberger D. and Vivanco J. M. (2022). The rhizosphere microbiome: plant–microbial interactions for resource acquisition. *Journal of Applied Microbiology*, **133**(5), 2864–2876, https://doi.org/10.1111/jam.15686

Petousi I., Thomaidi V., Kalogerakis N. and Fountoulakis M. S. (2023). Removal of pathogens from greywater using green roofs combined with chlorination. *Environmental Science and Pollution Research*, **30**(9), 22560–22569, https://doi.org/10.1007/s11356-022-23755-6

Qiu G. Y., Zou Z., Li W., Li L. and Yan C. (2022). A quantitative study on the water-related energy use in the urban water system of Shenzhen. *Sustainable Cities and Society*, **80**, 103786, https://doi.org/10.1016/j.scs.2022.103786

Rasul M. G. and Arutla L. K. R. (2020). Environmental impact assessment of green roofs using life cycle assessment. *Energy Reports*, **6**, 503–508, https://doi.org/10.1016/j.egyr.2019.09.015

Rayner J. P., Farrell C., Raynor K. J., Murphy S. M. and Williams N. S. G. (2016). Plant establishment on a green roof under extreme hot and dry conditions: the importance of leaf succulence in plant selection. *Urban Forestry & Urban Greening*, **15**, 6–14, https://doi.org/10.1016/j.ufug.2015.11.004

Razzaghmanesh M., Beecham S. and Kazemi F. (2014). The growth and survival of plants in urban green roofs in a dry climate. *Science of The Total Environment*, **476–477**, 288–297, https://doi.org/10.1016/j.scitotenv.2014.01.014

Regmi R., Nitisoravut R. and Ketchaimongkol J. (2018a). A decade of plant-assisted microbial fuel cells: looking back and moving forward. *Biofuels*, **9**(5), 605–612, https://doi.org/10.1080/17597269.2018.1432272

Regmi R., Nitisoravut R., Charoenroongtavee S., Yimkhaophong W. and Phanthurat O. (2018b). Earthen pot–plant microbial fuel cell powered by vetiver for bioelectricity production and wastewater treatment. *CLEAN – Soil, Air, Water*, **46**(3), 1700193, https://doi.org/10.1002/clen.201700193

Roesch L. F. W., Fulthorpe R. R., Riva A., Casella G., Hadwin A. K. M., Kent A. D., Daroub S. H., Camargo F. A. O., Farmerie W. G. and Triplett E. W. (2007). Pyrosequencing enumerates and contrasts soil microbial diversity. *The ISME Journal*, **1**(4), 283–290, https://doi.org/10.1038/ismej.2007.53

Roy H., Rahman T. U., Tasnim N., Arju J., Rafid M. M., Islam M. R., Pervez M. N., Cai Y., Naddeo V. and Islam M. S. (2023). Microbial fuel cell construction features and application for sustainable wastewater treatment. *Journal of Renewable and Sustainable Energy*, **13**(5), 490, https://doi.org/10.3390/membranes13050490

Rusyn I. (2021). Role of microbial community and plant species in performance of plant microbial fuel cells. *Renewable and Sustainable Energy Reviews*, **152**, 111697, https://doi.org/10.1016/j.rser.2021.111697

Shahmohammad M., Hosseinzadeh M., Dvorak B., Bordbar F., Shahmohammadmirab H. and Aghamohammadi N. (2022). Sustainable green roofs: a comprehensive review of influential factors. *Environmental Science and Pollution Research*, **29**(52), 78228–78254, https://doi.org/10.1007/s11356-022-23405-x

Shaikh R., Rizvi A., Quraishi M., Pandit S., Mathuriya A. S., Gupta P. K., Singh J. and Prasad R. (2021). Bioelectricity production using plant–microbial fuel cell: present state of art. *South African Journal of Botany*, **140**, 393–408, https://doi.org/10.1016/j.sajb.2020.09.025

Silva C. M., Gomes M. G. and Silva M. (2016). Green roofs energy performance in Mediterranean climate. *Energy and Buildings*, **116**, 318–325, https://doi.org/10.1016/j.enbuild.2016.01.012

Singh D. K. and Tayal S. (2021). Assessment of regional water demand for coal-based power plants in India: exploring its regional impact on other cross-sectoral water stress. *Environmental Monitoring and Assessment*, **193**(8), 503, https://doi.org/10.1007/s10661-021-09298-2

Sophia A. C. and Sreeja S. (2017). Green energy generation from plant microbial fuel cells (PMFC) using compost and a novel clay separator. *Sustainable Energy Technologies and Assessments*, **21**, 59–66, https://doi.org/10.1016/j.seta.2017.05.001

Sorgato A. C., Jeremias T. C., Lobo F. L. and Lapolli F. R. (2023). Microbial fuel cell: interplay of energy production, wastewater treatment, toxicity assessment with hydraulic retention time. *Environmental Research*, **231**, 116159, https://doi.org/10.1016/j.envres.2023.116159

Speak A. F., Rothwell J. J., Lindley S. J. and Smith C. L. (2014). Metal and nutrient dynamics on an aged intensive green roof. *Environmental Pollution*, **184**, 33–43, https://doi.org/10.1016/j.envpol.2013.08.017

Strik D. P. B. T. B., Hamelers H. V. M., Snel J. F. H. and Buisman C. J. N. (2008). Green electricity production with living plants and bacteria in a fuel cell. *International Journal of Energy Research*, **32**(9), 870–876, https://doi.org/10.1002/er.1397

Tapia N. F., Rojas C., Bonilla C. A. and Vargas I. T. (2017). Evaluation of *Sedum* as driver for plant microbial fuel cells in a semi-arid green roof ecosystem. *Ecological Engineering*, **108**, 203–210, https://doi.org/10.1016/j.ecoleng.2017.08.017

Tapia N. F., Rojas C., Bonilla C. A. and Vargas I. T. (2018). A new method for sensing soil water content in green roofs using plant microbial fuel cells. *Sensors*, **18**(1), 71, https://doi.org/10.3390/s18010071

Thomaidi V., Petousi I., Kotsia D., Kalogerakis N. and Fountoulakis M. S. (2022). Use of green roofs for greywater treatment: role of substrate, depth, plants, and recirculation. *Science of The Total Environment*, **807**, 151004, https://doi.org/10.1016/j.scitotenv.2021.151004

Timmers R. A., Rothballer M., Strik D. P., Engel M., Schulz S., Schloter M., Hartmann A., Hamelers B. and Buisman C. (2012). Microbial community structure elucidates performance of *Glyceria maxima* plant microbial fuel cell. *Applied Microbiology and Biotechnology*, **94**(2), 537–548, https://doi.org/10.1007/s00253-012-3894-6

Tongphanpharn N., Chou C.-H., Guan C.-Y. and Yu C.-P. (2021). Plant microbial fuel cells with *Oryza rufipogon* and *Typha orientalis* for remediation of cadmium contaminated soil. *Environmental Technology & Innovation*, **24**, 102030, https://doi.org/10.1016/j.eti.2021.102030

Ujang F. A., Roslan A. M., Osman N. A., Norman A., Idris J., Farid M. A. A., Halmi M. I. E., Gozan M. and Hassan M. A. (2021). Removal behaviour of residual pollutants from biologically treated palm oil mill effluent by *Pennisetum purpureum* in constructed wetland. *Scientific Reports*, **11**(1), 18257, https://doi.org/10.1038/s41598-021-97789-0

Vaz Monteiro M., Blanuša T., Verhoef A., Richardson M., Hadley P. and Cameron R. W. F. (2017). Functional green roofs: importance of plant choice in maximising summertime environmental cooling and substrate insulation potential. *Energy and Buildings*, **141**, 56–68, https://doi.org/10.1016/j.enbuild.2017.02.011

Venkata Mohan S., Mohanakrishna G. and Chiranjeevi P. (2011). Sustainable power generation from floating macrophytes based ecological microenvironment through embedded fuel cells along with simultaneous wastewater treatment. *Bioresource Technology*, **102**(14), 7036–7042, https://doi.org/10.1016/j.biortech.2011.04.033

Vijayaraghavan K. (2016). Green roofs: a critical review on the role of components, benefits, limitations and trends. *Renewable and Sustainable Energy Reviews*, **57**, 740–752, https://doi.org/10.1016/j.rser.2015.12.119

Vo N. X. P., Dang Nguyen Hoang D., Doan Huu T., Doan Van T., Lam Pham Thanh H. and Vo Nguyen Xuan Q. (2023). Performance of vertical up-flow-constructed wetlands integrating with microbial fuel cell (VFCW-MFC) treating ammonium in domestic wastewater. *Environmental Technology*, **44**(12), 1822–1837, https://doi.org/10.1080/09593330.2021.2014574

Wagner F., Nusrat F., Thiem L. and Akanda A. S. (2022). Assessment of urban water–energy interactions and heat island signatures in Rhode Island. *Energy Nexus*, **7**, 100093, https://doi.org/10.1016/j.nexus.2022.100093

Wakeel M. and Chen B. (2016). Energy consumption in urban water cycle. *Energy Procedia*, **104**, 123–128, https://doi.org/10.1016/j.egypro.2016.12.022

Wang C., Guo L., Li Y. and Wang Z. (2012). Systematic comparison of C3 and C4 plants based on metabolic network analysis. *BMC Systems Biology*, **6**(2), S9.

Wang F., Gu Y., O'Brien JP. ., Yi S. M., Yalcin S. E., Srikanth V., Shen C., Vu D., Ing N. L., Hochbaum A. I., Egelman E. H. and Malvankar N. S. (2019). Structure of microbial nanowires reveals stacked hemes that transport electrons over micrometers. *Cell*, **177**(2), 361–369.e310, https://doi.org/10.1016/j.cell.2019.03.029

Wetser K., Liu J., Buisman C. and Strik D. (2015). Plant microbial fuel cell applied in wetlands: spatial, temporal and potential electricity generation of *Spartina anglica* salt marshes and *Phragmites australis* peat soils. *Biomass and Bioenergy*, **83**, 543–550, https://doi.org/10.1016/j.biombioe.2015.11.006

Wetser K., Dieleman K., Buisman C. and Strik D. (2017). Electricity from wetlands: tubular plant microbial fuels with silicone gas-diffusion biocathodes. *Applied Energy*, **185**, 642–649, https://doi.org/10.1016/j.apenergy.2016.10.122

Wu W., Maier H. R., Dandy G. C., Arora M. and Castelletti A. (2020). The changing nature of the water–energy nexus in urban water supply systems: a critical review of changes and responses. *Journal of Water and Climate Change*, **11**(4), 1095–1122, https://doi.org/10.2166/wcc.2020.276

Xu B., Ge Z. and He Z. (2015). Sediment microbial fuel cells for wastewater treatment: challenges and opportunities. *Environmental Science: Water Research & Technology*, **1**(3), 279–284, https://doi.org/10.1039/C5EW00020C

Yan Z., Jiang H., Cai H., Zhou Y. and Krumholz L. R. (2015). Complex interactions between the macrophyte *Acorus calamus* and microbial fuel cells during pyrene and benzo[a]pyrene degradation in sediments. *Scientific Reports*, **5**(1), 10709, https://doi.org/10.1038/srep10709

Zhang X., Huang H., Tu K., Li R., Zhang X., Wang P., Li Y., Yang Q., Acerman A. C., Guo N. and Liu Y. (2024). Effects of plant community structural characteristics on carbon sequestration in urban green spaces. *Scientific Reports*, **14**(1), 7382, https://doi.org/10.1038/s41598-024-57789-2

doi: 10.2166/9781789065015_0175

Chapter 9

Co-designing water-sensitive suburbs through blue-green infrastructure planning by research, municipal and housing association partners

J. Friesen[1]*, *G. Khurelbaatar*[1], *B. Plaul*[2], *D. Despot*[1], *M. van Afferden*[1], *R. A. Müller*[1] and *M. Breulmann*[1]

[1]*Department of Systemic Environmental Biotechnology, Helmholtz Centre for Environmental Research GmbH – UFZ, Leipzig, Germany*
[2]*Leipziger Wohnungs- und Baugesellschaft mbH (LWB), Leipzig, Germany*
**Corresponding author: jan.friesen@ufz.de*

ABSTRACT

Cities around the world are struggling with water management in the face of climate change and extreme events. To combat this, we are using blue-green infrastructure planning. This involves combining an urban block approach for stormwater management with irrigation demands during dry summer months. We are implementing practical infrastructure planning for the refurbishment of municipal housing. In a network of researchers and city partners such as water utilities and a municipal housing company, we co-designed blue-green scenarios for the city of Leipzig (Germany). The co-designed water infrastructure scenarios were calculated using stormwater and irrigation demand modelling. The results show that more than 90% of a 100-year stormwater event can be retained and that courtyard irrigation can be largely covered by roof runoff under different climatic conditions.

Keywords: Block approach, blue-green infrastructure, building complexes, co-design, community housing, existing buildings, urban planning.

9.1 INTRODUCTION

Urban water management and blue-green infrastructure (BGI) planning are key to managing stormwater in urban areas and developing water-sensitive cities. Although greenfield development projects allow planners and municipalities to design water-sensitive building complexes from scratch, existing suburbs and buildings still remain a challenge. Our approach addresses this issue by developing and evaluating BGI scenarios through a co-design process between researchers, a communal housing association and landscape architects. The co-designed water infrastructure scenarios include constraints such as available green space, but also constraints related to refurbishment costs, and groundwater levels.

The Leipzig BlueGreen project (see box 'Leipzig BlueGreen' and www.ufz.de/leipzigerblaugruen) investigated a co-design process on a pilot scale in the preliminary planning rounds of a greenfield investment project, L416 (www.leipzig416.de). L416 (former Eutritzscher Freiladebahnhof, Leipzig, Germany) is planned as a new city suburb with housing, parks, a school and sports facilities close to

© 2025 IWAP. This is an Open Access book chapter distributed under the terms of the Creative Commons Attribution License (CC BY-NC-ND 4.0) which permits copying and redistribution for non-commercial purposes with no derivatives, provided the work is properly cited (https://creativecommons.org/licenses/by-nc-nd/4.0/). The chapter is from the book *Nature-Based Solutions for Urban Sustainability*, Piet N.L. Lens and Xuan-Thanh Bui (Editors).

Leipzig's city centre. Concepts were developed for both the local management of storm events and for the long-term creation of a sustainable water balance through BGI. Finding BlueGreen solutions for existing neighbourhoods is more challenging than for new development areas; the Leipzig BlueGreen research focuses on the needs of the city of Leipzig for existing neighbourhoods. Challenges in existing buildings are the fixed space allocation between buildings and open spaces as well as structural building constraints, ownership structure and refurbishment cost caps amongst many others, compared to greenfield development. The implementation of water-efficient climate adaptation measures in existing buildings requires the integration of existing building structures and management systems. Current practice is the integration of individual or 'decoupled' blue-green measures (e.g., green roofs and infiltration trenches) as part of the modernization of existing buildings.

An integrated use of the multifunctionality of the overall system in terms of sewer drainage, cooling and vegetation vitality is often not taken into account. The approaches developed in the often large-scale pilot projects, which are presented in the guidelines for water-sensitive urban development of the Bavarian State Ministry of the Environment and Consumer Protection (StMUV, 2020), are to be translated into simple, cost-effective construction methods with evaluated performance.

The development of standardized procedures for cascading BGI systems reduces the development and planning costs for individual projects. This means that even more complex systems, such as those combining green roofs with retention areas, storage tanks and above-ground infiltration systems, can be implemented in smaller projects. Without proven standard workflows, the design and planning effort for many projects including BGI is currently higher than standard refurbishments. With the development of easy-to-use blue-green building blocks, their use can also be made mandatory through specifications in bylaws and requirements for building owners. In this way, the measures developed can be applied universally and enable the existing infrastructure to be optimized in many places.

Research project: Leipzig BlueGreen (LeipzigerBlauGrün)

The Leipzig BlueGreen research project focuses on the development of resource-efficient urban districts (LBG II, 2023). In a first phase, a concept for climate-adapted water and energy management was developed using the Leipzig 416 greenfield site. Together with the city of Leipzig, an investor, business enterprises and scientific institutions, a resource-efficient urban district was developed and granted planning permission.

In a second project phase, Leipzig BlueGreen expanded its efforts to meet the city of Leipzig's needs to include existing neighbourhoods in their resource planning. In response to the climate emergency, the Leipzig City Council decided to increase the resilience of existing neighbourhoods to the local impacts of climate change, in addition to new buildings. In cooperation with the Leipziger Wohnungs-und Baugesellschaft (LWB), blue-green modernization concepts for the existing portfolio are to be developed and jointly implemented. As a model city for resource-efficient urban neighbourhoods, Leipzig has declared a 'climate emergency' due to a particularly critical development and has therefore outlined measures to implement '[...] climate-friendly and energy-efficient neighbourhoods' in order to meet the city's climate targets. A major challenge now is to translate the political goal and resolutions into municipal action in the context described. Concrete measures must be defined for the decision-makers, implemented and consolidated in the transformation process.

In a transdisciplinary process, the results can be incorporated into the redevelopment measures. For this purpose, blue-green scenarios adapted to the existing structures will be developed, which allow the complete decoupling of rainwater discharge into the combined sewer system (sustainable water balance) and irrigation for the local urban greenery.

With the aim of providing both a reduction of stormwater runoff and a balanced use of water for irrigation demands, different BGI scenarios have been developed to support the planning process of building complex refurbishments. The results provide the detailed planners of the communal housing association with tangible and plausible pre-planning scenarios that can be used to make informed decisions about the different alternatives and streamline both tendering and detailed planning. For the municipality and water utilities, these scenarios also provide a clear assessment of the impact of BGI on stormwater mitigation and water management. Preliminary results for a number of existing municipal building complexes in a suburb of the city of Leipzig (Germany) show that both stormwater retention (90% of a 100-year stormwater event) and irrigation are feasible for different climatic conditions through cost-effective BGI measures.

9.2 BLUE-GREEN REFURBISHMENT SCENARIOS

As part of the co-design processes with specialist planners, representatives from the scientific community and the investor 'Leipziger Wohnungs- und Baugesellschaft' (LWB), the following main objectives and scenarios were agreed for the BGI conversion:

- Objective 1: Complete decoupling in the event of heavy rainfall events (100-year design storm) for the building complexes in Kolonnadenviertel through the designing of BGI.
- Objective 2: Approaching a water balance through the management of BGI.

The objectives were part of a co-design process with the city of Leipzig, the Leipzig water utilities and the communal housing company, LWB. LWB owns and operates ~10% of the city's housing market; amongst those is the city quarter 'Kolonnadenviertel' in which the LWB has scheduled 14 building complexes for refurbishment. The refurbishment of the LWB building complexes (Figure 9.1) focuses on existing buildings, which poses challenges not only in terms of available space, but also in terms of rent structure, historic preservation, refurbishment cost limits and building structure, to name just a few. Although a variety of BGI measures are available on the market, the scenario development helps to narrow down a wide range of BGI to a set of plausible scenarios. The resulting scenarios for objectives 1 and 2 are detailed in Sections 9.2.1 and 9.2.2.

9.2.1 Objective 1: dealing with heavy rainfall events

On the basis of the scenario development, the complete retention of rainwater from a 100-year design storm, and the interest of the urban drainage system (disconnection of 20–25% of the connected areas), the potential of BGIs for the management of heavy rainfall events in the Kolonnadenviertel should be demonstrated. Therefore, the infiltration swale will be evaluated in combination with two different types of green roofs (Figure 9.2). To be able to use the swale area in other ways, depths of 20 and 10 cm were also investigated in addition to the 30 cm depth suggested by DWA 138 (DWA, 2005). The focus was initially placed on the infiltration swale, so that the green roof was only considered when the capacity of the infiltration swale was no longer sufficient for the assumed precipitation event. To achieve full retention of the design storm in building complexes with insufficient available BGI (swale) area in the inner courtyard, both the area and the storage height of the green roofs were increased.

On the basis of objective 1, four scenarios were identified:

- S_{sw_0}: Status quo

  This scenario shows the initial situation of each building complex in the event of a heavy rainfall event. The amount of surface runoff per building complex is determined as a target for the design of the BGI.

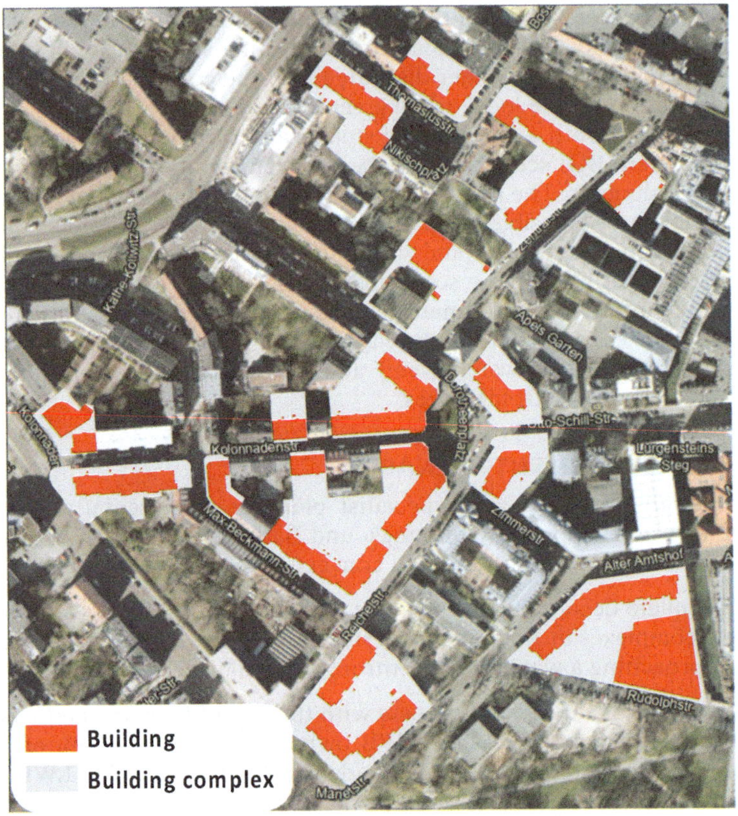

Figure 9.1 Communal housing company buildings and building complexes near the city centre of Leipzig (Germany).

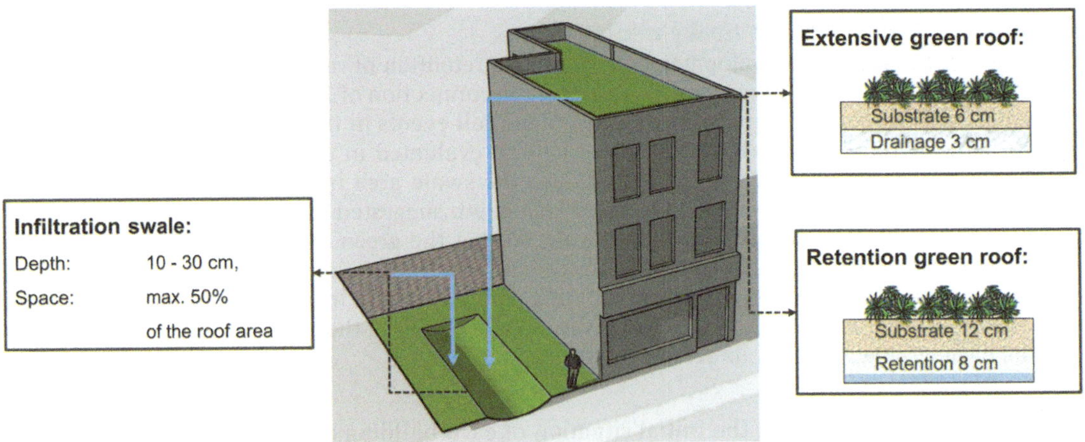

Figure 9.2 BGIs and cascading scheme.

- S_{sw_1}: Swale
  Scenario S_{sw_1} considers swales of different depths – 30, 20 and 10 cm – for stormwater management. Surface runoff is collected in a surface swale and infiltrated through a planted layer of soil.
- S_{sw_2}: Swale and extensive green roof
  In addition to S_{sw_1}, the retention volume of the building complexes is increased by an extensive green roof. Runoff of the roof and other sealed areas within the building complex feeds into the swale. Extensive green roofs have been added in cases where the swale is unable to retain all the total stormwater runoff – that is, for small inner courtyards that may limit the size of the swale.
- S_{sw_3}: Swale and retention green roof
  For S_{sw_3}, a retention green roof was considered instead of an extensive green roof (S_{sw_2}) in order to represent the spatial availability for nature-based solution (NBS) location and the retention capacity. Retention green roofs were included in case the swale was not able to retain all stormwater runoff.

9.2.2 Objective 2: towards a near-natural water balance

The protection, restoration and permanent safeguarding of the near-natural water balance is anchored as one of the 10 strategic topics in the German National Water Strategy (BMUV, 2023). The near-natural water balance is defined in DWA 102 as reference values for an area characterized by cultivated land use without settlement (DWA, 2022; Schmit *et al.*, 2023) and traffic areas ('undeveloped state'). The reference values for components of the water balance are: direct runoff, groundwater recharge and evaporation are to be defined as long-term mean values.

To achieve a near-natural water balance state, (i) the implementation of BGI and (ii) the maintenance of green spaces through irrigation are assessed. For the existing building complexes this was formulated into five scenarios:

- S_{irr_0}: Status quo
  The swale (S_{sw_1}) is considered in the sense that all storm/rain runoff is retained within the building complex. Irrigation of the swale or green spaces within the building complexes or green roofs is not considered.
- S_{irr_1}: Green roofs
  Scenario S_{irr_1} includes retention green roofs in the annual water balances. Irrigation is not included.
- S_{irr_2}: Courtyard irrigation
  In scenario S_{irr_2}, the courtyard is irrigated. No green roofs are included.
- S_{irr_3}: Courtyard irrigation and green roofs
  Scenario S_{irr_3} includes retention green roofs in the annual water balances and courtyard irrigation (S_{irr_2}).
- S_{irr_4}: Courtyard and green roof irrigation
  Scenario S_{irr_4} includes irrigation for both courtyard and the retention green roofs.

9.3 STORMWATER AND ANNUAL WATER BALANCE METHODS

For the assessment of different scenarios (S_{sw_0-3} and S_{irr_0-4}) a block-based approach, MUST-B, was used (Khurelbaatar *et al.*, 2021). MUST-B approach is based on the principle of dividing the city into residential blocks separated by roads. The urban blocks provide a basis for aggregating relevant data (e.g., climate data, soil data, buildings, green spaces and degree of sealing) and, based on these data, calculating the impact of different infrastructure and operational scenarios on the hydrological cycle. To apply the approach to existing building complexes of a communal housing company, the block area was set to the building complex footprint instead of the street boundaries. Each building complex is seen as a modelling unit for both stormwater and annual water balance assessment. Another key element of the MUST-B approach is the assumption that all rainwater on each urban block is managed (infiltrated and re-used for irrigation) within the block in order to minimize water transport.

For the short-term simulation, storm water management model (SWMM) (Rossman, 2015, Version 5.2) was used to illustrate the management of storm events (objective 1, Section 9.2.1) and to calculate the corresponding scenarios. For the long-term water balance components such as direct runoff, groundwater recharge and evaporation (objective 2, Section 9.2.2), both SWMM and the blue-green irrigation model, B-GRIIN, were used in combination (Breulmann *et al.*, 2024).

9.3.1 Climate data

For stormwater design events, the KOSTRA-DWD-2010 data (Malitz & Ertel, 2015), provided by the German Weather Service (DWD) were used for the short-term simulation. The KOSTRA-DWD-2010 dataset 'Coordinated heavy precipitation regionalisation and evaluation' is used to provide stormwater design events as a function of precipitation duration and annularity for Germany. On the basis of German regulations, DWA-A 138 (DWA, 2005) specifies that the technical measures for the management of precipitation water, such as conventional urban drainage systems (e.g., sewers and retention basins), should be designed for a 5-year storm event. Due to the increasing number of extreme weather events with high rainfall intensity and the associated risk of flooding, even at a block level, the 100-year heavy rainfall event was considered in the potential analysis in accordance with DIN 1986-100 (DIN, 2016).

Daily weather data from a DWD climate station (#2928) in Leipzig-Holzhausen (Germany) were used for the long-term simulation. Precipitation (DWD, 2021) and potential evapotranspiration (according to Penman Monteith, DWD, 2019) were taken for a dry year (2018), a wet year (1994) and for 10 randomly selected years from 1990 to 2020 (1993, 1999, 2000, 2003, 2005, 2008, 2012, 2015, 2016 and 2019).

9.3.2 Soil and vegetation

The permeability coefficient (k_f) of the soil is of crucial importance for the planning and design of urban water management infrastructure for precipitation management. In addition, soil values such as the permanent wilting point and field capacity are essential parameters for calculating irrigation requirements. The permeability coefficient in the study area is between 10^{-5} and 10^{-8} m/s (NATURWB, Schmit *et al.*, 2023). Its permanent wilting point is 14.2 mm and the field capacity is 33.2 mm. A k_f value of 10^{-5} was assumed for the scenario assessment.

For vegetation, grassland was assumed for irrigation estimation using a crop factor (k_c) of 0.8 (see Breulmann *et al.*, 2024). Although courtyard vegetation can include trees, grassland irrigation was assumed to allow trees to survive drought years in the form of deficit irrigation.

9.3.3 Urban blocks

For meaningful modelling of rainwater management systems and water balance components, an accurate description of the soil sealing is essential. These data are often derived from aerial photographs and geospatial data. For the studied building complexes, data provided by the communal housing company were used – buildings and sealed and unsealed areas. Unsealed areas within a block/building complex and building (roof) areas were considered for the placement of potential BGI (Table 9.1).

As not all of the green space within a building complex is usually available for the construction of BGI, it was assumed that the area required for swales would occupy a maximum of 50% of the unsealed area. For green roofs, it was assumed that up to 100% of the roof areas could be used.

9.3.4 Stormwater modelling

The SWMM stormwater model was used to dimension the BGI – swale, extensive and retention green roofs – specified in the scenarios S_{sw_0-3} (see Section 9.2.1). In accordance with the German rainwater

Table 9.1 Area composition of the communal housing building complexes.

Building Complex	Sealed Surface (m^2)	Open Surface (m^2)	Buildings (m^2)	Total Area (m^2)
1	510	1.029	457	1.996
2	1.460	2.227	1.512	5.199
3	368	1.156	453	1.978
4	1.610	4.091	1.608	7.309
5	78	0	477	555
6	909	2191	1.399	4.499
7	234	560	404	1.198
8	577	1.272	848	2.696
9	185	164	453	803
10	109	377	372	858
11	547	3.153	1.319	5.019
12	296	979	183	1.458
13	362	1.520	809	2691
14	821	3.270	1.455	5.546
Total sum	8.065	21.990	11.749	41.805

infiltration regulations (DWA, 2005), the inflow volumes of precipitation events with a return period of 100 years were calculated in various duration stages (in 5-min increments) up to 72 h.

The swale was designed and modelled in each building complex so that the 30, 20 or 10 cm storage height were exhausted, but no overflow occurred. In the event of an overflow, the swale area was increased and the simulation repeated until the swale area attained 50% of the green area of the courtyard.

Once the 50% of the green area was exhausted, two different types of green roofs (extensive and retention) were modelled on the roof area as additional retention space in case the building complexes still had residual runoff. The green roof area was progressively increased until either the modelling showed no further runoff from the building complex or the proportion of the green roof attained 100% of the total roof area. Figure 9.3 shows the SWMM modelling including the assumptions for different infrastructures and areas.

9.3.5 Irrigation modelling

For the long-term simulation of water availability and water demand a water balance approach was coupled with an irrigation module as applied by Breulmann *et al.* (2024). Different water demands were calculated based on the scenarios S_{irr_0-4} (see Section 9.2.2). The BGI parameterization was taken from the stormwater scenarios S_{sw_0-3} (see Section 9.2.1). All simulations were performed per building complex (Table 9.1). The cascading water flows of the different surfaces and infrastructures within the building complexes are detailed in Figure 9.2.

The water balance is based on the temporal variation (Δt) of the water inflow, outflow and change in soil storage:

$$\frac{\Delta Q_{in}}{\Delta t} - \frac{\Delta Q_{out}}{\Delta t} - \frac{\Delta (S_0 - S_1)}{\Delta t} = 0 \tag{9.1}$$

where Q_{in} is the inflow (precipitation and irrigation), Q_{out} is the outflow (evaporation, surface runoff, infiltration and interception) and $S_0 - S_1$ is the change in soil water storage. For irrigation, soil water

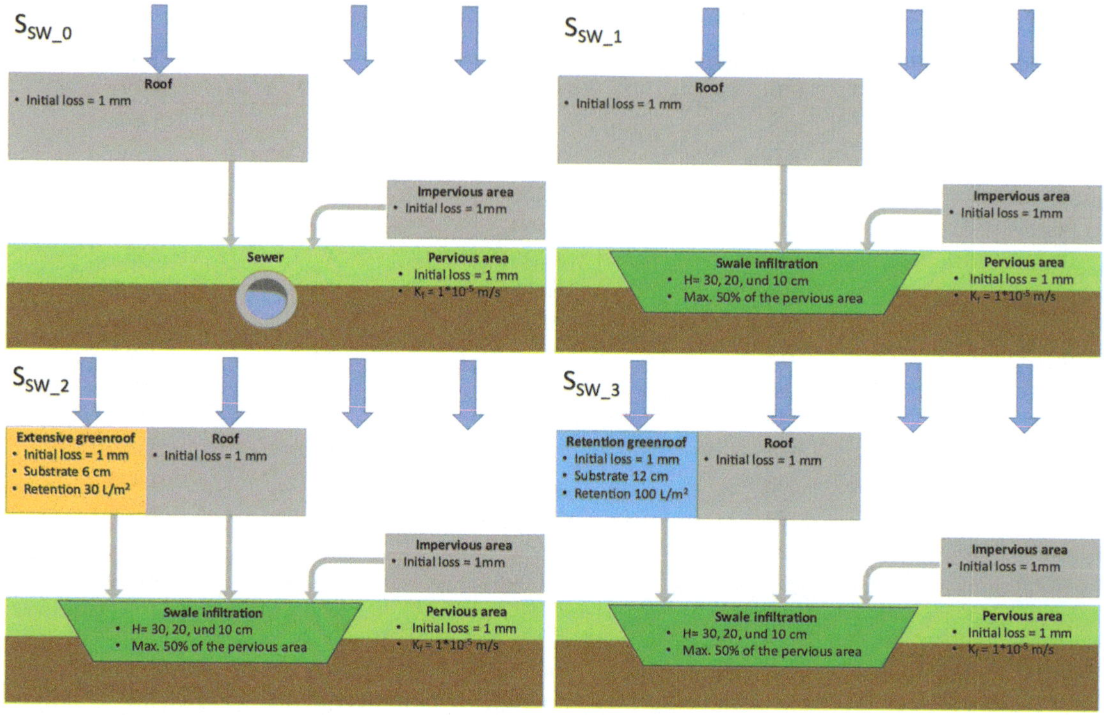

Figure 9.3 SWMM parameterization and infrastructure dimensioning.

storage is calculated as plant available water capacity (AWC). The irrigation amount was set to 20 mm and irrigation was applied at or below 30% AWC.

Inflow (Q_{in}) is defined as the change in water inflow over time. The water influx to the compartments of the building complexes (inner courtyard, retention green roof and sealed areas) was considered as the sum of precipitation and irrigation water. For irrigation, root zone water storage determines the irrigation amount and timing of irrigation. The total substrate thickness of the inner courtyard and the retention area was defined as the root zone. It was also specified that the AWC of the plant should not be less than 30% and not more than 80%. Individual irrigation amounts were set at 20 mm and the substrate was assumed to be a soil with $k_f = 10^{-5}$ with a field capacity of 35% and a wilting point of 10%.

Outflow (Q_{out}) is defined as the water runoff. The sum of potential evapotranspiration, interception, groundwater infiltration and runoff from multifunctional areas was considered as discharge. Daily precipitation and potential evapotranspiration data for different years were taken as input from the DWD (see Section 9.3.1). For the calculation of actual evapotranspiration, an annual mean value of the plant coefficient (k_c) for grass of 0.8 was used. Actual evapotranspiration was calculated by multiplying the plant coefficient by the potential evapotranspiration. Grass normally has a k_c of ~1 and by choosing 0.8 we take into account the inactive winter months. The model allows for a constant water loss by interception of 1 mm per precipitation day, which reduces the effective precipitation. The surface runoff from sealed surfaces is calculated as the total precipitation minus the interception and flows into the courtyard. If the maximum water storage capacity of the substrate of the retention green roof and the inner courtyard is exceeded, the excess water infiltrates directly into the groundwater.

The storage change ($S_0 - S_1$) is defined as the sum of the amount of water stored in the courtyard substrate and the infrastructure storage tank over time. The model initialization assumes that the substrate of the courtyard root zone and the green roof are saturated on day 1 ($t = 0$). The total available substrate storage capacity is therefore 43 mm for the retention green roof and 268 mm for the unsealed courtyard area.

No initial source or storage is assumed for irrigation water. The model results can be used to estimate a sustainable irrigation using groundwater as a virtual cistern or an actual cistern dimensioned by model results.

9.4 STORMWATER AND IRRIGATION SCENARIO RESULTS

9.4.1 BGI dimensioning for stormwater mitigation

The simulation of decentralized rainwater management in the Kolonnadenviertel was carried out for a 100-year storm event (122.5 mm within 72 h). A total of 5121 m^3 of rainwater falls on the 14 building complexes (~41,000 m^2). Four BGI scenarios (S_{sw_0-3}), including a status quo scenario, were assessed to determine whether the surface runoff of a 100-year stormwater event could potentially be decoupled from the combined sewer system through the decentral placement of BGI measures (Figure 9.4).

S_{sw_0} reflects the status quo of the study area, resulting in a surface runoff of 2545 m^3/72 h (50% of the rainwater), which is currently overloading the existing sewer network.

Overall, the results of the scenario S_{sw_1} show that an almost complete (98%) stormwater retention with a residual runoff of 120 m^3 (~2% of the stormwater volume) can be achieved with a 20 cm deep swale infiltration in most building complexes. The total area of the swales is 8354 m^2 (20% of the total area or 38% of the open area) (Figure 9.5). Only four building complexes were found to have residual runoff and therefore a need for additional retention through green roofs. By adding extensive green roofs to the four building complexes (S_{sw_1}), with a total roof area of 2591 m^2, 99% of the stormwater was retained. The retention green roof (S_{sw_2}) achieved the same retention (99%) that was achieved with a slightly smaller retention roof area (1875 m^2).

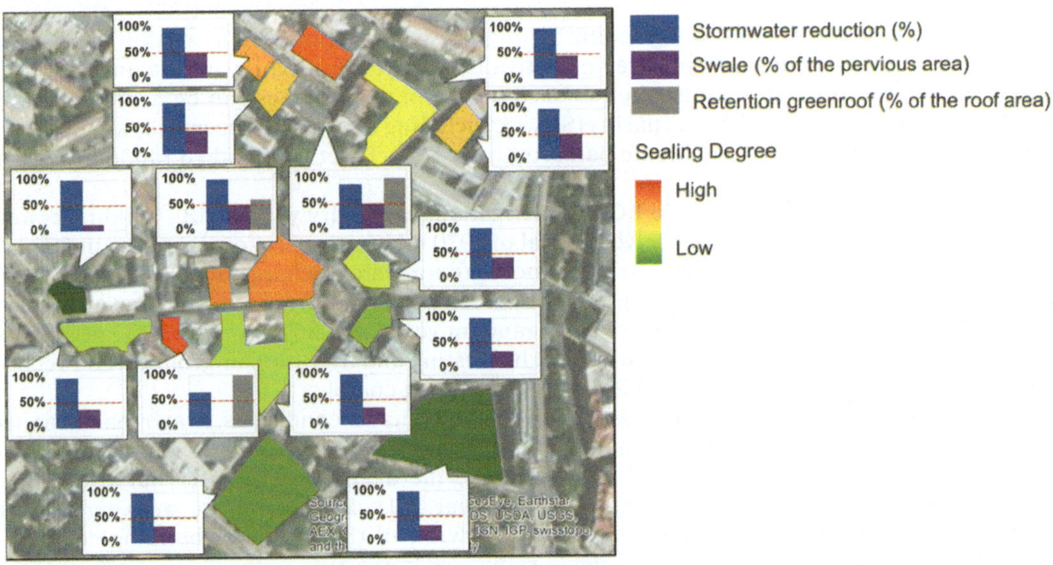

Figure 9.4 Stormwater mitigation potential for all building complexes under scenarios S_{sw_1} (swale) and S_{sw_3} (swale and retention green roofs).

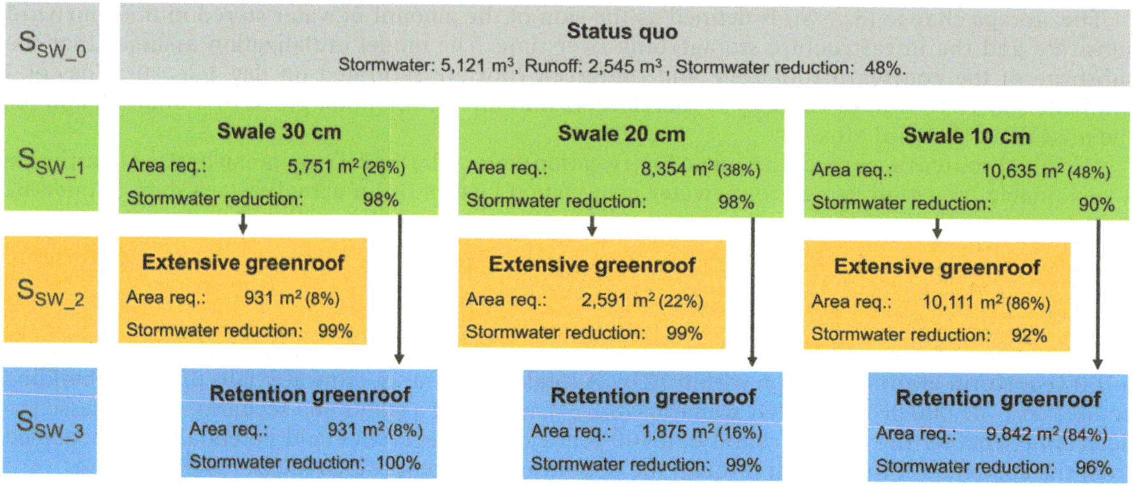

Figure 9.5 Stormwater mitigation results for scenarios S_{SW_0-4} considering all communal housing building complexes in the Kolonnadenviertel quarter of Leipzig.

The swale area (S_{sw_1}) was reduced from 8354 to 5751 m^2 by increasing the depth of the swale from 20 to 30 cm. In this case, the roof area required for the extensive (S_{sw_2}) and retention green roof (S_{sw_2}) was also halved (Figure 9.5). The opposite was observed when the swale depth was reduced to 10 cm. The area required for both the swales and the green roof increased. In this case, the swale area doubled and the green roof area quadrupled to achieve a slightly lower performance compared to the deeper swales.

9.4.2 Sustainable urban water management through BGIs and rainwater irrigation

In addition to stormwater mitigation, a key objective is to shift the urban water balance towards a near-natural water balance. To assess whether such a shift in water fluxes can be achieved through BGI and irrigation, the effect of five scenarios (S_{irr_0-4}), including a status quo scenario, was estimated. Each scenario was forced by three climate datasets, 1 dry year, 1 wet year and 10 randomly selected years.

The status quo (S_{irr_0}) did not include BGI and irrigation. In the dry year (2018), building complex 2 (Box 'Building complex factsheet') received a total of 1970 m^3 of precipitation, resulting in 880 m^3 of water available for irrigation (e.g., as subsurface runoff, groundwater recharge or collected in cisterns). In the wet year (1994), 3930 m^3 of precipitation was received and 2172 m^3 of water was available for reuse. The average precipitation received over 10 random years was 2574 m^3 with 1117 m^3 available water (Table 9.2). The water available for rainwater harvesting and irrigation or groundwater recharge in different years thus ranges from 880 to 2172 m^3 for building complex 2.

Taking into account the climate forcing for different climatic conditions, water balances were modelled for different building complexes and it was shown how BGIs and water management affect the water balance. Considering all building complexes, the available rainwater harvesting (S_{irr_0}) corresponds to ~25–45% (of precipitation) in the dry year 2018, 23–55% in the wet year 1994 and 23–46% in the average years (see Table 9.2 for an overview of the results of building complex 2). Considering the green roofs (S_{irr_1}), the available rainwater harvesting reduced to ~25% in the dry year and 23% in average wet years.

Building complex factsheet

Clear communication of scientific results in co-design processes is essential. To this end, factsheets were designed that summarize the scenario results for specific building complexes.

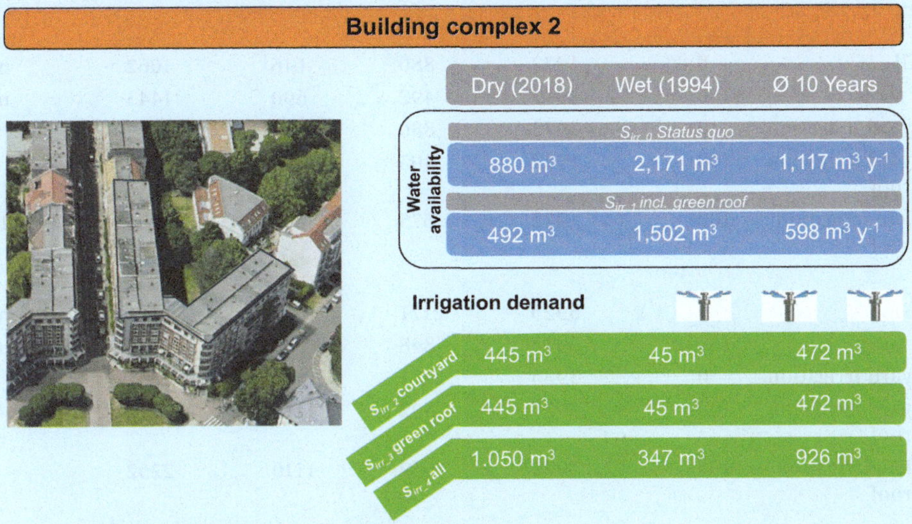

The factsheet highlights the results for the different irrigation scenarios under three climate conditions (dry year, wet year, and 10-year average). Depending on the co-designed infrastructure scenarios, water availability through rainwater harvesting and water demand through different irrigation scenarios are shown.

The factsheet highlights the results for the different irrigation scenarios under three climate conditions (dry year, wet year and 10-year average). Depending on the co-designed infrastructure scenarios, water availability through rainwater harvesting and water demand through different irrigation scenarios are shown.

The objectives for scenarios S_{irr_2-4} were drainage-free building complexes and an optimally irrigated green inner courtyard, even in dry years. The collected rainwater would be used for irrigation. When modelling the blue-green scenarios (S_{irr_2-4}), different water flows were taken into account (Figure 9.2). Taking into account the precipitation data, it is possible to achieve a sustainable water balance with the BGI and provide irrigation water for the inner courtyard or for the green roof. As no runoff is discharged to the sewer, the entire cost of discharging the rainwater into the combined sewer system from the current situation (S_{irr_0}) is saved.

By installing green roofs (S_{irr_1-4}) – in our simulation we initially assume 100% roof area – a large part of the rainwater is already consumed by evapotranspiration and thus significantly reduces the available water compared to the status quo scenario S_{irr_0} (see Table 9.2). For example, the available water in the dry year in building complex 2 is reduced from 45 to 25%.

Looking at all building complexes in the suburb 'Kolonnadenviertel', results show that sustainable irrigation (Figure 9.6) varies with climatic conditions and available space. The installation of green roofs on existing buildings may not always be possible due to the existing challenges such as building statics of the building (roof load), heritage protection or roof shapes. Therefore, S_{irr_2} has considered

Table 9.2 Building complex 2.

Scenario/Climate	Green Roof (% Roof Area)	Rainfall[a] (m^3/year)	Reuse Potential (m^3/year)	Soil Water (m^3/year)	Evaporation (m^3/year)	Irrigation Demand[b] (% Reuse Potential)
Dry						
S_{irr_0} no BGI	0	1972	880	616	1062	n.a.
S_{irr_1} green roof	100	1972	492	690	1443	n.a.
S_{irr_2} courtyard irrigation	0	1972	880	806	1329	51
S_{irr_3} courtyard irrigation and green roof	100	1972	493	877	1710	90
S_{irr_4} complete irrigation and green roof	100	1972	575	878	2231	183
Wet						
S_{irr_0} no BGI	0	3925	2171	1012	1341	n.a.
S_{irr_1} green roof	100	3925	898	958	2016	n.a.
S_{irr_2} courtyard irrigation	0	3925	2171	1057	1341	2
S_{irr_3} courtyard irrigation and green roof	100	3925	1502	1116	2016	3
S_{irr_4} complete irrigation and green roof	100	3925	1569	1110	2252	22
Average						
S_{irr_0} no BGI	0	2574	1117	635	882	n.a.
S_{irr_1} green roof	100	2574	598	638	1404	n.a.
S_{irr_2} courtyard irrigation	0	2574	1117	626	1336	42
S_{irr_3} courtyard irrigation and green roof	100	2574	598	656	1858	79
S_{irr_4} complete irrigation and green roof	100	2574	664	659	2244	139

All scenarios are initiated with full soil storage at field capacity.
[a]In addition to the annual rainfall, all scenario calculations are initialized with soil water at field capacity.
[b]Irrigation demand is calculated independent of available water, in case irrigation demand is above 100%, irrigation is not sustainable in view of the water balance.

a scenario that only takes into account the optimal irrigation of the inner courtyard. Only in the extremely dry year 2018 is the water supply insufficient for optimal irrigation of the inner courtyard (Figure 9.6). If, in addition to the green roofs, an optimal irrigation of the inner courtyard (S_{irr_3}) or of the inner courtyard and the green roof (S_{irr_4}) is targeted, the water supply is often insufficient, especially in dry years, taking into account our assumptions (Figure 9.6).

Scenarios including irrigation (S_{irr_2-4}) further lead to a transition of the water balance towards a near-natural state. This transition, however, can only be considered in case the irrigation demand is provided through available water (Figure 9.6, blue boxes). Looking at evaporation – under sustainable conditions – results further show that BGIs have an increased evaporation. Averaged over all building complexes evaporation increases from 43 to 59% (dry year), from 32 to 46% (wet year) and from 37 to 58% (random years). Near-natural water balance estimates that simulate the undeveloped state show even higher evaporation (83% dry year, 51% wet year, 68% random years). However, introducing BGI measures clearly closes the gap towards the near-natural water balance.

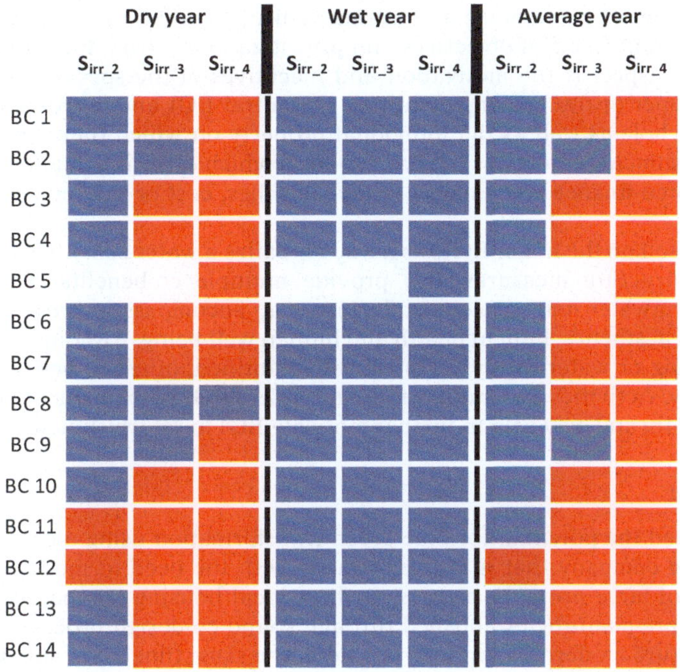

Figure 9.6 Sustainable irrigation potential per building complex (BC 1–14). Blue boxes indicate scenarios and climate conditions under which the irrigation demand is covered by annual precipitation. Red boxes indicate insufficient precipitation for the irrigation demand (scenario details are given in Section 9.2.2). Blank boxes indicate scenarios that were not applicable to the building complex.

9.4.3 Potential for BGI

The assessed scenarios show the potential for blue-green refurbishments within existing suburbs. Depending on the specific objectives, for example, set by the communal housing company, local regulations, water utilities, the most appropriate scenarios can form the basis of an informed stakeholder process or be used directly to tender for detailed planning.

9.5 DISCUSSION

Transforming cities into resource-efficient and climate-sensitive urban spaces is an ongoing and pressing challenge. Although this is often considered and well planned for greenfield developments, it remains a challenge for existing suburbs and buildings.

In response to this challenge, concepts and models were simulated that could be incorporated into the planning of the modernization of several existing building complexes in the suburb 'Kolonnadenviertel' of the city of Leipzig. When modernizing existing buildings with BGI, stakeholder interaction is essential in order to cope with local specificities and to respect restrictions or goals set by the property owners. At the same time, large-scale goals such as the United Nation's sustainable development goal 11 'Sustainable cities & communities' (UN DESA, 2023), the European Union (EU) green deal (EC DG RTD, 2021) with its mission to increase urban green spaces, as well as national strategies to transition urban water balances towards near-natural states (BMUV, 2023) need to be kept in mind. Considering the municipality level specific goals set, for example, by water utilities, to reduce the stormwater pressure onto the sewer systems need consideration. Our approach to transition urban spaces towards more sustainable and attractive urban conditions is to use BGI. Although we can consider the different levels of objectives and prioritizations – from local to national or regional (EU) – an important aspect is the motivation and incentives of the specific refurbishment project, that is, the house owner or the housing company. Decentral BGI and its co-benefits affect both the city or suburb as a whole as well as the individual property on which they are located. Financially, however, the refurbishment project has to bear the costs which often is a challenge. Options to target this challenge are to co-finance these measures by subsidies, cost or tax reductions related to water use and treatment.

Although we focused on the use of BGIs as a way to realize a water-sensitive urban transformation, the proposed infrastructure measures also provide multiple co-benefits. The multifunctionality of BGIs beyond stormwater management and reuse may encompass improvements for air quality, biodiversity, cooling, aesthetics, human health and human well-being. Although water-related benefits linked to water saving or water utility costs can be quantified, other co-benefits can often only be assessed qualitatively. Activities and studies to better assess BGI co-benefits are currently under way (Moeller, 2024a, 2024b) and may help to further motivate BGI in refurbishment or greenfield projects.

9.6 CONCLUSIONS

We have demonstrated the potential for blue-green transformation using a co-design approach with a communal housing company that is regularly faced with the need to modernize all its buildings. A major advantage of large property owners, such as local housing companies, municipalities and industry, is that the pre-planning established through sound co-design processes can be more easily incorporated into future modernization efforts, and at the same time allow for significant up-scaling of water-sensitive modernization across multiple buildings.

Our focus was on two main objectives – stormwater mitigation and sustainable water use. Both objectives can be achieved, depending on the local conditions, through the implementation of BGI and appropriate water management schemes. To provide users with a range of plausible infrastructure measures rather than a single solution, scenario-based modelling was chosen. Scenario-based modelling allows users and stakeholders to design or co-design scenarios and assess their impact on defined objectives. Scenarios can be prescribed, for example, by climate-sensitive municipal regulations, or co-designed with stakeholders, taking into account their constraints and objectives. The ultimate aim is to provide a sound comparison of different scenarios before a detailed planning process begins, thereby reducing planning costs and optimizing planning processes.

For the suburb 'Kolonnadenviertel', we were able to show that multifunctional blue-green water infrastructure and sustainable irrigation concepts are feasible and can be used to (i) mitigate the effect of stormwater and (ii) move towards a near-natural water balance. Although this may not always be feasible for all climatic conditions and every building complex within a suburb, it does provide a pre-planning overview of how much can be achieved through water-sensitive infrastructure planning.

In summary, stormwater mitigation, even for an extreme event, such as a 100-year design event, it was shown that it is possible to reduce stormwater runoff using relatively low-cost infrastructure measures such as swales. Even with the use of swales alone, stormwater runoff to the combined sewer system could be reduced by at least 90% – with the additional use of green roofs, even greater reductions can be achieved. The reuse of available roof runoff for courtyard and green roof irrigation is, of course, highly dependent on climatic conditions from dry to wet or average years. Depending on the irrigated areas, that is courtyard or courtyard and green roof, it is not always possible to irrigate with roof water runoff. However, in many years and for most building complexes, a courtyard irrigation could be reached. Modernization towards a near-natural water balance not only meets the overall objective of many recent policies, but also results in a more liveable urban environment and has an impact on co-benefits such as cooling or biodiversity.

ACKNOWLEDGEMENTS

The work presented in this book chapter was supported by grant 033W110AN 'Leipziger BlauGrün II – Blau-grüne Quartiersentwicklung in Leipzig' as part of the funding initiative Resource-efficient urban districts (RES:Z) by the German Federal Ministry of Education and Research (BMBF). Further support was provided by the MULTISOURCE and WATERUN projects, which have received funding from the European Union's Horizon 2020 Program under grant agreements 101003527 and 101060922, respectively.

REFERENCES

BMUV (2023) National Water Strategy. Federal Ministry for the Environment, Nature Conservation, Nuclear Safety and Consumer Protection, Berlin, Germany. https://www.bmuv.de/fileadmin/Daten_BMU/Download_PDF/Binnengewaesser/nationale_wasserstrategie_2023_en_bf.pdf (accessed 13 February 2024).

Breulmann M., Müller R. A. and van Afferden M. (2024) Modeling urban stormwater and irrigation management with coupled blue-green infrastructure in the context of climate change. *Blue-Green Systems*, **6**(1), 100–113. https://doi.org/10.2166/bgs.2024.101

DIN (2016) DIN 1986-100:2016-12 Entwässerungsanlagen für Gebäude und Grundstücke – Teil 100: Bestimmungen in Verbindung mit DIN EN 752 und DIN EN 12056. https://doi.org/10.31030/2586228

DWA (2005) DWA-A 138E – Planning, Construction and Operation of Facilities for the Percolation of Precipitation Water. Deutsche Vereinigung für Abwasserwirtschaft, Abwasser und Abfall e. V. (DWA): Hennef, Germany.

DWA (2022) DWA-M 102-4/BWK-M 3-4 – Grundsätze zur Bewirtschaftung und Behandlung von Regenwetterabflüssen zur Einleitung in Oberflächengewässer – Teil 4: Wasserhaushaltsbilanz für die Bewirtschaftung des Niederschlagswassers. Deutsche Vereinigung für Abwasserwirtschaft, Abwasser und Abfall e. V. (DWA): Hennef, Germany.

DWD (2019) Calculated daily values for different characteristic elements of soil and crops. Version v19.3. German Weather Service (DWD). Climate Data Center (CDC). Available online at: https://opendata.dwd.de/climate_environment/CDC/derived_germany/soil/daily/ (accessed 15 November 2022)

DWD (2021) Historical Daily station observations (temperature, pressure, precipitation, sunshine duration, etc.) for Germany. Version v21.3. German Weather Service (DWD). Climate Data Center (CDC). Available online at: https://opendata.dwd.de/climate_environment/CDC/observations_germany/climate/daily/kl/ (accessed 15 November 2022).

EC DG RTD (2021) European Green Deal: research & innovation call. Publications Office of the European Union. European Commission, Directorate-General for Research and Innovation. https://data.europa.eu/doi/10.2777/33415

Khurelbaatar G., van Afferden M., Ueberham M., Stefan M., Geyler S. and Müller R. A. (2021) Management of urban stormwater at block-level (MUST- B): A new approach for potential analysis of decentralized stormwater management systems. *Water*, **13**(3), 378, https://doi.org/10.3390/w13030378

LBG II (2023) Blau-grüne Quartiersentwicklung in Leipzig 'Leipziger BlauGrün II'. https://www.ufz.de/leipzigerblaugruen/index.php?en=46672 (accessed 13 September 2023)

Malitz G. and Ertel H. (2015) KOSTRA-DWD-2010. Starkniederschlagshöhen für Deutschland. Deutscher Wetterdienst: Offenbach am Main, Germany, 2015.

Moeller L. (2024a) PhD college CLEANER – Fostering pollutant-sink functions of blue-green infrastructures: Towards local urban water cycles in climate-resilient cities. Available online at: https://www.ufz.de/index.php?en=49715 (accessed 26 February 2024).

Moeller L. (2024b) The UFZ research green roof. Available online at: https://www.ufz.de/index.php?en=44676 (accessed 26 February 2024).

Rossman L. (2015) Storm water management model reference manual volume I – Hydrology I. U.S. Environmental Protection Agency, Office of Research and Development, National Risk Management Laboratory, Cincinnati.

Schmit M., Steinbrich A., Leistert H. and Weiler M. (2023) Webtool zur Bestimmung der naturnahen urbanen Wasserbilanz (NatUrWB) als deutschlandweiter, einheitlicher Referenzzustand. Tagungsband/Forum zur Hydrologie und Wasserbewirtschaftung, 44.23, DWA. https://doi.org/10.6094/UNIFR/237475. Available online at: https://freidok.uni-freiburg.de/data/237475 (accessed 20 December 2023).

StMUV (2020) Wassersensible Siedlungsentwicklung in Bayern- Empfehlungen für ein zukunftsfähiges und klimaangepasstes Regenwassermanagement in Bayern. Bayerisches Staatsministerium für Umwelt und Verbraucherschutz, Munich, Germany. Available online at: https://www.bestellen.bayern.de/application/eshop_app000006?SID=1571021679&ACTIONxSESSxSHOWPIC(BILDxKEY:%27stmuv_wasser_018%27,BILDxCLASS:%27Artikel%27,BILDxTYPE:%27PDF%27) (accessed 26 February 2024).

UN DESA (2023) The Sustainable Development Goals Report 2023: Special Edition – July 2023. New York, USA: UN DESA.

doi: 10.2166/9781789065015_0191

Chapter 10

Mangrove–eco-geomorphic interactions modelling for nature-based solutions

S. M. Beselly[1,2]*, *M. van Der Wegen*[1,3] *and D. Roelvink*[1,3,4]

[1]IHE Delft Institute for Water Education, Coastal and Urban Risk & Resilience, Delft, the Netherlands
[2]Water Resources Engineering, Brawijaya University, Malang, Indonesia
[3]Deltares, Delft, the Netherlands
[4]Department of Hydraulic Engineering, Delft University of Technology, Delft, the Netherlands
**Corresponding author: s.beselly@un-ihe.org*

ABSTRACT

Mangrove ecosystems are widely recognized for having highly valued multiple ecosystem services. These services, however, are often overlooked because of the lack of understanding of mangrove's species-specific and associated eco-geomorphological dynamics. Therefore, it will lead to a limited quantification and valuation of mangrove's functional and structural attributes. A mangrove ecosystem model capable of mechanistically simulating the feedback loop between mangrove stands and physical–environmental drivers is essentially important, specifically in the strategy of integrating mangroves as nature-based solutions for climate change adaptation and mitigation. The main objectives of this chapter are to gain a better insight into complex mangrove ecosystem and eco-geomorphic interactions to predict their trajectories, the possibility of modelling those utilizing a process-based model, and explore the interactions of mangrove, mudflat, and physical–environmental drivers. Following that, this chapter introduces a new hybrid model, so-called DFMFON, achieved by coupling mangrove individual-based and landscape-scale hydro-morphodynamic models which is capable of reproducing mangrove forest dynamics and morphodynamic delta features. To conclude, the application, limitations, and future development of process-based mangrove modelling for nature-based solutions are discussed.

Keywords: Conservation, eco-hydro-morphological modelling, ecosystem services, mangrove coasts, mangrove dynamics, mangrove modelling, mangrove–eco-geomorphic, mechanistic mangrove–morphodynamic, nature-based solutions, restoration.

10.1 INTRODUCTION

10.1.1 Values and functions of mangrove forests

Mangrove forests, in the past, were often misunderstood to be associated with a wasteland (Dahdouh-Guebas *et al.*, 2020) with a smelly, rotten-egg-like environment and a swarm of mosquito-borne diseases (Yeo *et al.*, 2021). That undervalued perspective is somewhat of a justification for clearing

© 2025 IWAP. This is an Open Access book chapter distributed under the terms of the Creative Commons Attribution License (CC BY-NC-ND 4.0) which permits copying and redistribution for non-commercial purposes with no derivatives, provided the work is properly cited (https://creativecommons.org/licenses/by-nc-nd/4.0/). The chapter is from the book *Nature-Based Solutions for Urban Sustainability*, Piet N.L. Lens and Xuan-Thanh Bui (Editors).

mangrove swamps, primarily in favour of aquaculture and agriculture (Hagger *et al.*, 2022) or other developments. Between 2000 and 2016, a satellite remote-sensing analysis study revealed that 62% of global losses of mangrove area were human-driven through land conversion (Goldberg *et al.*, 2020). This decline mostly happened in developing countries, where the mangrove forests were cleared for other high economic value activities. Mangroves have been declining at an alarming rate, ~1–2% annually, even faster than coral reefs or tropical rainforests (Alongi, 2002; Duke *et al.*, 2007). These factors make mangroves one of the threatened ecosystems.

In contrast to the previous justifications for mangrove forest deforestation, the loss of mangrove forests, in reality, deprives the population of the high intrinsic value of mangrove forests. The intrinsic values are derived from the high capacity of primary productivity, terrestrial ecosystem supports, and marine food webs (Dahdouh-Guebas *et al.*, 2022). The presence of mangrove forests has proven to be an effective buffer for storm surges (De Dominicis *et al.*, 2023). The effectivity of the wave attenuation function is non-linear, dependent on the width, density, and fragmentation of the forest (De Dominicis *et al.*, 2023; van Zelst *et al.*, 2021), where scientists have acknowledged wave height reduction between 13 and 66% over 100 m mangrove forest width (McIvor *et al.*, 2012). Another benefit can be correlated with how communities perceive their relation with mangrove ecosystems in cultural non-materialistic interrelationships. Here, mangrove forests serve as places for recreational or intangible spiritual purposes for people (Das *et al.*, 2022). These benefits have been recognized as the mangrove ecosystem services, which can be divided into habitat, regulating services (e.g., regulating climate and disaster risk reduction), provisioning services (e.g., source of food), and cultural services (Das *et al.*, 2022).

10.1.2 Mangrove forests under threat

Over-exploitation and loss of mangroves significantly impact biodiversity and the surrounding ecosystems (Carugati *et al.*, 2018). A dramatic impact of mangrove forest loss to biodiversity can be seen, for example, in Indonesia – home of the largest mangrove forests, ~20% of the global mangrove area. The United Nations Environment Programme (UNEP) (2023) reports that large-scale mangrove land conversion to oil palm plantations, rice fields, and aquaculture is directly linked to habitat destruction and has endangered 41 bird species unique to mangrove forests. Winterwerp *et al.* (2020) report a direct relationship between mangrove loss and the severe erosion of the muddy coast on Java's north coast, comprising ~44% of Java coastlines. This is an alarming situation for the community, whose lives to a major extent depend on mangrove ecosystems. In Demak, Central Java, aquaculture establishment close to the waterline and mangrove loss has resulted in self-accelerated coastal erosion (World Bank Group, 2018). Land subsidence and mismatch of infrastructure development aggravated the coastline retreat up to 1.5 km, where 70,000 people have been affected (Damastuti *et al.*, 2022). That caused significant income loss for almost 80% of the village inhabitants (Winterwerp *et al.*, 2016).

10.1.3 Impacts of climate change

The recent sixth assessment of the Intergovernmental Panel on Climate Change (IPCC) reports that human activities have caused 1°C global warming with already widespread and intensifying impacts (IPCC, 2022). The primary evidence of climate change impacts on coastal areas, among others, is sea-level rise (SLR) and more frequent and stronger storms. The current estimate shows that 23% of the world population has been exposed to flood inundation of over 0.15 m in a 100-year return period (Rentschler *et al.*, 2022). This figure may increase following the SLR scenario, where the frequency and intensity will become higher. When compounded with the changing patterns of precipitation in the hinterland, it increases the depth and duration of flooding in urbanized coastal cities worldwide, where more than 600 million people live in these low-lying regions (Magnan *et al.*, 2022). The exposure to coastal flooding will increase with the growing population and social-economic activities, where the destruction effect will be multiplied.

10.1.4 Mangroves for climate mitigation

Sustainable approaches to mitigation and adaptation measures are critical to address the impacts posed by climate change. The goal of mitigation is to reduce the amount of greenhouse gases in the atmosphere by reducing the sources or enhancing the carbon sink. As climate change is inevitable, followed by the global target of achieving net zero emissions by 2050 (Saintilan *et al.*, 2023), the high potential of ocean and coastal ecosystems referred to as blue carbon ecosystems (BCE) has been proposed as the natural climate solution (Macreadie *et al.*, 2021). BCE consists of tidal marshes, seagrass meadows, and includes mangroves.

Owing to its co-benefits of carbon sequestration and climate adaptation, the potential of BCE for climate mitigation has been the subject of extensive research (Kusumaningtyas *et al.*, 2022). BCE has a disproportionally large carbon storage, storing half of the carbon sequestration in the ocean while only occupying 0.5% of the sea floor (Macreadie *et al.*, 2021). Even though the sequestering capacity varies highly among sites (Sidik *et al.*, 2023), among the other ecosystems, mangrove forests have the highest rate of carbon sequestration (Kusumaningtyas *et al.*, 2022). The mangrove area, nevertheless, can shift the role from carbon sink to carbon source when deforested. It is estimated, with global annual deforestation of 0.4% (Hamilton & Casey, 2016), that the carbon emitted due to the loss of mangroves will contribute to 3–19% of the total carbon emission (Donato *et al.*, 2011; Pendleton *et al.*, 2012). Between 1996 and 2020, the global carbon stocks net reduction associated with mangrove loss was estimated to equal 139 megatonnes (Mt) (United Nations Environment Programme, 2023). This net reduction equals four times the carbon produced globally in 2018, originating from fossil fuels burning and cement manufacturing. On the contrary, when rehabilitation attempts to increase the mangrove extent can be achieved, the potential climate benefit can reach more than 424 Mt carbon by 2030 (Sasmito *et al.*, 2023).

10.1.5 Mangroves for climate adaptation

Climate adaptation involves measures to reduce the risk and vulnerability to the unavoidable climate change impacts due to the past and current emissions. Adaptation measures include building coastal defences, early warning systems, or setting the setback area. Traditionally, coastal defences are approached from an engineering perspective and comprise man-made or 'grey' structures. The grey structures are designed to interact and mitigate a specific part of the physical processes, for example, waves, sediment transport, or water level (Coastal Engineering Manual, 2002; Shore Protection Manual, 1973). Given the continuous physical environment actions, grey structures require careful operations and likely costly maintenance to fulfil their design function and reach the expected service life (Duarte *et al.*, 2013). In mangrove–muddy coastal systems, the solutions using conventional permanent structures could lead to fragmentation and disruption of ecological connectivity of mangrove ecosystem services (Borsje *et al.*, 2011) or can induce scour due to the reflective surface in muddy coasts (Winterwerp *et al.*, 2020). Several studies mentioned the benefits of incorporating vegetation or ecosystem-based approaches have been approved to be more sustainable and cost-effective compared to hard structure measures (Tiggeloven *et al.*, 2022; van Zelst *et al.*, 2021). The approaches utilizing natural elements for coastal hazard mitigation, such as mangroves are commonly termed nature-based solutions (Narayan *et al.*, 2016). Mangroves can aid climate adaptation with their coastal protection function (Temmerman *et al.*, 2023), with the highest economic value per hectare (Macreadie *et al.*, 2019). The role of mangroves as coastal defence is well appreciated. Many studies have shown mangrove functionality, such as a buffer from tsunamis (Dahdouh-Guebas *et al.*, 2005) or attenuating wind waves and storm surges (Marois & Mitsch, 2015; Montgomery *et al.*, 2018).

10.1.6 Coordinated actions

Acknowledging a wide range of ecosystem services, mangroves have been considered one of the high priorities with coordinated global communities for mangrove conservation (Friess *et al.*, 2020). Countries have committed to their National Determination Contribution to conserve mangrove forests as a derivation of the Paris Agreement. To date, countries and international organizations have been advocating green belt policies (Su *et al.*, 2021), for example the United Nations (UN) Framework Convention on Climate Change (UNFCCC. Secretariat, 2021), Australia (Kelleway *et al.*, 2017; Morris *et al.*, 2021), China (Fu *et al.*, 2021), Engineering With Nature in the USA (Bridges *et al.*, 2018; Nature-Based Solutions Resource Guide, 2022) or the US National Academy of Sciences agenda (Board and National Academies of Sciences, 2019), and Indonesia (Sasmito *et al.*, 2023). Globally, mangrove restoration and conservation are in urgent need as targeted by the UN Decades of Ecosystem Restoration 2021–2030, Sustainable Development Goals, and Aichi Biodiversity Targets. The global conservation community has committed to increasing the global mangrove cover by 20% in 2030 (Global Mangrove Alliance, 2021; Sasmito *et al.*, 2023). Given the close deadline of both targets in climate change mitigation–adaptation and mangrove conservation, prompt actions in recent years are critical. When it is successful, conserving the remaining and restoring degrading mangroves not only could mitigate climate change but also reduce the impacts to 296 million vulnerable people in tropical coastal regions (Sasmito *et al.*, 2023).

Altogether, recognizing mangroves as nature-based solutions will include conservation, restoration, and ecosystem creation (Temmerman *et al.*, 2023). Despite global commitment and optimism, conserving and rehabilitating mangroves should be ecologically sound. As an illustration, of the reported mangrove restoration projects, 80–90% experienced failures (Lewis & Brown, 2014). The reasons for failures are mainly a mismatch of mangrove species, lack of understanding of the eco-geomorphological characteristics, and the mangrove species-specific ecological requirements (Ellison *et al.*, 2020; Kodikara *et al.*, 2017; Lewis & Brown, 2014; Primavera & Esteban, 2008). Additionally, it requires more than a decade for mangrove forests to reach their optimal forest capacity for mitigation–adaptation purposes. To achieve that objective, mid-course adjustment is often required, emphasizing the need for operation and maintenance.

10.1.7 Mangrove restoration

Global targets have mandated mangrove restoration and conservation as one of the high priorities and are in urgent need (Friess *et al.*, 2020) – recognizing their multifunctionalities for climate mitigation and adaptation. Those ambitious targets have promoted a push for slowing down mangrove deforestation from an annual average loss of 0.21% during 1996–2010 to 0.04% during 2010–2020 (The State of the World's Mangroves 2022, 2022). Nevertheless, many past restoration attempts have failed due to the misunderstanding of the mangrove wetlands' ecological–physical processes (Lee *et al.*, 2019) and land (social) conflicts (Sasmito *et al.*, 2023). At first, ensuring mangrove restoration success requires a mechanistic understanding of the ecological–physical processes, where current knowledge is segregated.

Several guidelines exist, either in mangrove restoration (Lewis, 2005; SER (Society for Ecological Restoration Science and Policy Working Group), 2002; Zimmer *et al.*, 2022) or focusing on the practical design (engineering) and management (procedures to achieve) on the mangroves' integration in the coastal protection (Bridges *et al.*, 2021; Van Wesenbeeck *et al.*, 2021; World Bank, 2016). Those guidelines have resulted from ongoing learning-by-doing processes (Aerts *et al.*, 2014; Wilms *et al.*, 2021; Winterwerp *et al.*, 2020). Some guidelines remain quite site-specific, so that duplication of such approaches should be conducted carefully. However, given the dynamic nature of mangroves, it requires a system understanding from the tree, forest, and ecosystem levels. The system understanding requires field observation, periodic monitoring and assessment, and quantification approaches with modelling.

10.1.8 Need for prediction tools

Understanding the urgent needs mandated by global targets coupled with complexities and the state of the current knowledge mentioned above reassures us that advancing new knowledge in mechanistic understanding of mangrove–intertidal flat interactions is critical. Field observation and monitoring techniques have been experiencing rapid growth, for instance, the Global Mangrove Alliance initiative (www.mangrovealliance.org) with their mangrove restoration tracker and global mangrove watch. However, tools that can mechanistically assess and predict the process components and interactions of mangrove forest evolution are yet underdeveloped. Examples of these modelling tools resolving eco-geomorphological interactions are spatial and statistical models, followed by process-based and conceptual models (Rivera-Monroy *et al.*, 2022). All of these approaches express mangrove dynamics as a top-down hierarchy. In this hierarchy, the response of mangroves is averaged or pre-defined based on specific environmental conditions.

In light of climate change uncertainty, the pre-defined response of mangroves, as in a top-down approach, may not be valid under the changing physical–environmental stressors. Pretzsch (2009) suggests the system's understanding may be built upon integrating all known processes, components, and their interactions from the bottom-up. As in our work (Beselly, 2024), we could mechanistically model forest expansion, retreat, and colonization influenced by physical–environmental drivers. In this regard, this research study will contribute to obtaining more realistic projections of forest structure and eco-geomorphological change that account for the dynamic environmental conditions, for example driven by climate change.

10.2 INTERACTIONS BETWEEN MANGROVES AND THE ENVIRONMENT

10.2.1 Understanding mangrove dynamics

By definition, mangroves are plants thriving in the wedge of the intertidal zone (Spalding *et al.*, 2010). The global presence of mangroves is limited within the (sub)tropics and warm temperate climates, between 30°N and 37°S (Mukherjee *et al.*, 2014), and specifically located in a sheltered waterlogged environment (Krauss *et al.*, 2014; Sharma *et al.*, 2022). Mangroves can only thrive in specific eco-geomorphological characteristics determined by their species-specific tolerance. For instance, the latitudinal range of mangroves is limited by the tolerance of mangroves to cold temperatures due to the efficient temperature range of photosynthesis (Ball & Sobrado, 1999) and limited capability to assimilate CO_2 in cold temperatures (Ward *et al.*, 2016). Regional climate variability, such as precipitation and evapotranspiration, can affect mangrove zonation. Under high net evaporation conditions, dieback within the interior of a mangrove forest can occur due to an increase in salinity (Cortés, 2019). In contrast, during events of high precipitation, mangrove productivity increases (Gilman *et al.*, 2007), and subsequent increase in the mangrove area due to landward migration into salt marsh zones because of the decrease in porewater salinity and sulfate concentration (Ward *et al.*, 2016). More locally, hydrological forcings, for example, river discharge, tides, and waves, determine the lateral expansion of mangrove forest, whereas soil biogeochemistry determines the structure, for example, basal area and tree height (Jennerjahn *et al.*, 2017). The regional scale geophysical description of coastal environments guided the ecological classification of mangroves into fringe, basin, scrub, or riverine mangroves (Twilley *et al.*, 2017).

Utilizing mangroves as a measure in the face of climate change uncertainty requires well-defined predictability of their functional capacity (Morris *et al.*, 2018) and the potential to persistently restore after disturbance (Temmerman *et al.*, 2023). However, mangrove capacity in attenuating waves, for example, varies over locations. It depends on the mangroves' biophysical characteristics (frontal area, density, and composition) and wave conditions (wave height and wave period), where the latter may benefit from highly detailed local observation or global hindcast. However, predicting the wave attenuation capacity, for instance, leads to a logistical challenge for forest inventory to carry out on the ground (Bispo *et al.*, 2019) as it is labour intensive. Studies have used different proxies to characterize the forest, for example, forest age (Maza *et al.*, 2021), characterization of the mangrove root structure (Mori *et al.*, 2022), and remote sensing (Beselly *et al.*, 2021). Attempts to approximate short-period wave propagation through vegetation fields have been formulated in the earliest original work by Dalrymple *et al.* (1984) and adapted in several recent works, for example Maza *et al.* (2019)

and Mendez and Losada (2004). The formulation requires an estimation of the vegetation bulk drag coefficient (C_D) where it is typically obtained through field measurements (Horstman *et al.*, 2021) or flume experiments (van Hespen *et al.*, 2021). Another approach is to estimate C_D as a function of mangrove projected area and volume (van Maanen *et al.*, 2015).

Mangroves have species-specific critical thresholds on the physical–environmental forcings that control their establishment or collapse. As an effect, these species-specific thresholds influence how we can predict the mangrove forest evolution and, therefore, measure persistence capacity. To add complexity, the critical thresholds are also varied over the life-stages (propagule, seedling, sapling, mature). The natural successful establishment would require propagule availability (Lewis, 2005), where the parental trees could be from the vicinity or adjacent as the propagules can float up to several months and be transported by hydrodynamic processes within short/long distances (Shih *et al.*, 2022). Successful seedling recruitment requires propagules to survive the window of opportunity (WoO), where external disturbances remain within their critical threshold. This WoO concept (Balke *et al.*, 2011) determines the set of species-specific thresholds on (1) the inundation-free phase, where a propagule can germinate and anchor the root within the low tide period, (2) a calm wave period, where waves remain low enough to prevent the seedling dislodgment, and (3) limited accretion–erosion, where the accretion will lead to burial and suffocate the seedling and erosion can lead to toppling and dislodgement (Balke *et al.*, 2013, 2014). Seedling establishment is an important factor in the forest's lateral (horizontal) expansion (Shih *et al.*, 2022; van Hespen *et al.*, 2022a). The species-specific optimum mangrove (vertical) growth is reduced by their local environmental conditions, that is, salinity, pH, and hydrogen sulfide concentration (Berger *et al.*, 2008; Chen & Twilley, 1998; Grueters *et al.*, 2014) in addition to the competition with the neighbouring trees. The interspecific competition to sources (space, nutrients, light, and by environmental forcings) limits the growth and, hence, leads to mortality.

10.2.2 Mangrove–mudflat interactions

Mangroves interact with the environment and provide feedback (Figure 10.1) dependent on the three factors in physical processes, that is, nutrients as the resources, salinity as the regulator, and sea level/hydroperiod (Grueters *et al.*, 2014; Krauss *et al.*, 2014; Wimmler *et al.*, 2021). These factors

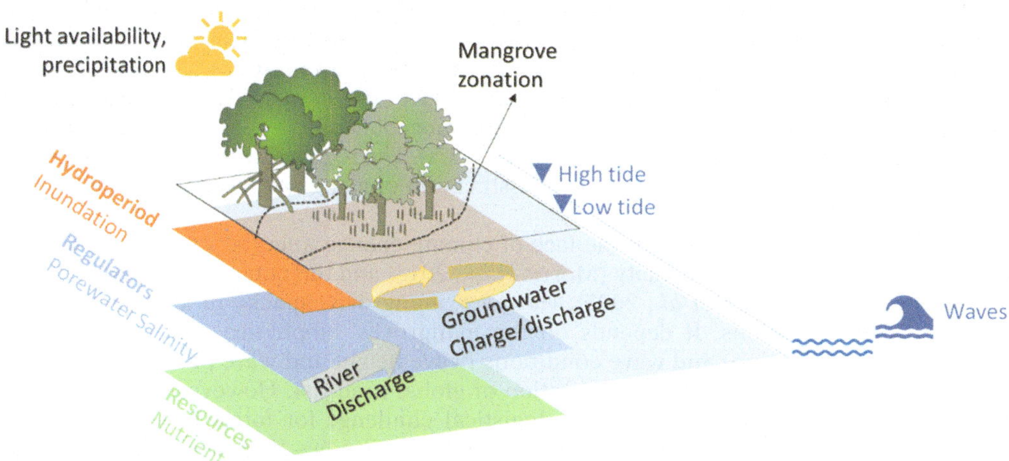

Figure 10.1 Schematic of hydrological factors (tides, waves, groundwater, and river discharge) and the regulating-resources factors along with the hydrological connectivity between mangrove forest and coastal waters (*source*: adapted from Rivera-Monroy *et al.*, 2022).

are intertwined and affect the growth and productivity of mangroves by regulating the zonation of mangrove forests as an influence on the porewater salinity distribution (Lovelock *et al.*, 2006; Piou *et al.*, 2006), limiting the growing capacity due to the availability of chemical/biochemical nutrients (Lovelock *et al.*, 2006; Reef *et al.*, 2010), and defining the spatial expansion dependent on the tidal flooding and duration (Balke *et al.*, 2015; Lovelock *et al.*, 2015). These factors determine the mangroves' growth, dieback, and seaward expansion (through tree recruitment).

Spatially explicit processes and feedback between mangroves and soil can be explained in three vertical soil layers, as shown in Figure 10.2. The relatively stable layer that comprises bedrock is located at the very bottom of the soil. In this layer, the movement is merely related to geologic processes. Thus, it is assumed to remain stable along the life of mangrove stands (Krauss *et al.*, 2014). Above the deep layer, subsurface layer processes are mainly related to the mangroves' rooting system (McIvor *et al.*, 2013). The root production contributes to the increase in soil volume and the subsurface elevation gain.

On the contrary, following the decomposition and compression of the roots after dieback, shallow subsidence may occur as root remnants take less space, decrease soil porosity, and promote soil collapse. Moreover, the shrink–swell of soil can be referred to as the increasing and decreasing soil water content due to the water absorption of the roots. In short, the subsurface processes comprise biological factors due to root production-decomposition and the physical factors in soil compaction and shrink/swell (Roskoden *et al.*, 2020). The top layer, as the most active, related to the surface processes that dynamically interact with the seawater. Here, the active processes are governed by hydrodynamics due to waves, tides, storm surges, and interactions with the mangrove's roots, stem, and canopy. This layer represents the transport and fate of materials in the form of sediment transport, nutrient exchange, salinity mixing, and mangroves' seedlings dispersal. The materials can be derived from outside of the mangrove forest (allochthonous) or inside the mangrove forest (autochthonous) (McIvor *et al.*, 2013).

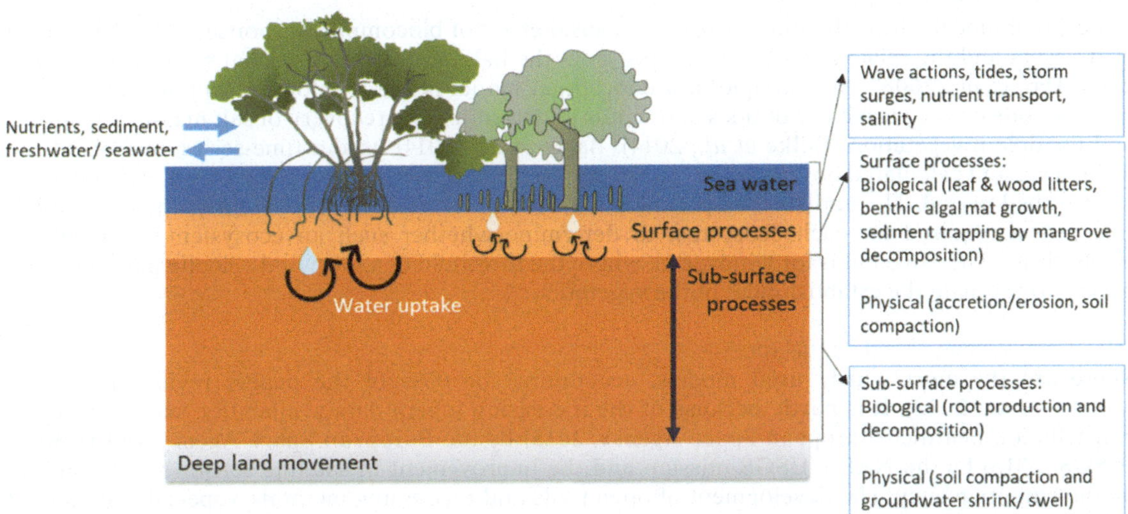

Figure 10.2 Illustration of the conceptual model mangrove–soil interactions. The interactions in soils are presented in three layers: the static and stable deep land movement layer, the subsurface layer related with root growth and decay processes, and the active layer associated with the hydro-morphodynamic processes. Mangrove trees respond to growth, dieback, and tree recruitment, thus providing new environmental feedback (*source*: adapted from Krauss *et al.*, 2014).

10.2.3 Modelling mangrove–mudflat interactions

10.2.3.1 Eco-geomorphic hierarchical approach

Due to the high variety of mangrove ecosystems, hierarchical approaches based on the biophysical characteristics describing their abiotic controls (hydroperiod, resources, and regulators, see Section 10.2.2) have been adopted and recommended (Krauss *et al.*, 2008; Rivera-Monroy *et al.*, 2004). This approach can be adopted, for example, at the geomorphological level, such as assessing mangrove vulnerability due to SLR (Lovelock & Ellison, 2007), at the forest level such as in ecological mangrove restoration techniques (Lewis & Brown, 2014) that provide structured restoration attempts by considering wetland's geomorphology and hydrology, and a detailed tree-level investigation on nutrition competition and hydroperiod effect on growth and change in community structure (Rivera-Monroy *et al.*, 2004). It shows how mangrove ecosystems interact and operate on different spatiotemporal and life-stage scales. Any disturbance on each hierarchy would alter ecosystem processes and may cascade into the higher hierarchical level and, in turn, affect the interactions and traits of the abiotic–biotic component at the lower levels. For example, if the change in sea level exceeds the sediment build-up, it may lead to the collapse and reorganization of mangrove zonation. To this end, no single model is capable of representing and simulating the processes encompassing all levels; they were designed to explain processes occurring at their own level (Rivera-Monroy *et al.*, 2022).

10.2.3.2 Model classifications

A mangrove ecosystem model is essential to assess and predict forest structure development. Still, ecological models need to parameterize interactions and processes in mangrove wetlands because it is impossible to cover the entire spectrum of spatial scales and time scales involved in mangrove biocomplexity. In other words, a model should be complex enough to pay tribute to the complexity it aims to describe but simple enough to understand what is going on. Mangrove ecosystem models can be classified into conceptual-, statistical-, spatial-, and process-based (Rivera-Monroy *et al.*, 2022).

10.2.3.2.1 Conceptual models

Conceptual models draw the intercorrelation causal effect of biocomplex responses with the change of stressors and the related attributes varying from the habitat unit to the global scale (Davis *et al.*, 2005; Day *et al.*, 2008). An example of a conceptual model is the windows of opportunity during calm periods on the variability of tides and winds to determine the recovery of salt marsh, mangrove, and floodplain vegetation (Balke *et al.*, 2014). Balke *et al.* (2014) used a time-series analysis of the water level and calculated how long the low water level lasted with respect to bed level, defined as an undisturbed period. The undisturbed period is then compared with the threshold of the inundation-free period required for each ecosystem to determine whether such an ecosystem is capable of establishing. The same holds for windspeed, where the threshold of a low wind velocity period should prevail to estimate the establishment of the vegetation.

10.2.3.2.2 Statistical and spatial models

These are the most widely used models, accounting for 60% of the papers reviewed (Rivera-Monroy *et al.*, 2022). It is partly because of the increasing (open) data availability, such as satellite constellation Sentinel (European Space Agency, 2015) by the European Space Agency and Landsat (USGS, 2013) by the NASA/USGS mission and the improvement in satellite sensor resolution and monitoring frequency. The development of open tools and processing methods, especially attributed to the cloud computing analysis, for instance Google Earth Engine (Gorelick *et al.*, 2017), allows for global and near-daily to weekly analysis of satellite imagery. The dominant usage of those big spatial datasets is to update mangrove inventory maps (United Nations Environment Programme, 2023). When complemented with field-based observations and statistical models, these maps can be used, for example, to evaluate mangrove spatial distribution (Bunting *et al.*, 2022) and assess carbon stocks (Murdiyarso *et al.*, 2015). The statistical models utilize the curated information of either remotely sensed or field-based datasets to infer the relationship between environmental variables and mangrove biophysical characteristics. For instance, an estimate of the global mangrove above-ground biomass based on the climate, assuming mangrove biomass corresponds to temperature and precipitation (Hutchison *et al.*, 2014). Although the application of statistical and spatial models has played an important role in explicitly explaining, for example, the drivers of mangrove loss and gain (Hagger *et al.*, 2022) or impacts of climate change on mangrove carbon stocks and fluxes (Alongi, 2022), what empowers this approach is also the major limiting factor, which is the datasets. The available dataset is rarely long enough or lacks consistency (Macreadie *et al.*, 2019; Sharma *et al.*, 2022; Zimmer *et al.*, 2022). The information gap may add complication in validating the prediction of the fate of the mangrove ecosystems in the climate uncertainty scenarios.

10.2.3.2.3 Process-based models

Process-based models, sometimes known as mechanistic models (Cox *et al.*, 2006), are numerical representations based on explicit causal mechanisms or processes on how the systems work, grounded in proven scientific knowledge (Cuddington *et al.*, 2013; Roelvink & Reniers, 2012). The explicit nature of process-based models in representing reality and its transparency on the assumptions of the processes provide more confidence in understanding phenomena either in hindcast or forecast simulations (Best *et al.*, 2018). In this chapter, we define process-based for two subclasses, that is, hydro-morphodynamic and vegetation dynamic models. Process-based hydro-morphodynamic models have long been used in wetland analysis, for example, modelling estuarine hydrodynamics (Thanh *et al.*, 2017), wave attenuation (Yoshikai *et al.*, 2023), tidal circulation (Horstman *et al.*, 2015), and sediment dynamics (Willemsen *et al.*, 2016). Another subclass, vegetation dynamic models, aims to simulate the mangrove forest trajectory.

10.2.3.2.3.1 Process-based: hydro-morphodynamics A process-based hydro-morphodynamic model is identical to simulating abiotic processes, for example, hydro-morphodynamics (Roelvink & Reniers, 2012). We can find examples of hydro-morphodynamic models in the application of mangrove wetlands, such as the investigation of tidal circulation within the wetland (Horstman *et al.*, 2015), wave attenuation function of mangrove forests (De Dominicis *et al.*, 2023; Pelckmans *et al.*, 2023), and effect on hydrodynamic and sediment exchange (Bryan *et al.*, 2017; Nardin *et al.*, 2016). The simulation exercises in the references mentioned previously assumed that physical processes occur in a static vegetation state, ignoring forest structure changes. Those studies commonly include detailed spatially varying hydro-morphodynamic–ecology interactions in a relatively limited vegetation time scale (seconds to weeks) where vegetation development within that period can be assumed to not significantly affect hydrodynamics (Bryan *et al.*, 2017; Friess *et al.*, 2012; Nardin *et al.*, 2016). Therefore, the model structure does not allow the inclusion of the feedback loop mechanisms of environmental (abiotic) changes to vegetation growth (biotic).

10.2.3.2.3.2 Process-based: vegetation dynamics The detailed physical processes involved in hydro-morphodynamic models have proven to be capable of elucidating mangroves' functional capacity on the physical drivers (van Hespen *et al.*, 2022a). However, the persistence or capacity to recover in mangrove forests is reflected in their ecological processes, which occur on a longer time scale, and encompass life stages from seedling to mature trees (Wang *et al.*, 2014). Over time, mangrove forest structure and composition will change in response to environmental conditions. Evidence can be obtained from past studies observing the vegetation dynamics based on satellite imagery or aerial photographs, which take years to decades of observation. For example, a study by Kleinhans *et al.* (2019) shows the migration and succession of riparian vegetation due to the river meandering process and the impacts of mangrove extent variation to estuarine hydro-morphodynamics in Waikaraka Estuary, New Zealand (Glover *et al.*, 2022). With this in mind, introducing the co-evolution of morphology and vegetation has an important role in providing a process understanding of the interactions of vegetation and physical drivers. We describe vegetation dynamic models in two types: population dynamic and individual-based models (IBMs).

(a) Vegetation dynamics – population dynamic
One approach to simulate vegetation dynamics is a population dynamic model, assuming the composition of individuals who share similar traits (e.g., species and biophysical properties) are grouped on a grid/plot and behave under a set of rules (Cappuccino, 1995). This grid-based population model can change simultaneously depending on the endogenous (e.g., density, above-belowground biomass) and exogenous (e.g., inundation frequency, wave energy) factors. To our knowledge, many studies focused more on the dynamics of mangrove primary productivity, for example, the phenology of *Rhizophora apiculata* on seasonal climate variation (Christensen & Wium-Andersen, 1977), the effect of nutrient availability on *Rhizophora mangle* (Onuf *et al.*, 1977), biomass production of *R. apiculata* (Christensen, 1978), and effect of nutrient and irradiance to seedling growth of *Ceriops australis* and *Ceriops decandra* (Ball, 2002). Exception is in Thi Ha *et al.* (2003), where they investigate the seasonality of seedling growth and production rates of *Kandelia candel*. The main issue in population dynamic models, specifically in mangroves, is not being able to treat propagule dispersal. Instead, considered individuals within populations can follow homogenous behaviour without individual variation (Jørgensen & Fath, 2011).

(b) Vegetation *dynamics – individual-based*
Another approach is the IBM, where populations and communities are composed of discrete individuals and emerge by following the feedback loop of individual interactions with their environment (DeAngelis & Grimm, 2014). Individual organisms (biotic) are considered explicitly with their variability, local interactions, and specific adaptive behaviour in their physical environment (Grimm & Railsback, 2005), where they distinguish themselves from each other. Individuals can have different growth, survival probability, or reproduction capacities due to their relative position with other individuals (competition/endogenous factors) and with regards to the environmental conditions (exogenous factor). In this approach, individuals have self-directed motivation and adapt or modify their environment through their actions (Jørgensen & Fath, 2011). The main challenges in IBM modelling are their extensive computational requirement to run in large spatial scales (>1 km^2) and the need for long-term species-specific biophysical characteristics datasets. With the growing number of field datasets available and advances in IBM architecture, there is a possibility to improve IBM performance.

10.2.3.2.3.3 Integration of vegetation dynamics and hydro-morphodynamics In recent years, few attempts have been made to integrate two-way couplings of biotic–abiotic processes. In wetland modelling, several studies have attempted to include complex interactions on ecological, hydrodynamic, and morphological changes. This eco-morphodynamic modelling ranges from riparian (van Oorschot *et al.*, 2016) and salt marshes (Best *et al.*, 2018; Brückner *et al.*, 2019) to mangroves (van Maanen *et al.*, 2015; Xie *et al.*, 2020). Those models were based on well-appreciated knowledge of hydro-morphodynamic processes. They assumed an abundant supply of seedlings to fill the numerical grid when the inundation threshold allowed, indicating colonization. Meanwhile, the numerical grid area is a proxy to explicitly define above-ground competition for resources (maximum density or biomass capacity), which drives growth and mortality. Eventually, the models do not consider the dispersal mechanism in a smaller time scale (Friess *et al.*, 2012; Piercy *et al.*, 2023) that is closely correlated with hydrodynamic processes (Duke *et al.*, 1998; Shih *et al.*, 2022), which determines vegetation establishment and lateral expansion. On the contrary, the development of IBMs has incorporated abiotic interactions in a somewhat limited way. One example is MANGA (Bathmann *et al.*, 2020), which includes mangrove's effect on groundwater salinity by coupling IBMs with OpenGeoSys.

Integrating IBMs with abiotic processes is still mainly occurring in plot scale and over a short period. Despite already including the complex abiotic–biotic interactions, all integrated models discussed still lack representation of feedback loop processes at each life stage. Thus, it can be a challenge when we want to assess the wetland responses on non-stationary drivers such as variation in wave climate, sediment supply, and multispecies vegetation dynamics.

10.3 RESEARCH GAPS IN MODELLING MANGROVE ECO-HYDRO-MORPHODYNAMICS

10.3.1 Detailed mangrove dynamics

Despite the increasing number of numerical models evaluating the physical–ecological processes, those models still focus on the detailed abiotic interactions and use 'prescribed' ecological interactions, and thus may not reflect the actual mangrove dynamics, losing the important interactions at the lower level. Examples are models that consider interactions of the wave, tidal, and sediment trapping efficiency in mangrove forests but do not include dynamic ecological processes (Willemsen *et al.*, 2016), simulation of morphological evolution in sandy tidal embayments with aggregated mangrove feedback (van Maanen *et al.*, 2015), and mangrove response to SLR with pre-defined species composition concerning inundation depth (Buffington *et al.*, 2021; Rodríguez *et al.*, 2017; Xie *et al.*, 2022). Mechanistic interactions should consider the full life stages of the mangrove from propagule, seedling, and sapling to mature tree. An important note is given by Zainol *et al.* (2022) on the role of propagule dispersion in determining colonization and regeneration of mangrove forests, which is affected by hydrodynamic processes (Shih *et al.*, 2022).

10.3.2 Optimizing restoration strategies

There have been explicit approaches in mangrove restoration attempts; examples are ecological restoration (Lewis & Brown, 2014) and assisted rehabilitation or planting (Primavera *et al.*, 2011). Even though these approaches have considered the hydro-morphology and ecological system in the planning, they lack mechanistic understanding as time progresses. Therefore, restoration practice is still site-specific and results from long trial-and-error processes (Primavera & Esteban, 2008; Wilms *et al.*, 2021). Although the (long-term) prediction of abiotic systems has been well understood, for example, in Roelvink *et al.* (2020) and van der Wegen *et al.* (2011), when a biotic system is involved, either the spatiotemporal scale may be reduced to include details (Bryan *et al.*, 2017) or such simplification in vegetation dynamics should be required (Brückner *et al.*, 2020). The lack of spatial-explicit understanding of biotic–abiotic systems encompassing local to ecosystem scales may hinder successful restoration attempts. When looking at the global targets, the approaching deadline to cut carbon emissions and climate actions requires prompt solutions in developing restoration strategies. Therefore, a full life stage process-based model may be a helping hand to understand the interactions and the important factors determining a successful restoration.

10.4 MANGROVE–MUDFLAT PROCESS-BASED ECO-HYDRO-MORPHODYNAMIC MODELLING

10.4.1 Simulating mangrove–mudflat dynamics with a hybrid eco-hydro-morphodynamic model

Mechanistic mangrove–mudflat interactions are modelled with the two-way coupling of biotic–abiotic processes. It is approached by enabling the feedback loop of landscape-scale process-based hydro-morphodynamics with an individual-based mangrove dynamic model. DFMFON model (Beselly *et al.*, 2023) approached this by integrating the hydro-morphodynamic Delft3D-Flexible Mesh (DFM) and individual-based MesoFON mangrove model (MFON). The coupling of the two model paradigms allows for resolving the feedback loop of both short-term variation in local scale abiotic factors (flow, waves, sediment availability, and salinity), including the propagule dispersals, including the long-term bio-geomorphic mangrove forest dynamics.

As shown in the model schematic (Figure 10.3), the feedback loop between mangrove and the physical–environmental drivers is represented in two loops, that is, the inner and outer loops. The simulation is initialized by parameterizing the mangrove stand properties, that is species, diameter at breast height at 130 cm above the ground (D_{130}), vegetation density, and vegetation height in each DFM numerical cell. DFM applies the modified Baptist equation (Baptist *et al.*, 2007) to include mangrove-induced resistance λ (Equation (10.1)) and bare bed roughness C momentum equation (Equation (10.2)) by considering vegetation height h_v (m), density n (number of trees/m^2), bulk drag coefficient C_D and unvegetated bed roughness C_b to water depth h, κ is the dimensionless von Kármán constant, and D is the stem diameter (m). In DFMFON, we predict C_D (Equation (10.3)) as the sum of the total drag coefficient of the bare surface $C_{D,no}$ and the ratio between the constant e and vegetation length L (Equation (10.4)). L is defined by considering the species-specific mangrove root and stem volume V_M varying over the water depth, by comparing it with the volume of water in a numerical cell (V).

$$\lambda = \frac{1}{2} C_D n D \frac{h_v}{h} \frac{C_b^2}{C^2} \tag{10.1}$$

$$C = C_b + \frac{\sqrt{g}}{\kappa} \ln\left(\frac{h}{h_v}\right) \sqrt{1 + \frac{C_D n D h_v C_b^2}{2g}} \tag{10.2}$$

$$C_D = C_{D,no} + \frac{e}{L} \tag{10.3}$$

$$L = \frac{V - V_M}{A} \tag{10.4}$$

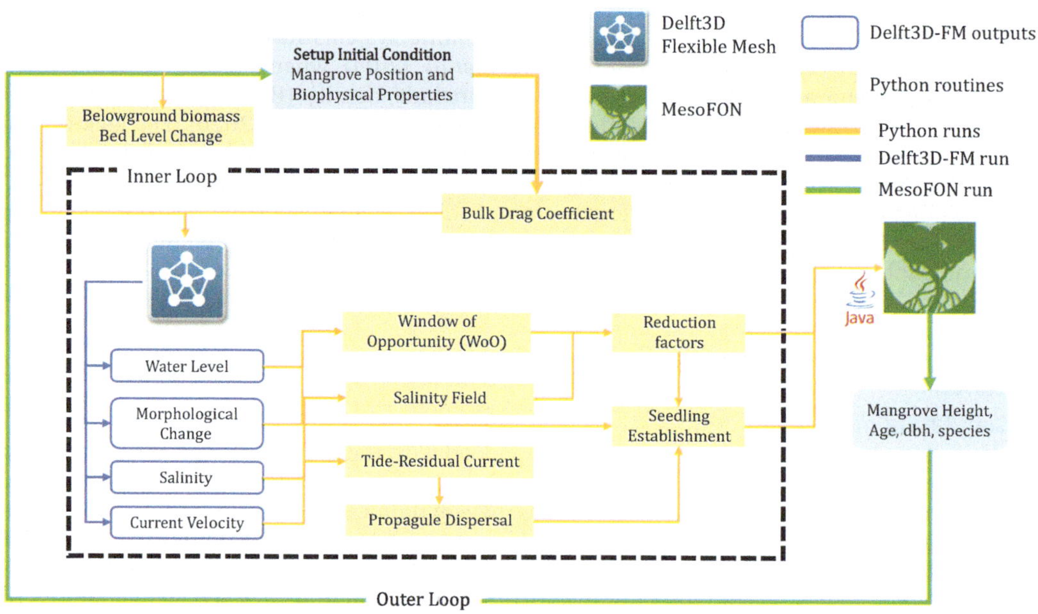

Figure 10.3 Schematic of the DFMFON model (*source*: adapted from Beselly *et al.*, 2023).

In the inner loop, DFM uses the vegetation information to simulate the hydro-morphodynamics and salinity for 90 morphological days. The reason for choosing this period is to allow the model to consider the seasonal fluctuation of the local climate condition and incorporate the mangrove's reproduction, that is, flowering, fruiting, and seedling production. The resulting DFM variables (water level, salinity, and bed level) are then converted as a WoO (Balke *et al.*, 2011) map, salinity distribution, and the tide-residual current. The WoO map is applied to evaluate the seedling establishment. In the model, we define a minimum of 3 inundation-free days and 5 inundation-free days for *Avicennia* sp. and *Rhizophora* sp., respectively. When the cell is inundation-free on the minimum required days, the propagules are allowed to establish. We assume that the propagules are buoyant and able to float irrespective of the salinity value. The magnitude and direction of the average tidal current over 2 weeks determined the final location of the propagules' spatial distribution over the numerical cells. In addition to the establishment due to inundation frequency, the effect of sediment disturbance on propagules, that is, burial and uprooting due to erosion is also considered. Eventually, the surviving propagules are allowed to grow into seedlings and afterwards into saplings when reaching 2 years of age. With this technique, the model closes the life cycle by considering the propagules production, seedlings establishment, saplings, and adult trees. DFMFON with this regard explicitly considers a successful seedling establishment as the factor driving the forest expansion.

Retrieving the information from the 90 morphological days inner loop simulations, the MFON considers the physical–environmental drivers that influence the mangrove dynamics. The salinity field simulated in the inner loop affects the growing capacity of mangroves and the number of propagule production. The surviving seedlings will be transferred into the outer loop once they have reached the sapling stage. MFON utilizes the field of neighbourhood (FON) approach (Berger & Hildenbrandt, 2000) to represent the strength of influence one tree exerts on another. In addition to the distance, the FON strength depends on their diameter and is subject to calibration for each species. The level of competition with the neighbouring trees is sensed via the FON value both for the above- and below-ground. With this, the mangrove growth is governed by the competition for resources and salinity field. The mortality occurs when the mangrove is indicated to not grow for 5 years sequentially (Grueters *et al.*, 2014). Currently, the model only considers the mortality as a result of the tree-to-tree competition for resources in the sapling-to-adult stage and burial-uprooting in the propagule stage. At the end of the outer loop processes, the simulated distribution and mangrove biophysical properties are fed to the inner loop as input, where the below-ground biomass contributes to the positive bed-level development.

10.4.2 Biocomplex interactions of the individual mangroves and physical–environmental stressors

The advantages of the DFMFON model are the two-dimensional horizontal characterization of the physical–environmental drivers and the capability of representing multiple species of mangroves and their individual interactions. This approach makes it possible to investigate forest expansion, retreat, and colonization, including the feedback loop with varying physical–environmental drivers. The model is applied with a case study in the Porong Delta, East Java Province, Indonesia. We investigated the model's capability and limitation in reproducing the mangrove extent development, the mangrove age–height relationship, and the morphodynamic delta features.

Mangrove forests in a mixed fluvial-deltaic with predominantly muddy sediment of the Porong Delta have seen a rapid expansion due to the unprecedented sediment load from the hinterland. The high sediment load originated from the mud volcanic eruption, which since the first burst in 2006 still delivers mud from the surface, albeit has declined almost to one-third of the first eruption (Karyadi *et al.*, 2012). This unprecedented flow forced thousands of residents to evacuate (Mazzini, 2018). To reduce the damage to the community and environment, the discharge of mud is contained in a reservoir. During the wet season, the mud is diluted and pumped into the river, increasing sediment load and concentration in the Porong River. The Porong River is regulated by a series of barrages and floodgates upstream, during the dry season the river flow is reduced to almost zero as it is diverted into the nearby metropolitan, in contrast to the high flow as the barrage is opened to prevent flooding in the city. As a result, within 11 years of the operation, a study by Besselly *et al.* (2021) shows the rapid delta expansion and seasonal mangrove dynamics in the estuary. The study combines multiple sources of satellite imageries to derive a 6-to-3 monthly mangrove extent map; which is more frequent than the globally available annual mangrove map. The study unravels the seasonal characteristic of the mangrove dynamic, where the mangrove forests tend to have successional patterns on the transition during the wet to dry season and dieback during the dry to wet season. The wet season correlates with the fruiting season and high freshwater discharge with additional mud disposal operation from the reservoir. The propagules are then dispersed and established on the newly deposited mud towards seawards. A relatively low saline condition and soft elevated mud layer increase the establishment opportunities. In the middle-to-end of the dry season, the increasing competition for resources and space leads to the dieback. More specifically for the Porong Delta, the dieback is pronounced likely because the massive sediment load is deposited further in the margin of the forest. Thus, it is transformed into the basin mangrove type in certain areas. The seedlings that are more sensitive to salt and drought might die under this condition in the basin-type areas by the end of the dry season.

The unique characteristics of the Porong Delta and the associated mangrove dynamics are simulated in DFMFON by creating a schematized funnel-shaped estuary with scenarios representing the seasonal high–low discharge and sediment load operation. In total, eight scenarios were tested to investigate the response of the system on variation in salinity (high, medium, low) on the condition of high, medium, and poor sediment availability (see the detailed scenarios in Beselly *et al.*, 2023). The model is schematized into a 1500 m length × 500 m width funnel-shaped estuary with a delta lobe in the middle. It starts with 140 m width of the river channel bifurcates into two branches encircling the 500 m × 640 m delta lobe at the upstream and the river branches are connected at the downstream. The mangrove population is characterized by consisting of only a single species of *Avicennia marina* that is dominating the Porong Delta. The mangroves are initialized as saplings at the age of 1-year-old with uniform biophysical properties and randomly placed in the upstream part of the delta lobe with a density of $0.03/m^2$, similar to the density found in the study area. The model is forced with two open boundaries, that is, fresh river discharge that also supplies sediment and tidal water level in the seaward boundary, and is simulated for 60 years.

The simulation results (Figure 10.4) show the realistic spatiotemporal mangrove–mudflat dynamics in prograding delta settings. The schematized model reproduces the expansion pattern of mangrove forest with the elongated shape towards downstream similar to the observation. However, the model could not reproduce the expansion rate similar to the case study. This has been expected considering the source of propagules in the model is originating from one source and one species, that is, *A. marina* in the delta lobes. Meanwhile, the case study has multiple sources from mangrove forests in the vicinity of the delta. In addition, the fruiting season for most of the mangrove species occurs during the wet season, where in this case, the river discharge and local precipitation could transport propagules over longer distances. The canopy height development shows a good comparison with the mean absolute error of 1.105 m (Figure 10.4d). The simulation realistically represents the cyclical seasonal pattern (Figure 10.4c) with the expansion during the wet season and retreat during the dry season.

The findings demonstrate the dependency of the mangrove ecosystem on the sediment supply and the gradual shift of mangroves from colonizers into ecosystem engineers. Understanding these interactions and functions is relevant, especially the latter, considering the function of mangroves as nature-based solutions. The capacity of mangrove stands in influencing hydrodynamics is varied over space and time. We observed three stages on the interplays of mangrove and mudflat, indicated by the mangrove canopy area as a proxy. The first stage occurs during the first 15 years of development indicating mangroves as colonizers, the seedlings establish and fill the allowable space until reaching the physical limit (mean sea level elevation). The high sediment supply and river discharge assisted in dispersing the propagules. As the established mangroves are relatively sparse, the effect on hydrodynamics tends to be limited and mostly dominated by local topography. Thus,

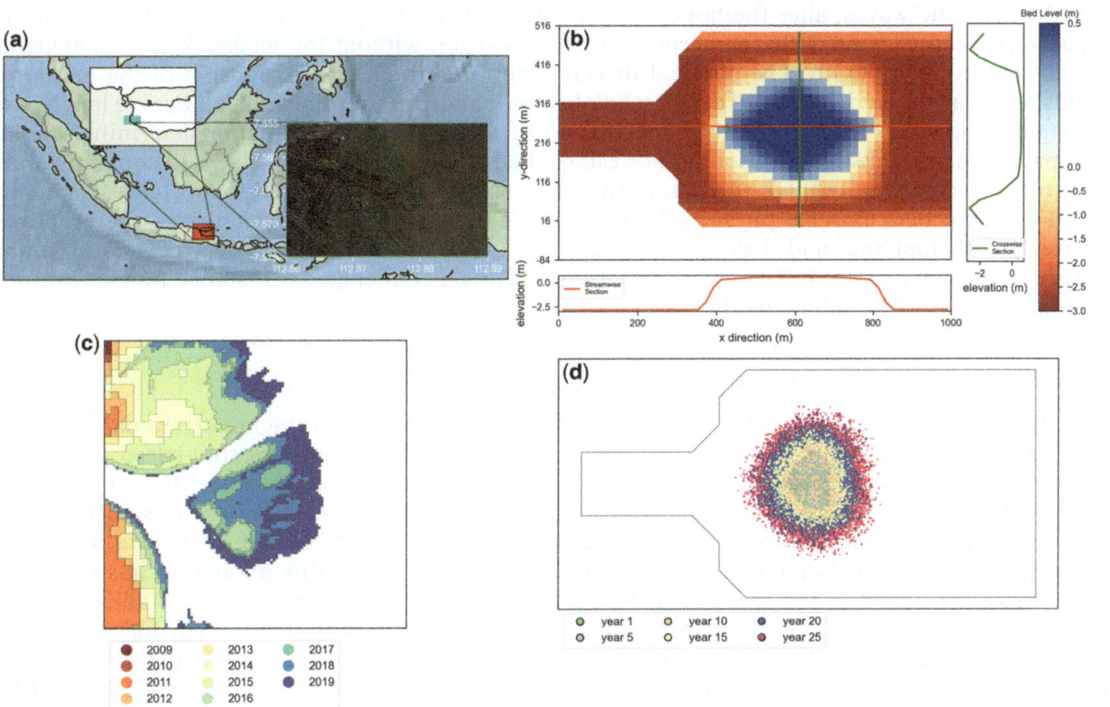

Figure 10.4 Depiction of real case study located in (a) Porong Delta, East Java Province, Indonesia, with focus on the northern delta lobe, (b) in which schematized as funnel-shaped delta with initial delta lobe in the centre, the schematized model is run for six different scenarios by varying the sediment concentration and fluvial discharge, here (c) a snapshot of simulated mangrove forest after 60 years for the base scenario (scenario A: high salinity and rich sediment concentration condition) is depicted, and (d) comparison of the simulated and observed mangrove canopy height (*source*: adapted from Beselly *et al.*, 2023).

there is not so much difference in morphology between no- and with-mangrove scenarios. In the second stage representing years 15 to 30, the expansion rate is increasing, given the initial population has become adult and produces more propagules. Here, we see the gradual distribution of mangrove stands with the older trees in the interior and younger trees at the fringe. The allowable space in the mudflat platform is seen to be rapidly filled with saplings. In this stage, we start to see the effect of mangroves in hydrodynamics, particularly by the large and dense older stands in the interior near the upstream of delta lobes. Their presence alters the flow and settles the sediment downstream, increasing the establishment probability. Hence, we would argue mangroves during the second stage act as ecosystem engineers, although somewhat limited and dependent on the sediment supply. After 30 years of simulation, the majority of the population has reached the reproductive age, and adding the number of propagules in the system. Here, in stage three, the propagules have been dispersed outside the delta and colonized the mudflat along the northern and southern channels.

The modelling experiment unravels the benefit of including full life-cycle mangrove dynamics and the feedback loops with the physical–environmental drivers. DFMFON accommodates the possibility of representing mangrove establishment and mangrove dispersal by considering hydrodynamic conditions. The capability to include this process is important since the establishment and dispersal are critical in determining the persistence of mangrove forests (Van der Stocken *et al.*, 2019).

The present development of the DFMFON model (Beselly *et al.*, 2023), however, considers the hydrodynamics as the main physical driver where mangroves act as obstacles. The extra obstacles can attenuate the waves, alter the flow, and eventually affect the soil surface development. Salinity is parameterized to control the mangrove growth, however, without the feedback from mangroves on the salinity distribution. The physical drivers determine the lateral expansion of the mangrove stands, whereas the environmental driver regulates the mangrove tree diameter, mangrove height, and propagule production. The tidal water level, wave force, and erosion-sedimentation determine whether a propagule can establish into a sapling. The established seedlings and adult trees will compete for the resources (space and nutrients), as a result, the high competition would lead to mortality. Currently, DFMFON considers extreme physical drivers (e.g., sedimentation and erosion) into the seedlings' failure (burial, toppling, and dislodgement). Estimation of the extreme physical drivers of young-mature trees such as stem breakage or uprooting remains a challenge, for instance how to estimate the mortality threshold following the breakage.

10.5 FUTURE DEVELOPMENTS OF MANGROVE MODELLING

Our modelling attempts have contributed to an ongoing exploration on the important parameters of mudflat–mangrove system dynamics. Currently, our knowledge of important trait-based parameters is limited to a few mangrove tree species. Extending it to other species and different regions is deemed necessary. An example is the WoO approach to describe seedling establishment that is valid for *Avicennia alba* in Singaporean mangroves (Balke *et al.*, 2011) and *A. marina* in Firth of Thames, New Zealand (Balke *et al.*, 2015). Interest in investigating this parameter has increased, for instance, as shown in the study of hydro-morphodynamic factors in seedling establishment for eight mangrove species in China (van Hespen *et al.*, 2022b). Replicating such a study for an extended database of mangroves and regions is possible. Current growth functions such as the one in Berger and Hildenbrandt (2000) initiates the growth parameter when the mangroves have reached the sapling stage, equal to breast height (137 cm). It is worth investigating the mangrove's growth after a successful seedling establishment into the sapling stage.

The presence of mangroves has important impacts on hydrodynamics, for example, attenuating waves and altering the flow. This is a relevant research topic, considering mangroves function as sustainable solution in disaster risk reduction. However, the mangrove capacity in attenuating waves, for example, varies between locations. It depends on the mangroves' biophysical characteristics (frontal area, density, and composition) and wave conditions (wave height and wave period). To approximate wave propagation, formulations require a function of mangrove projected area and volume on different water depth. Idealized vegetation configurations are commonly applied, such as for *Avicennia* sp. (van Maanen *et al.*, 2015) and *Rhizophora* sp. (Ohira *et al.*, 2013). All of these assume water depth will inundate the whole mangrove structure below the canopy. The hydrodynamic estimation will benefit most if such a database describing the function of (prop, pencil, cone, buttress) root, stem, and canopy exists. The correlations on species-specific root dimension would be better to be built based on the commonly used key parameters such as trunk diameter at breast height.

Although the chapter focuses more on the primary production and surface contributions, the contribution of below ground biomass to bed level is noticed, especially in the capability of mangrove areas to adjust to SLR (Krauss *et al.*, 2014). Similar to that, the role of, for example, mangrove litter and detritus to sediment accumulation in the forest interior is considerable (Ding & Fu, 2021). However, the challenge remains how to estimate their contribution and the relationship to determine the magnitude in a dynamic way in the model.

Mortality in the model is controlled by the competition between sapling-mature trees and during seedling establishment stage. Limited literature is available investigating mechanical tree damage, for instance due to the exposure to extreme wind and waves. This will be relevant, especially to understand the role of mangrove as coastal protection (Morris *et al.*, 2019; World Bank, 2017). Such a mechanistic limit on the extreme conditions leading to tree mortality, once it is known, will be useful to investigate mangroves' persistence.

The process-based mangrove hydro-morphodynamic model presented in this chapter is validated, focusing on Porong Delta, Indonesia. As of any other model, for site- or species-specific use, this coupled model requires to be validated with field datasets, with the emphasis on mangrove attributes trajectories, particularly, the growth parameter and seedlings production. More efforts are needed to gather local- to forest-scale datasets in mangrove wetlands, as these available datasets will be useful to further validate the model. There are increasing remotely sensed monitoring efforts in mangrove ecosystems, such as Global Mangrove Watch (Bunting *et al.*, 2022), and several field-based observations, for example, in Guyana (Best *et al.*, 2022) and New Zealand (Gijsman *et al.*, 2023). Although remotely sensed observations can provide a larger extent in understanding the dynamics of mangrove forests in their area development or vegetation phenology, a field-based dataset is limited to its temporal period and location. Additionally, as mentioned previously, the field dataset collected in the study area requires state-of-the-art equipment that needs trained experts to operate and can be costly – considering the remoteness of the mangrove forests. Off-the-shelf tools such as unoccupied aerial vehicles demonstrated in Beselly *et al.* (2021) can be a promising solution to provide very high-resolution imagery to complement satellite-based tools with lower resolution (10–30 m). Remotely sensed datasets in this way can only detect changes above the canopy and, hence, estimate the dynamics with canopy height or leaf index as a proxy; observing processes occurring under the canopy remains a challenge to investigate. Innovative low cost and autonomous monitoring tools are being developed, such as laser-based surface elevation dynamics sensors (Hu *et al.*, 2020) to monitor daily bed level changes or Mini Buoy (Balke *et al.*, 2021) to observe water level and current velocity in mangrove wetlands. Such low-cost observation equipment can potentially provide an extended period and a larger extent of the study.

There are roughly two types of process-based, eco-hydro-morphodynamic models (see Section 10.1). Each type of model serves the specific needs of the simulations. The question remains, how much complexity should we include in the models? Such as to what extent does including the effect of light availability, precipitation, or including interactions with subsurface hydrodynamics in the model affect the accuracy. Applying these tools requires more considerations than just adding physical and ecological processes. Although more datasets will provide opportunities to improve and include more processes in the model, the type of model applied depends also on the study objectives and research questions. In future research, a systematic comparison between individual-based and population dynamic vegetation models coupled with hydro-morphodynamics can be conducted, considering, for example, geomorphic and sedimentary settings, single or multiple tree species, or based on short- or long-term physical–environmental changes.

REFERENCES

Aerts J. C. J. H., Wouter Botzen W. J., Emanuel K., Lin N., De Moel H. and Michel-Kerjan E. O. (2014). Evaluating Flood Resilience Strategies for Coastal Megacities. *Science*, **344**(6183), 473–75, https://doi.org/10.1126/science.1248222

Alongi D. M. (2002). Present state and future of the world's mangrove forests. *Environmental Conservation*, **29**, 331–349, https://doi.org/10.1017/S0376892902000231

Alongi D. M. (2022). Impacts of climate change on blue carbon stocks and fluxes in mangrove forests. *Forests*, **13**, 149, https://doi.org/10.3390/f13020149

Balke T., Bouma T., Horstman E., Webb E., Erftemeijer P. and Herman P. (2011). Windows of opportunity: thresholds to mangrove seedling establishment on tidal flats. *Marine Ecology Progress Series*, **440**, 1–9, https://doi.org/10.3354/meps09364

Balke T., Webb E. L., van den Elzen E., Galli D., Herman P. M. J. and Bouma T. J. (2013). Seedling establishment in a dynamic sedimentary environment: a conceptual framework using mangroves. *Journal of Applied Ecology*, **50**, 740–747, https://doi.org/10.1111/1365-2664.12067

Balke T., Herman P. M. J. and Bouma T. J. (2014). Critical transitions in disturbance-driven ecosystems: identifying Windows of Opportunity for recovery. *Journal of Ecology*, **102**, 700–708, https://doi.org/10.1111/1365-2745.12241

Balke T., Swales A., Lovelock C. E., Herman P. M. J. and Bouma T. J. (2015). Limits to seaward expansion of mangroves: translating physical disturbance mechanisms into seedling survival gradients. *Journal of Experimental Marine Biology and Ecology*, **467**, 16–25, https://doi.org/10.1016/j.jembe.2015.02.015

Balke T., Vovides A., Schwarz C., Chmura G. L., Ladd C. and Basyuni M. (2021). Monitoring tidal hydrology in coastal wetlands with the 'Mini Buoy': applications for mangrove restoration. *Hydrology and Earth System Sciences*, **25**, 1229–1244, https://doi.org/10.5194/hess-25-1229-2021

Ball M. C. (2002). Interactive effects of salinity and irradiance on growth: implications for mangrove forest structure along salinity gradients. *Trees*, **16**, 126–139, https://doi.org/10.1007/s00468-002-0169-3

Ball M. and Sobrado M. (1999). Ecophysiology of mangroves: challenges in linking physiological processes with patterns in forest structure. In Physiological Plant Ecology. British Ecological Society. Blackwell Science, London, pp. 331–346.

Baptist M. J., Babovic V., Rodríguez Uthurburu J., Keijzer M., Uittenbogaard R. E., Mynett A. and Verwey A. (2007). On inducing equations for vegetation resistance. *Journal of Hydraulic Research*, **45**, 435–450, https://doi.org/10.1080/00221686.2007.9521778

Bathmann J., Peters R., Naumov D., Fischer T., Berger U. and Walther M. (2020). The MANgrove–GroundwAter feedback model (MANGA) – describing belowground competition based on first principles. *Ecological Modelling*, **420**, 108973, https://doi.org/10.1016/j.ecolmodel.2020.108973

Berger U. and Hildenbrandt H. (2000). A new approach to spatially explicit modelling of forest dynamics: spacing, ageing and neighbourhood competition of mangrove trees. *Ecological Modelling*, **132**, 287–302, https://doi.org/10.1016/S0304-3800(00)00298-2

Berger U., Rivera-Monroy V. H., Doyle T. W., Dahdouh-Guebas F., Duke N. C., Fontalvo-Herazo M. L., Hildenbrandt H., Koedam N., Mehlig U., Piou C. and Twilley R. R. (2008). Advances and limitations of individual-based models to analyze and predict dynamics of mangrove forests: a review. *Aquatic Botany*, **89**, 260–274, https://doi.org/10.1016/j.aquabot.2007.12.015

Beselly S. (2024). The Mangrove Tale: Mechanistic modelling of mangrove–eco-geomorphic interactions (Dissertation, TU Delft). Mech. Model. Mangrove–eco-geomorphic interact. IHE Delft Institute for Water Education, Delft, the Netherlands, https://doi.org/10.4233/uuid:17a9d384-5416-40d0-a714-306b1e284976

Beselly S. M., van der Wegen M., Grueters U., Reyns J., Dijkstra J. and Roelvink D. (2021). Eleven years of mangrove–mudflat dynamics on the mud volcano-induced prograding Delta in East Java, Indonesia: integrating UAV and satellite imagery. *Remote Sensing*, **13**, 1084, https://doi.org/10.3390/rs13061084

Beselly S. M., Grueters U., Van Der Wegen M., Reyns J., Dijkstra J. and Roelvink D. (2023). Modelling mangrove–mudflat dynamics with a coupled individual-based-hydro-morphodynamic model. *Environmental Modelling & Software*, **169**, 105814, https://doi.org/10.1016/j.envsoft.2023.105814

Best Ü. S. N., Van der Wegen M., Dijkstra J., Willemsen P. W. J. M., Borsje B. W. and Roelvink D. J. A. (2018). Do salt marshes survive sea level rise? Modelling wave action, morphodynamics and vegetation dynamics. *Environmental Modelling & Software*, **109**, 152–166, https://doi.org/10.1016/j.envsoft.2018.08.004

Best Ü. S. N., van der Wegen M., Dijkstra J., Reyns J., van Prooijen B. C. and Roelvink D. (2022). Wave attenuation potential, sediment properties and mangrove growth dynamics data over Guyana's intertidal mudflats: assessing the potential of mangrove restoration works. *Earth System Science Data*, **14**, 2445–2462, https://doi.org/10.5194/essd-14-2445-2022

Bispo P. D. C., Pardini M., Papathanassiou K. P., Kugler F., Balzter H., Rains D., dos Santos J. R., Rizaev I. G., Tansey K., dos Santos M. N. and Spinelli Araujo L. (2019). Mapping forest successional stages in the Brazilian Amazon using forest heights derived from TanDEM-X SAR interferometry. *Remote Sensing of Environment*, **232**, 111194, https://doi.org/10.1016/j.rse.2019.05.013

Board O.S., National Academies of Sciences E. and Medicine (2019). Negative emissions technologies and reliable sequestration: a research agenda.

Borsje B. W., van Wesenbeeck B. K., Dekker F., Paalvast P., Bouma T. J., van Katwijk M. M. and de Vries M. B. (2011). How ecological engineering can serve in coastal protection. *Ecological Engineering*, **37**, 113–122, https://doi.org/10.1016/j.ecoleng.2010.11.027

Bridges T., Bourne E. M., King J., Kuzmitski H., Moynihan E. and Suedel B. (2018). Engineering with nature: An Atlas. Environmental Laboratory (U.S.), https://doi.org/10.21079/11681/27929

Bridges T., King J., Simm J., Beck M., Collins G., Lodder Q. and Mohan R. (2021). International guidelines on natural and nature-based features for flood risk management. *Engineer Research and Development Center (U.S.)*, 1–989, https://doi.org/10.21079/11681/41946

Brückner M. Z. M., Schwarz C., Dijk W. M., Oorschot M., Douma H. and Kleinhans M. G. (2019). Salt marsh establishment and eco-engineering effects in dynamic estuaries determined by species growth and mortality. *Journal of Geophysical Research: Earth Surface*, **124**, 2962–2986, https://doi.org/10.1029/2019JF005092

Brückner M. Z. M., Braat L., Schwarz C. and Kleinhans M. G. (2020). What came first, mud or biostabilizers? Elucidating interacting effects in a coupled model of mud, saltmarsh, microphytobenthos, and estuarine morphology. *Water Resources Research*, **56**(9), 1–24, https://doi.org/10.1029/2019WR026945

Bryan K. R., Nardin W., Mullarney J. C. and Fagherazzi S. (2017). The role of cross-shore tidal dynamics in controlling intertidal sediment exchange in mangroves in Cù Lao Dung, Vietnam. *Continental Shelf Research*, **147**, 128–143, https://doi.org/10.1016/j.csr.2017.06.014

Buffington K. J., MacKenzie R. A., Carr J. A., Apwong M., Krauss K. W. and Thorne K. M. (2021). Mangrove species' response to sea-level rise across Pohnpei, Federated States of Micronesia (Report No. 2021-1002), Open-File Report. Reston, VA, https://doi.org/10.3133/ofr20211002

Bunting P., Rosenqvist A., Hilarides L., Lucas R. M. and Thomas N. (2022). Global mangrove watch: updated 2010 mangrove forest extent (v2.5). *Remote Sensing*, **14**, 1034, https://doi.org/10.3390/rs14041034

Cappuccino N. (1995). Chapter 1 – Novel approaches to the study of population dynamics. In: Population Dynamics, N. Cappuccino and P. W. Price (eds), Academic Press, San Diego, pp. 3–16, https://doi.org/10.1016/B978-012159270-7/50002-6

Carugati L., Gatto B., Rastelli E., Lo Martire M., Coral C., Greco S. and Danovaro R. (2018). Impact of mangrove forests degradation on biodiversity and ecosystem functioning. *Scientific Reports*, **8**, 13298, https://doi.org/10.1038/s41598-018-31683-0

Chen R. and Twilley R. R. (1998). A gap dynamic model of mangrove forest development along gradients of soil salinity and nutrient resources. *Journal of Ecology*, **86**, 37–51, https://doi.org/10.1046/j.1365-2745.1998.00233.x

Christensen B. (1978). Biomass and primary production of *Rhizophora apiculata* Bl. in a mangrove in southern Thailand. *Aquatic Botany*, **4**, 43–52, https://doi.org/10.1016/0304-3770(78)90005-0

Christensen B. and Wium-Andersen S. (1977). Seasonal growth of mangrove trees in Southern Thailand. I. The phenology of *Rhizophora apiculata* Bl. *Aquatic Botany*, **3**, 281–286, https://doi.org/10.1016/0304-3770(77)90030-4

Coastal Engineering Manual (2002). The Official United States Army Corps of engineers manual. Coastal Hydraulics Laboratory Publication.

Cortés I. M. (2019). Exploring the Role of Evaporation and Precipitation Rates on Mangrove Island Morphology. Montclair State University, New Jersey, USA.

Cox G. M., Gibbons J. M., Wood A. T. A., Craigon J., Ramsden S. J. and Crout N. M. J. (2006). Towards the systematic simplification of mechanistic models. *Ecological Modelling*, **198**, 240–246, https://doi.org/10.1016/j.ecolmodel.2006.04.016

Cuddington K., Fortin M.-J., Gerber L. R., Hastings A., Liebhold A., O'Connor M. and Ray C. (2013). Process-based models are required to manage ecological systems in a changing world. *Ecosphere*, **4**, 1–12, https://doi.org/10.1890/ES12-00178.1

Dahdouh-Guebas F., Jayatissa L. P., Di Nitto D., Bosire J. O., Lo Seen D. and Koedam N. (2005). How effective were mangroves as a defence against the recent tsunami? *Current Biology*, **15**, R443–R447, https://doi.org/10.1016/j.cub.2005.06.008

Dahdouh-Guebas F., Ajonina G. N., Amir A. A., Andradi-Brown D. A., Aziz I., Balke T., Barbier E. B., Cannicci S., Cragg S. M., Cunha-Lignon M., Curnick D. J., Duarte C. M., Duke N. C., Endsor C., Fratini S., Feller I. C., Fromard F., Hugé J., Huxham M., Kairo J. G., Kajita T., Kathiresan K., Koedam N., Lee S. Y., Lin H.-J., Mackenzie J. R., Mangora M. M., Marchand C., Meziane T., Minchinton T. E., Pettorelli N., Polanía J., Polgar G., Poti M., Primavera J., Quarto A., Rog S. M., Satyanarayana B., Schaeffer-Novelli Y., Spalding M., Van Der Stocken T., Wodehouse D., Yong J. W. H., Zimmer M. and Friess D. A. (2020). Public perceptions of mangrove forests matter for their conservation. *Frontiers in Marine Science*, **7**, 603651, https://doi.org/10.3389/fmars.2020.603651

Dahdouh-Guebas F., Friess D. A., Lovelock C. E., Connolly R. M., Feller I. C., Rogers K. and Cannicci S. (2022). Cross-cutting research themes for future mangrove forest research. *Nature Plants*, **8**, 1131–1135, https://doi.org/10.1038/s41477-022-01245-4

Dalrymple R. A., Kirby J. T. and Hwang P. A. (1984). Wave diffraction due to areas of energy dissipation. *Journal of Waterway, Port, Coastal, and Ocean Engineering*, **110**, 67–79, https://doi.org/10.1061/(ASCE)0733-950X(1984)110:1(67)

Damastuti E., De Groot R., Debrot A. O. and Silvius M. J. (2022). Effectiveness of community-based mangrove management for biodiversity conservation: A case study from Central Java, Indonesia. *Trees, Forests and People*, **7**, 100202, https://doi.org/10.1016/j.tfp.2022.100202

Das S. C., Das S. and Tah J. (2022). Mangrove ecosystems and their services. In: Mangroves: Biodiversity, Livelihoods and Conservation, S. C. Das Pullaiah and E. C. Ashton (eds), Springer Nature Singapore, Singapore, pp. 139–152, https://doi.org/10.1007/978-981-19-0519-3_6

Davis S. M., Childers D. L., Lorenz J. J., Wanless H. R. and Hopkins T. E. (2005). A conceptual model of ecological interactions in the mangrove estuaries of the Florida Everglades. *Wetlands*, **25**, 832–842, https://doi.org/10.1672/0277-5212(2005)025[0832:ACMOEI]2.0.CO;2

Day J. W., Christian R. R., Boesch D. M., Yáñez-Arancibia A., Morris J., Twilley R. R., Naylor L., Schaffner L. and Stevenson C. (2008). Consequences of climate change on the ecogeomorphology of coastal wetlands. *Estuaries and Coasts*, **31**, 477–491, https://doi.org/10.1007/s12237-008-9047-6

De Dominicis M., Wolf J., van Hespen R., Zheng P. and Hu Z. (2023). Mangrove forests can be an effective coastal defence in the Pearl River Delta, China. *Communications Earth & Environment*, **4**, 13, https://doi.org/10.1038/s43247-022-00672-7

DeAngelis D. L. and Grimm V. (2014). Individual-based models in ecology after four decades. *F1000Prime Reports*, **6**, 1–6, https://doi.org/10.12703/P6-39

Ding L. and Fu S. (2021). Effects of different types of undecomposed surface litter on the sediment transport capacity. *Geoderma*, **385**, 114842, https://doi.org/10.1016/j.geoderma.2020.114842

Donato D. C., Kauffman J. B., Murdiyarso D., Kurnianto S., Stidham M. and Kanninen M. (2011). Mangroves among the most carbon-rich forests in the tropics. *Nature Geoscience*, **4**, 293–297, https://doi.org/10.1038/ngeo1123

Duarte C. M., Losada I. J., Hendriks I. E., Mazarrasa I. and Marbà N. (2013). The role of coastal plant communities for climate change mitigation and adaptation. *Nature Climate Change*, **3**, 961–968, https://doi.org/10.1038/nclimate1970

Duke N. C., Ball M. C. and Ellison J. C. (1998). Factors influencing biodiversity and distributional gradients in mangroves. *Global Ecology and Biogeography Letters*, **7**, 27–47, https://doi.org/10.2307/2997695

Duke N. C., Meynecke J.-O., Dittmann S., Ellison A. M., Anger K., Berger U., Cannicci S., Diele K., Ewel K. C., Field C. D., Koedam N., Lee S. Y., Marchand C., Nordhaus I. and Dahdouh-Guebas F. (2007). A world without mangroves? *Science*, **317**, 41–42, https://doi.org/10.1126/science.317.5834.41b

Ellison A. M., Felson A. J. and Friess D. A. (2020). Mangrove rehabilitation and restoration as experimental adaptive management. *Frontiers in Marine Science*, **7**, 327, https://doi.org/10.3389/fmars.2020.00327

European Space Agency. (2015). Sentinel-2 user handbook, Issue 1, Rev 2nd ed., ESA Standard Document. European Space Agency, Paris.

Friess D. A., Krauss K. W., Horstman E. M., Balke T., Bouma T. J., Galli D. and Webb E. L. (2012). Are all intertidal wetlands naturally created equal? Bottlenecks, thresholds and knowledge gaps to mangrove and saltmarsh ecosystems. *Biological Reviews*, **87**, 346–366, https://doi.org/10.1111/j.1469-185X.2011.00198.x

Friess D. A., Yando E. S., Abuchahla G. M. O., Adams J. B., Cannicci S., Canty S. W. J., Cavanaugh K. C., Connolly R. M., Cormier N., Dahdouh-Guebas F., Diele K., Feller I. C., Fratini S., Jennerjahn T. C., Lee S. Y., Ogurcak D. E., Ouyang X., Rogers K., Rowntree J. K., Sharma S., Sloey T. M. and Wee A. K. S. (2020). Mangroves give cause for conservation optimism, for now. *Current Biology*, **30**, R153–R154, https://doi.org/10.1016/j.cub.2019.12.054

Fu X.-M., Tang H.-Y., Liu Y., Zhang M.-Q., Jiang S.-S., Yang F., Li X.-Y. and Wang C.-Y. (2021). Resource status and protection strategies of mangroves in China. *Journal of Coastal Conservation*, **25**, 42, https://doi.org/10.1007/s11852-021-00800-z

Gijsman R., Horstman E. M., Swales A., MacDonald I. T., Bouma T. J., Van Der Wal D. and Wijnberg K. M. (2023). Mangrove forest drag and bed stabilisation effects on intertidal flat morphology. *Earth Surface Processes and Landforms*, **49**, 1117–1134, https://doi.org/10.1002/esp.5758

Gilman E., Ellison J. and Coleman R. (2007). Assessment of mangrove response to projected relative sea-level rise and recent historical reconstruction of shoreline position. *Environmental Monitoring and Assessment*, **124**, 105–130, https://doi.org/10.1007/s10661-006-9212-y

Global Mangrove Alliance (Ed.) (2021). Global Mangrove Alliance Roadmap 2021.

Glover H. E., Stokes D. J., Ogston A. S., Bryan K. R. and Pilditch C. A. (2022). Decadal-scale impacts of changing mangrove extent on hydrodynamics and sediment transport in a quiescent, mesotidal estuary. *Earth Surface Processes and Landforms*, **47**, 1287–1303, https://doi.org/10.1002/esp.5317

Goldberg L., Lagomasino D., Thomas N. and Fatoyinbo T. (2020). Global declines in human-driven mangrove loss. *Global Change Biology*, **26**, 5844–5855, https://doi.org/10.1111/gcb.15275

Gorelick N., Hancher M., Dixon M., Ilyushchenko S., Thau D. and Moore R. (2017). Google earth engine: planetary-scale geospatial analysis for everyone. *Remote Sensing of Environment*, **202**, 18–27, https://doi.org/10.1016/j.rse.2017.06.031

Grimm V. and Railsback S. F. (2005). Individual-Based Modeling and Ecology. Princeton Series in Theoretical and Computational Biology. Princeton University Press, Princeton.

Grueters U., Seltmann T., Schmidt H., Horn H., Pranchai A., Vovides A. G., Peters R., Vogt J., Dahdouh-Guebas F. and Berger U. (2014). The mangrove forest dynamics model MesoFON. *Ecological Modelling*, **291**, 28–41, https://doi.org/10.1016/j.ecolmodel.2014.07.014

Hagger V., Worthington T. A., Lovelock C. E., Adame M. F., Amano T., Brown B. M., Friess D. A., Landis E., Mumby P. J., Morrison T. H., O'Brien K. R., Wilson K. A., Zganjar C. and Saunders M. I. (2022). Drivers of global mangrove loss and gain in social-ecological systems. *Nature Communications*, **13**, 6373, https://doi.org/10.1038/s41467-022-33962-x

Hamilton S. E. and Casey D. (2016). Creation of a high spatio-temporal resolution global database of continuous mangrove forest cover for the 21st century (CGMFC-21): CGMFC-21. *Global Ecology and Biogeography*, **25**, 729–738, https://doi.org/10.1111/geb.12449

Horstman E. M., Dohmen-Janssen C. M., Bouma T. J. and Hulscher S. J. M. H. (2015). Tidal-scale flow routing and sedimentation in mangrove forests: combining field data and numerical modelling. *Geomorphology*, **228**, 244–262, https://doi.org/10.1016/j.geomorph.2014.08.011

Horstman E. M., Bryan K. R. and Mullarney J. C. (2021). Drag variations, tidal asymmetry and tidal range changes in a mangrove creek system. *Earth Surface Processes and Landforms*, **46**, 1828–1846, https://doi.org/10.1002/esp.5124

Hu Z., Zhou J., Wang C., Wang H., He Z., Peng Y., Zheng P., Cozzoli F. and Bouma T. J. (2020). A novel instrument for bed dynamics observation supports machine learning applications in mangrove biogeomorphic processes. *Water Resources Research* **56**(7), 1–14, https://doi.org/10.1029/2020WR027257

Hutchison J., Manica A., Swetnam R., Balmford A. and Spalding M. (2014). Predicting global patterns in mangrove forest biomass. *Conservation Letters*, **7**, 233–240, https://doi.org/10.1111/conl.12060

IPCC (2022). Climate Change 2022 – impacts, adaptation and vulnerability: Working Group II contribution to the Sixth Assessment Report of the Intergovernmental Panel on Climate Change, 1st ed., Summary for Policymakers. Cambridge University Press, Cambridge, UK and New York, USA, https://doi.org/10.1017/9781009325844

Jennerjahn T. C., Gilman E., Krauss K. W., Lacerda L. D., Nordhaus I. and Wolanski E. (2017). Mangrove ecosystems under climate change. In: Mangrove Ecosystems: A Global Biogeographic Perspective: Structure, Function, and Services, V. H. Rivera-Monroy, S. Y. Lee, E. Kristensen and R. R. Twilley (eds), Springer International Publishing, Cham, pp. 211–244, https://doi.org/10.1007/978-3-319-62206-4_7

Jørgensen S. E. and Fath B. D. (2011). Chapter 9 – Individual-based models. In: Developments in Environmental Modelling, S. E. Jørgensen and B. D. Fath (eds), Elsevier, Amsterdam, pp. 291–308, https://doi.org/10.1016/B978-0-444-53567-2.00009-0

Karyadi, Soegiarto and Harnanto A. (2012). Pengaliran Lumpur Sidoarjo ke Laut Melalui Kali Porong. Bayumedia Publishing, Malang, Indonesia.

Kelleway J., Serrano O., Baldock J., Cannard T., Lavery P., Lovelock C., Macreadie P., Masqué P., Saintilan N. and Steven A. (2017). Technical review of opportunities for including blue carbon in the Australian Government's Emissions Reduction Fund. Canberra ACT CSIRO.

Kleinhans M., Douma H. and Addink E. A. (2019). Fate of pioneering vegetation patches in a dynamic meandering river. *Earth Surface Processes and Landforms*, **44**, 1618–1632, https://doi.org/10.1002/esp.4596

Kodikara K. A. S., Mukherjee N., Jayatissa L. P., Dahdouh-Guebas F. and Koedam N. (2017). Have mangrove restoration projects worked? An in-depth study in Sri Lanka: evaluation of mangrove restoration in Sri Lanka. *Restoration Ecology*, **25**, 705–716, https://doi.org/10.1111/rec.12492

Krauss K. W., Lovelock C. E., McKee K. L., López-Hoffman L., Ewe S. M. L. and Sousa W. P. (2008). Environmental drivers in mangrove establishment and early development: a review. *Aquatic Botany*, **89**, 105–127, https://doi.org/10.1016/j.aquabot.2007.12.014

Krauss K. W., McKee K. L., Lovelock C. E., Cahoon D. R., Saintilan N., Reef R. and Chen L. (2014). How mangrove forests adjust to rising sea level. *New Phytologist*, **202**, 19–34, https://doi.org/10.1111/nph.12605

Kusumaningtyas M. A., Kepel T. L., Solihuddin T., Lubis A. A., Putra A. D. P., Sugiharto U., Ati R. N. A., Salim H. L., Mustikasari E., Heriati A., Daulat A., Sudirman N., Suryono D. D. and Rustam A. (2022). Carbon sequestration potential in the rehabilitated mangroves in Indonesia. *Ecological Research*, **37**, 80–91, https://doi.org/10.1111/1440-1703.12279

Lee S. Y., Hamilton S., Barbier E. B., Primavera J. and Lewis R. R. (2019). Better restoration policies are needed to conserve mangrove ecosystems. *Nature Ecology & Evolution*, **3**, 870–872, https://doi.org/10.1038/s41559-019-0861-y

Lewis R. R. (2005). Ecological engineering for successful management and restoration of mangrove forests. *Ecological Engineering*, **24**, 403–418, https://doi.org/10.1016/j.ecoleng.2004.10.003

Lewis R. R. and Brown B. (2014). Ecological mangrove rehabilitation. A field manual for practitioner.

Lovelock C. E. and Ellison J. (2007). Chapter 9: Vulnerability of mangroves and tidal wetlands of the Great Barrier Reef to climate change, climate change and the Great Barrier Reef: a vulnerability assessment. The Great Barrier Reef Marine Park Authority.

Lovelock C. E., Ball M. C., Feller I. C., Engelbrecht B. M. J. and Ling Ewe M. (2006). Variation in hydraulic conductivity of mangroves: influence of species, salinity, and nitrogen and phosphorus availability. *Physiologia Plantarum*, **127**, 457–464, https://doi.org/10.1111/j.1399-3054.2006.00723.x

Lovelock C. E., Cahoon D. R., Friess D. A., Guntenspergen G. R., Krauss K. W., Reef R., Rogers K., Saunders M. L., Sidik F., Swales A., Saintilan N., Thuyen L. X. and Triet T. (2015). The vulnerability of Indo-Pacific mangrove forests to sea-level rise. *Nature*, **526**, 559–563, https://doi.org/10.1038/nature15538

Macreadie P. I., Anton A., Raven J. A., Beaumont N., Connolly R. M., Friess D. A., Kelleway J. J., Kennedy H., Kuwae T., Lavery P. S., Lovelock C. E., Smale D. A., Apostolaki E. T., Atwood T. B., Baldock J., Bianchi T. S., Chmura G. L., Eyre B. D., Fourqurean J. W., Hall-Spencer J. M., Huxham M., Hendriks I. E., Krause-Jensen D., Laffoley D., Luisetti T., Marbà N., Masque P., McGlathery K. J., Megonigal J. P., Murdiyarso D., Russell B. D., Santos R., Serrano O., Silliman B. R., Watanabe K. and Duarte C. M. (2019). The future of blue carbon science. *Nature Communications*, **10**, 3998, https://doi.org/10.1038/s41467-019-11693-w

Macreadie P. I., Costa M. D. P., Atwood T. B., Friess D. A., Kelleway J. J., Kennedy H., Lovelock C. E., Serrano O. and Duarte C. M. (2021). Blue carbon as a natural climate solution. *Nature Reviews Earth & Environment*, **2**, 826–839, https://doi.org/10.1038/s43017-021-00224-1

Magnan A. K., Oppenheimer M., Garschagen M., Buchanan M. K., Duvat V. K. E., Forbes D. L., Ford J. D., Lambert E., Petzold J., Renaud F. G., Sebesvari Z., van de Wal R. S. W., Hinkel J. and Pörtner H.-O. (2022). Sea level rise risks and societal adaptation benefits in low-lying coastal areas. *Scientific Reports*, **12**, 10677, https://doi.org/10.1038/s41598-022-14303-w

Marois D. E. and Mitsch W. J. (2015). Coastal protection from tsunamis and cyclones provided by mangrove wetlands – a review. *International Journal of Biodiversity Science, Ecosystem Services & Management*, **11**, 71–83, https://doi.org/10.1080/21513732.2014.997292

Maza M., Lara J. L. and Losada I. J. (2019). Experimental analysis of wave attenuation and drag forces in a realistic fringe *Rhizophora* mangrove forest. *Advances in Water Resources*, **131**, 103376, https://doi.org/10.1016/j.advwatres.2019.07.006

Maza M., Lara J. L. and Losada I. J. (2021). Predicting the evolution of coastal protection service with mangrove forest age. *Coastal Engineering*, **168**, 103922, https://doi.org/10.1016/j.coastaleng.2021.103922

Mazzini A. (2018). 10 years of Lusi eruption: lessons learned from multidisciplinary studies (LUSI LAB). *Marine and Petroleum Geology*, **90**, 1–9, https://doi.org/10.1016/j.marpetgeo.2017.12.025

McIvor A., Möller I., Spencer T. and Spalding M. (2012). Reduction of wind and swell waves by mangroves (Natural Coastal Protection Series: Report 1), Cambridge Coastal Research Unit Working Paper 40. The Nature Conservancy and Wetlands International, Cambridge, UK.

McIvor A., Spencer T., Möller I. and Spalding M. (2013). The response of mangrove soil surface elevation to sea level rise (Natural Coastal Protection Series: Report 3), Cambridge Coastal Research Unit Working Paper 42. The Nature Conservancy and Wetlands International, Cambridge, UK.

Mendez F. J. and Losada I. J. (2004). An empirical model to estimate the propagation of random breaking and nonbreaking waves over vegetation fields. *Coastal Engineering*, **51**, 103–118, https://doi.org/10.1016/j.coastaleng.2003.11.003

Montgomery J., Bryan K., Horstman E. and Mullarney J. (2018). Attenuation of tides and surges by mangroves: contrasting case studies from New Zealand. *Water*, **10**, 1119, https://doi.org/10.3390/w10091119

Mori N., Chang C.-W., Inoue T., Akaji Y., Hinokidani K., Baba S., Takagi M., Mori S., Koike H., Miyauchi M., Suganuma R., Sabunas A., Miyashita T. and Shimura T. (2022). Parameterization of mangrove root structure of *Rhizophora stylosa* in coastal hydrodynamic model. *Frontiers in Built Environment*, **7**, 782219, https://doi.org/10.3389/fbuil.2021.782219

Morris R. L., Konlechner T. M., Ghisalberti M. and Swearer S. E. (2018). From grey to green: efficacy of eco-engineering solutions for nature-based coastal defence. *Global Change Biology*, **24**, 1827–1842, https://doi.org/10.1111/gcb.14063

Morris R., Strain E. M. A., Konlechner T. M., Fest B. J., Kennedy D. M., Arndt S. K. and Swearer S. E. (2019). Developing a nature-based coastal defence strategy for Australia. *Australian Journal of Civil Engineering*, **17**, 167–176, https://doi.org/10.1080/14488353.2019.1661062

Morris R. L., Bishop M. J., Boon P., Browne N. K., Carley J. T., Fest B. J., Fraser M. W., Ghisalberti M., Lovelock C. E., Lowe R. J., Rogers A. A., Simpson V., Strain E. M. A., van Rooijen A. A., Waters E. and Swearer S. E. (2021). The Australian guide to nature-based methods for reducing risk from coastal hazards, Earth Systems and Climate Change Hub Report No. 26. NESP Earth Systems and Climate Change Hub, Australia.

Mukherjee N., Sutherland W. J., Khan M. N. I., Berger U., Schmitz N., Dahdouh-Guebas F. and Koedam N. (2014). Using expert knowledge and modeling to define mangrove composition, functioning, and threats and estimate time frame for recovery. *Ecology and Evolution*, **4**, 2247–2262, https://doi.org/10.1002/ece3.1085

Murdiyarso D., Purbopuspito J., Kauffman J. B., Warren M. W., Sasmito S. D., Donato D. C., Manuri S., Krisnawati H., Taberima S. and Kurnianto S. (2015). The potential of Indonesian mangrove forests for global climate change mitigation. *Nature Climate Change*, **5**, 1089–1092, https://doi.org/10.1038/nclimate2734

Narayan S., Beck M. W., Reguero B. G., Losada I. J., van Wesenbeeck B., Pontee N., Sanchirico J. N., Ingram J. C., Lange G.-M. and Burks-Copes K. A. (2016). The effectiveness, costs and coastal protection benefits of natural and nature-based defences. *PLoS ONE*, **11**, e0154735, https://doi.org/10.1371/journal.pone.0154735

Nardin W., Edmonds D. A. and Fagherazzi S. (2016). Influence of vegetation on spatial patterns of sediment deposition in deltaic islands during flood. *Advances in Water Resources*, **93**, 236–248, https://doi.org/10.1016/j.advwatres.2016.01.001

Nature-Based Solutions Resource Guide. (2022). White House Council on Environmental Quality, White House Office of Science and Technology Policy, White House Office of Domestic Climate Policy, Washington D.C.

Ohira W., Honda K., Nagai M. and Ratanasuwan A. (2013). Mangrove stilt root morphology modeling for estimating hydraulic drag in tsunami inundation simulation. *Trees*, **27**, 141–148, https://doi.org/10.1007/s00468-012-0782-8

Onuf C. P., Teal J. M. and Valiela I. (1977). Interactions of nutrients, plant growth and herbivory in a mangrove ecosystem. *Ecology*, **58**, 514–526, https://doi.org/10.2307/1939001

Pelckmans I., Belliard J.-P., Dominguez-Granda L. E., Slobbe C., Temmerman S. and Gourgue O. (2023). Mangrove ecosystem properties regulate high water levels in a river delta. *Natural Hazards and Earth System Sciences*, **23**, 3169–3183, https://doi.org/10.5194/nhess-23-3169-2023

Pendleton L., Donato D. C., Murray B. C., Crooks S., Jenkins W. A., Sifleet S., Craft C., Fourqurean J. W., Kauffman J. B., Marbà N., Megonigal P., Pidgeon E., Herr D., Gordon D. and Baldera A. (2012). Estimating global 'blue carbon' emissions from conversion and degradation of vegetated coastal ecosystems. *PLoS ONE*, **7**, e43542, https://doi.org/10.1371/journal.pone.0043542

Piercy C. D., Charbonneau B. R., Russ E. R. and Swannack T. M. (2023). Examining the commonalities and knowledge gaps in coastal zone vegetation simulation models. *Earth Surface Processes and Landforms*, **49**, 24–48, https://doi.org/10.1002/esp.5565

Piou C., Feller I. C., Berger U. and Chi F. (2006). Zonation patterns of Belizean offshore mangrove forests 41 years after a catastrophic hurricane. *Biotropica*, **38**, 365–374, https://doi.org/10.1111/j.1744-7429.2006.00156.x

Pretzsch H. (2009). Forest Dynamics, Growth and Yield: From Measurement to Model. Springer Berlin Heidelberg, Berlin, Heidelberg, https://doi.org/10.1007/978-3-540-88307-4

Primavera J. H. and Esteban J. M. A. (2008). A review of mangrove rehabilitation in the Philippines: successes, failures and future prospects. *Wetlands Ecology and Management*, **16**, 345–358, https://doi.org/10.1007/s11273-008-9101-y

Primavera J., Rollon R. and Samson M. (2011). The pressing challenges of mangrove rehabilitation: pond reversion and coastal protection. In: Ecohydrology and Restoration, in Treatise on Estuarine and Coastal Science, E. Wolanski and D. McLusky, Academic Press, Waltham, 217–244, http://www.mangroverestoration.com/pdfs/Primavera2011-ThePressingChallengesofMangroveRehabilitation.pdf.

Reef R., Feller I. C. and Lovelock C. E. (2010). Nutrition of mangroves. *Tree Physiology*, **30**, 1148–1160, https://doi.org/10.1093/treephys/tpq048

Rentschler J., Salhab M. and Jafino B. A. (2022). Flood exposure and poverty in 188 countries. *Nature Communications*, **13**, 3527, https://doi.org/10.1038/s41467-022-30727-4

Rivera-Monroy V. H., Twilley R. R., Bone D., Childers D. L., Coronado-Molina C., Feller I. C., Herrera-Silveira J., Jaffe R., Mancera E. and Rejmankova E. (2004). A conceptual framework to develop long-term ecological research and management objectives in the wider Caribbean region. *BioScience*, **54**, 843–, https://doi.org/10.1641/0006-3568(2004)054[0843:ACFTDL]2.0.CO;2

Rivera-Monroy V. H., Zhao X., Wang H. and Xue Z. G. (2022). Are existing modeling tools useful to evaluate outcomes in mangrove restoration and rehabilitation projects? A minireview. *Forests*, **13**, 1638, https://doi.org/10.3390/f13101638

Rodríguez J. F., Saco P. M., Sandi S., Saintilan N. and Riccardi G. (2017). Potential increase in coastal wetland vulnerability to sea-level rise suggested by considering hydrodynamic attenuation effects. *Nature Communications*, **8**, 16094, https://doi.org/10.1038/ncomms16094

Roelvink J. A. and Reniers A. (2012). A Guide to Modeling Coastal Morphology, Advances in Coastal and Ocean Engineering. World Scientific, Hackensack, NJ, Singapore.

Roelvink D., Huisman B., Elghandour A., Ghonim M. and Reyns J. (2020). Efficient modeling of complex sandy coastal evolution at monthly to century time scales. *Frontiers in Marine Science*, **7**, 535, https://doi.org/10.3389/fmars.2020.00535

Roskoden R. R., Bryan K. R., Schreiber I. and Kopf A. (2020). Rapid transition of sediment consolidation across an expanding mangrove fringe in the Firth of Thames New Zealand. *Geo-Marine Letters*, **40**, 295–308, https://doi.org/10.1007/s00367-019-00589-9

Saintilan N., Horton B., Törnqvist T. E., Ashe E. L., Khan N. S., Schuerch M., Perry C., Kopp R. E., Garner G. G., Murray N., Rogers K., Albert S., Kelleway J., Shaw T. A., Woodroffe C. D., Lovelock C. E., Goddard M. M., Hutley L. B., Kovalenko K., Feher L. and Guntenspergen G. (2023). Widespread retreat of coastal habitat is likely at warming levels above 1.5°C. *Nature*, **621**, 112–119, https://doi.org/10.1038/s41586-023-06448-z

Sasmito S. D., Basyuni M., Kridalaksana A., Saragi-Sasmito M. F., Lovelock C. E. and Murdiyarso D. (2023). Challenges and opportunities for achieving sustainable development goals through restoration of Indonesia's mangroves. *Nature Ecology & Evolution*, **7**, 62–70, https://doi.org/10.1038/s41559-022-01926-5

SER (Society for Ecological Restoration Science and Policy Working Group) (2002). The SER primer on ecological restoration. Science Policy Working Group 9.

Sharma S., Suwa R., Ray R., Rozaimi M., MacKenzie R. A. and Nakaoka M. (2022). Preface: blue carbon studies in Asia-Pacific regions: current status, gaps, and future perspectives. *Ecological Research*, **37**, 5–8, https://doi.org/10.1111/1440-1703.12290

Shih S.-S., Huang Z.-Z. and Hsu Y.-W. (2022). Nature-based solutions on floodplain restoration with coupled propagule dispersal simulation and stepping-stone approach to predict mangrove encroachment in an estuary. *Science of The Total Environment*, **851**, 158097, https://doi.org/10.1016/j.scitotenv.2022.158097

Shore Protection Manual (1973). U.S. Army Coastal Engineering Research Center; For sale by the Superintendent of Documents, U.S. Government Printing Office, Fort Belvoir, VA, Washington.

Sidik F., Lawrence A., Wagey T., Zamzani F. and Lovelock C. E. (2023). Blue carbon: A new paradigm of mangrove conservation and management in Indonesia. *Marine Policy*, **147**, 105388, https://doi.org/10.1016/j.marpol.2022.105388

Spalding M., Kainuma M. and Collins L. (2010). World Atlas of Mangroves. Taylor & Francis Group, London.

Su J., Friess D. A. and Gasparatos A. (2021). A meta-analysis of the ecological and economic outcomes of mangrove restoration. *Nature Communications*, **12**, 5050, https://doi.org/10.1038/s41467-021-25349-1

Temmerman S., Horstman E. M., Krauss K. W., Mullarney J. C., Pelckmans I. and Schoutens K. (2023). Marshes and mangroves as nature-based coastal storm buffers. *Annual Review of Marine Science*, **15**, 95–118, https://doi.org/10.1146/annurev-marine-040422-092951

Thanh V. Q., Reyns J., Wackerman C., Eidam E. F. and Roelvink D. (2017). Modelling suspended sediment dynamics on the subaqueous delta of the Mekong River. *Continental Shelf Research*, **147**, 213–230, https://doi.org/10.1016/j.csr.2017.07.013

The State of the World's Mangroves 2022 (2022). 2022nd ed, Global Mangrove Alliance.

Thi Ha H., Duarte C. M., Tri N. H., Terrados J. and Borum J. (2003). Growth and population dynamics during early stages of the mangrove *Kandelia candel* in Halong Bay, North Viet Nam. *Estuarine, Coastal and Shelf Science*, **58**, 435–444, https://doi.org/10.1016/S0272-7714(03)00109-4

Tiggeloven T., De Moel H., Van Zelst V. T. M., Van Wesenbeeck B. K., Winsemius H. C., Eilander D. and Ward P. J. (2022). The benefits of coastal adaptation through conservation of foreshore vegetation. *Journal of Flood Risk Management*, **15**, e12790, https://doi.org/10.1111/jfr3.12790

Twilley R. R., Castañeda-Moya E., Rivera-Monroy V. H. and Rovai A. (2017). Productivity and carbon dynamics in mangrove wetlands. In: Mangrove Ecosystems: A Global Biogeographic Perspective: Structure, Function, and Services, V. H. Rivera-Monroy, S. Y. Lee, E. Kristensen and R. R. Twilley (eds), Springer International Publishing, Cham, pp. 113–162, https://doi.org/10.1007/978-3-319-62206-4_5

UNFCCC. Secretariat (2021). Nationally determined contributions under the Paris Agreement. Synthesis report by the secretariat (Synthesis reports No. FCCC/PA/CMA/2021/8), Glasgow Climate Change Conference – October/November 2021. Glasgow.

United Nations Environment Programme (2023). Decades of mangrove forest change: What does it mean for nature, people and the climate?, Technical Report. UNEP, Nairobi, Kenya.

USGS (2013). Landsat – A global land-imaging mission. (U.S. Geological Survey Fact Sheet), Fact Sheet.

Van der Stocken T., Wee A. K. S., De Ryck D. J. R., Vanschoenwinkel B., Friess D. A., Dahdouh-Guebas F., Simard M., Koedam N. and Webb E. L. (2019). A general framework for propagule dispersal in mangroves: A general framework for propagule dispersal in mangroves. *Biological Reviews*, **94**, 1547–1575, https://doi.org/10.1111/brv.12514

van der Wegen M., Jaffe B. E. and Roelvink J. A. (2011). Process-based, morphodynamic hindcast of decadal deposition patterns in San Pablo Bay, California, 1856–1887: morphodynamic hindcast in San Pablo Bay. *Journal of Geophysical Research: Earth Surface*, **116**, 1–22, https://doi.org/10.1029/2009JF001614

van Hespen R., Hu Z., Peng Y., Borsje B. W., Kleinhans M., Ysebaert T. and Bouma T. J. (2021). Analysis of coastal storm damage resistance in successional mangrove species. *Limnology and Oceanography*, **66**, 3221–3236, https://doi.org/10.1002/lno.11875

van Hespen R., Hu Z., Borsje B., De Dominicis M., Friess D. A., Jevrejeva S., Kleinhans M., Maza M., van Bijsterveldt C. E. J., Van der Stocken T., van Wesenbeeck B., Xie D. and Bouma T. J. (2022a). Mangrove forests as a nature-based solution for coastal flood protection: biophysical and ecological considerations. *Water Science and Engineering*, **16**, 1–13, https://doi.org/10.1016/j.wse.2022.10.004

van Hespen R., Hu Z., Peng Y., Zhu Z., Ysebaert T. and Bouma T. J. (2022b). Identifying trait-based tolerance to sediment dynamics during seedling establishment across eight mangrove species. *Limnology and Oceanography*, **67**, 2281–2295, https://doi.org/10.1002/lno.12202

van Maanen B., Coco G. and Bryan K. R. (2015). On the ecogeomorphological feedbacks that control tidal channel network evolution in a sandy mangrove setting. *Proceedings of the Royal Society A: Mathematical, Physical and Engineering Sciences*, **471**, 20150115, https://doi.org/10.1098/rspa.2015.0115

van Oorschot M., Kleinhans M., Geerling G. and Middelkoop H. (2016). Distinct patterns of interaction between vegetation and morphodynamics: distinct patterns of interaction between vegetation and morphodynamics. *Earth Surface Processes and Landforms*, **41**, 791–808, https://doi.org/10.1002/esp.3864

Van Wesenbeeck B., van Rees Cronin F., Tonneijck F. K. and Winterwerp H. (2021). System understanding: Building with nature to restore eroding tropical muddy coasts. Ecoshape technical report. Ecoshape International, Dordrecht, The Netherlands.

van Zelst V. T. M., Dijkstra J. T., van Wesenbeeck B. K., Eilander D., Morris E. P., Winsemius H. C., Ward P. J. and de Vries M. B. (2021). Cutting the costs of coastal protection by integrating vegetation in flood defences. *Nature Communications*, **12**, 6533, https://doi.org/10.1038/s41467-021-26887-4

Wang W. J., He H. S., Fraser J. S., Thompson F. R., Shifley S. R. and Spetich M. A. (2014). LANDIS PRO: a landscape model that predicts forest composition and structure changes at regional scales. *Ecography*, **37**, 225–229, https://doi.org/10.1111/j.1600-0587.2013.00495.x

Ward R. D., Friess D. A., Day R. H. and Mackenzie R. A. (2016). Impacts of climate change on mangrove ecosystems: a region by region overview. *Ecosystem Health and Sustainability*, **2**, e01211, https://doi.org/10.1002/ehs2.1211

Willemsen P. W. J. M., Horstman E. M., Borsje B. W., Friess D. A. and Dohmen-Janssen C. M. (2016). Sensitivity of the sediment trapping capacity of an estuarine mangrove forest. *Geomorphology*, **273**, 189–201, https://doi.org/10.1016/j.geomorph.2016.07.038

Wilms T., van der Goot F., Tonneijck F., Nurhabni F. and Sembiring L. (2021). Building with nature approach. Building with Nature to restore eroding tropical muddy coasts. Ecoshape technical report. Ecoshape International, Dordrecht, The Netherlands.

Wimmler M.-C., Bathmann J., Peters R., Jiang J., Walther M., Lovelock C. E. and Berger U. (2021). Plant–soil feedbacks in mangrove ecosystems: establishing links between empirical and modelling studies. *Trees*, **35**, 1423–1438, https://doi.org/10.1007/s00468-021-02182-z

Winterwerp H., Wilms T., Siri H., Vries J., Noor Y., Van Wesenbeeck B., Cronin K., Van Eijk P. and Tonneijck F. (2016). Building with nature: sustainable protection of mangrove coasts. *Terra Aqua*, **1**, 5–15.

Winterwerp J. C., Albers T., Anthony E. J., Friess D. A., Mancheño A. G., Moseley K., Muhari A., Naipal S., Noordermeer J., Oost A., Saengsupavanich C., Tas S. A. J., Tonneijck F. H., Wilms T., Van Bijsterveldt C., Van Eijk P., Van Lavieren E. and Van Wesenbeeck B. K. (2020). Managing erosion of mangrove–mud coasts with permeable dams – lessons learned. *Ecological Engineering*, **158**, 106078, https://doi.org/10.1016/j.ecoleng.2020.106078

World Bank (2016). Managing coasts with natural solutions: Guidelines for measuring and valuing the coastal protection services of mangroves and coral reefs.

World Bank (2017). Implementing nature based flood protection: Principles and implementation guidance (Working Paper). World Bank, Washington, DC, USA.

World Bank Group (2018). Building with nature (English) (Brief No. 126688), West Africa Coastal Areas Management Program, case study no. 4. World Bank Group, Washington, D.C.

Xie D., Schwarz C., Brückner M. Z. M., Kleinhans M. G., Urrego D. H., Zhou Z. and van Maanen B. (2020). Mangrove diversity loss under sea-level rise triggered by bio-morphodynamic feedbacks and anthropogenic pressures. *Environmental Research Letters*, **15**, 114033, https://doi.org/10.1088/1748-9326/abc122

Xie D., Schwarz C., Kleinhans M. G., Zhou Z. and van Maanen B. (2022). Implications of coastal conditions and sea-level rise on mangrove vulnerability: a bio-morphodynamic modelling study. *Journal of Geophysical Research: Earth Surface*, **127**, e2021JF006301, https://doi.org/10.1029/2021JF006301

Yeo D., Srivathsan A., Puniamoorthy J., Maosheng F., Grootaert P., Chan L., Guénard B., Damken C., Wahab R. A., Yuchen A. and Meier R. (2021). Mangroves are an overlooked hotspot of insect diversity despite low plant diversity. *BMC Biology*, **19**, 202, https://doi.org/10.1186/s12915-021-01088-z

Yoshikai M., Nakamura T., Herrera E. C., Suwa R., Rollon R., Ray R., Furukawa K. and Nadaoka K. (2023). Representing the impact of *Rhizophora* mangroves on flow in a hydrodynamic model (COAWST_rh v1.0): the importance of three-dimensional root system structures. *Geoscientific Model Development*, **16**, 5847–5863, https://doi.org/10.5194/gmd-16-5847-2023

Zainol Z., Peris A. D., Akhir M. F., Rahim N. H. A., Satyanarayana B. and Dahdouh-Guebas F. (2022). Mangrove propagule dispersal in a shallow and narrow coastal lagoon: a simulation-based assessment of the Setiu Wetlands, Malaysia. *Forests*, **13**, 1525, https://doi.org/10.3390/f13091525

Zimmer M., Ajonina G. N., Amir A. A., Cragg S. M., Crooks S., Dahdouh-Guebas F., Duke N. C., Fratini S., Friess D. A., Helfer V., Huxham M., Kathiresan K., Kodikara K. A. S., Koedam N., Lee S. Y., Mangora M. M., Primavera J., Satyanarayana B., Yong J. W. H. and Wodehouse D. (2022). When nature needs a helping hand: different levels of human intervention for mangrove (re-)establishment. *Frontiers in Forests and Global Change*, **5**, 784322, https://doi.org/10.3389/ffgc.2022.784322

Part 4

Integration of nature-based solutions into urban biorefineries

doi: 10.2166/9781789065015_0219

Chapter 11

Integrated aquaculture for fish production coupled with water purification in urban areas: Indonesian cases

*Agus Waluyo[1], Awalina Satya[1], Ika A. Satya[2], Kukuh Nirmala[3], Yuni P. Hastuti[3], Tjandra Chrismadha[1], Novi Mayasari[1] and Tjandra Setiadi[4]**

[1]Research Center for Limnology and Water Resources, National Research and Innovation Agency (BRIN), KST Soekarno, Jl. Raya Bogor-Jakarta km46, Cibinong, Bogor 16911, Indonesia
[2]Research Center for Computation, National Research and Innovation Agency (BRIN), Komplek Cisitu, Jl. Sangkuriang, Coblong, Bandung 40135, Indonesia
[3]Aquaculture Department Fisheries and Marine Sciences Faculty, IPB University, Bogor, Indonesia
[4]Department of Chemical Engineering, Institut Teknologi Bandung, Jl. Ganesha No. 10, Bandung 40132, Indonesia
**Corresponding author: tjandra.setiadi@itb.ac.id*

ABSTRACT

Adverse detrimental impacts of water pollution on the health of urban people have driven a shift in lifestyle toward cleaner and natural resources, especially in food production, consumption, and spatial needs for agriculture. Urban-integrated aquaculture for producing fish and purifying domestic polluted water is a challenging concept for being implemented in developing countries such as Indonesia. In this chapter, the basic concept of urban-integrated aquaculture will be discussed. The advantages relate to biomass productivity, environment impacts, and socio-economic aspects. Furthermore, different types of aquaculture models are overviewed, followed with some related cases in Indonesia. We concluded that the challenges faced by urban areas for achieving sustainable food security through this concept can be effectively managed by proper planning, management, and sharing knowledge and skills on a regional level.

Keywords: Fish production, food security, integrated aquaculture, sustainable food, urban farming, water purification.

11.1 INTRODUCTION

Urbanization and environmental pollution have become important issues in developing countries including Indonesia. Urbanization speeds up economic development but at the same time also causes loss of agricultural lands and increases pollutant emissions. Environmental pollution, such as water, air, and land contamination are unavoidable almost everywhere (Zhang & Cai, 2022). Water contamination in urban areas generally originates from human activities, such as housing, industries, and aquaculture. Urban contaminated water is hazardous for human health and the environment (Ayilara & Babalola, 2023).

© 2025 IWAP. This is an Open Access book chapter distributed under the terms of the Creative Commons Attribution License (CC BY-NC-ND 4.0) which permits copying and redistribution for non-commercial purposes with no derivatives, provided the work is properly cited (https://creativecommons.org/licenses/by-nc-nd/4.0/). The chapter is from the book *Nature-Based Solutions for Urban Sustainability*, Piet N.L. Lens and Xuan-Thanh Bui (Editors).

Urban wastewater treatment becomes a challenge in many countries as the development of infrastructure and regulation have not kept in line with the population growth and urbanization. At the same time, the urban community has started to be selective in fulfilling their needs, particularly food and water, so as to make a change in life style to be more hygienic and put efforts to produce own food utilizing readily available resources around them, including in aquaculture activities (Roan *et al.*, 2019).

The world aquaculture activities have increased from year to year, as seen from the increasing demand for fishery products including aquaculture. According to the FAO (2022) report, the world per capita fish consumption will amount to 21.2 kg in 2030, increasing from 20.5 kg in 2020 (Figure 11.1). This is also true in Indonesia, where the per capita fish consumption increased by 5.47%, from 54.5 kg in 2019 to 566.59 kg in 2020 (Harianto *et al.*, 2022). The drivers of this strong increase in fish consumption for food are a combination of numerous factors, including income development and urbanization, more extensive fish production, improvement in distribution network, product innovations, high price of livestock meat, as well as better people appreciation of fish as a source of nutritious and healthy food. This increasing trend is estimated to further speed up fish consumption in the next decades (Campanati *et al.*, 2022).

To meet fish demand, aquaculture in Indonesia has extended to urban areas and tends to become a more intensive activity (Sambodo *et al.*, 2023; Susanti *et al.*, 2023). Escalation of fish production is generally associated with larger amounts of input, particularly feed per unit of pond (Henriksson *et al.*, 2018). The culture density affects the daily amount of feed to be delivered, in which the higher density aquaculture tends to elevate the amount of waste materials (Dauda *et al.*, 2019). The high nutrient content of aquaculture waste along with various contaminants from other domestic activities leads to a heavier pollution burden on the water environment. In urban areas that are close to coastal zones, it causes nutrient contamination to coastal areas and becomes a serious threat to the sustainability of marine and coastal ecosystems and exerts a direct impact on human healthiness as well as economic activities including tourism and aquaculture (Desbonnet & Costa-Pierce, 2005). This potential pressure on the aquatic environment leads to a call for immediate measures for pollution mitigation as well as efforts to utilize the waste so as to minimize the amount that is disposed to the environment.

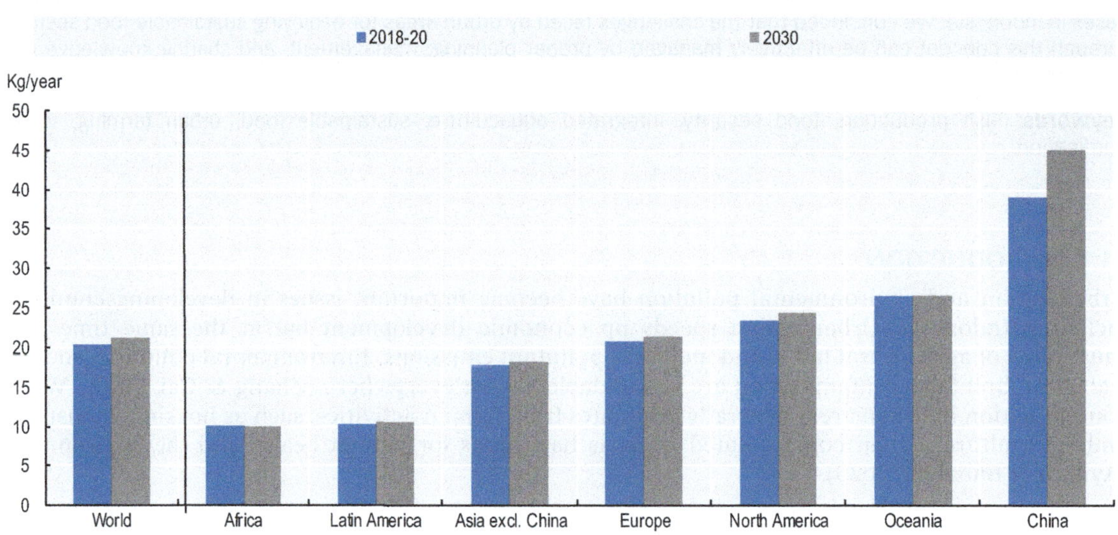

Figure 11.1 World distribution of the per capita fish consumption – 2018–2020 vs 2030 (FAO, 2022).

Aquaculture waste has a high content of various nutrients required by plants to grow, and plants can be harvested and can become valuable input for conjunction production systems. Aquaculture waste comprises fish metabolite products and the decomposed residual feed which are mainly composed of nitrogenous (N) and phosphorous (P) compounds (Boyd & Massaut, 1999). These two elements are very important for protein synthesis in plants, but at high concentrations in water they can lead to a contamination hazard. The aquaculture waste nutrients originate from degradation of proteins, carbohydrates, and lipid compounds by microorganisms and the fish metabolism itself. Some formed nutrients that, among others, contain the elements N and P in the form of ammonia (NH_3-N), nitrite (NO_2^- - N), nitrate (NO_3^- - N), and phosphate (PO_4^{3-} - P), in which the concentration is dependent on the feed ratios, type and size of fish, and the aquaculture management (Novriadi *et al.*, 2020). One of the waste utilization ways is by bioconversion processes (Ayilara & Babalola, 2023).

Through bioconversion processes, aquaculture wastes can potentially be reused by means of transformation from organic waste into energy (Xu *et al.*, 2022). There has also been some reutilization ways as media for growing silkworm (*Bombyx mori*), water flea (*Daphnia magna*), and for some plants. Enhancing the role of other organisms apart from the main cultivated fish is an important element of the integrated aquaculture concept, in which the aquaculture waste can be turned into useful means as it contains various nutrients that can be expected to have added value by increasing production efficiency, improving economic performance, and reducing the negative impact on the environment (Alameen *et al.*, 2023; Sahubawa *et al.*, 2020; Yadav *et al.*, 2023).

Research efforts to develop integrated aquaculture are carried out worldwide, including in Indonesia. They involved some organisms, both animals and plants that potentially provide added value while having the capability to control waste contamination and convert it into a source of energy or in the form of feed available to be reused for aquaculture.

Integrated aquaculture can become a model for any sustainable food production system that applies the principle that the waste disposal from any biological system can be used for nutrition in any other concurrent biological system, so as the incomposition of those two biological systems can result in a more efficient polyculture system, because it has a particular biological water self-purification system that enables employing a closed water recirculating aquaculture (Yadav *et al.*, 2023). Through integrated aquaculture, urban farming and aquaculture activities can be expected to provide urban communities a production system and accessibility for local healthy food as well as for improving their economy.

11.2 URBAN AQUACULTURE WASTEWATER

Urban aquaculture has been considered a part of food resilience in city areas, which currently grow faster accelerated by better accessibility of information and markets as well as water resource availability (Bunting & Little, 2015). In Indonesia, rural aquaculture development has been hindered by land-use change, issues of regulation to control aquaculture activities due to overloaded waste into the environment, and other factors, such as distribution of ponds over a wide area that leads to some difficulties in managing the aquaculture activities resulting in imbalance of ecosystems that lead to disease outbreaks and decline of the production rate. This factor stimulates urban areas as the main consumer of aquaculture products, to develop self-production systems utilizing any available lands as well as the house yards as aquaculture sites.

Different from conventional aquaculture, urban aquaculture is generally conducted in populated areas with limitation of land and ecological resources. Various types of urban aquaculture are currently employed around the world, which can be classified according to the function and resource utilization, including land-based aquaculture, multifunctional wetlands, ponds, lakes, and floating cage aquaculture (Bunting & Little, 2015; Roan *et al.*, 2019)

The major characteristic of urban aquaculture in Asia is concerned with wastewater and its reuse (Bunting & Little, 2015). Dense population and high number of waste-producing industries combined with land and water limitation makes that the management of urban aquaculture wastes requires serious attention. In particular those urban aquaculture activities close to the sea, where the waste is easily discharged to the coastal and sea water. A report stated there was an increase in nutrition input from aquaculture to the coastal water around the world by six folds, from 0.43×10^9 kg N/year in 1985 up to 2.60×10^9 kg N/year in 2005 (Malone and Newton, 2020). Aquaculture waste contamination can be naturally eliminated by microbial activities, but when there is a continual waste production without any controlling measures, moreover combined with input of industrial and anthropogenic wastes and sub-optimal environmental conditions to support microbial degradation processes, the waste contamination will accumulate in the natural water and causes eutrophication (Ayilara and Babalola, 2023; Streicher *et al.*, 2021).

There has been a concern on water contamination caused by aquaculture activities, which is becoming more intense as there is also an increase in water resource demand (Tom *et al.*, 2021). Globally, more than 4 trillion m^3 fresh water is being utilized for daily life activities on a yearly basis. In addition to 70% freshwater consumption for agriculture including aquaculture, there is also 20% available water resource used to fulfill industrial needs and the other 10% is allocated for household activities (Lawnstarter, 2023). Indonesian statistics reported that the freshwater consumption in the country was 222.6 billion m^3 in 2018 (Anonymous, 2022). By considering this water consumption, it can be calculated how much wastewater would be produced from the aquaculture sector. Research on efforts to improve aquaculture productivity as well as control and utilization of freshwater resources has been widely conducted in Indonesia. There have also been technological developments, including biofloc aquaculture, recirculating aquaculture systems, and integrated aquaculture (Sagita *et al.*, 2014).

11.3 URBAN-INTEGRATED AQUACULTURE PRINCIPLE

Integrated aquaculture is a fishery culture system that is combined with other production systems, in particular agriculture or animal husbandry, in which an integration among the sub-systems is maintained. In this case, one particular waste product from one sub-system is considered a valuable resource that is useful to support other relevant production sub-systems in such an arrangement to make the entire available land and water resources mutually employed in order to attain maximum productivity, cost efficiency, and product diversification with considerable added values. Integrated aquaculture is not only applicable for the industrial scale, but also for household scale activities. The culture arrangement can be adapted in particular to anticipate land limitation problems in urban areas (Zajdband, 2011).

Aquaculture intensification has led to higher input requirements followed by worse impacts on the environment and increasing risks especially for small-scale aquaculturists. Meanwhile, there is an expectation that the improvement in aquaculture products does not sacrifice water ecosystems, but decreases water decontamination and increases animal welfare, biodiversity, and social balances. Therefore, it is considered a need to formulate a new strategy toward an efficient and sustainable aquaculture development (FAO, 2022; Murshed-E-Jahan & Pemsl, 2011). Integrated aquaculture has been recommended by some studies as a sustainable aquaculture technology. It is an interesting alternative because of wasted nutrient utilization for producing other valuable and marketable products as well as enhancing the total land productivity (Hasimuna *et al.*, 2023; Rodrigues *et al.*, 2019a; Yadav *et al.*, 2023).

It has been shown by some studies that cultivation of Nile tilapia fish (*Oreochromis niloticus*) in combination with Amazonian river prawn (*Macrobrachium amazonicum*) resulted in an additional prawn production without affecting the fish growth (Rodrigues *et al.*, 2019). At the same time, a mixed culture of African cat fish (*Clarias* sp.), Nile tilapia (*O. niloticus*), and minute duckweed (*Lemna perpusilla*) enhanced minute duckweed biomass production, which is very useful as alternative natural feed for both fish species (Chrismadha, 2020; Chrismadha *et al.*, 2016, 2021). In Indonesia which belongs to developing countries, 70% of aquaculturists are in the category of small-scale enterprises with limitation in land area as well as capital and technology accessibility, and most of them habitat in coastal urban areas (Harianto *et al.*, 2022). As those in other developing countries, such as Zambia and Bangladesh which also have the majority of their aquaculture carried out at small-scale activities, providing training on long-term integrated aquaculture has led to a significant improvement in the total productivity (Hasimuna *et al.*, 2023; Murshed-E-Jahan & Pemsl, 2011).

11.4 IMPACT ON THE ENVIRONMENT

Some studies have elaborated the role of integrated aquaculture in mitigation of water contamination. Wastewater is particularly rich in nutrient elements, such as N and P, which are required as macro-nutrients by plants. These can be utilized by preselected plants to attain both nutrient elimination and the plant biomass production (Campanati *et al.*, 2022; Ezziddine & Liltved, 2021).

Incorporation of phytoremediation processes into a closed water recirculated aquaculture system is able to maintain the water conditions suitable for the aquaculture purpose, so as to support both the aquaculture sustainability as well as water conservation. A study has revealed this advantage where integration of the aquatic plant minute duckweed (*L. perpusilla* Torr) into a closed recirculated cat fish aquaculture could reduce the water consumption up to 85% compared to conventional ones (Chrismadha, 2020). Therefore, it supports water conservation programs in anticipation of water scarcity conditions due to environmental pollution and climate change (Correia *et al.*, 2020; Dauda *et al.*, 2019).

Some water eutrophication has been associated with continuous water pollution. This will then lead to water ecosystem deterioration followed by reduction in water productivity, biodiversity loss, and water shortage. Integrated aquaculture has been developed to aim the conventional aquaculture to utilize wastes before or without disposal to the open water environment. Integrated aquaculture develops an artificial ecosystem in which aquaculture, agriculture, and animal husbandry activities are entirely organized to have some mutual functional relations among them (Nenciu *et al.*, 2022).

11.5 SOCIO-ECONOMIC BENEFITS

In economical respect, integrated aquaculture can be considered as production cost reduction when the aquaculture waste is utilized back for fish natural feed. Some duckweed species and water ferns are examples of plants that can potentially be used for fish natural feed as well as eliminating the nutrient contamination from aquaculture waste. Minute duckweed (*L. perpusilla*) can be used as supplemental feed for cat fish to reduce the artificial pellet feed consumption up to 40% (Chrismadha *et al.*, 2021).

Aquaculture waste can also be employed as fertilizer in agricultural activities so as to enhance diversification of valuable products which yields additional income or self-food resilience for farmers. Under land shortage conditions in urban areas, adaptation is conducted by employing various materials for pond formation, including plastic tarpaulin, glass fiber tanks, and various drums. It actually makes it easier to combine ponds with other culture activities in order to enhance the integrated aquaculture purposes, in which the design and size can be arranged according to the available land and financial sources. It then enables the urban society to provide a part of their own food, while the intention to gain natural and healthy food is also accomplished, which also increases their wealth. Integrated aquaculture also supports the national blue economy program in the aquaculture sector, in which aquaculture is designed to develop the urban economy without any consequences of environment pollution.

11.6 TYPES OF INTEGRATED AQUACULTURE MODELS

11.6.1 Integrated multitrophic aquaculture

The concept of integrated multitrophic aquaculture (IMTA) has been developed a long time ago, originating from China and Egypt which featured pond development and incorporation of fish culture with other cultivation systems. A book from the year 1639 titled Nong Zheng Quan Shu (The

Complete Book on Agriculture) explained many topics of agriculture, such as irrigation and rotation production systems between fish and plant cultivation, utilization of manures as fertilizer in mulberry culture and paddy field, and integrated aquaculture (Azhar and Memiş, 2023; Nissar *et al.*, 2023). Modern IMTA has started to be developed as John Ryther via a project named 'Integrated Waste-recycling Marine Polyculture System' recalled the IMTA concept in the modern era by developing a method to employ sea grass and bivalves for intensive treatment of wastewater disposal. Around the year 2000, the method was applied in the aquaculture field with various modifications, such as polyculture, integrated mariculture, and ecological-based aquaculture. In 2004, the term of integrated multitrophic culture appeared that was initiated by Thierry Chopin and Jack Taylor (Figure 11.2). The basic concept of IMTA systems is the conversion of nutrition produced from aquatic culture activities through species diversification (Azhar and Memiş, 2023; Chopin, 2013).

The IMTA concept has received worldwide attention as it has been shown to be an ecosystem-based approach that is potentially effective at enhancing more advantageous and sustainable aquaculture (Correia *et al.*, 2020; Kim *et al.*, 2022). IMTA is a culture system that employs various feeding habits or trophic levels in one particular place. One of the species functions to process the wasted nutrients and convert them into biomass. The species is placed together in one culturing compartment or separated in a wastewater channel. Today, IMTA has been developed for fresh water, brackish water, and mariculture with various approaches suitable for local environments and socio-economic conditions (Azhar and Memiş, 2023).

Many studies have been conducted concerning the species combination in IMTA. Although different from that in rural areas, in urban areas design of IMTA has to take into account the available landscape, materials for pond construction, prospective species commodities, and production costs. Species selection and the combination ratio according to the available landscape are very important

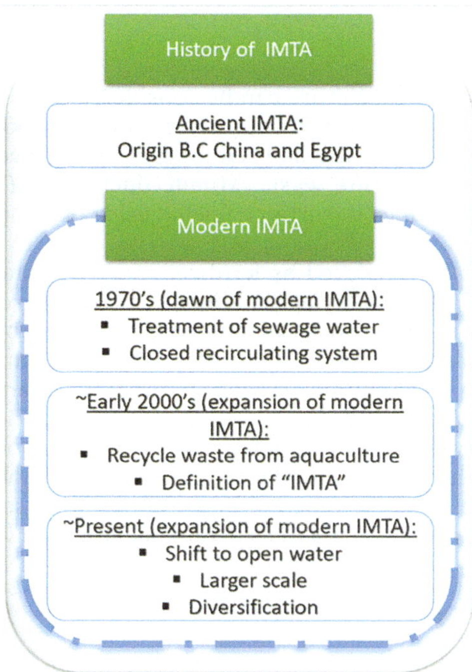

Figure 11.2 History of IMTA practices (*source*: redrawn from Kodama, 2022).

for IMTA (Yadav *et al.*, 2023). At the same time, IMTA design has to be directed for cutting the water usage in anticipation of water shortage which generally occurs in urban areas.

There is an example of IMTA development, particularly in freshwater aquaculture in Indonesia, which is a combination of catfish (*Clarias* sp.) culture with the floating aquatic plant of minute duckweed (*L. perpusilla*). The plant is able to serve as fishery natural feed, and the study revealed that the integration of these two commodities can reduce the production cost up to 30% (Gustiano *et al.*, 2009; Minich and Michael, 2023; Paolacci *et al.*, 2021). Although there are some other potential plant species involved, *Wolffia* and *Spirodela*, the minute duckweed has more potential, as it has a better growth performance, high protein content, is tolerant to high light intensity, and high temperature. Considering these advantages, minute duckweed can potentially be easily cultured for purpose of IMTA in urban areas.

Demand for catfish in Indonesia is constantly increasing. The fish currently has become the most popular commodity among aquaculturists, especially in urban areas. Some factors that stimulate the catfish culture development are low initial investment, easy maintenance, low operational costs, and fast marketing. Technology of catfish cultivation can be easily adopted by aquaculturists, while the marketplace is widely opened and the required investment is relatively low. The culture period is relatively short: it takes only 2.5–3.0 months per production cycle from juvenile up to the consumable size. Catfish aquaculture currently can be carried out at a high density of 2000–3000 fish/m^3, depending on the harvesting size. IMTA is rarely conducted in earthen ponds as aquaculturists tend to change to ponds made of tarpaulin or glass fiber materials. It enables the IMTA design to be more adjustable following the available landscape.

There is no absolute pattern for construction of IMTA systems; the design is generally set according to the biomass production target, which particularly determines the width and compartment number. There are three types of IMTA design: those with additional water recirculation systems, without water recirculation systems, and floating systems. Some Indonesian aquaculturists have implemented IMTA by integration of fish and duckweed or azolla cultures. Design of IMTA with recirculating water systems consists of one compartment for fish culture and some compartments for minute duckweed culture (Figure 11.3). The size of the minute duckweed culture compartment determines the quantity of biomass production. It has to be considered with the need of continual daily supplemental feed delivery in order to enhance the feed efficiency. Chrismadha *et al.* (2021) showed that the culture combination of catfish and minute duckweed can reduce the artificial feed requirements up to 40%. Application of a water recirculation system in IMTA systems helps distribute nutrients from the waste and speed up nitrification processes which convert NH_4^+ into NO_3^- available for absorption by the minute duckweed, and to maintain the dissolved oxygen content in the water during cultivation (Figure 11.4).

The design of IMTA without water recirculation systems has more application in large-scale culture with a large pond size and continual production (Figure 11.5). IMTA without water recirculation systems only channels the wastewater disposed from the regular fish culture pond water into a

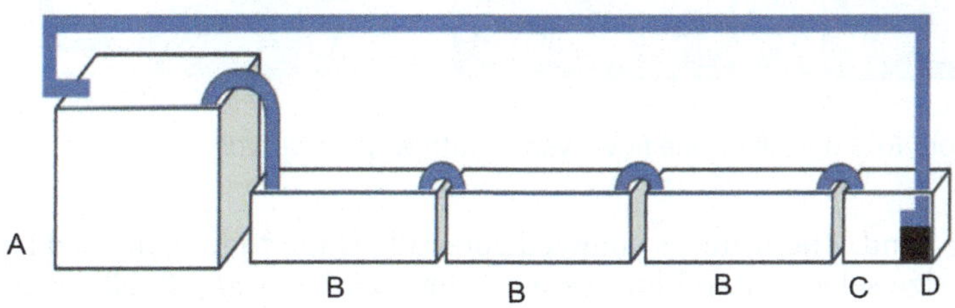

Figure 11.3 IMTA design connected to Recirculatory Aquaculture System (RAS). A, fish tank, B, duckweed compartment, C, water reservoir, D, submersible pump.

Figure 11.4 IMTA ponds combined with RAS in KST Soekarno-Cibinong West of Java (*source*: Chrismadha *et al.*, 2024).

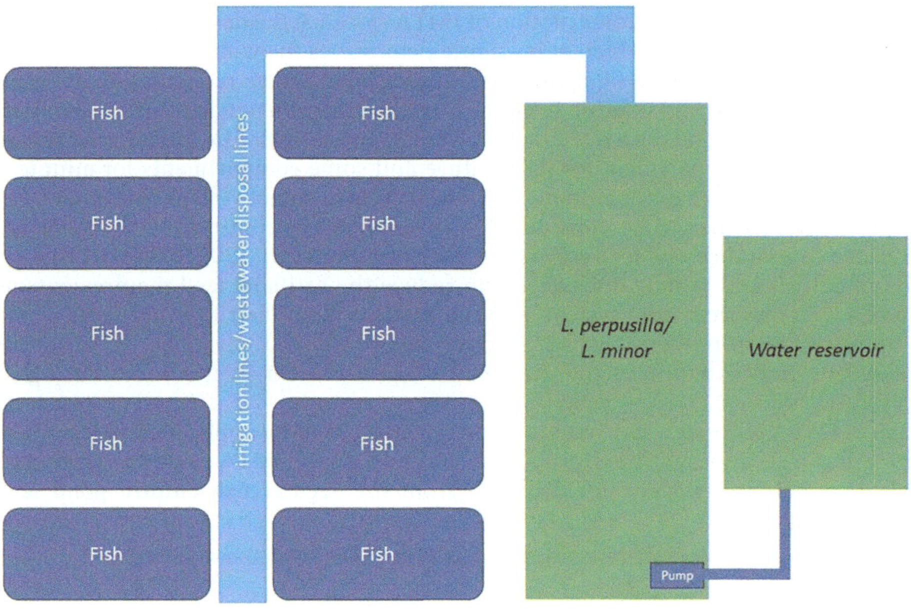

Figure 11.5 IMTA without RAS design or one flow system with water reservoir.

particular collecting pond. The water is changed generally three to four times during the production period, in line with size selection and harvesting (Chrismadha *et al.*, 2024). In the collecting pond, minute duckweed is grown to function as water bioremediatory, whereas the biomass is harvested for supplemental feed. The remedial water is then pumped into a reservoir pond as the water stock available for reuse (Satya *et al.*, 2022).

Figure 11.6 IMTA pond at an aquaculturist hatchery in Bogor Regency-West of Java (Indonesia): (a) tarpaulin-lined pond for culturing catfish, (b) an outlet of the pond connected to the wastewater discharging channel, and (c) wastewater reservoir for growing minute duckweed functioning as phytoremediator, it was harvested daily for fulfilling supplementary catfish feed.

Figure 11.6 shows a series of tarpaulin-lined ponds used for catfish and Nile tilapia cultures in an Indonesian urban area which applies minute duckweed as the supplemental feed. The design can be modified according to the number of fish culture ponds and the number of the minute duckweed culture ponds in order to maintain the continuous supply of biomass for supplemental feed and efficiency state of the nutrient removal process for water purification.

The simplest IMTA design is a floating cage system (Figure 11.7), in which the duckweed-rearing compartment is floated on the surface of the fish cultivation pond. The compartment is made of dragnet arranged like a fishery floating cage with wood materials or polyvinyl chloride pipes as the float but it only has a shallow depth of 10–15 cm, while the width suits the pond size.

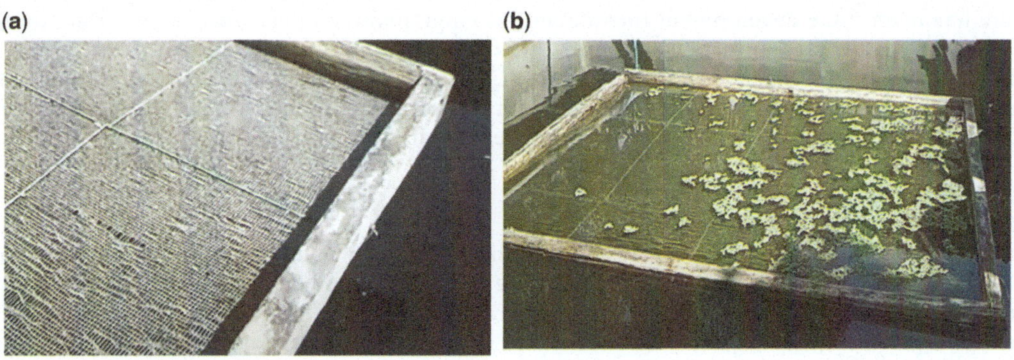

Figure 11.7 Floating IMTA with floating cage for minute duckweed (a) or azolla (b) culture.

Integrated agriculture–aquaculture (IAA) is a development of IMTA, in which the aquaculture nutrient waste is more directed toward various products of horticulture which have economic value or as the additional products. Similar to IMTA, IAA can also be designed with various modifications according to factors of landscape, production capacity, and cost. This technology provides resolution in terms of aquaculture wastewater treatment, water conservation, fertilizer efficiency, and product diversification that is very useful to improve human wealth. Another example is a cultivation system using cages that were set in netting brackish water ponds for growing *Gracilaria* seaweed, milk fish (*Chanos chanos*), tilapia (*O. niloticus*), shrimp (*Penaeus vannamei*), sea weed (*Gracilaria verrucosa*), and green mussel (*Perna viridis*) (Rejeki *et al.*, 2016).

IAA systems have been developed since a long time ago where the integration was mainly conducted at a large area as the agriculture and aquaculture activities generally occur in rural areas. In line with technology development, however, both agriculture and aquaculture activities are nowadays practiced in urban areas. Fish culturing together with growing horticulture plants, such as vegetables and fruits, can be performed by urban societies at a household scale (Sastro, 2016).

IAA enables to use water efficiently, which is an important issue particularly in arid and semi-arid areas. As that of the urban water efficiency principle, IAA enables to reduce the requirement of water and fertilizer for plants while improving the water productivity. Integrating agriculture and aquaculture can improve the fishpond water quality, reduce water and fertilizer consumption, reduce the environmental impact of nutrient-rich water disposal, diversify agricultural products, and improve revenue (Abdul-Rahman *et al.*, 2011).

The most recent developed IAA which is considered the most applicable in urban areas is aquaponic, which is a cultivation system that combines fish culture and hydroponic, a water-based horticulture. Hydroponics have currently become more favorable as it makes it easier to find the source of healthy food by own production of vegetables that are cleaner and free of pesticides. This hydroponic application, however, has been an obstacle by the high fertilizer costs. IAA, with the aquaponic models in particular, provides a technology to resolve this problem. Utilization of aquaculture waste by recycling it through an aquaponic system with proper modification to suit the available resources will enable the society to produce fish and horticultural goods on their own yard.

Aquaponics is very suitable for the application in urban areas which generally has limitation in landscape and water resources. Besides producing fish, this system also produces valuable food products and accordingly increases the revenue. It is generally designed by constructing a water recirculation system that channels fishery ponds into plant cultivation containers. Basic aquaponic design involves a series of pipes connecting the fish culture compartments with the horticulture compartments, so that the aquaculture wastewater can be channeled into the plant container and then flow back to the aquaculture compartment (Figure 11.8). The wastewater nutrient content is designed to eliminate NH_4^+ by the plants so that the water can be recycled back to the fish compartment.

There has been some development introduced into aquaponic models, and in Indonesia these are generally applied for aquaculture wastewater utilization in cultivation of catfish (*Clarias* sp.), Nile tilapia (*Oreochromis niloticus*), and common carp (*Cyprinus carpio*). There are some methods or types of aquaponics available for application, such as nutrient film techniques (NFTs), deep water culture (DWC) or floating raft systems, drip systems, and aeroponic systems.

11.6.2 Aquaponic systems

11.6.2.1 Nutrient film techniques

In a nutrient film technique (NFT) aquaponic system, a plant is grown in a series of narrow channels with a continuous flow of a thin water layer at the bottom to provide water, nutrients, and oxygen for the plant (Chakravartty *et al.*, 2017) as can be seen in Figure 11.9. Some studies revealed that utilization of grass crap (*Ctenopharyngodon idella*) culture wastewater for growing leafy green plant species, such as lettuce (*Lactuca sativa* L.), dill (*Anethum graveolens* L.), rocket (*Eruca sativa*), coriander (*Coriandrum sativum* L.), and parsley (*Petroselinum crispum*) under an NFT scheme yields performance as good as that of fertilized hydroponics (Lennard & Ward, 2019).

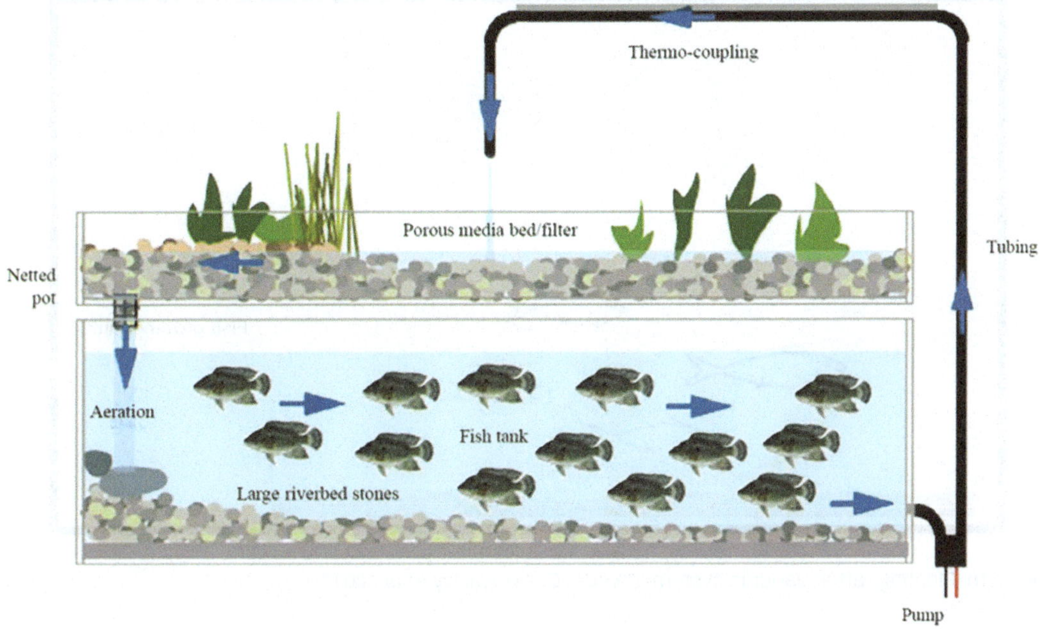

Figure 11.8 Basics of an aquaponic system (*source*: Al-Kodmany, 2018).

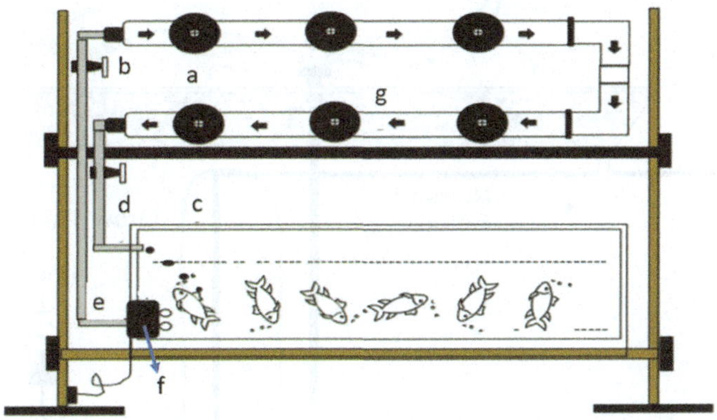

Figure 11.9 Schematic diagram of the nutrient film technique of an aquaponics system: (a) plant growing chamber, (b) water-flow controller, (c) fish growing tank, (d) inlet, (e) outlet, (f) a water pumping motor, and (g) water flow direction (*source*: Alameen *et al.*, 2023).

11.6.2.2 Deep water culture/floating raft systems

DWC method or floating raft system is also known as deep channel system (Figure 11.10). In this system, plants are grown on the rafts made from a polystyrene shelf. This system makes a maximum exposure of the plant roots into the aquaculture water below and avoids clogging.

11.6.2.3 Drip systems

A drip system provides continual or regular trickles of aquaculture wastes dripping into a tray of plants placed above the aquaculture compartment (Figure 11.11), so the excess water can be directed

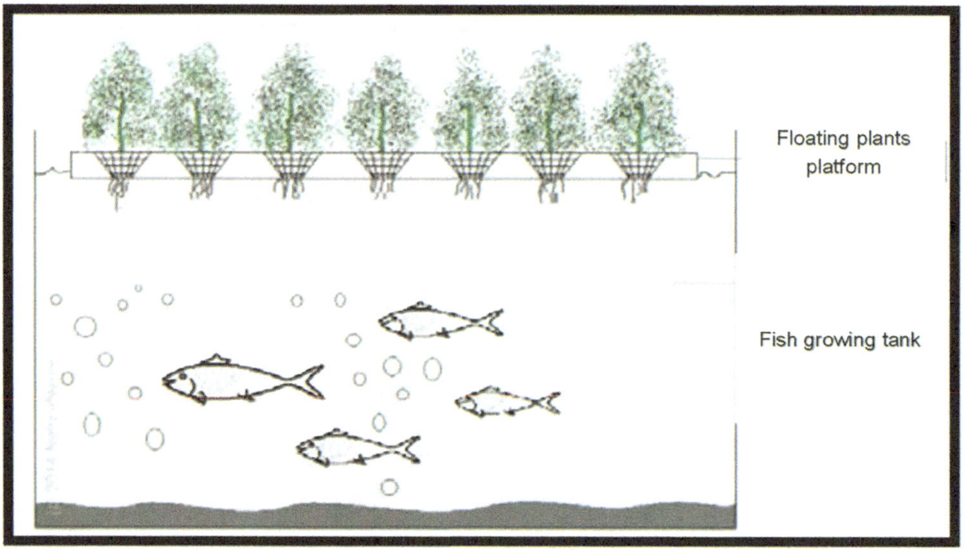

Figure 11.10 Floating raft aquaponic system (*source*: Chakravartty *et al.*, 2017).

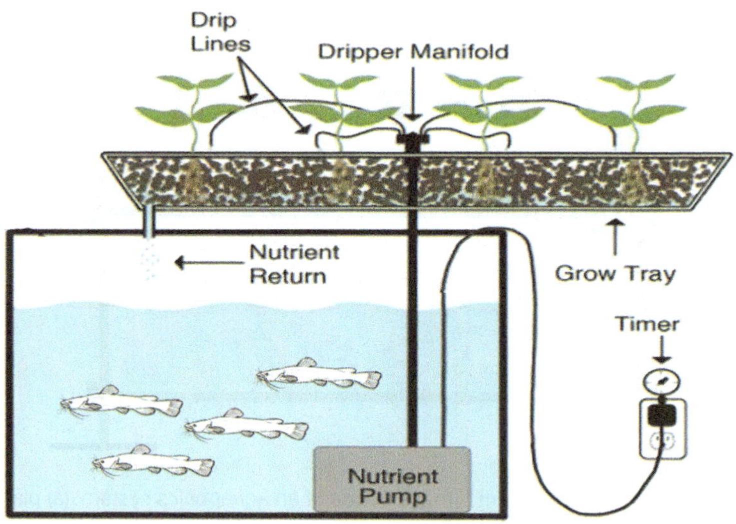

Figure 11.11 Drip aquaponic system (*source*: modified from https://nanobubble.id/).

back to the aquaculture body. The plant cultivation is generally conducted on media of gravel, root tile fragments, or other small-sized stones.

11.6.2.4 Aeroponics systems

Aeroponics are aquaponic systems employing relatively high technology. The growth media is air provided with hazing of nutrients from aquaculture wastewater (Figure 11.12). The hazing is generally

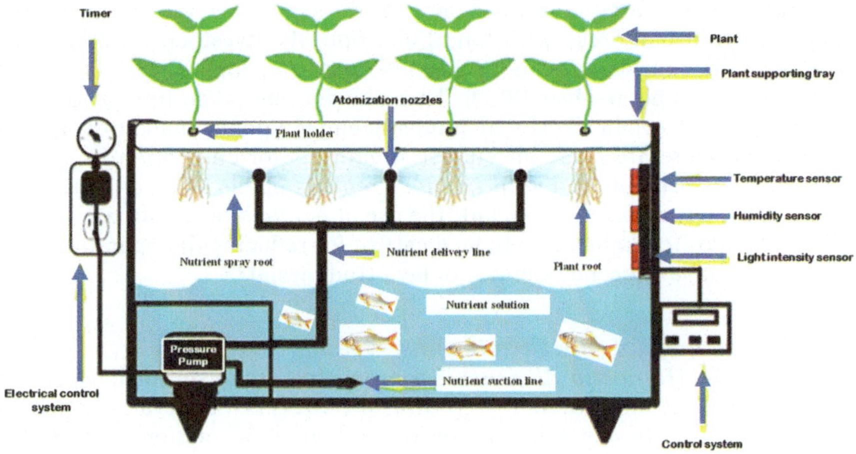

Figure 11.12 Principle of aeroponic system (*source*: modified from Lakhiar *et al.*, 2018).

conducted by water nozzles every few minutes. As the roots are completely exposed to air, it has a risk of drying when the hazing fails. This method is rarely employed as it requires a more expensive operation. This method also has disadvantages when it is applied to an aquaponic system, such as nozzle clogging, improper fogging time, and slow nutrient absorption that makes the wastewater purification more time consuming.

Among the above four methods, Indonesian urban aquaculturists tend to select NFT and floating raft systems, because of easy installation, operation, and maintenance, low-cost construction materials, as well as the better performance. Apart from these aquaponic systems, aquaculture wastewater can also be directly used in a manual way for liquid organism fertilizer (LOF) which is applicable instantly for any plant grown both on the farm land or in the backyard, or involving treatment with sugar or molasses and the addition of dispersed microorganism (EM4) solution (Putro *et al.*, 2015). LOF can be produced and packed for long time storage and marketed (Syamsuddin and Isyanita, 2022).

11.6.3 Expanded aquaculture–animal husbandry

An integrated aquaculture–animal husbandry (IAAH) system is an expanded IAA with additional connection to animal husbandry activities. It is developed based on the idea that 'there is no waste', in which waste is considered a misplaced resource and can be turned into any other valuable product. It is also directed toward a mutual incorporation between animal husbandry and conservation activities, including utilization of all the animal dung (Gebru, 2022).

Water biological productivity is generally manifested in terms of the planktonic biotic abundance which functions as natural fish food. Animal waste can be used as organic fertilizer to stimulate the biological productivity in pond water which in turn will increase the fish production rate. At the same time, it will also accelerate the pond wastewater production that can be employed as fertilizer in cultivation of various plants potentially useful for animal feed. It is a mutual activity that will generate a sustainable aquaculture activity.

A proper selection of cultured animals, both for fishery and husbandry, determines the success of an integrated process intended to be developed. The most consumable fish with good efficiency in feeding on phytoplankton and zooplankton as well as other macrophytic plants become a criterion for the fish selection to employ for integrated aquaculture. In Indonesia, common carp, cat fish, Nile tilapia, gourami, and pangasius are the favorite ones. At the same time, the livestock also determines the required land size. In urban areas with land limitation, the livestock selection and production target have to be carefully considered. A pond has also not to be located in a flooding hazard area and has a yearlong water supply (Gebru, 2022). These have to be taken into account in developing an integrated aquaculture in urban areas which mostly employ ponds made of acrylic, concrete, or tarpaulin. In Indonesia, there are some alternative organisms for cultivation in an IAAH scheme, such as *Clarias gariepinus* fish cultured with *Ipomoea reptans* plants using an aquaponic biofiltered pond (Sagita *et al.*, 2014). Other examples include the use of *Azolla* biomass mixed with quail dung (as pellets) (Sudadi, 2017), and *Azolla* biomass to feed catfish for reducing the operational cost in catfish farms located on Karanganyar Regency Central of Java (Indonesia).

11.6.3.1 Fish, azolla, and maggot – black soldier fly systems

Maggot of black soldier fly (*Hermetia illucens*) is a promising organism for livestock and fish feed as it has a protein content of more than 19% (Indrasari, 2021). This larva is largely employed for waste treatment because of its capability to decompose organic matter, including poultry and livestock manures, livestock innards, fishery wastes, kitchen disposal, fruits, and vegetables (Siddiqui *et al.*, 2022). Maggot lives in humid places that are avoided from direct sun light with the optimum temperature of 24–30°C (Wahyuni *et al.*, 2020). Close placement to the ponds maintains the humidity of the maggot growth media. One of the feed sources is water fern (*Azolla*). Besides the feed source, azolla is also known to function as a phytoremediator on wastewater. Integration of fish, azolla, and maggot is carried out without any water recirculating system or cultivated in separate containers. First of all, the disposed aquaculture wastes are channeled into the waste collecting container where the azolla is cultivated. Then the harvested azolla is used for supplemental feed in maggot culture, or can be directly used for feeding fish. Furthermore, the harvested maggot can be marketed or delivered as additional feed for fishery (Sajuri, 2018).

To attain a good synergy among the culture systems, the capacity, compartment size, and growth period have to be carefully considered, as the culture period of those three organisms is different, as shown in Table 11.1. By taking into account the cultivated biotic culture production cycle of all the cultivated organisms, it can be estimated when to start each cultivation and the minimum production size to ensure synergy in the integrated cultivation can be sustainable. However, there is no standard design for this integrated livestock and fish farming in Indonesia. It can be adjusted to the conditions of the available ponds or other cultivation sites (Figure 11.13). Urban people can employ their properties, and make some modifications to suit the available landscape.

11.6.3.2 Fish, azolla, and poultry systems

From the integration of fishery and azolla as previously explained in Section 11.6.3.1, it can be further developed by incorporation of an additional poultry culture, including chicken or duck. In this case,

Table 11.1 Production cycle of some biota for aquaculture in Indonesia.

Biota	Production Cycle	Growth Rate
Azolla	5–15 days (depends on the pond density)	35%/day
Lemna	3–10 days (depends on the pond density)	32–45%/day
Maggot	15–21 days	
Catfish	2.5–4 months	
Nile tilapia	4–6 months	

Source: Chrismadha *et al.* (2012); Dewi *et al.* (2021), and Widianingrum (2021).

Figure 11.13 Culture pond for Nile tilapia equipped with a raft for culturing maggot (a), rack of maggots' cultivation trays (b), and azolla delivery system to maggot (c) in Bogor Regency-West of Java, Indonesia.

azolla is cultivated in fishery waste-based media to produce high protein feed for birds; after the azolla consumption, birds defecate organic waste that with a little treatment of drying they can be recycled back to nourish the azolla growth media. Water fern (*Azolla*) cultivation can be carried out on trays, containers, or tarpaulin ponds. It can employ IMTA with a water recirculation system as shown in Figure 11.4 or floating cages as shown in Figure 11.14; meanwhile the application for poultry feeding is conducted separately after harvesting.

11.6.4 Role of urban aquaculture in Indonesian economy

Climate change has a negative influence on global weather, increased storm frequency, and sea level as well as temperature rise, all of which affect the survival of aquatic ecosystems and the aquatic economy. Indonesia, with 6.4 million km^2 of water, has a huge potential for fisheries income and commercial value. However, sustainable fisheries development must be based on development of principles that help the current generation while also considering future generations. Blue economy strategies and initiatives have emerged as the most appropriate and effective approach to fisheries development, encouraging the optimal and sustainable use and exploitation of fishery resources.

Figure 11.14 Simple floating cage for azolla and duckweed culture in Bogor Regency-West of Java, Indonesia.

However, Indonesia" fisheries sector has yet to reach its full potential. Aquaculture, like other aspects of the sector, has not been managed optimally. Although aquaculture growth must focus on excellent management practices, it is also necessary to develop more innovative, responsible, sustainable, and profitable technologies that are more ecologically efficient, environmentally friendly, product diverse, and useful to society. Maintaining sustainability is important not only from an environmental standpoint, but also from an economic, social, and technical standpoint. It has become a key issue as consumer awareness grows, demanding quality, traceability, and fair working conditions. As a result, it is vital to reassess the idea and execution of IMTA in the blue aquaculture. This chapter showed that implementing the IMTA scheme's aquaculture idea provides both ecological and economic benefits while remaining sustainable. IMTA approach is particularly important to the Priority Program of the Indonesian Ministry of Marine Affairs and Fisheries to encourage the development of blue aquaculture in Indonesia.

11.7 CONCLUSION

Integrated aquaculture has several advantages, both from the fish production and environmental conservation sectors, as well as the socio-economic development sector especially in urban areas and potentially improves the food production efficiency. Application of integrated aquaculture in the forms of IMTA, IAA, and IAAH has to be encouraged, particularly for small-scale aquaculturists and farmers. To achieve this goal, it needs a support of counseling services that are capable of delivering appropriate knowledge to aquaculturists and farmers and research for determination of the optimum cultivation ratio in a particular unit area. In urban areas where the landscape is limited, IMTA with a water recirculation system or aquaponic are the best and easiest techniques for application. Integrated aquaculture systems harmonize the interaction among fisheries, agriculture, and animal husbandries for achieving efficient agroecosystems that in turn will improve fish and other food production, socio-economic welfare, and natural resource conservation.

REFERENCES

Abdul-Rahman S., Saoud I. P., Owaied M. K., Holail H., Farajalla N., Haidar M. and Ghanawi J. (2011). Improving water use efficiency in semi-arid regions through integrated aquaculture/agriculture. *Journal of Applied Aquaculture*, **23**(3), 212–230, https://doi.org/10.1080/10454438.2011.600629

Alameen A. M., Devi K. N., Kumar S. D., Gunabal S., Krishnaveni N., Gowthami A., Sinduja S., Roopavathy J., Santhanam P., Jebapriya G. R., Saranya M., Aravinth A., Dhanasundaram S. and Perumal P. (2023). A sustainable utilization of aquaculture wastewater for the production of commercially important tilapia fish and plants (mint and chickpea) in improved integrated aqua–agriculture system. *Bioresource Technology Reports*, **21**(October 2022), 101313, https://doi.org/10.1016/j.biteb.2022.101313

Al-Kodmany K. (2018). The vertical farm: a review of developments and implications for the vertical city. *Buildings*, **8**(2), 24, https://doi.org/10.3390/buildings8020024

Anonymous (2022). Green Economy Mendorong Terciptanya Pembangunan Ekonomi yang Inklusif dan Berkelanjutan. Kementerian Koordinator Bidang Perekonomian Republik Indonesia, Siaran Pers. Jakarta, 24 April 2022.

Ayilara M. S. and Babalola O. O. (2023). Bioremediation of environmental wastes: the role of microorganisms. *Frontiers in Agronomy*, **5**(May), 1–15, https://doi.org/10.3389/fagro.2023.1183691

Azhar M. H. and Memiş D. (2023). Application of the IMTA (integrated multi-trophic aquaculture) system in freshwater, brackish and marine aquaculture. *Aquatic Sciences and Engineering*, **38**(2), 106–121, https://doi.org/10.26650/ASE20231252136

Boyd C. E. and Massaut L. (1999). Risks associated with the use of chemicals in pond aquaculture. *Aquacultural Engineering*, **20**, 113–132, https://doi.org/10.1016/S0144-8609(99)00010-2

Bunting S. W. and Little D. C. (2015). Urban aquaculture for resilient food systems. *Cities and Agriculture: Developing Resilient Urban Food Systems, March*, 312–335, https://doi.org/10.4324/9781315716312-18

Campanati C., Willer D., Schubert J. and Aldridge D. C. (2022). Sustainable intensification of aquaculture through nutrient recycling and circular economies: more fish, less waste, blue growth. *Reviews in Fisheries Science and Aquaculture*, **30**(2), 143–169, https://doi.org/10.1080/23308249.2021.1897520

Chakravartty D., Mondal A., Raychowdhury P., Bhattacharya S. B., Mitra A. and Bikash Bhattacharya S. (2017). Role of aquaponics in the sustenance of coastal India-Aquaponics is a solution for modern agriculture in ecologically sensitive Indian mangrove Sundarbans: a review. *International Journal Fisheries and Aquatic Studues*, **5**(2), 441–448. www.fisheriesjournal.com

Chopin T. (2013). Integrated Multi-Trophic Aquaculture. Ancient, adaptable concept focuses on ecological integration. *Global Aquaculture Advocate*, **16**, 16–19.

Chrismadha T. (2020). Phytotechnology for eutrophic waters: ecological approach to increase benefits. A review. *IOP Conference Series: Earth and Environmental Science*, **535**(1), 12011, https://doi.org/10.1088/1755-1315/535/1/012011

Chrismadha T., Sulawesty F., Satya A., Mardiati Y., Mulyana E. and Widoretno M. R. (2012). Use of duckweed (*Lemna perpusilla* Torr.) for natural feed and fitoremedial agent in aquaculture: Profit improvement and sustainability enhancement. International Seminar on New Paradigm and Innovation on Natural Science and Its Application (ISNPINSA), pp. 1–5.

Chrismadha T., Sulawesty F., Awalina A., Mardiati Y., Mulyana E. and Widoretno M. R. (2016). Growth performance of Minute duckweed (*Lemna perpusilla*) in an integrated common carp (*Cyprinus carpio*) closed recirculation aquaculture. *Aquacultura Indonesiana*, **15**(1), 15–26, https://doi.org/10.21534/ai.v15i1.21

Chrismadha T., Tanjung L. R. and Sutrisno (2021). Use of minute duckweed (*Lemna perpusilla*) for supplemental feed in catfish (*Clarias* sp.) culture: determination of the optimal proportion using Powersim simulation. *E3S Web of Conferences*, **322**, 2015, https://doi.org/10.1051/e3sconf/202132202015

Chrismadha T., Satya A., Satya I. A., Satya A. D. M., Yazdi S. K. and Mardawati E. (2024). An overview on managing minute duckweed (*Lemna perpusilla* Torr) cultivation for fish feed purpose BT – Biomass conversion and sustainable biorefinery: Towards circular bioeconomy. In: M. A. R. Lubis, S. H. Lee, E. Mardawati, S. Rahimah, P. Antov, R. Andoyo, Ľ. Kršťák and B. Nurhadi (eds), Springer Nature, Singapore, pp. 171–191, https://doi.org/10.1007/978-981-99-7769-7_8

Correia M., Azevedo I. C., Peres H., Magalhães R., Oliva-Teles A., Almeida C. M. R. and Guimarães L. (2020). Integrated multi-trophic aquaculture: a laboratory and hands-on experimental activity to promote environmental sustainability awareness and value of aquaculture products. *Frontiers in Marine Science*, **7**(March), 1–12, https://doi.org/10.3389/fmars.2020.00156

Dauda A. B., Ajadi A., Tola-Fabunmi A. S. and Akinwole A. O. (2019). Waste production in aquaculture: sources, components and managements in different culture systems. *Aquaculture and Fisheries*, **4**(3), 81–88, https://doi.org/10.1016/j.aaf.2018.10.002

Desbonnet A. and Costa-Pierce B. (2005). Aquaculture in future urban ecosystems. *Urban Aquaculture*, pp. 267–277, https://doi.org/10.1079/9780851998299.0267

Dewi R. K., Ardiansyah F., Fadhlil R. C. and Wahyuni (2021). Maggot BSF: Kualitas Fisik dan Kimianya. In *Litbang Pemas Unisla*. http://fapet.unisla.ac.id/wp-content/uploads/2021/07/Revisi-Layout-Maggot-Ok-104hlm-15-x-23-cm-2.pdf

Ezziddine M. and Liltved H. (2021). Nutrients recovery from aquaculture waste for use as fertilizer in soilless growth systems. *Acta Horticulturae*, **1305**, 399–406, https://doi.org/10.17660/ActaHortic.2021.1305.52

FAO (2022). The State of World Fisheries and Aquaculture, Towards Blue Transformation, Rome, FAO, https://doi.org/10.4060/cc0461en

Gebru T. (2022). Integrated aquaculture with special reference to fish integration with animal husbandry to enhance production and productivity. *Journal of Agriculture and Aquaculture*, **3**(2), 1–6.

Gustiano R., Prakoso V. A., Iswanto B., Radona D., Kusmini I. I. and Ath-thar M. H. F. (2020). Biodiversitas, Status, dan Tren Budidaya Ikan Lele. [Biodiversity, Status and Culture Trend of Catfish in Indonesia]. IPB Press. ISBN: 978-623-256-214-1, Kota Bogor, Indonesia, pp. 1–106.

Harianto E., Supriyono E., Budiardi T., Affandi R., Hadiroseyani Y. and Sabilu K. (2022). Water and land efficiency in eel (*Anguilla bicolor bicolor*) rearing in development of urban aquaculture through vertical aquaculture system. *IOP Conference Series: Earth and Environmental Science*, **1033**(1), 012013, https://doi.org/10.1088/1755-1315/1033/1/012013

Hasimuna O. J., Maulu S., Nawanzi K., Lundu B., Mphande J., Phiri C. J., Kikamba E., Siankwilimba E., Siavwapa S. and Chibesa M. (2023). Integrated agriculture–aquaculture as an alternative to improving small-scale fish production in Zambia. *Frontiers in Sustainable Food Systems*, **7**, 1–13, https://doi.org/10.3389/fsufs.2023.1161121

Henriksson P. J. G., Belton B., Jahan K. M. and Rico A. (2018). Measuring the potential for sustainable intensification of aquaculture in Bangladesh using life cycle assessment. *Proceedings of the National Academy of Sciences of the United States of America*, **115**(12), 2958–2963, https://doi.org/10.1073/pnas.1716530115

Indrasari M. (2021). Maggot BSF cultivation development strategy as economic resilience during pandemic. *Sinergi: Jurnal Ilmiah Ilmu Manajemen*, **11**(2), 18–20, https://doi.org/10.25139/sng.v11i2.4153

Kim D. Y., Shinde S. K., Kadam A. A., Saratale R. G., Saratale G. D., Kumar M., Syed A., Bahkali A. H. and Ghodake G. S. (2022). Advantage of species diversification to facilitate sustainable development of aquaculture sector. *Biology*, **11**(3), 368, https://doi.org/10.3390/biology11030368

Kodama M. (2022). Overview and history of IMTA, from ancient to modern times. JIRCAS-SEAFDEC/AQD Joint Workshop on IMTA Research: Understanding Current Challenges and Future Prospects, pp. 2–5.

Lakhiar I. A., Gao J., Syed T. N., Chandio F. A. and Buttar N. A. (2018). Modern plant cultivation technologies in agriculture under controlled environment: a review on aeroponics. *Journal of Plant Interactions*, **13**(1), 338–352, https://doi.org/10.1080/17429145.2018.1472308

Lawnstarter (2023). water-consumption-statistics/#https://www.lawnstarter.com/. https://www.lawnstarter.com/blog/statistics/

Lennard W. and Ward J. (2019). A comparison of plant growth rates between an NFT hydroponic system and an NFT aquaponic system. *Horticulturae*, **5**(2), 27, https://doi.org/10.3390/horticulturae5020027

Malone T. C. and Newton A. (2020). The globalization of cultural eutrophication in the coastal ocean: causes and consequences. *Frontiers in Marine Science*, https://doi.org/10.3389/fmars.2020.00670

Minich J. J. and Michael T. P. (2023). A review of using duckweed (Lemnaceae) in fish feeds. Research Square Platform LLC, https://doi.org/10.21203/rs.3.rs-3340590/v1

Murshed-E-Jahan K. and Pemsl D. E. (2011). The impact of integrated aquaculture–agriculture on small-scale farm sustainability and farmers' livelihoods: experience from Bangladesh. *Agricultural Systems*, **104**(5), 392–402, https://doi.org/10.1016/j.agsy.2011.01.003

Nenciu F., Voicea I., Cocarta D. M., Vladut V. N., Matache M. G. and Arsenoaia V. N. (2022). 'Zero-Waste' food production system supporting the synergic interaction between aquaculture and horticulture. *Sustainability (Switzerland)*, **14**(20), 13396, https://doi.org/10.3390/su142013396

Nissar S., Bakhtiyar Y., Arafat M. Y., Andrabi S., Mir Z. A., Khan N. A. and Langer S. (2023). The evolution of integrated multi-trophic aquaculture in context of its design and components paving way to valorization via optimization and diversification. *Aquaculture*, **565**, 739074, https://doi.org/10.1016/j.aquaculture.2022.739074

Novriadi R., Wijayanti E., Agustatik S. and Hartanto W. (2020). Potensi dan interaksi limbah nitrogen dan fosfor dari pakan pada sistem produksi marikultur. Jurnal Pengawasan Budidaya, December.

Paolacci S., Stejskal V. and Jansen M. A. K. (2021). Estimation of the potential of *Lemna minor* for effluent remediation in integrated multi-trophic aquaculture using newly developed synthetic aquaculture wastewater. *Aquaculture International*, **29**(5), 2101–2118, https://doi.org/10.1007/s10499-021-00736-z

Putro S. P., Widowati S. and Muhammad F. (2015). The application of integrated multi trophic aquaculture (IMTA) using stratified double net rounded cage (SDFNC) for aquaculture sustainability. *International Journal of Science and Engineering*, **9**(2), 85–89.

Rejeki S., Ariyati R. W. and Widowati L. L. (2016). Application of integrated multi tropic aquaculture concept in an abraded brackish water pond. *Jurnal Teknologi* **78**(4-2), 227–232, https://doi.org/10.11113/jt.v78.8213

Roan E. H., Tiu L., Yanong R. P. E., Dimaggio M. A. and Patterson J. T. (2019). Overview of urban aquaculture 1 types of urban aquaculture systems. pp. 1–6.

Rodrigues C. G., Garcia B. F., Verdegem M., Santos M. R., Amorim R. V. and Valenti W. C. (2019). Integrated culture of Nile tilapia and Amazon River prawn in stagnant ponds, using nutrient-rich water and substrates. *Aquaculture*, **503**, 111–117, https://doi.org/10.1016/j.aquaculture.2018.12.073

Sagita A., Wicaksana S. N., Primasaputri N. R., Prakoso K., Afifah F. N., Nugraha A. and Hastuti S. (2014). Pengembangan Teknologi Akuakultur Biofilter-Akuaponik (Integrating fish and plant culture) Sebagai Upaya Mewujudkan Rumah Tangga Tahan Pangan. Seminar Nasional Tahunan Ke-IV Hasil-Hasil Penelitian Perikanan Dan Kelautan, November 2014, pp. 353–361. https://www.researchgate.net/profile/Andi_Sagita/publication/322150482_PENGEMBANGAN_TEKNOLOGI_AKUAKULTUR_BIOFILTER-AKUAPONIK_Integrating_Fish_And_Plant_Culture_SEBAGAI_UPAYA_MEWUJUDKAN_RUMAH_TANGGA_TAHAN_PANGAN/links/5a484078aca272d2945fdaa0/PENGEMBANGAN-

Sahubawa L., Triyatmo B. and Ambarwati E. (2020). Bioconversion and bioeconomic of wastewater from red tilapia aquaculture on the aquaponics system as source of nutrient in green mustard growth. *E3S Web of Conferences*, **147**, 1013, https://doi.org/10.1051/e3sconf/202014701013

Sajuri (2018). Potensi Tepung Pakan Alternatif dari Maggot dan Azolla (Malla) sebagai Bahan Baku Pakan Ternak dengan Kandungan Protein Tinggi Alternative Feed Flour Potential from Maggot and Azolla (Malla) as animal feed raw materials with high protein content. *Biofarm Jurnal Ilmiah Pertanian*, **14**(1), 35–40.

Sambodo L. A. A. T., Pane D. D. P., Pertamawa L. H., Mafukhah S., Firdaus I. T., Wikapuspita T., Harianto S. K., Collin S., Muhammad T., Mangunsong W. M. C., Suardhin M., Narjoko D., Fauzi A., Nurrahmani S. M., Nugrahaeni T. A., Mukhairiq M. T., Alfarisy M. A. R., M., R. A., Manullang and … Tamadhika R. (2023). Indonesia blue economy roadmap executive team to the minister of national development planning for leading sectors and infrastructure development. *Indonesia Blue Economy Roadmap*, **2**, 3145374.

Sastro Y. (2016). Teknologi Akuaponik Mendukung Pengembangan Urban Farming. Balai Pengkajian Teknologi Pertanian (BPTP). Jakarta.

Satya A., Chrismadha T., Satya A. D. M. and Satya I. A. (2022). Optimizing crude protein production from minute duckweed (*Lemna perpusilla* Torr) grown in varied NPK based medium. *IOP Conference Series: Earth and Environmental Science*, **1062**(1), 12008, https://doi.org/10.1088/1755-1315/1062/1/012008

Siddiqui S. A., Ristow B., Rahayu T., Putra N. S., Widya Yuwono N., Nisa' K., Mategeko B., Smetana S., Saki M., Nawaz A. and Nagdalian A. (2022). Black soldier fly larvae (BSFL) and their affinity for organic waste processing. *Waste Management*, **140**, 1–13, https://doi.org/10.1016/j.wasman.2021.12.044

Sudadi S. (2017). Keterpaduan Antara Beternak Puyuh, Lele Dan Azolla Dalam Mengatasi Limbah Puyuh Dan Mahalnya Pakan Lele. *PRIMA: Journal of Community Empowering and Services*, **1**(1), 11, https://doi.org/10.20961/prima.v1i1.35148

Streicher M. D., Reiss H. and Reiss K. (2021). Impact of aquaculture and agriculture nutrient sources on macroalgae in a bioassay study. *Marine Pollution Bulletin*, **173**, 113025, https://doi.org/10.1016/j.marpolbul.2021.113025

Susanti R., Sugianto S., Asih T., Sasi F. A., Widiatningrum T., Kenarni N., Mukaromah L., Nisa F. and Jaya A. (2023). Penerapan Teknologi Akuaponik pada pemberdayaan masyarakat kelurahan Kalisegoro. Prosiding Semnas Biologi XI Tahun 2023 FMIPA Universitas Negeri Semarang, pp. 70–75.

Syamsuddin R. and Isyanita (2022). Indonesia Menuju Akuakultur Berkelanjutan Indonesia. *Prosiding Simposium Nasional IX Kelautan Dan Perikanan*, **9**, 1–14, https://journal-old.unhas.ac.id/index.php/proceedingsimnaskp/article/view/22567

Tom A. P., Jayakumar J. S., Biju M., Somarajan J. and Ibrahim M. A. (2021). Aquaculture wastewater treatment technologies and their sustainability: A review. *Energy Nexus*, **4**, 100022, https://doi.org/10.1016/j.nexus.2021.100022

Wahyuni W., Fadhlil R. C. and Sholikha R. (2020). Kualitas Fisik Maggot BSF (*Hermetia illucens*) yang Dipelihara Pada Media Limbah Buah Dan Eceng Gondok Terfermentasi. *International Journal of Animal Science*, **3**(4), 129–133, https://doi.org/10.30736/ijasc.v3i04.61

Widianingrum D. C. (2021). Pengembangan Budidaya Azolla Mycrophilla Sebagai Alternatif Pakan Ternak dan Pemanfaatannya Sebagai Pupuk Bio Organik di Wilayah Masyarakat Desa Baletbaru, Sukowono. *Jurnal Abdimas Madani Dan Lestari (JAMALI)*, **3**(1), 11–19, https://doi.org/10.20885/jamali.vol3.iss1.art2

Xu M., Yang M., Sun H., Gao M., Wang Q. and Wu C. (2022). Bioconversion of biowaste into renewable energy and resources: a sustainable strategy. *Environmental Research*, **214**(P2), 113929, https://doi.org/10.1016/j.envres.2022.113929

Yadav M., Yadav N., Chauhan R. and Khati A. (2023). Chapter – 3 Integrated Aquaculture Systems: A Sustainable Farming Approach. February, pp. 37–60.

Zhang Y. and Cai Q. (2022). Impact mechanism of new urbanization on environmental pollution: empirical analysis based on spatial panel model. *Frontier Public Health*, **10**, 928100, https://doi.org/10.3389/fpubh.2022.928100

Zajdband A. D. (2011). Integrated agri–aquaculture systems. In: Genetics, Biofuels and Local Farming Systems, Sustainable Agriculture Reviews, Vol. **7**, E. Lichtfouse (eds), Springer, Dordrecht, pp. 87–127, https://doi.org/10.1007/978-94-007-1521-9_4

doi: 10.2166/9781789065015_0239

Chapter 12

Nature-based solutions in the solid waste management framework in San Carlos City, Philippines

Jeanger P. Juanga-Labayen[1], Ildefonso V. Labayen[2] and Qiuyan Yuan[3]*

[1]*Carlos Hilado Memorial State University, Talisay City, Philippines*
[2]*Technological University of the Philippines Visayas, Talisay City, Philippines*
[3]*Civil Engineering Department, University of Manitoba, Manitoba, Canada*
**Corresponding author: jeanger.labayen@chmsu.edu.ph*

ABSTRACT

Solid waste management (SWM) in many cities in the Philippines remains an escalating challenge, even after the 'Ecological Solid Waste Management Act of 2000' or Republic Act (RA) 9003 came into effect. The San Carlos City is one of the local government units in the Negros Occidental province, which is committed to its solid waste management plan (SWMP) and RA 9003. The city successfully surpassed the minimum requirement of a 25% municipal solid waste diversion rate from landfills by utilizing low-cost technology and community involvement. This study utilizes the awareness, action, and advancement framework on the top priority areas of SWM in San Carlos City: the information, education, and communication and the SWMP including the eco-center. The community's awareness saturation is crucial during the initial stages of the law implementation. Action and advancement analysis reveals that the SWMP should adopt innovative nature-based solutions through scaling up the windrow composting and vermicomposting facilities and landfill gas recovery. Other potential green technologies such as fermentation, anaerobic digestion, and bioremediation transform organic waste into valuable resources. The city's environmental management system should consider green infrastructures inspired by nature-based solutions to enhance its sustainable SWM program.

Keywords: solid waste, environmental management, San Carlos City, eco-centre, sustainability.

12.1 INTRODUCTION

Nature-based solutions offer practical ways to build sustainable green infrastructure that safeguards economy, society, and environment. Integrating nature-based remedies into the city's environmental management system (EMS) establishes a comprehensive framework for reviving and restoring damaged ecosystems. A paramount problem of uncontrolled waste generation, exacerbated by escalating population and urbanization in many developing nations, causes detrimental concerns on municipal solid waste (MSW) management. In the Philippines, challenging strategies on MSW management such as waste prevention, reduction, reuse, recycling, recovery, treatment, and safe disposal are the core policy of Republic Act (RA) 9003, known as the Ecological Solid Waste Management Act of 2000 (Romallosa & Kraft, 2017).

© 2025 IWAP. This is an Open Access book chapter distributed under the terms of the Creative Commons Attribution License (CC BY-NC-ND 4.0) which permits copying and redistribution for non-commercial purposes with no derivatives, provided the work is properly cited (https://creativecommons.org/licenses/by-nc-nd/4.0/). The chapter is from the book *Nature-Based Solutions for Urban Sustainability*, Piet N.L. Lens and Xuan-Thanh Bui (Editors).

The RA 9003 created the National Solid Waste Management Commission to oversee the implementation of this Act and crafted the national solid waste management framework (NSWMF) (PHILJA, 2011). This framework adopts a holistic program on ecological solid waste management (SWM) that promotes waste avoidance, reduction, reuse, recycling, and recovery of other materials (Llanes, 2015; Romallosa & Kraft, 2017). The law mandates the vital role of the local government units (LGUs) in the MSW management as the key to the administration of the provisions of this Act (EMB, 2000). The law also specifies a mandatory waste diversion rate of at least 25% of all solid waste from disposal sites within 5 years of this Act taking effect in 2001. Furthermore, the law imposed the closure of existing open dumpsites by 2006 and the establishment of siting criteria for sanitary landfills. Moreover, the NSWMF encourages LGUs to promote village-based composting on biodegradable waste and to establish a material recovery facility (MRF) for recyclable waste fractions (Acosta *et al.*, 2012).

The Philippines' LGUs primarily ensure the mandatory segregation and diversion of solid waste from landfills by establishing an MRF and other resource recovery facilities. The RA 9003 is considered a major government-driven breakthrough in restructuring MSW management in the Philippines; however, its implementation at the LGU level is still minimal (Premakumara *et al.*, 2014). Compliance with RA 9003 policies remains challenging for many LGUs due to financial, technical, and political constraints (Acosta *et al.*, 2012). Despite this legislation, inadequate SWM practices persist, posing significant environmental and public health risks (Adhikari *et al.*, 2006; Hanuschke *et al.*, 2013; Wynne *et al.*, 2018). This chapter aims to provide an overview of the best practices of San Carlos City's SWM strategy, ensuring successful compliance to surpass the mandated 25% solid waste diversion rate. Besides, the application of the awareness, action, and advancement (AAA) framework to demonstrate the sustainability of the SWM practices in San Carlos City is presented.

12.2 MATERIALS AND METHODS

12.2.1 Description of the study area

San Carlos City is located in the northeast of the Negros Occidental province in the Philippines, having a population of 132,536 in 2015 (PSA, 2015) and 132,650 people in 2020 reflecting an annual growth rate of 0.02% (PhilAtlas, 2024). The city consists of 18 barangays or villages, out of which 15 are classified as urban and three as rural. San Carlos has a land area of 45,150 hectares, of which 52 and 44% can be classified as agricultural and forest areas, respectively (CEMO, 2015). The city has an excellent potential for various types of developments due to its abundant natural resources and available land for development. The city developed a San Carlos eco-zone area where the three major companies, San Carlos Bioenergy Incorporated, San Carlos Solar Energy Incorporated, and San Carlos Bio-Power, engaged in renewable energy production.

Primary data on MSW composition and the percentage of waste diversion from landfills were taken from the San Carlos City Environmental Management Office. The AAA framework was utilized in the city's two priority areas of SWM: the information, education, and communication (IEC) and the solid waste management plan (SWMP) including the eco-center.

12.2.2 Overview of the AAA system

The EMS oversees environmental aspects, satisfies compliance requirements, and counteracts risks and opportunities (ISO 14001, 2015). The EMS primarily aims to protect the environmental conditions of an organization by preventing and mitigating adverse environmental impacts and assisting the organization in fulfilling compliance obligations (ISO 14001, 2015). On the basis of the study conducted by Zeng *et al.* (2005), the benefits of EMS include standardized environmental management procedures, conserving resources and reducing waste, enhancing corporate image for marketing effects, and improving the environmental supply chain. Moreover, Zeng *et al.* (2005) highlighted the critical factors that can affect the implementation of the standards, including the environmental consciousness of the top and intermediate management, as well as the transparent commitment to environmental management, the legal system, and its enforcement.

Figure 12.1 AAA in environmental management scheme (*source*: modified from Sambataro & Hughey, 2006).

According to Sambataro and Hughey (2006), the AAA model in Environmental Management and Reporting System unified the vital elements of ISO 14001 and sustainability reporting in a cyclic scheme. Figure 12.1 presents the cyclic process of the AAA management and reporting system. Organizational *Awareness* considers the environmental impact and creates environmental policy. Then, the organization can proceed to *Action* based on the set goals and targets. The organization prepares performance assessments and generates programs on environmental management to attain the set targets. The *Advancement* aspect is for continuous improvement of the organization's environmental sustainability performance, which can be conducted through evaluation, review processes, and environmental audit. The AAA model in Environmental Management and Reporting Systems provides a strategic framework to managing environmental performance, ensuring compliance, and continual improvement leading to a more sustainable future by taking action and consistently enhancing sustainability initiatives. This chapter considers the two priority areas: (1) the IEC campaign and (2) the SWMP and eco-center of San Carlos City. Herein, the focus was on these priority areas to explore the sustainability of the city's SWM practices.

12.3 AAA FRAMEWORK OF SWM IN SAN CARLOS CITY

12.3.1 Awareness

The Philippine Ecological Solid Waste Management Act was constituted to provide a legal framework for controlling the worsening problem of solid waste. Moreover, the RA 9003 valued the significant roles of LGUs in dealing with their assorted solid wastes. Every city was obliged to create a Solid Waste Management Board (SWMB), which has the tasks to prepare an SWMP, monitor the implementation of the SWMP, adopt economic measures to foster the sustainability of SWMP, recommend measures to safeguard the environment against pollution, and suggest efficient systems for managing the waste from the collection up to disposal. Meanwhile, the SWMP of every city should take the appropriate steps in implementing the 3Rs (reduce, reuse/ and recycle), including waste characterization, collection and transfer, and final disposal (CEMO, 2015).

12.3.1.1 Information, education, and communication

In response to the call of RA 9003, the Integrated Waste Management Pollution Control Division under the City Environment Management Office (CEMO) was tasked to be mainly accountable for the overall management of solid waste in San Carlos City. The CEMO created SWMB in 2002 and prepared its 10-year SWMP covering 2000–2010. One of the initial programs established by SWMP was the IEC, which started in 2002. The Local Government of San Carlos City, in partnership with the Global Environment and Nature Ecosystems Society (GENESYS) Foundation, implemented a massive 5-year city-wide IEC program. The IEC campaign was conducted to support the implementation of SWMP. The local government also recognized that the people's involvement and participation mainly determine the success of SWM. The stakeholders' active participation remains the initial catalyst for change to comply with waste segregation at the source. Moreover, it is considered indispensable to attain better compliance with SWM law. This includes but is not limited to, the private/industrial/agricultural sector, the schools, the 18 barangays, and all households (CEMO, 2015).

12.3.1.2 SWMP and eco-center

CEMO is responsible for the city's SWM from the point of collection to final disposal, encompassing the residential and non-residential waste streams. Proper waste characterization and understanding the nature and quantity of waste is a crucial step toward waste management. Figure 12.2 presents the MSW composition in San Carlos City. It shows that the biodegradable waste fraction predominates in the solid waste stream, with 72% in the market sector, followed by 60% of the residential waste, and 53% of the institutional waste. Generally, biodegradable waste consists of tree trimmings, garden waste, kitchen waste, as well as waste from the market and slaughterhouses, which have the potential for biological processing in windrow composting to convert waste into compost or soil conditioner. In addition, selected waste from the market and kitchen can be subjected to vermicomposting to produce valuable organic fertilizer (Hanuschke *et al.*, 2013). In 2004, the city imposed a 'No Segregation, No Collection' policy to implement segregated waste collection. In 2008, a Solid Waste Management Office under the city mayor's office was created to enhance the implementation of SWMP.

Furthermore, the city's SWMP (2000–2010) indicated the establishment of an eco-center, an integrated SWM facility with a sanitary landfill, leachate treatment, and resource recovery facility. The construction of a sanitary landfill within the city's eco-center was prioritized to cease solid waste disposal in the open dumpsite. The sanitary landfill initiated its operation in 2007, whereas the closure and rehabilitation of the open dumpsite was initiated in 2008. In 2010, the closure and rehabilitation of the dumpsite was completed.

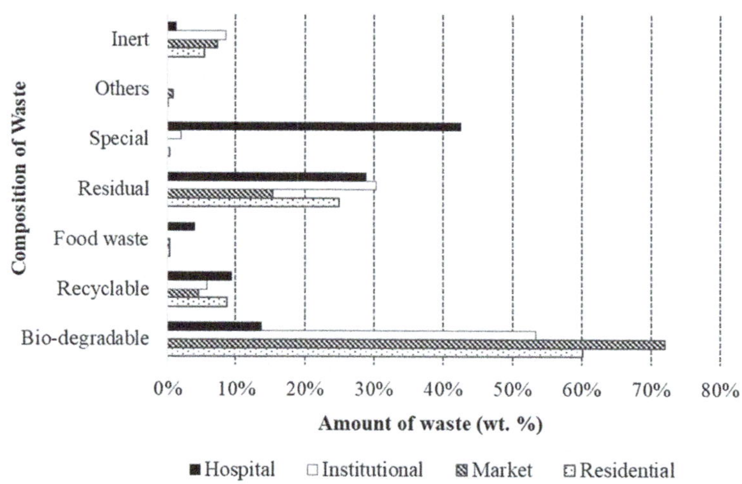

Figure 12.2 Solid waste composition by sector in San Carlos City (*source*: CEMO, 2015).

Kuitunen *et al.* (2008) proposed the Rapid Impact Assessment Matrix (RIAM) as a valuable tool for planning, evaluating, and presenting the sustainability results of the environmental impact assessment (EIA). Taheri *et al.* (2014) conducted an EIA study on four MSW options (i.e., open dumping, sanitary landfill, composting, and recycling) using the RIAM. The results revealed that open dumping is the least preferred option because of its significant adverse effects, whereas composting with a sanitary landfill can be regarded as the most favorable alternative due to the high biodegradability of MSW. Tables 12.1 and 12.2 show the results of the RIAM analysis for the three preferred alternatives and the classification of environmental scores, respectively. This shows that a composting facility with a sanitary landfill has lower negative impacts than recycling because of the high fraction of biodegradable waste in the MSW stream (Taheri *et al.*, 2014). These studies show that the SWM program implemented by San Carlos City has a positive environmental impact.

12.3.2 Action and advancement

12.3.2.1 Information, education, and communication

The IEC program was implemented in 2002 until 2007 by the GENESYS Foundation and has been continuously supported by the city because of its integral role in stakeholders' capacity building. An integrated SWM IEC and outreach program was implemented through training, workshops, and demonstrations in rural and urban villages. The villagers were educated about the significance of waste reduction at source, sorting, composting, and residual disposal. Incentives were awarded to the best-performing villages to motivate the residents' active participation. The impact of the IEC was translated into the successful SWM performance of the city.

Table 12.1 RIAM analysis results for the three proposed alternatives.

Options	Environmental Scores	Range Bands	Description
Sanitary landfill	−48/+16	−D/+B	Significantly negative/positive
Recycling	−67/+18	−D/+B	Significantly negative/positive
Composting	−45/+20	−D/+C	Significantly negative/moderately positive

Source: Taheri et al. (2014).

Table 12.2 Description of environmental scores used in RIAM analysis.

Environmental Scores	Range Bands	Description
+72 to +108	+E	Major positive
+36 to +71	+D	Significantly positive
+19 to +35	+C	Moderately positive
+10 to +18	+B	Positive
+1 to +9	+A	Slightly positive
0	N	No change
−1 to −9	−A	Slightly negative
−10 to −18	−B	Negative
−19 to −35	−C	Moderately negative
−36 to −71	−D	Significantly negative
−72 to −108	−E	Major negative

Source: Pastakia (1998); Pastakia and Jensen (1998).

12.3.2.2 SWMP and eco-center

The city's SWMP encompasses a holistic, SWM and disposal program approach. The CEMO has updated its SWMP for another 10 years, from 2015 up to 2024. The updated SWMP shows the 10-year waste diversion target of at least 1% annually. Hence, it is estimated that the waste diversion in 2015 of 65% will attain 74% by 2024 (CEMO, 2015). The city's SWMB aims to find better waste treatment and recycling alternatives to further improve the waste diversion rate. The CEMO strives to strengthen the existing partnership with Holcim Cement in co-processing residual plastics and other alternative fuels and raw materials. Plastic bags, laminates, and other waste materials accepted by the waste-to-energy plant were separated and adequately baled for shipment to the Holcim Cement factory for co-processing (CEMO, 2015).

Figure 12.3 presents the existing and advanced technologies for sustainable SWM. The segregated organic fraction of MSW can be subjected to biological treatment to generate valuable resources. A portable in-vessel composting, vermicomposting, and sequential batch anaerobic digester can be considered to foster resource recovery from biodegradable organic waste. These decentralized technologies can be situated near market places where market waste is generated, thereby limiting the cost of waste transfer. Landfill gas from a sanitary landfill can be utilized as renewable energy source. Another potential option is the utilization of organic waste for biofuel production. Green energy or bio-based fuel is an emerging powerful nature-based solution that offers the dual benefit of sustainable waste management and reducing greenhouse gas (GHG) emissions to combat global warming (Emmanuel *et al.*, 2024). Important emerging green technologies that utilize microorganisms include fermentation, anaerobic digestion, vermicomposting, and bioremediation. These technologies are able to transform organic waste fractions into valuable eco-friendly products such as bioplastic, biofuel, and biofertilizer (Labayen-Juanga *et al.*, 2024).

The implementation of the integrated eco-center in 2007 was made possible with technical support from the German Society for International Cooperation (GIZ). Figure 12.4 shows some photos taken at the eco-center in San Carlos City. The GIZ identified the city's local soil as a potential landfill liner. This led to significant savings as the sanitary landfill was built for 7.5 million PHP (Philippine Peso), which has an equivalent amount of 150,000 USD, instead of the initially predicted amount of 20 million PHP (400,000 USD). The eco-center consists of a sorting plant, a windrow and vermicomposting

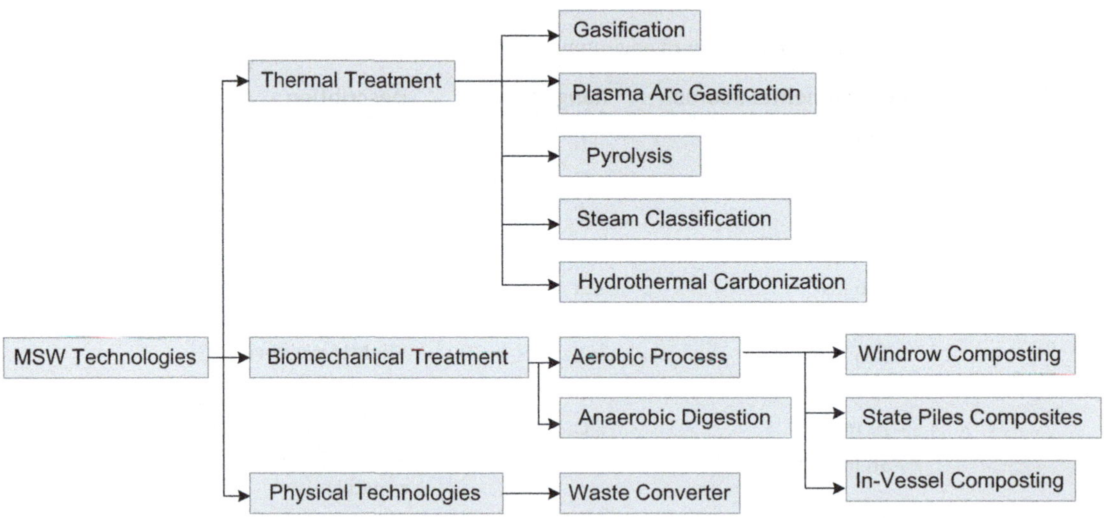

Figure 12.3 Classification of SWM technologies (*source*: modified from Celik *et al.*, 2016).

Figure 12.4 Photos taken at the eco-center in San Carlos City: (a) novel gravity-driven MRF, (b) vermiculture composting area, and (c) windrow composting area.

system, and a landfill site equipped with a leachate treatment system that was developed by employing practical and economical technologies (Paul *et al.*, 2012). All incoming recyclable waste at the eco-center has been segregated at the source and is further segregated at the waste sorting facility by passing the waste along several slides driven by gravity (Hanuschke *et al.*, 2013). Figure 12.5 presents the amount of waste diverted from landfills and disposed of in landfills. A large fraction of the solid

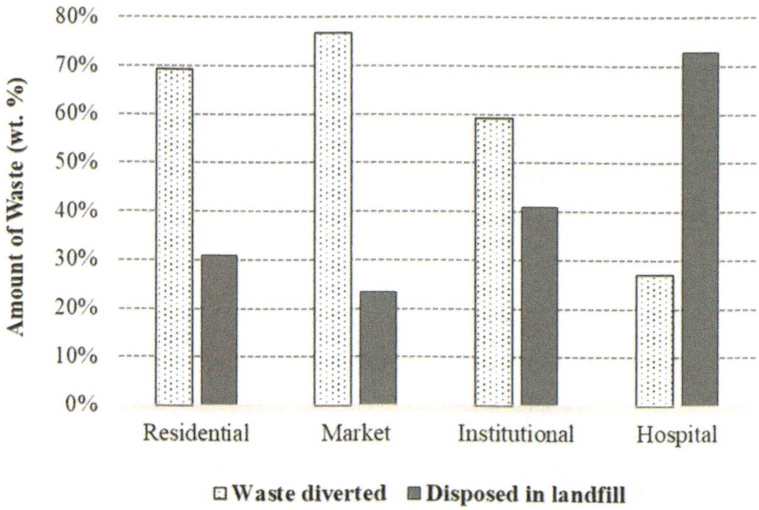

Figure 12.5 Percentage of waste diverted from landfill and disposed in landfill (*source*: CEMO, 2015).

waste streams from residential (69%), market (77%), and institutional (59%) sectors has been diverted from landfilling. However, a large fraction of hospital waste is disposed of in landfills. Hospitals in the city currently have placenta digesters that treat infectious waste before landfill disposal. Consequently, the city's SWMB should continue to make a strategic plan to minimize the amount of hospital waste being landfilled.

Paul *et al.* (2012) reported that 64% of the waste has been diverted from landfill disposal, surpassing the RA 9003 target of at least a 25% waste diversion rate. Moreover, a climate calculator for GHG assessment developed for the SWM sector ascribed that the increased material recovery at the eco-center resulted in a reduction of GHG emissions by around 7600 tons CO_2-eq/year (net) if compared with the baseline scenario in 2005 when all collected wastes were disposed at a dumpsite and material recovery was associated with waste pickers. Table 12.3 summarizes the activities involved in the

Table 12.3 AAA framework towards environmental sustainability of SWM in San Carlos City.

Objectives	Awareness	Action	Advancement
1. Effective IEC	Integral role of stakeholders' participation in SWMP Intensive IEC campaign through 'Lifestyle Change Program'	IEC training, workshops, and demonstrations Formulation and implementation of policies for SWMP	IEC campaign reaching out to 88% of the city population Dissemination of updated/revised training/workshop manual on SWM
2. Execution and evaluation of SWMP	Strategic SWMP prepared by CEMO. Application of 3R principle for biodegradable waste management	Closure and rehabilitation of the old dumpsites Establishment of sanitary landfill with leachate treatment	Periodic review and assessment of SWMP Enhancement of innovative waste management with landfill gas recovery
3. Enhancement of eco-center for improved waste diversion rate	Sustainable operation of eco-center and MRF	Improvement of 'Gravity MRF', an eco-center's sorting facility	Conversion of organic waste into sustainable resources via anaerobic digestion, windrow composting, and vermicomposting system

AAA framework utilizing the SWM program of San Carlos City. Key objectives are highlighted with emphasis on AAA. To foster the environmental sustainability of SWM in San Carlos City, regularly upgrading the IEC and SWMP program and engaging in low-cost innovative technologies to enhance the eco-center operation is required.

12.4 CONCLUSIONS

The execution of a comprehensive SWMP involves holistic, active participation from different stakeholders. The San Carlos City's approach to environmental sustainability requires political will, technological innovativeness, public engagement, and consultation, as well as transparency. The successful compliance of the city to the RA 9003 is due to the combined efforts of the San Carlos LGUs, NGOs (GENESYS Foundation and GIZ), and the community. However, an optimal SWMP involves periodic revising and updating as the city's local conditions change. The AAA framework serves as a valuable tool for monitoring and evaluating the performance indicators of the organization's priority areas. The application of the AAA framework on SWMPs and the EIA study using RIAM analysis are highly recommended. Importantly, it is of utmost importance to integrate nature-based inspired solutions into a comprehensive SWMP, because these provide cost-effective approaches leading to sustainable green infrastructure solutions to protect the society, environment, and economy. Waste treatment technologies driven by microorganisms are innovative nature-based solutions for the biodegradable fraction of MSW. Green technologies infused by nature-based solutions embedded into the city's EMS create a strategic mechanism for restoring and revitalizing the degraded urban ecosystem.

ACKNOWLEDGEMENTS

We thank Mr. Arthur Alfaro Batomalaque and acknowledge the CEMO of San Carlos City for providing valuable data and information.

REFERENCES

Acosta V., Paul J., Lao C., Aguinaldo E. and Valdez M. D. C. (2012). Development of the Philippines national solid waste management strategy 2012–2016. Proceedings of the 7th International Conference on Waste Management and Technology. *Procedia Environmental Sciences*, **16**, 9–16, https://doi.org/10.1016/j.proenv.2012.10.003

Adhikari R., Juanga J. P. and Visvanathan C. (2006). Anaerobic digestion of municipal solid waste in thermophilic sequential batch process. Proceedings of the 5th International Conference ORBIT, Welmar, Germany.

Celik N., Yasar D. and Bastani M. (2016). Assessment and Evaluation of Advanced Solid Waste Management Technologies for Improved Recycling Rates. University of Miami, Department of Industrial Engineering, Gainesville, Florida, United States.

City Environment Management Office (CEMO) (2015). Solid Waste Management Plan (updated) 2015. San Carlos City, Negros Occidental, Philippines.

Emmanuel O., Rozina and Ezeji T. C. (2024). Utilization of biomass-based resources for biofuel production: A mitigating approach towards zero emission. *Sustainable Chemistry One World*, **2**, 100007, https://doi.org/10.1016/j.scowo.2024.100007

Environmental Management Bureau (EMB) (2000). Republic Act 9003. Ecological Solid Waste Management Act of 2000. Republic of the Philippines, Congress of the Philippines, 11th congress, 3rd regular session.

Hanuschke K., Romallosa A. R. D., Paul J. G., Sanchez L. C. and Batomalaque A. (2013). Evaluation of waste diversion effects through a v-cost, gravity-driven waste sorting plant in San Carlos City, Philippines. Proceedings of the International Conference on Solid Waste Management, Hong Kong.

International Organization for Standard (ISO) 14001 (2015). Environmental management systems – Requirements with guidance for use. ISO 2015, Published in Switzerland.

Kuitunen M., Jalava K. and Hirvonen K. (2008). Testing the usability of the rapid impact assessment matrix (RIAM) method for comparison of EIA and SEA results. *Environmental Impact Assessment Review*, **28**, 312–320, https://doi.org/10.1016/j.eiar.2007.06.004

Labayen-Juanga J., Labayen I. V., Montalbo-Lomboy M., Ngo M. T. T. and Doan P. V. M. (2024). Emerging green technologies for organic waste management. In: Waste Management in Climate Change and Sustainability Perspectives: Organic and Medical Waste, 1st Edn, S. Varjani, I. Haq, A. Pandey, V. K. Gupta and X.-T. Bui (eds), CRC Press, Boca Raton, Florida, United States, 3–26, https://doi.org/10.1201/9781003386902

Llanes Jr. C. A. (2015). Environmental Investments in the Philippines for Blue Economy Development. The East Asian Seas (EAS) Congress, Danang, Vietnam.

Pastakia C. M. R. (1998). The rapid impact assessment matrix (RIAM) – A new tool for environmental impact assessment. In: Environmental Impact Assessment Using the Rapid Impact Assessment Matrix (RIAM), K. Jensen (ed.), Olsen & Olsen, Fredensborg, pp. 8–17.

Pastakia C. M. R. and Jensen A. (1998). The rapid impact assessment matrix (RIAM) for EIA. *Environmental Impact Assessment Review*, **18**, 461–482, https://doi.org/10.1016/S0195-9255(98)00018-3

Paul J. G., Sanchez L., Batholomaque A. and Hanuschke K. (2012). Increase of bio-waste recovery with low-cost technologies in San Carlos City, Philippines. Proceedings of the 8th International Conference ORBIT, Rennes, France.

PhilAtlas (2024). Western Visayas Region, Negros Occidental. Available from https://www.philatlas.com/visayas/r06/negros-occidental.html (accessed 15 May 2024)

Philippine Judicial Academy (PHILJA) (2011). Access to environment justice: A sourcebook on environmental rights and legal remedies. http://www.ombudsman.gov.ph/UNDP4/access-to-environmental-justice-a-sourcebook-on-environmental-rights-and-legal-remedies/index.html.

Philippines Statistics Authority (PSA) (2015). Population of Negros Island Region (Based on the 2015 Census of Population). Available from https://psa.gov.ph/content/population-negros-island-region-based-2015-census-population (accessed 10 May 2024).

Premakumara D. G. J., Canete A. M. L., Nagaishi M. and Kurniawan T. A. (2014). Policy implementation of the Republic Act (RA) No. 9003 in the Philippines: a case study of Cebu city. *Waste Management*, **34**, 971–979, https://doi.org/10.1016/j.wasman.2013.10.040

Romallosa A. R. D. and Kraft E. (2017). Feasibility of biomass briquette production from municipal waste streams by integrating the informal sector in the Philippines. *Resources*, **6**, 12, https://doi.org/10.3390/resources6010012

Sambataro III J. P. and Hughey K. F. D. (2006). The awareness, action and, advancement (AAA) environmental management and reporting system: moving from the household to small and medium-size enterprises. *Journal of Cleaner Production*, **14**, 1017–1027, https://doi.org/10.1016/j.jclepro.2005.11.046

Taheri M., Gholamalifard M., Ghazizade M. J. and Rahimoghli S. (2014). Environmental impact assessment of municipal solid waste disposal site in Tabriz, Iran using rapid impact assessment matrix. *Impact Assessment and Project Appraisal*, **32**(2), 162–169. https://doi.org/10.1080/14615517.2014.896082

Wynne A. L., Nieves P. M., Vulava V. M., Qirko H. N. and Callahan T. J. (2018). A community-based approach to solid waste management for riverine and coastal resource sustainability in the Philippines. *Ocean & Coastal Management*, **151**, 36–44, https://doi.org/10.1016/j.ocecoaman.2017.10.028

Zeng S. W., Tam C. M., Tam V. W. Y. and Deng Z. M. (2005). Towards implementation of ISO 14001 environmental management systems in selected industries in China. *Journal of Cleaner Production*, **13**, 645–656, https://doi.org/10.1016/j.jclepro.2003.12.009

doi: 10.2166/9781789065015_0249

Chapter 13

Organic fertilizer production from agriculture by-products for sustainable agricultural systems

Chau Thi Da[1]*, *Ngoc Tuan Nguyen*[1], *Tran Van Dung*[2], *Mai-Nhu Hoang*[3] *and Xuan-Thanh Bui*[3]

[1]*Faculty of Applied Sciences, Ton Duc Thang University, 19 Nguyen Huu Tho street, District 7, Ho Chi Minh City 700000, Vietnam*
[2]*Faculty of Soil Science, College of Agriculture, Can Tho University, Can Tho, 94100 Vietnam*
[3]*VNU-HCM Key Laboratory of Advanced Waste Treatment Technology, Ho Chi Minh City University of Technology (HCMUT), Ho Chi Minh City, Vietnam*
**Corresponding author: chauthida@tdtu.edu.vn*

ABSTRACT

Vietnam is one of the leading countries in the world for exporting rice, agricultural foods, seafood, and aquaculture products. However, in the past few decades, this country has primarily focused on maximizing productivity for export, often neglecting the management of surplus inputs in the production process and over application of organic fertilizers. These fertilizers are crucial for enhancing soil nutrition, improving soil structure and fertility, protecting biodiversity, and increasing the value of organic and safe products for export. This chapter evaluates the practical benefits of reusing and converting local organic wastes and agricultural by-products into a renewable resource within a sustainable agricultural model. The high-speed centrifugal technology of VanTech was used to collect effluent and sludge from fish ponds, combined with Bioway Vietnam's advanced AT-6H rapid fermentation technology, to convert organic waste from agricultural by-products into organic fertilizer. These fertilizers were then used for rice production and the cultivation of key local vegetables with export potential. The results demonstrated a reduced reliance on inorganic fertilizers, potentially lowering production costs and minimizing environmental pollution. The research findings indicated that the nutrient quality of organic fertilizer using the advanced AT-6H rapid fermentation technology met Vietnam's national standards for organic fertilizers. Furthermore, the application of these fertilizers in crop models such as rice, cucumber, green been, water spinach, malabar spinach, *Amaranthus cruentus* vegetables, soybean vegetables, and *Brassica integrifolia*, allowed for a 25–75% reduction in chemical fertilizers while still achieving higher yields and profits compared to traditional farming methods.

Keywords: Advanced technology application, innovative technology solution, organic fertilizer, organic wastes, organic waste treatment, reusing of agriculture by-products.

13.1 INTRODUCTION

The world population has significantly increased from 3.75 billion in 1970 to over 7.87 billion in 2021, and it is projected to exceed 9 billion by 2050 (United Nations, 2022; Worldometer, 2022). This population growth poses major challenges for agriculture and food production, along with the issue

© 2025 IWAP. This is an Open Access book chapter distributed under the terms of the Creative Commons Attribution License (CC BY-NC-ND 4.0) which permits copying and redistribution for non-commercial purposes with no derivatives, provided the work is properly cited (https://creativecommons.org/licenses/by-nc-nd/4.0/). The chapter is from the book *Nature-Based Solutions for Urban Sustainability*, Piet N.L. Lens and Xuan-Thanh Bui (Editors).

of agricultural waste and by-products, which are forecasted to continue to increase. The trend toward more intensive agricultural production to ensure food security for the growing global population has led to a large amount of organic and solid wastes, agricultural by-products, and crops, as well as a significant increase in the use of synthetic fertilizers, chemicals, pesticides, and herbicides (Berg *et al.*, 2023; Da *et al.*, 2021; Livsey *et al.*, 2021). According to the Ministry of Agriculture and Rural Development in Vietnam (MARD, 2022a), agricultural production in Vietnam has mainly focused on increasing productivity in the past few decades, without paying attention to the excess input of the production process or to the application of organic fertilizers to replenish soil nutrients, enhance soil structure and fertility, as well as protect biodiversity. The overuse of inorganic/chemical fertilizers (synthetic fertilizer), and pesticides in crop cultivation, along with the increase in waste from the livestock and aquaculture sectors threatens environmental quality and public health, seriously impacts ecological environments, leading to the loss of many valuable natural resources (MARD, 2022a). The Food and Agriculture Organization (FAO) of the United Nations has highlighted the global issue of land degradation, warning about the disaster caused by excessive use of chemical fertilizers and pesticides, and over-exploitation over the past 50 years. About 25–30% of the world's land area is currently 'severely degraded' with soil erosion, water resource depletion, and loss of natural ecosystems. About 1.2–1.3 billion hectares of land have lost vegetation cover and 10–20% of land has turned into desert (FAO, 1990, 2002, 2024).

Vietnam currently ranks 4th among the top 10 countries in the world for the export of rice, agricultural products, seafood, and aquaculture products. Agricultural and aquaculture production accounts for about 7.2% of the country's total export value, with more than 70 types of agricultural products exported to over 130 countries worldwide (MARD, 2022a; VASEP, 2022). The Crop Production Department of Vietnam reported that the total volume of agricultural by-products or agricultural wastes was over 156.8 million tons in 2021, including 88.9 million tons of post-harvest by-products from rice, crops, and fruits processing in the crop production sectors (accounting for 56.7%), about 61.4 million tons (39.1%) of livestock and poultry manure, 5.5 million tons (3.5%) from the forestry sector; and nearly 1 million tons from the aquaculture and fisheries (seafood) sector (10.6%) (GSO, 2022). Moreover, it is estimated that the farmed aquaculture sector, which produced approximately 5,163,700 tons in 2022, with *Pangasius* catfish and shrimp production achieving 1,216,000 tons and 1,080,600 tons, respectively, generated from 510 to 665 million tons of solid waste, and sludge/sediment discharged into the local environment (Anh *et al.*, 2010, 2011; GSO, 2022). MARD (2022a) estimated that the total emissions from agricultural production and by-products in Vietnam account for more than 15% of the total emissions, with projection reaching about 120 million tons of CO_2 by 2030 if there is no intervention. Up to half of these emissions come from rice production (GSO, 2022; MARD, 2022b; Truc, 2024). The by-products and organic wastes from the key agriculture and aquaculture sectors mentioned above emit millions of tons of organic waste and solid wastes into the environment each year. These sources of organic by-products are renewable resources that can be reused in agricultural production and other fields, potentially bringing economic benefits to local communities. They also offer a potential solution for producing organic fertilizers for economically valuable crops, which can be exported to promote a circular economy in agriculture towards future green development. If utilized and managed effectively, these resources can provide economic benefits while contributing to environmental protection, improving soil fertility, supplying nutrients to soil and crops, and reducing greenhouse gas emissions. However, these organic resources are currently not effectively utilized for the right purposes to add value to the agricultural production value chain (Koul *et al.*, 2022; Westerman & Bicudo, 2005).

Given the overload of agricultural by-products in many areas of Vietnam and the limitations of some traditional composting or organic fertilizer production technologies, which cannot completely process these agricultural by-products, there is a need to seek and apply advanced technological solutions and innovative technologies for collecting solid wastes and sludge/sediment, processing these and utilizing organic agricultural by-products for organic fertilizer production. This can reduce the use of synthetic fertilizer, environmental pollution, operation cost, and enhance quality of production. The main goal is to convert agricultural by-products and organic wastes into high-quality organic fertilizers for agricultural crops production and renewable bioenergy sources, thereby reducing agricultural by-products and organic wastes and bringing their value to the local communities in the future. The organic fertilizers derived from natural sources such as agricultural by-products, organic wastes, and sludge/sediment of fish ponds, have become increasingly important in sustainable agricultural production systems. The shift from chemical fertilizers (synthetic fertilizers) to organic alternatives is driven by the need to maintain soil health and soil fertility, reduce the amount of chemical fertilizers, minimize environmental pollution, enhance biodiversity, improve the value of agriculture production, and increase income for local people and local livelihood. This chapter explores and investigates the current situation of generation and utilization of agricultural by-products and organic wastes in Vietnam, and application of advanced technological solutions and innovative technologies to produce organic fertilizers from locally available agricultural by-products, organic wastes mixed with effluent and sludge from fish ponds. It also assesses the benefits of organic fertilizers use for soil nutrient and fertility improvement, reduce synthetic fertilizers, enhance yield of vegetables and rice cultivation, and their potential in promoting sustainable agriculture and green agriculture.

13.2 OVERVIEW OF CURRENT GENERATION AND UTILIZATION OF AGRICULTURAL BY-PRODUCTS AND WASTES IN VIETNAM

According to a report of the General Statistics Office of Vietnam, and the Department of Crop Production of the Ministry of Agriculture and Rural Development, Vietnam, the total estimated volume of agricultural by-products in Vietnam is approximately 156.8–200 million tons per year (GSO, 2022; MARD, 2022b; Truc, 2024) (Table 13.1). Among the four key agricultural production sectors, the highest proportion of by-product emissions comes from crop production (rice, vegetables, and fruit trees) and livestock, with 88.9 and 61.4 million tons, respectively, compared to forestry and fisheries (Figure 13.1).

The amount of post-harvest by-products in the fields is about 88.9 million tons, including straw, rice husks, rice bran, corn stalks, vegetables, fruits, sugarcane bagasse, corn cobs, coffee husks, and cassava peels. However, the collection and reuse rate of crop by-products is only 52.2%, with the utilization rate of straw being just 56.6%. This utilization includes making fertilizers, biological pads, mushroom substrates, handicrafts, fruit wrapping, field mulching, livestock feed, and bedding materials (Table 13.1). Total farmed aquaculture production was 8.4 million tons in 2020, by-products in seafood processing amount to about 1 million tons per year. Currently, about 90% of these by-products are collected from processing plants and transformed into useful products, while the remaining portion is discharged into the environment. If the by-products from the seafood and aquaculture sectors were reused, they could generate 4–5 billion USD (World Bank, 2023).

Table 13.1 Amount of Vietnamese agricultural by-products and their usage rates by sector in 2022.

Agriculture	By-products (million tons)	Ratio (%)	Reuse Ratio (%)	Potential Values
Rice crop	88.9	56.6	52.3	Approximately 43 million tons of organic fertilizer, 87 million liters of bioethanol
Livestock	61.4	39.1	23	Fertilizer, biogas
Aquaculture	1.0	3.5	90	4–5 billion USD
Forestry	5.5	0.8	15	Export of compressed wood and sawdust
Total	156.8	100	0	

Source: Lam and Son (2024), Department of Crop Production of Vietnam (2022), MARD (2022b).

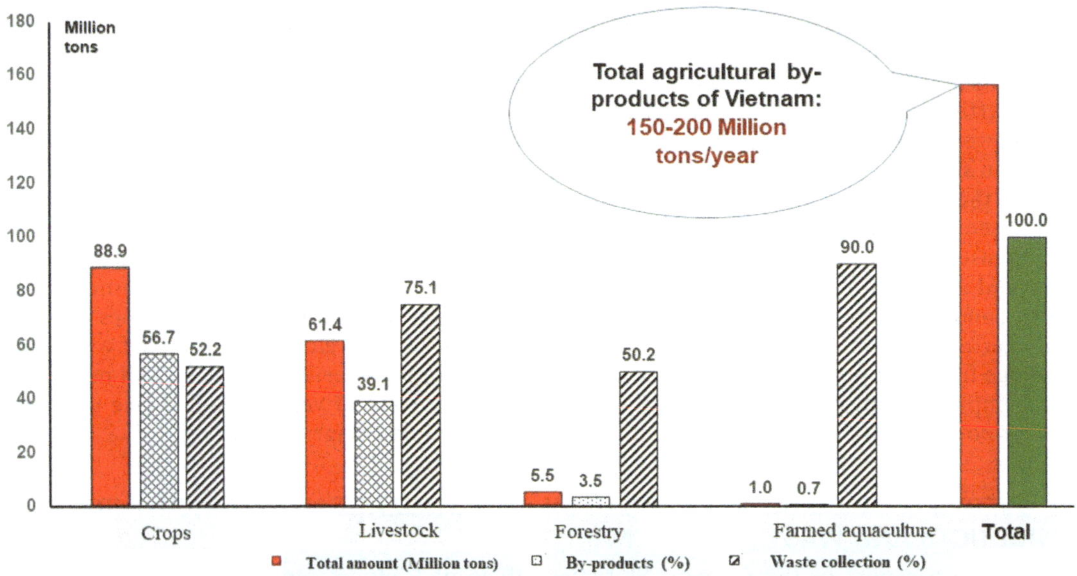

Figure 13.1 Total production of by-products of agricultural activities: crop production, livestock, forestry, and fisheries in Vietnam in 2022 (*Source*: GSO, 2022; MARD, 2022a).

Vietnam's agriculture annually produces 45 million tons of dry rice straw, 8 million tons of rice husks, and 30–50 million tons of other plant by-products. Reports indicate that a large amount of straw is wasted: 46% is burned in the fields, 29% is used as animal feed, 8.6% is buried in the fields, 5% is composted, 4.1% is used for mulching, and 7% is used as bedding material and raw material for handicrafts (Figure 13.2).

A study reported by Thinh *et al.* (2022) from the International Rice Research Institute shows that 60% of rice straw in Vietnam is burned, resulting in soil nutrient loss, greenhouse gas emissions, and environmental pollution from smoke and dust, which reduces biodiversity. However, directly burying rice straw in the fields can increase CH_4 emissions and cause black root disease in rice plants. Previous studies have identified the main reasons for low efficiency in the use of agricultural by-products: inconsistent policy mechanisms, lack of technology, low market prices for agricultural by-products, and insufficient knowledge about the carbon credit market. In the agricultural product processing sector, Vietnam has utilized only 50% of the rice husk for fuel, animal feed, husk pellets, and fertilizer, whereas about 30% of rice bran is used for oil extraction. The annual processing volume of fresh cassava is approximately 11 million tons, generating about 700,000 tons of cassava peels and 4 million tons of cassava pulp, which can be reused for bioethanol production and animal feeds. The sugar industry annually produces 7–8 million tons of sugarcane plants, generating 2.3 million tons of bagasse, 360,000 tons of molasses, and 400,000 tons of filter mud and furnace ash. The reuse rate of bagasse is about 28%, whereas 41% of molasses is used for animal feed, 20% for food flavoring, and 12% for monosodium glutamate production (GSO, 2022; MARD, 2022a).

In 2020, the amount of solid waste produced by livestock and poultry cultivation (including bedding material and leftover or spilled feed) was approximately 61.4 million tons (39.1%). This waste included around 290 million m^3 of liquid waste (urine and cleaning water) and tens of millions of tons of bedding material, with only 23% being reused. According to a report by the Department of Livestock Production in 2022, 48.2% of households managed solid waste, whereas 96.1% of large livestock

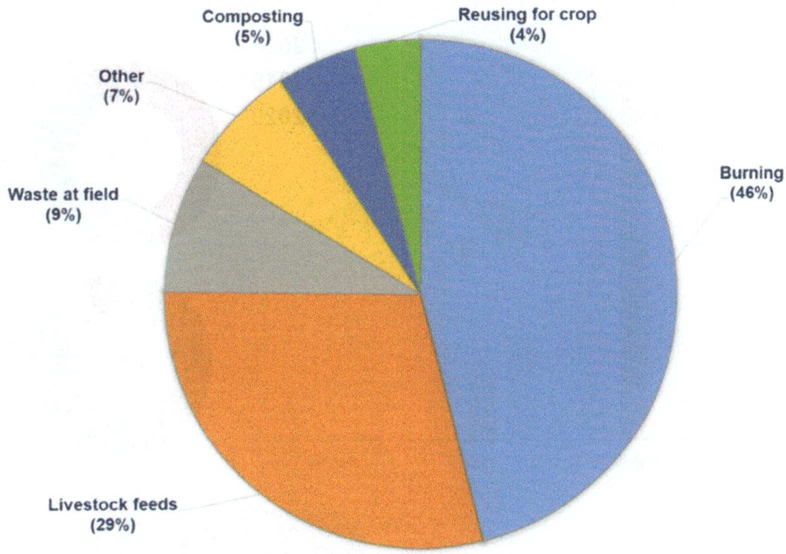

Figure 13.2 Processing ratio of agricultural by-products in the field of rice crop production in Vietnam (*Source*: GSO, 2022; MARD, 2022a).

farms effectively managed and treated their solid waste (Figure 13.3A). Animal manure contains high levels of nitrogen, phosphorus, zinc, copper, lead, arsenic, nickel (heavy metals), and other harmful microorganisms. These pollutants not only contaminate the air but also the soil, disrupting soil fertility and polluting both surface and groundwater. The estimated amount of livestock waste shows an increasing trend over the years: 62 million tons (2021), 68.15 million tons (2025), and 71.92 million tons (2030). The highest emissions come from cattle farming (31.8–34.0 million tons), followed by buffalo farming (8.5–8.9 million tons), pig farming (7.1–10.0 million tons), and poultry farming (0.8–1.0 million tons) (Figure 13.3A). Livestock waste is managed in various ways, including composting, treatment with microbial products, biogas plants, bioenergy production, and the direct use of fresh manure as fertilizer (Department of Livestock Production, 2022). Additionally, meat and bone meal, meat meal, poultry meal, feather meal, and animal fat are the main products of the processing industry. Slaughter by-products are also being used to produce animal feed. However, the Department of Livestock Production's 2022 report indicates that only 48.2% of households and livestock facilities apply waste treatment technologies. The adoption rates for these technologies among households remain modest: bio-bedding at 1.8%, biogas at 6.8%, and composting at 31.1%.

Regarding the forestry sector, according to the General Department of Forestry 2022 report, Vietnam produces approximately 30 million m^3 of roundwood annually, generating 16 million m^3 of firewood and 8.6 million m^3 of sawdust, wood shavings, and wood chips (Figure 13.3B). These by-products can be used for pellet production, industrial ethanol production, bioelectricity generation, wood drying fuel, bio-bedding in livestock, organic fertilizer, and growing media for soilless agriculture. Vietnam is the second-largest exporter of wood chips and wood pellets in the world, with an export volume of 3.2 million tons and revenue of 363 million USD. However, the utilization rate of forestry by-products for pellet production is only about 15%. Currently, Vietnam does not have specific policies or mechanisms to support the processing of wood shavings and sawdust. There are also no guiding criteria, and a lack of connectivity among production units leads to unhealthy competition among suppliers of forestry by-products.

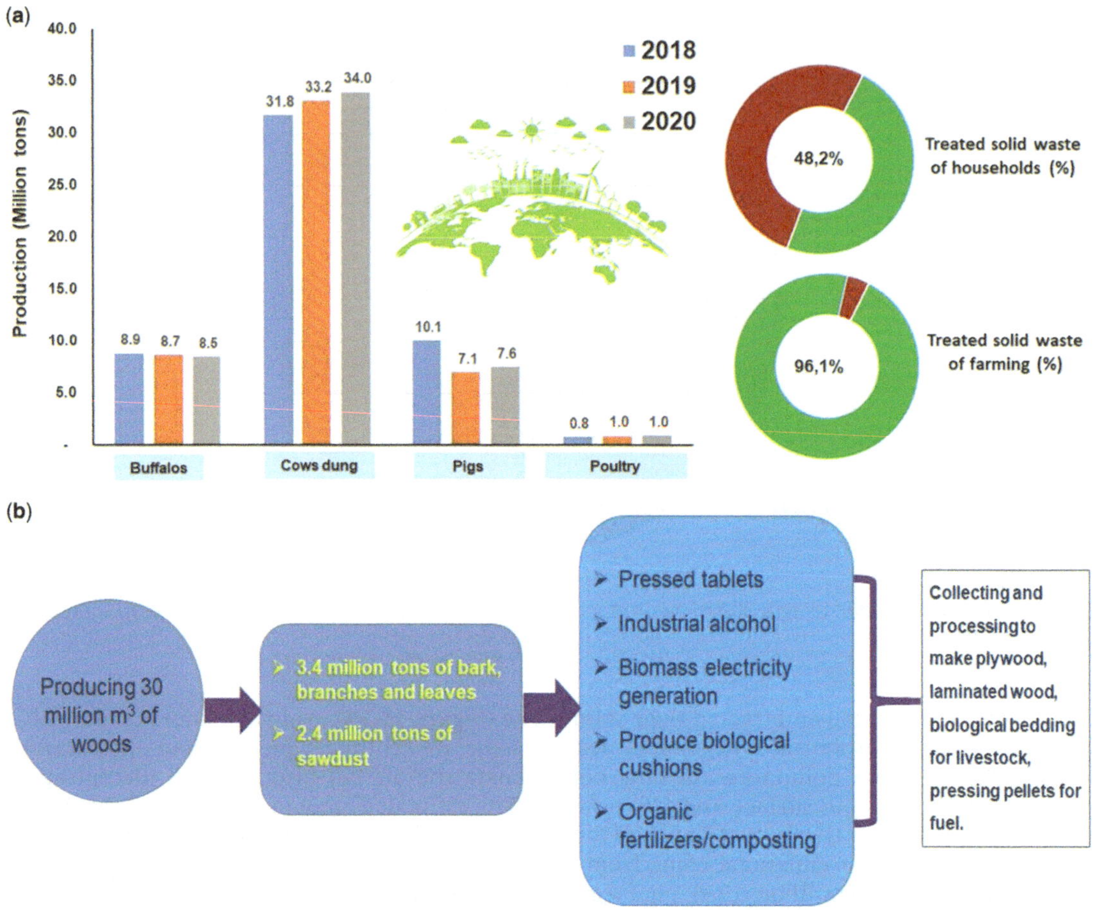

Figure 13.3 Management of solid waste and by-products from the livestock (A) and forestry (B) sector in Vietnam (*Source*: GSO, 2022; MARD, 2022a).

13.3 OVERVIEW OF ORGANIC FERTILIZERS IN VIETNAM

Fertilizers are often described as 'food or nutrition' provided by humans to support plant growth and they contain many essential nutrients for plants. The primary nutrients in fertilizers are nitrogen (N), phosphorus (P), and potassium (K), which are known as macronutrients. In addition, fertilizers also contain micronutrients. On the basis of the methods used in manufacture, fertilizers are divided into three main groups: (1) organic fertilizers (including animal-based organic fertilizers, plant-based organic fertilizers, and mineral based organic fertilizers); (2) chemical (inorganic) fertilizers or synthetic (including nitrogen fertilizers, phosphorus fertilizers, potassium fertilizers, and NPK fertilizers); and (3) microbial fertilizers (biofertilizers), which are products containing living microorganisms selected at a density that meets established standards. Through their biological activities, these microorganisms produce nutrients that plants can utilize (e.g., N, P, K, S, and Fe). Annually, the agricultural sector uses approximately 10 million tons of fertilizers, of which inorganic and synthetic fertilizers are about 75%. Noteworthy, Vietnam uses fertilizers at a much higher rate than many other countries, nearly three times the global average. Synthetic fertilizers are applied in an imbalanced manner, failing to adhere to recommended principles, leading to low utilization efficiency: only 40–45% for nitrogen fertilizers, 25–30% for phosphorus fertilizers, and 55–60% for potassium fertilizers. Moreover, this imbalance not only leads to waste but also causes soil and water pollution (both surface water and groundwater), increases greenhouse gas emissions and affects the quality of agricultural products. Recently, bio-organic fertilizer that is defined as a new type of fertilizer with both microbial fertilizer and organic fertilizer effect is showing a great potential for use in sustainable agricultural processes.

Vietnam has significant opportunities in both domestic and international markets. However, effectively competing will depend on the ability of farmers and firms to deliver products with reliability and assurances relating to quality, safety, and sustainability. Over the past 30 years, Vietnam's agriculture has primarily relied on inorganic fertilizers due to the pressure of intensive farming to increase productivity for national food consumption, food security, and international export. However, most farmers in Vietnam use fertilizers in an unreasonable and imbalanced manner, often applying synthetic fertilizers in quantities much higher than recommended (MARD, 2022a). The country has lost a significant portion of its arable land due to erosion and pollution in the past few years. Continuous plowing of fields, combined with heavy synthetic fertilizer use, has degraded and contaminated soils around the country. Currently, Vietnam is facing a reduction in arable land, prompting farmers to shift toward organic farming to meet the food demand driven by the population growth. As decreasing arable land results in a smaller area available for cultivation, productivity needs to be improved in this area, and the use of organic fertilizers can enhance plant growth and yield. Furthermore, with the increasing awareness and government initiatives related to organic fertilizers, the market is anticipated to grow over the next 20 years (World Bank, 2023). Recently, the use of organic fertilizers has become one of the Vietnam government's priorities because the country has favorable conditions for organic fertilizer production and application. The government of Vietnam promotes the production and use of organic fertilizers and has established policies to develop this sector. According to MARD (2022b), over the past 3 years (2019–2021), the number of registered organic fertilizers has increased tremendously by 500%, whereas the use of synthetic fertilizers is on a downward trend. In 2020, it was estimated that Vietnam produced about 16.8 million tons of organic fertilizers by farmer households. This increased to 18.36 million tons in 2021 and surpassed 21 million tons in 2022, but it only met about 10–25% of the demand (MARD, 2022b; World Bank, 2023). According to the approved plan, the Ministry of Agriculture and Rural Development (MARD, 2022b) aims to increase the share of organic fertilizer products permitted for circulation to over 30% of the total fertilizers products by 2030, with a production capacity of organic fertilizers reaching 5 million tons per year. The use of industrial organic fertilizers will account for at least 30% of the total fertilizers used in Vietnamese agricultural production.

Many previous studies have reported that if organic fertilizers are widely used in agricultural production, they not only provide stable nutrients for crops but also improve the physical, chemical, and biological properties of the soil, reduce erosion, increase soil permeability and water retention, and enhance the drought resistance of crops (Da *et al.*, 2020, 2021; Nuri *et al.*, 2021; Saioof *et al.*, 2019). Meanwhile, Vietnam has significant potential in raw materials for producing organic fertilizers, with approximately 150–200 million tons of organic wastes generated annually from livestock, animals, rice crop, vegetables, fruits, forestry, aquaculture, and other processing industries. These are great sources of organic raw materials with high nutrient contents and substantial humus, ideal for organic fertilizer production plants. However, the potential biomass by-products from these major crops have not yet been fully utilized or reused to produce organic fertilizers, which could reduce the pressure on the use of chemical fertilizers/synthetic fertilizers. Given the overwhelming wastes from agricultural by-products in many regions of Vietnam, coupled with the limitations in some composting technologies and traditional organic fertilizer production methods, it is challenging to manage these agricultural by-products effectively. Therefore, it is necessary to explore and apply advanced and innovative technologies in processing and utilizing organic agricultural by-products to convert them into high-quality organic fertilizers for crops and renewable energy sources that bring value to the local people.

13.4 APPLYING ADVANCED TECHNOLOGY FOR ORGANIC FERTILIZER PRODUCTION FROM AGRICULTURAL BY-PRODUCTS AND WASTES

In recent decades, Vietnam has imported a vast amount of chemical fertilizers annually but has not yet fully utilized the potential of biomass by-products from major crops such as rice, corn, sugarcane, various fruits, and vegetables. These by-products could potentially provide the equivalent of approximately 43 million tons of organic fertilizers, 10.8 million tons of urea nitrogen, 1.6 million tons of single superphosphate, and 2.2 million tons of potassium sulfate. According to a report by the Department of Crop Production of Vietnam (Thinh *et al.*, 2022), households produce 17 million tons of organic fertilizers annually from agricultural by-products, achieving about 40% of the potential. The excessive accumulation of agricultural by-product wastes in many regions of Vietnam, along with the limitations of some traditional composting and organic fertilizer production methods, poses significant challenges to the complete processing of these by-products. Therefore, it is essential to explore and implement advanced technological solutions for effectively processing and utilizing organic agricultural by-products. This approach can enable the conversion of these resources into high-quality organic fertilizers for crops and renewable energy sources, delivering real value to local communities.

This chapter identifies two different advanced technological solutions for reusing agricultural by-products to produce organic fertilizers: (1) the use of high-speed centrifugal technology (by VanTech, Japan) to treat and collect sludge and sediment from pond bottoms of freshwater aquaculture, as well as the effluent and suspended solid wastes to clean the aquaculture environment. The centrifuged sludge of fish ponds can then be reused as one of the input materials for organic fertilizers by mixing it with various agricultural by-products and locally available agricultural products to produce organic fertilizers. (2) The application of the AT-6H rapid fermentation technology (from Bioway Hitech, Vietnam) using high-temperature-resistant fermentation bacteria to convert agricultural by-products and organic waste into organic fertilizers.

13.4.1 Process of reusing agricultural by-products for organic fertilizer production

Reusing agricultural by-products for organic fertilizer production is an eco-friendly approach to improve soil health and reduce waste. Figure 13.4 illustrates the comprehensive process for producing organic fertilizer from agricultural by-products and waste. This production process consists of five key steps.

(1) *Material collection*: Identify and gather abundant, suitable raw materials while understanding the nutrient composition of the agricultural by-products. (2) *Fertilizer formulation*: Transport materials such as fish pond sludge, manure, and carcasses to the production plant using specialized vehicles to ensure cleanliness. Develop fertilizer formulations that comply with Vietnam's organic standards. (3) *Composting mixture*: Collect fish pond sludge (40–50% moisture) using high-speed centrifugal technology and mix it with local organic wastes and agricultural by-products, including manure, carcasses, and decomposed fruits and vegetables. The moisture content of these by-products should be 65% or lower, ideally around 45%, with a material size not exceeding 1.2 cm. (4) *Fertilizer production*: Combine all fertilizer materials (pond sludge, organic wastes, and agricultural by-products) with AT-6H high-temperature-resistant fermentation bacteria. Process the mixture in the Bioway AT-6H rapid fermentation machine for 45 min to 1 h, completing the bio-organic fertilizer production in about 6 h. (5) *Fertilizer quality testing and economic evaluation*: Collect samples of the bio-organic fertilizer for quality testing. Conduct trials on vegetables and rice crops to assess the fertilizer's effects, evaluate economic efficiency, and scale up the production model (Figure 13.4).

13.4.2 Organic fertilizer produced from wastes from agriculture by-products

13.4.2.1 Collection and chemical content of fertilizer amendments

In the past few decades, Vietnam has annually imported large amounts of chemical fertilizers, yet the country has not fully utilized the potential of biomass by-products from major crops such as rice, corn, sugarcane, and various vegetables. These by-products could provide million tons of organic fertilizer, urea nitrogen fertilizer, single superphosphate, and potassium sulfate for plants.

Step 1
Material collection

- Identify and collect sources of abundant and potential agricultural by-products and wastes to use as raw materials for organic fertilizer production.
- Material preparation (by-products are sorted to remove non-organic or unwanted materials; moisture is adjusted...).
- Analyze the nutrient composition of all raw materials as a basis for formulating fertilizer mixtures.

Step 2
Fertilizer formulation

- Fertilizer materials (such as fish pond sludge, livestock and poultry manure, fish/animal carcasses, etc.) need to be transported using specialized vehicles to ensure cleanliness during transportation to the material collection site at the fertilizer production plant.
- Develop organic fertilizer formulations that meet the standards of Vietnam's bio-organic fertilizers.

Step 3
Composting mixture

- Sludge from fish pond collected by high-speed centrifugal technology (40-50% of moisture contents) mixed with local organic wastes and agricultural by-products such as: poultry and livestock manure, fish/animal carcasses (whole or in parts), fish processing wastes, decomposed fruits, vegetables, and plants.
- The moisture content of agricultural by-products is about 65% or lower (the optimal moisture level is 45%). The size of materials is not exceeding 1.2 cm.

Step 4
Fertilizer production

- All materials of ferilizer (pond sluge, local organic wastes and agricultural by-products) mixed with AT-6H high-temperature resistant fermentation bacteria are loaded into the Bioway AT-6H rapid fermentation machine (This process takes about 45 minutes to 1 hour).
- Apply rapid fermentation technology of Bioway Hitech AT-6H (using high-temperature resistant fermentation bacteria) to process and produce bio-organic fertilizers (This process takes about 6 hours).

Step 5
Test quality of fertilizer and economic evaluation

- Take samples of the bio-organic fertilizer for quality testing.
- Conduct trials of the effects of bio-organic fertilizer on the vegetables and rice crops.
- Evaluate economic efficiency and scale up the model.

Figure 13.4 Processing scheme of organic fertilizer production from agricultural by-products and waste.

According to a report from the Department of Crop Production, households annually produce 17 million tons of organic fertilizer from agricultural by-products, meeting about 40% of the potential (MARD, 2022b). Recent studies have reported that the effluent and sludge/sediments of fish ponds (*Pangasius* catfish and snakehead fish farming), fish carcasses, seafood wastes (such as shrimp head), rice straw, rice husk, rice husk ash, water hyacinth residues, cow manure/urine, pig manure/urine, chicken manure, other livestock, slaughterhouse waste (blood and other residues), agriculture by-products (fruits and vegetable wastes) are enriched with high organic matter, essential nutrients, and micronutrients. These include protein, amino acids, fatty acids, nitrogen, phosphorus, potassium, and vitamins, making them valuable as continuously available fertilizer supplements and soil conditioners. Their use can enhance the soil environment for crop production, reduce environmental pollution, and increase crop yields (Da *et al.*, 2020, 2021; Karak *et al.*, 2013; Rahman & Yakupitiyage, 2006; Thanh *et al.*, 2015, 2023). The nutrient contents of the common organic sources of waste from agricultural by-products and other organic sources used for organic fertilizer production are presented in Tables 13.2 and 13.3.

Table 13.2 lists the total nitrogen (TN), phosphorus (TP), and potassium (TK) contents of various organic sources, showing that, in terms of total N, organic sources typically contain 10–100 times less N than urea. Meanwhile, rock phosphate and bone meal, which are accepted by the organic farming

Table 13.2 Total N, P, and K concentrations in common organic sources as compared with urea (46% N), phosphorous (20% P), and triple superphosphate (20% P), respectively.

Organic Sources	Total N (%)	Total P (%)	Total K (%)
1. Agricultural waste			
Poultry manure	1.5–3.0	0.5–1.5	–
Pig, horse, cow manure	0.3–0.6	–	–
Green manure	1.5–5.0	0.2–0.5	2–5
Compost	0.5–2.0	0.2–0.5	0.5–2.0
Fish waste/fish carcass	4.0–10.0	5–10	–
Blood (slaughterhouse)	10.0–12.0	–	–
Plant residues	1.0–1.5	0.15–0.2	1.0
Cattle manure	1.2–2.0	0.3–0.8	1.7
Poultry	1.8–4.1	1.5–3.3	1.5–3.2
Swine manure	1.2	0.4	0.6
2. Domestic waste			
Human urine/night soil	1.0–1.5	–	–
Sewage sludge	1.0–5.0	0.4–2.5	–
3. Treated waste			
Rock phosphate	–	17–26	–
Bone meal	–	20–30	–
Wood ash	–	2–5	5–10
Sul-Po-Mag [Mg^{2+}, K^+, SO_4^{2-}]	–	–	22–23
Polyhalite [Ca^{2+}, K^+, SO_4^{2-}]	–	–	10–15
Green sand	–	–	5–7

Source: Adapted from Sankar *et al.* (2017), Nick (1994), Rechcigl (1995), Stratton and Rechcigl (1998), Mills (1996).

Table 13.3 Chemical composition of effluent/sludge of fish pond, peanut shell, rice straw, water hyacinth, wastes of agroindustry food by-products.

Organic Sources	Organic Carbon (% C)	Total N (% N)	Total P (% P_2O_5)	Total K (% K_2O)
Effluent and sludge of *Pangasius* catfish pond	4.51	0.35	0.82	1.07
Effluent and sludge of snakehead fish pond	5.51	0.54	0.46	0.21
Peanut shell	21.20	5.00	0.11	4.32
Coir	9.80	0.55	0.32	0.40
Rice straw	47.31	0.81	–	2.24
Water hyacinth	40.19	2.32	0.05	1.46
Agroindustry fruit wastes from An Giang Food and Vegetables Joint Stock Company (ANTESCO)	13.1	0.27	0.05	0.63

Source: Adapted from Thanh *et al.* (2015, 2023), Da *et al.* (2020, 2021), Rahman and Yakupitiyage (2006), Karak *et al.* (2013).

community in most regions, contain relatively high levels of phosphorus, ranging from 20 to 30%. However, the phosphorus in these two sources is very insoluble, making it much less available to plants compared to phosphorus in triple superphosphate (Sankar *et al.*, 2017; Tilman *et al.*, 2002). Among the organic sources listed in Table 13.2, Sul-Po-Mag [Mg^{2+}, K^+, SO_4^{2-}] has the highest potassium content, followed by polyhalite [Ca^{2+}, K^+, SO_4^{2-}], compared to other organic sources. Nitrogen and phosphorus are also essential for normal plant development and are needed in substantial amounts. Additionally, potassium is critical for maintaining cell osmotic potential, facilitating water uptake from the soil, retaining water within plant tissues, and transporting water and nutrients over long distances in the xylem and phloem (Mills, 1996; Nick, 1994; Rechcigl, 1995; Sankar *et al.*, 2017; Stratton & Rechcigl, 1998). Chemical analysis of rice straw and dried water hyacinth found the highest content of organic carbon (% C), followed by peanut shell and agro-industrial fruit wastes. The organic carbon content was 2–10 times higher than that of effluents and sludge of fish ponds. The TN (% N), TP (% P_2O_5), and TK (% K_2O) concentration ranged from 0.27 to 2.32%, 0.05 to 0.82%, and 0.40 to 2.24%, respectively (Table 13.3). In general, the chemical composition and nutrient profiles of these organic sources can complement each other as input organic materials for the production of organic fertilizers for crops.

The increasing intensification of aquaculture and industrial agriculture production requires the development of strategies to reduce their environmental impacts such as pollution caused by the discharge of nutrient-rich sediments and agricultural waste into local water bodies and the environment. In fact, the growing risk of contamination from nutrients and toxic chemical compounds present in groundwater and surface waters leads to eutrophication and soil quality degradation (Sankar *et al.*, 2017). Facing with the overload of these wastes from agricultural by-products, the lack of cost-effective technologies, and improper disposal of organic waste, along with the limitations of some traditional composting technologies or organic fertilizer production methods, it is impossible to thoroughly process these agro-industrial wastes. Therefore, it is crucial to seek and implement advanced and innovative technologies to process and utilize organic agricultural by-products, converting them into high-quality organic fertilizers for crops and renewable energy sources that provide real value to local communities.

13.4.2.2 Organic fertilizer formulation

On the basis of the nutrient concentrations and availability of local raw materials for fertilizer amendments, a formulation for organic fertilizer production was developed to achieve a high-quality fertilizer and meet Vietnam's organic fertilizer standards. Two fertilizer formulations for the organic fertilizer production are presented in Figure 13.5. Fertilizer formulation (1) includes 50% pond sludge, mixed with 3% fish residue, 16.9% rice husk ash, 15% chicken manure, 1% molasses, and 10% base fertilizer containing temperature-resistant microorganisms. Fertilizer formulation (2) consists of 62.9% organic fertilizer produced from agriculture by-products (fruits and vegetable wastes), 2% *Pangasius* fish carcass, 10% rice husk, 15% chicken manure, 0.1% sugarcane molasses, and 10% original bio-microorganisms.

The detailed procedure for producing organic fertilizer comprises:

- *Mixing raw materials of formulation 1 or formulation 2*: The sludge and sediment of fish pond, as well as wastes from agricultural by-products that have undergone dewatering, should maintain optimal moisture content of 40–50%. These materials can be combined with additional materials such as poultry and livestock manure, whole or partial animal carcasses, seafood waste, and decomposable fruits, vegetables, and plants. All of these should have a moisture content of 65% or lower (with an optimal moisture content of 45%), and a particle size not exceeding 1.2 cm. The most common local raw materials should be selected depending on the different ecological regions to leverage regional advantages and reduce fertilizer production costs.

Fertilizer formulation 1: Organic fertilizer produced from sludge of *Pangasius* catfish (50%), *Pangasius* fish carcass (3%), rice husk (16.9%), chicken manure (15%), sugarcane molasses (0.1%) and orginal bio-microbiology (10%).

Fertilizer formulation 2: Organic fertilizer produced from agriculture by-products (fruits and vegetable wastes: 62.9%), *Pangasius* fish carcass (2%), rice husk (10%), chicken manure (15%), sugarcane molasses (0.1%) and orginal bio-microbiology (10%).

Figure 13.5 Formulation of organic fertilizer production from effluents and sludge of fish pond, and organic wastes from agroindustry food by-products. Left: Fertilizer formulation 1: organic fertilizer produced from sludge of *Pangasius* catfish (50%), *Pangasius* fish carcass (3%), rice husk (16.9%), chicken manure (15%), sugarcane molasses (0.1%), and original bio-microbiology (10%). Right: Fertilizer formulation 2: organic fertilizer produced from agriculture by-products (fruits and vegetable wastes: 62.9%), *Pangasius* fish carcass (2%), rice husk (10%), chicken manure (15%), sugarcane molasses (0.1%) and original bio-microbiology (10%).

- *Raw materials for organic fertilizer*: The production of organic fertilizers using Bioway AT-6H high rapid fermentation technology includes: (1) pond sludge or any agricultural by-products; (2) catfish carcasses or seafood processing by-products (such as shrimp shells), or by-products from livestock and poultry slaughterhouses; (3) rice husk, or rice husk ash, and rice straw; (4) manure from poultry, cattle, or buffalo; (5) sugarcane molasses; and (6) base fertilizers mixed high-temperature-resistant microorganisms.
- *Moisture and size requirement*: The moisture content of the mixture should be around 45%, and the size of the raw materials should not exceed 120 mm.
- *Fermentation process*: The next step involves adding the AT-6H microbial inoculant to the mixture and transferring it to the Bioway AT-6H fermentation machine (Figure 13.6).
- The AT-6H high-temperature fermentation bacteria can withstand temperatures up to 80–90°C, utilizing temperatures in the furnace above 100°C to eliminate viruses and harmful microorganisms in all organic waste materials.
- After 6 h in the Bioway AT-6H rapid fermentation machine, the semi-finished product is transferred to the secondary fermentation chamber, where the moisture content is stabilized over 24–36 h to achieve the desired quality (Figure 13.6). The secondary fermentation chamber system includes a mixing machine combined with an oxygen aeration system and is covered with a screen to prevent flies, mosquitoes, and ultraviolet light.
- After 2–3 days of fermentation, the product is screened to remove any non-decomposable components (if present), bagged in plastic for storage, and then tested for maturity and fertilizer quality parameters.

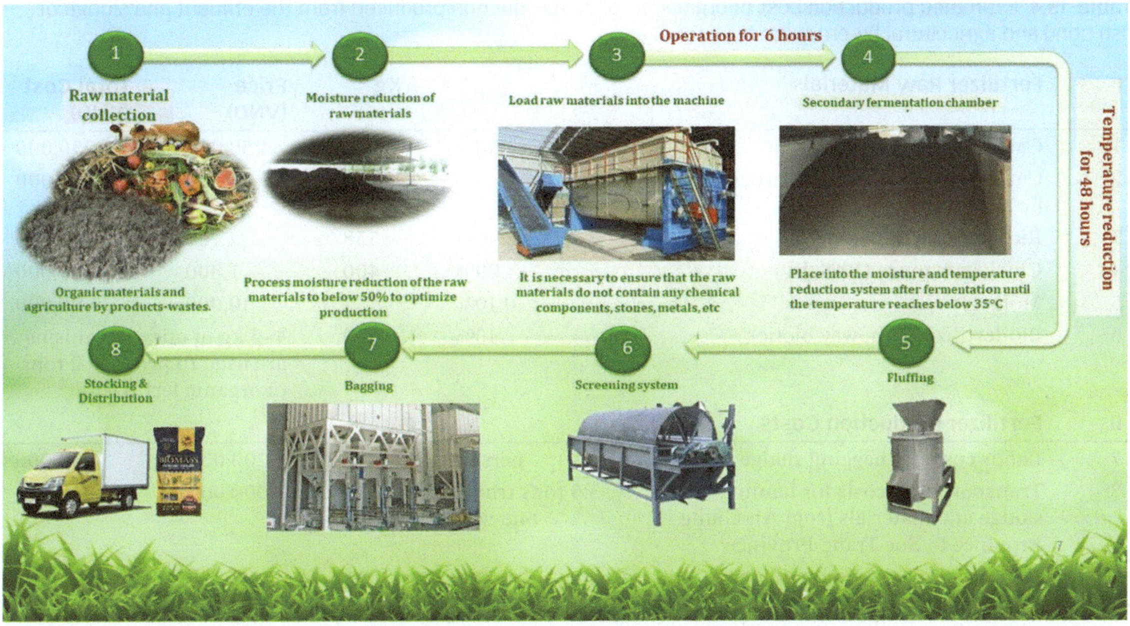

Figure 13.6 Processing diagram of organic fertilizer production by AT-6H rapid fermentation technology (high-temperature-resistant fermentation bacteria) of Bioway Hitech, Vietnam. Total time from material input to the ended product is about 56–60 h (*Source*: Bioway Vietnam).

The processing flowchart of organic fertilizer production by applying AT-6H rapid fermentation technology with high-temperature-resistant fermentation bacteria of Bioway Hitech, Vietnam is presented in Figure 13.6.

13.4.2.3 Cost estimation for organic fertilizer production using AT-6H rapid fermentation technology

The production cost for 2 tons of organic fertilizer, produced using AT-6H rapid fermentation technology, has been calculated (Table 13.4).

This fertilizer is made from various sources, including effluent and sludge of *Pangasius* catfish pond or wastes from agro-industrial by-products, fish carcasses (dead fish), chicken manure, rice husk ash, and other locally available wastes from agricultural wastes (see Figure 13.5). The detailed cost calculation consists of raw materials cost, transport costs, and operation costs of organic fertilizer production (covering equipment depreciation, labor, electricity, packaging, and storage). The total cost for producing 2 tons of organic fertilizer production is approximately 12,641,000 VND ($\approx$503 USD), equating to about 6230.5 VND/kg ($\approx$0.25 USD/kg) (Table 13.4). Currently, the market price for organic fertilizer ranges from 8000 to 12,000 VND, yielding a profit of 2000–5000 VND/kg of fertilizer produced from agricultural by-products. Moreover, applying rapid fermentation technology with high-temperature-resistant fermentation bacteria not only reduces production costs and enhances the value of agricultural by-products by converting them into high-quality products for crop use, but also prevents secondary pollution and significantly shortens the composting time compared to traditional methods/manual technologies.

Table 13.4 Estimated production cost of organic fertilizer production produced from the effluent and sludge of fish pond and agricultural by-products for 2 tons.

I	Fertilizer Raw Materials		Kg	Price (VND)	Total Cost (VND)
1	Catfish pond sludge	50%	1000	250,000	250,000
2	Catfish carcasses/Catfish processing by-products	3%	60	4,500	270,000
3	Rice husk ash	16.90%	338	1,700	574,600
4	Chicken manure (30% dry)	20%	400	1,800	720,000
5	Molasses	0.10%	2	10,000	20,000
6	Bio-fertilizers (Bioway biotic)	10%	200	4–5 kg of microorganisms are used to produce 2 tons of organic fertilizer	
II	**Fertilizer Production Costs**				
7	Labor costs for bagging sludge	person	2	200,000	400,000
8	Transportation costs for hauling sludge and materials from An Giang Province to Soc Trang Province	3.5 tons truck, one-way	1	3,000,000	3,000,000
9	Depreciation of equipment (cost based on the rental fee of a material composting machine for one day)	rental	1	3,000,000	3,000,000
10	Labor hire in the factory (day x number of workers x cost)	1 day * 3 people		500,000	1,500,000
11	Electricity cost for one production batch (2 tons of compost fertilizer)	80KWh/h	480	2,780	1,334,400
12	Packaging cost	bag	20	3,600	72,000
13	Other costs (sewing bags + sewing thread + storage)	–		1,500,000	1,500,000
III	**Total Cost of Producing Organic Bio-fertilizer (2 Tons)**	**100**	**2,000**	**268,000**	**12,641,000**
14	Total cost of producing organic bio-fertilizer per kilogram				**6,320.5**
15	Market price of organic bio-fertilizer				**8,500**
16	**Total net income**				**2,180**

13.4.2.4 Nutrient composition of organic fertilizer produced from the effluent and sludge of fish pond mixed with agricultural by-products

The physico-chemical properties of the organic fertilizer produced from wastes of agricultural by-products using different composting technologies are summarized in Table 13.5. Saldarriaga *et al.* (2019) stated that the physical and chemical properties of the composting products for organic fertilizer depend on the type of initial material, duration of the decomposition process, and the conditions during composting (e.g., temperature, moisture content, degree of aeration, pH, and C/N ratio, and the physical structure of the raw materials). As a result, it is explored that there would be some variation between our two batches of compost. Analysis of the fertilizer's nutritional parameters showed that the total organic carbon (%C), TN, total phosphorus (%P_2O_5), and total potassium (K_2O) content was approximately 46.50 ± 3.54 (44.0–49.0%), 1.07 ± 0.18 (0.94–1.20%), 1.56 ± 0.26 (1.37–1.74%), and 1.48 ± 0.48 (1.14–1.82%), respectively (Table 13.5). The nutrient content of the

Table 13.5 Physico-chemical properties of organic fertilizer produced from the effluent and sludge of fish pond mixed with agricultural by-products using traditional/manual technology and advanced technology application.

Parameters	Application of Traditional/ Manual Technology		Advanced Technology Application (AT-6H Rapid Fermentation Technology)	
pH	7.21 ± 0.25	6.85–7.38	8.06 ± 0.50	7.5–8.41
Organic carbon (% C)	12.01 ± 3.65	5.64–18.1	46.50 ± 3.54	44.0–49.0
TN (% N)	1.01 ± 0.23	0.65–1.33	1.07 ± 0.18	0.94–1.20
TP (% P_2O_5)	0.96 ± 0.41	0.60–1.80	1.56 ± 0.26	1.37–1.74
TK (% K_2O)	0.96 ± 0.86	0.22–2.14	1.48 ± 0.48	1.14–1.82
C/N ratio	11.92 ± 2.90	8.67–17.24	28.80 ± 12.46	40.8–46.8

Source: Thanh *et al.* (2015, 2023), Da *et al.* (2020, 2021), Rahman and Yakupitiyage (2006), Karak *et al.* (2013).

organic fertilizer produced using Bioway Vietnam's AT-6H rapid fermentation technology meets the requirements specified in the Vietnamese National Technical Regulation on Fertilizers. In general, the total organic carbon (%C) in organic fertilizer produced using Bioway Vietnam's AT-6H rapid fermentation technology is approximately three to four times higher than that in biofertilizers produced using traditional/manual technology. Additionally, the nutrient content (TN, TP, and TK) of organic fertilizer produced with advanced technology is approximately twice as high compared to fertilizers produced using traditional/manual technology (Table 13.5). The initial C/N ratio, which depends on the quality of the residues, affects the composting process and compost quality, whereas the final C/N ratio depends on the degree of decomposition (Zhu, 2007). The final C/N ratios of the organic fertilizer ranged between 8.67 and 17.24 for organic fertilizer produced applying traditional/manual technology, and between 40.8 and 46.8 for organic fertilizer produced with advanced technology (Table 13.5). A C/N ratio of around 25–50 is recommended for composition organic biofertilizers (Truc, 2024).

13.5 EFFECTIVENESS OF USING ORGANIC FERTILIZER PRODUCED FROM AGRICULTURAL BY-PRODUCTS ON VARIOUS VEGETABLES AND RICE CROP PRODUCTIONS

Organic fertilizer produced from agricultural by-products offer numerous benefits for crop production, particularly in sustainable farming. By recycling these by-products, these fertilizers reduce waste and minimize environmental impact, promoting sustainability. They also improve soil health by enhancing soil structure, increasing organic matter, and boosting microbial activity, which leads to more fertile and resilient soil. Moreover, the use of agricultural by-products as raw materials lowers production costs, making these fertilizers a cost-effective option for farmers. Organic fertilizers naturally supply essential nutrients, reducing the dependency on synthetic fertilizers, which can harm the environment and human health. This environmentally friendly approach also decreases the risk of water contamination from chemical runoff and helps to lower greenhouse gas emissions linked to synthetic fertilizer production. In terms of effectiveness, organic fertilizer offers a steady release of vital nutrients such as nitrogen, phosphorus, and potassium, ensuring a consistent supply for crops. This results in improved crop yield and quality, with enhancements in size, taste, and nutritional content. Additionally, these fertilizers also strengthen plant immunity, increasing resistance to diseases and pests, thus reducing the need for chemical pesticides. Furthermore, the improved soil conditions foster better root development, allowing plants to access water and nutrients more efficiently. Unlike synthetic fertilizers, which can degrade soil quality over time, organic fertilizer helps building and maintaining long-term soil fertility, ensuring ongoing agricultural productivity.

Previous studies have demonstrated that organic bio-fertilizers can replace or supplement synthetic fertilizers, positively impacting the yield of crops such as cucumber, green been, water spinach, malabar spinach, *Amaranthus cruentus* vegetables, soybean (*Glycine max*) vegetables, *Brassica integrifolia*, and rice production (Da *et al.*, 2015, 2020, 2021; Thanh *et al.*, 2015, 2023). The studies also indicated that the nutrients in organic fertilizer produced from effluent and sludge of fish ponds mixed with waste from agricultural by-products or agro-industrial food waste can replace 25–50%, or even up to 75% of chemical fertilizers/synthetic fertilizers as a nutrient source for certain vegetables, rice, and other crops (Da *et al.*, 2015, 2020, 2021; Karak *et al.*, 2013; Rahman & Yakupitiyage, 2006; Thanh *et al.*, 2015, 2023). These organic amendments satisfy the nutritional requirements of the plants and increased yields compared to using only inorganic fertilizers. Moreover, combining organic amendments with integrated aquaculture and agriculture farming creates a diversified production system that offers tangible environmental benefits, potentially increases farm income, and supports the development of a circular economy model for sustainable agriculture in the future.

Organic fertilizer derived from agricultural by-products provides an eco-friendly, cost-effective solution that enhances crop yield and quality while supporting sustainable farming practices. In addition to the effectiveness and benefits of reusing agricultural by-products and residues in agriculture, there are, however, still some limitations. The availability of input materials from these by-products can be unstable due to dependence on seasons and weather. The collection and storage of materials also face certain challenges, and the costs of collecting and transporting materials to fertilizer production plants remain relatively high. Moreover, we have not yet raised public awareness and understanding of the benefits and effectiveness of using organic fertilizers, so farmers have not yet fully embraced the use of organic fertilizers for their crops.

13.6 VIETNAM POLICY ON THE DEVELOPMENT AND USE OF ORGANIC FERTILIZER

The development of organic fertilizers is crucial for the future of Vietnamese agriculture, both in the short and long terms. Recently, various levels of governmental and local authorities have introduced multiple strategies to advance organic fertilizer use. Despite these efforts, fertilizer usage remains imbalanced, and the production and utilization of organic fertilizers do not fully align with the available domestic raw materials. To further promote the development of organic fertilizers and their role in fostering clean, sustainable agriculture, enhancing agricultural product quality, and protecting the environment, the Ministry of Agriculture and Rural Development outlined several key priorities in 2020:

- Disseminate and promote legal documents, standards, and technical regulations related to fertilizers.
- Support and prioritize new registrations, scale expansions, and capacity building in organic fertilizer production.
- Develop and implement effective models and supply chains for the production and use of organic fertilizers.
- Enhance research and application of high-quality organic fertilizers using Vietnam's available raw materials.
- Improve the distribution system to ensure on-site supply at reasonable prices.
- Allocate funding for science and technology initiatives and policies that support and encourage investment in organic fertilizer production and related business activities.
- Guide specialized agencies to collaborate with organizations and individuals in the fertilizer sector to promote organic fertilizer use and develop suitable, effective models for local applications.
- Strengthen inspections and oversight of fertilizer production, business practices, and usage; enforce strict penalties for the production, sale, and use of poor-quality or fake fertilizers that harm the environment and undermine public trust.

13.7 CONCLUSIONS

In Vietnam, the annual generation of agricultural by-products is substantial, amounting to approximately 150–200 million tons per year, including the crop sector (56.6%), livestock (39%), aquaculture (0.6%), and forestry (3.8%). Although Vietnam has many technologies and models for processing agricultural by-products, their application has not been highly effective: burning rice straw accounts for 45–60%, sawdust, wood shavings, and wood chips for about 15%, and livestock waste treatment for less than 50%. The obstacles to managing agricultural by-products include: (i) the form, scale, and level of production; (ii) the cost and labor required for by-product processing; (iii) production planning; (iv) inconsistent policies; and (v) environmental factors and diseases. Facing with the overload of these wastes from agriculture by-products, lack of cost-effective technologies and improper disposal of organic waste, along with the limitations of some traditional composting technologies or organic fertilizer production methods, it is impossible to thoroughly process these agroindustry wastes. Regarding technological solutions, the application of high-speed centrifugal technology for collecting sludge and treating aquaculture pond environments, combined with the application of Bioway Hitech AT-6H rapid fermentation technology to process and produce organic fertilizer from locally available agricultural by-products, has proven effective in converting and adding value to these by-products into valuable products and wealth such as organic fertilizers. The results of applying organic fertilizer in several crop models, including rice, soybean, and sweet mustard greens, showed that reducing chemical fertilizers by 25–50% up to 75% while using organic fertilizers still resulted in higher yields and profits compared to conventional farming. This chapter proposed several solutions to promote the effective management and use of agricultural by-products and organic waste, which should be based on: (i) policy synchronization, (ii) technology transfer, (iii) circular economy and building a sustainable society, and (iv) complementary solutions, including communication and training to raise awareness among those involved in agricultural production.

ACKNOWLEDGEMENTS

The authors acknowledge Navico Company, An Giang Food and Vegetables Joint Stock Company (ANTESCO), Department of Agriculture and Rural Development and local farmers of An Giang province for a great collaboration and providing the necessary facilities to conduct experiments, and appreciate technician staff to support and assist with the implementation of the research. The authors are grateful to VanTech Vietnam Co., Ltd, and Bioway Hitech Organic Fertilizera, Vietnam, for funding the research team from Ton Duc Thang University, HCMCT, Vietnam to carry out the research of organic fertilizer in An Giang province, Vietnam.

REFERENCES

Anh P. T., Kroeze C., Bush S. R. and Mol A. P. J. (2010). Water pollution by *Pangasius* production in the Mekong Delta, Vietnam: causes and options for control. *Aquaculture Research*, **42**(1), 108–128, https://doi.org/10.1111/j.1365-2109.2010.02578.x

Anh P. T., My Dieu T. T., Mol A. P. J., Kroeze C. and Bush S. R. (2011). Towards eco-agro industrial clusters in aquatic production: the case of shrimp processing industry in Vietnam. *Journal of Cleaner Production*, **19**(17), 2107–2118, https://doi.org/10.1016/j.jclepro.2011.06.002

Berg H., Lan T. H. P., Tam N. T., Trang D. H., Van P. H. T., Duc H. N. and Da C. T. (2023). An ecological economic comparison between integrated rice-fish farming and rice monocultures with low and high dikes in the Mekong Delta, Vietnam. *Ambio*, **52**(9), 1462–1474, https://doi.org/10.1007/s13280-023-01864-x

Da C. T., Phuoc L. H., Duc H. N., Troell M. and Berg H. (2015). Use of wastewater from striped catfish (*Pangasianodon hypophthalmus*) pond culture for integrated rice–fish–vegetable farming systems in the Mekong Delta, Vietnam. *Agroecology and Sustainable Food Systems*, **39**(5), 580–597, https://doi.org/10.1080/21683565.2015.1013241

Da C. T., Anh Tu P., Livsey J., Tang V. T., Berg H. and Manzoni S. (2020). Improving productivity in integrated fish–vegetable farming systems with recycled fish pond sediments. *Agronomy*, **10**(7), 1025, https://doi.org/10.3390/agronomy10071025

Da C. T., Vu T. H., Duy D. T., Ty N. M., Thanh D. T., Nguyen-Le M. T., Berg H., Nguyen Q. H. and Bui X. T. (2021). Recycled *Pangasius* pond sediments as organic fertilizer for vegetables cultivation: strategies for sustainable food production. *Clean Technologies and Environmental Policy*, **25**(2), 369, https://link.springer.com/article/10.1007/s10098-021-02109-9

Department of Crop Production of Vietnam (2022). Current status of by-products in crop production and proposed management and utilization solutions. Workshop proceeding on Agricultural by-products – A renewable resource. Workshop Proceeding. Ministry of Agriculture and Rural Development, pp. 46–48.

Department of Livestock Production of Vietnam (2022). The current status of by-products in livestock production and proposed waste management and utilization solutions. Workshop Proceeding on animal waste management and treatment (Vietnamese). Workshop Proceeding. Ministry of Agriculture and Rural Development, pp. 49–56.

FAO (1990). World agriculture: towards 2015/2030, Interim Report, FAO, Rome, Italy.

FAO (2002). Land degradation assessment in drylands – LADA project, World Soil Resources Reports, Food And Agriculture Organization of the United Nations.

FAO (2024). The State of World Fisheries and Aquaculture 2024 – Blue Transformation in Action. Food and Agricultural Organization of the United Nations, Rome, Italy. ISBN: 978-92-5-138763-4, **264**, https://doi.org/10.4060/cd0683en

GSO (2022). Production of Agriculture, Forestry and Fishing by Province. Statistical Yearbook of Vietnam 2022. Statistical Documentation and Service Centre – General Statistics Office of Vietnam, 54 Nguyen Chi Thanh street, Dong Da district, Hanoi, Vietnam, https://www.gso.gov.vn/default_en.aspx?tabid=778

Karak T., Bhattacharyya P., Paul R. K., Das T. and Saha S. K. (2013). Evaluation of composts from agricultural wastes with fish pond sediment as bulking agent to improve compost quality. *CLEAN – Soil, Air, Water*, **41**(7), 711–723, https://doi.org/10.1002/clen.201200142

Koul B., Yakoob M. and Shah M. P. (2022). Agricultural waste management strategies for environmental sustainability. *Environmental Research*, **206**, 112285, https://doi.org/10.1016/j.envres.2021.112285

Lam N. T. and Son C. T. (2024). Current status of agricultural by-products and management solutions in Vietnam. Proceedings of the National Scientific Conference on 'Agricultural and Rural Environment and Sustainable Development', pp. 159–168.

Livsey J., Da C., Scaini A., Lan T. H. P., Long T. X., Berg H. and Manzoni S. (2021). Floods, soil and food – interactions between water management and rice production within an Giang province, Vietnam. *Agriculture, Ecosystems & Environment*, **320**, 1, https://doi.org/10.1016/j.agee.2021.107589

MARD (2022a). Annual Reports of the Ministry of Agriculture and Rural Development, Vietnam. Unpublished documents (in Vietnamese).

MARD (2022b). Current status of by-products in crop production and proposed management and utilization solutions. The proceeding of Biomass Waste Agricultural By-Products – Renewable Resource, pp. 46–48.

Mills H. J. (1996). Plant Analysis Handbook II. Micromacro Publishing, Inc., Athens, GA.

Nick J. M. A. (1994). Growing Fruits and Vegetables Organically: The Complete Guide to a Great-Tasting, More Bountiful, Problem-Free Harvest. Rodale Press, Emmaus. Bradley, Fern Marshall [Editor]; Atthowe, Helen [Editor]. From Orion Tech, Kingwood, Texas, USA.

Nuri S. A., Hossean K. A. and Hamza K. I. (2021). Use of fertilizers or importance of fertilizers in agriculture. *International Journal of Advanced Academic Studies*, **3**(2), 52–57, https://doi.org/10.33545/27068919.2021.v3.i2a.770

Rahman M. and Yakupitiyage A. (2006). Use of fishpond sediment for sustainable aquaculture–agriculture farming. *International Journal of Sustainable Development and Planning*, **1**(2), 192–202, https://doi.org/10.2495/SDP-V1-N2-192-202

Rechcigl J. E. (1995). Soil Amendments and Environmental Quality. CRC Press, Lewis Publishers, Boca Raton, Florida 33431, United States.

Saioof H. K., Ghamry A. M. E., Fouda K. F. and Sally F. A. E. E. (2019). Organic fertilizers derived from farm by-products for sustainable agriculture – a review. *Journal of Soil Sciences and Agricultural Engineering*, **10**(12), 815–819, https://doi.org/10.21608/jssae.2019.79578

Saldarriaga J. F., Gallego J. L., López J. E., Aguado R. and Olazar M. (2019). Selecting monitoring variables in the manual composting of municipal solid waste based on principal component analysis. *Waste and Biomass Valorization*, **10**(7), 1811–1819, https://doi.org/10.1007/s12649-018-0208-y

Sankar G. K., Sundaramoorthy P., Nagarajan M. and Lawrence Xavier R. (2017). Chapter 7: Role of organic amendments in sustainable agriculture. In: Sustainable Agriculture Towards Food Security, A. Dhanarajan (ed.), Springer, Singapore, pp. 111–124, https://doi.org/10.1007/978-981-10-6647-4_7

Stratton M. L. and Rechcigl J. E. (1998). Agronomic Benefits of Agricultural, Municipal, and Industrial By-Products and Their Co-Utilization: An Overview. In: Beneficial Co-Utilization of Agricultural, Municipal and Industrial by-Products, S. Brown, J. S. Angle and L. Jacobs (eds), Springer, Netherlands, Dordrecht, pp. 9–34, https://doi.org/10.1007/978-94-011-5068-2_2

Thanh X. B., Hien V. T. M., Cong T. T., Da C. T. and Berg H. (2015). Reuse of sediment from catfish pond through composting with water hyacinth and rice straw. *Sustainable Environment Research*, **25**(1), 1–5.

Thanh D. T., Ty N. M., Berg H., Hien N. V., Nguyen T. K. O., Vu P. T., Minh V. Q. and Da C. T. (2023). Effects of organic fertilizers produced from fish pond sediment on growth performances and yield of *Malabar* and *Amaranthus* vegetables. *Frontiers in Sustainable Food Systems*, **7**, 31, https://doi.org/10.3389/fsufs.2023.1045592

Thinh D. N., Son V. N., Hien H., Phuong N. T., Hương T. T. D., Hang T. B. B., Scippo M. L., Quetin-Leclercq J., Kestemont P. and Phu T. (2022). The use of drugs, chemicals, herbs, and herbal extract products in grow-out farms of snakehead (*Channa striata*) and *Pangasius* catfish (*Pangasianodon hypophthalmus*) in the Mekong Delta, Vietnam. *Vietnam Journal of Agricultural Sciences*, **5**(1), 1336–1344, https://doi.org/10.31817/vjas.2022.5.1.03

Tilman D., Cassman K. G., Matson P. A., Naylor R. and Polasky S. (2002). Agricultural sustainability and intensive production practices. *Nature*, **418**, 671–677, https://doi.org/10.1038/nature01014

Truc N. T. N. (2024). Chapter 5. By-products and current treatment status of agricultural by-products in Vietnam. In: Book of Circular Agriculture, Agriculture Publishing House, Le Tan, Nguyen Thi Diem Yen, Nguyen Khanh Ha (eds), Vietnam, pp. 81–97

United Nations (2022). Department of economic and social affairs population division. World Population Prospects, 2022. https://population.un.org/wpp/

VASEP (2022). Vietnam's fisheries industry overview 2022 of Vietnam Association of Seafood Exporters and Producers. Vietnam Association of Seafood Exporters and Producers of Ministry of Agriculture and Rural Development, Vietnam.

Westerman P. W. and Bicudo J. R. (2005). Management considerations for organic waste use in agriculture. *Bioresource Technology*, **96**(2), 215–221, https://doi.org/10.1016/j.biortech.2004.05.011

World Bank (2023). Vietnam organic fertilizer market size and share analysis – growth trends & forecasts (2024–2029). Precise market intelligence and advisory. https://www.mordorintelligence.com/industry-reports/vietnam-organic-fertilizers-market. (accessed 09 August 2024)

Worldometer (2022). World population projections. https://www.worldometers.info/world-population/world-population-projections/

Zhu N. (2007). Effect of low initial C/N ratio on aerobic composting of swine manure with rice straw. *Bioresource Technology*, **98**(1), 9–13, https://doi.org/10.1016/j.biortech.2005.12.003

Part 5

Sustainability assessment of urban nature-based solutions

doi: 10.2166/9781789065015_0271

Chapter 14

Enhancing sustainability of constructed wetlands for agricultural water and nutrient management practices through social innovations

Tamara Avellan[1,2], Alba Canet-Martí[3], Eriona Canga[4], Valentina Guerrieri[5], Simone Amadori[5], Stevo Lavrnić[5], Marco Hartl[4], Nidhi Nagabhatla[2], Bishal Dahal[1], Kedar Ghag[1], Syed Mustafa[1,6] and Björn Klöve[1]*

[1]University of Oulu, Oulu, Finland
[2]UNU-CRIS, Bruges, Belgium
[3]University of Natural Resources and Life Sciences, Vienna (BOKU), Vienna, Austria
[4]alchemia-nova, Institute for Circular Economy & Nature-based Solutions, Vienna, Austria
[5]University of Bologna, Bologna, Italy
[6]Wageningen Research, Wageningen, The Netherlands
**Corresponding author: tamara.avellan@oulu.fi*

ABSTRACT

Constructed wetlands (CWs) have a long history of pollution control in urban but also in rural and agricultural settings. As one of the solutions to water and nutrient management issues in European agriculture, CWs were assessed in the European H2020 project WATERAGRI not just for their effectiveness but also for their role towards sustainability. In this chapter, we show different methods used to assess the effectiveness and sustainability of CWs by applying a rapid assessment, driver–pressure–state–impact–response analyses, and life-cycle thinking. We also demonstrate the need for an advanced inter- and transdisciplinary understanding of how sustainability, both as a concept and in its implications, is perceived by applying a Delphi analysis and an online questionnaire. The results show that CWs can be considered a sustainable measure for water and nutrient management in agricultural sites for their environmental benefits, which are the most highlighted by the scientific community and practitioners. However, findings also show that social benefits may indeed be the tipping point for these systems to be sustainable and that more research and insights are needed to highlight the role of social innovation.

Keywords: Nature-based solutions, Delphi approach, Life Cycle Thinking, Social sustainability, Policy recommendations.

14.1 INTRODUCTION

Intense agricultural practices are the leading cause of Europe's water, soil, and biodiversity degradation. Since the implementation of the Water Framework Directive 15 years ago, farm fertilizer and pesticide applications have been identified consistently as the leading cause of excess pollution loads both of surface waters and groundwater and for aquatic species loss (Bieroza *et al.*, 2021; Schäfer *et al.*, 2007). As per Dudley (2022), the massive loss of insects, both in species diversity as well as in absolute

© 2025 IWAP. This is an Open Access book chapter distributed under the terms of the Creative Commons Attribution License (CC BY-NC-ND 4.0) which permits copying and redistribution for non-commercial purposes with no derivatives, provided the work is properly cited (https://creativecommons.org/licenses/by-nc-nd/4.0/). The chapter is from the book *Nature-Based Solutions for Urban Sustainability*, Piet N.L. Lens and Xuan-Thanh Bui (Editors).

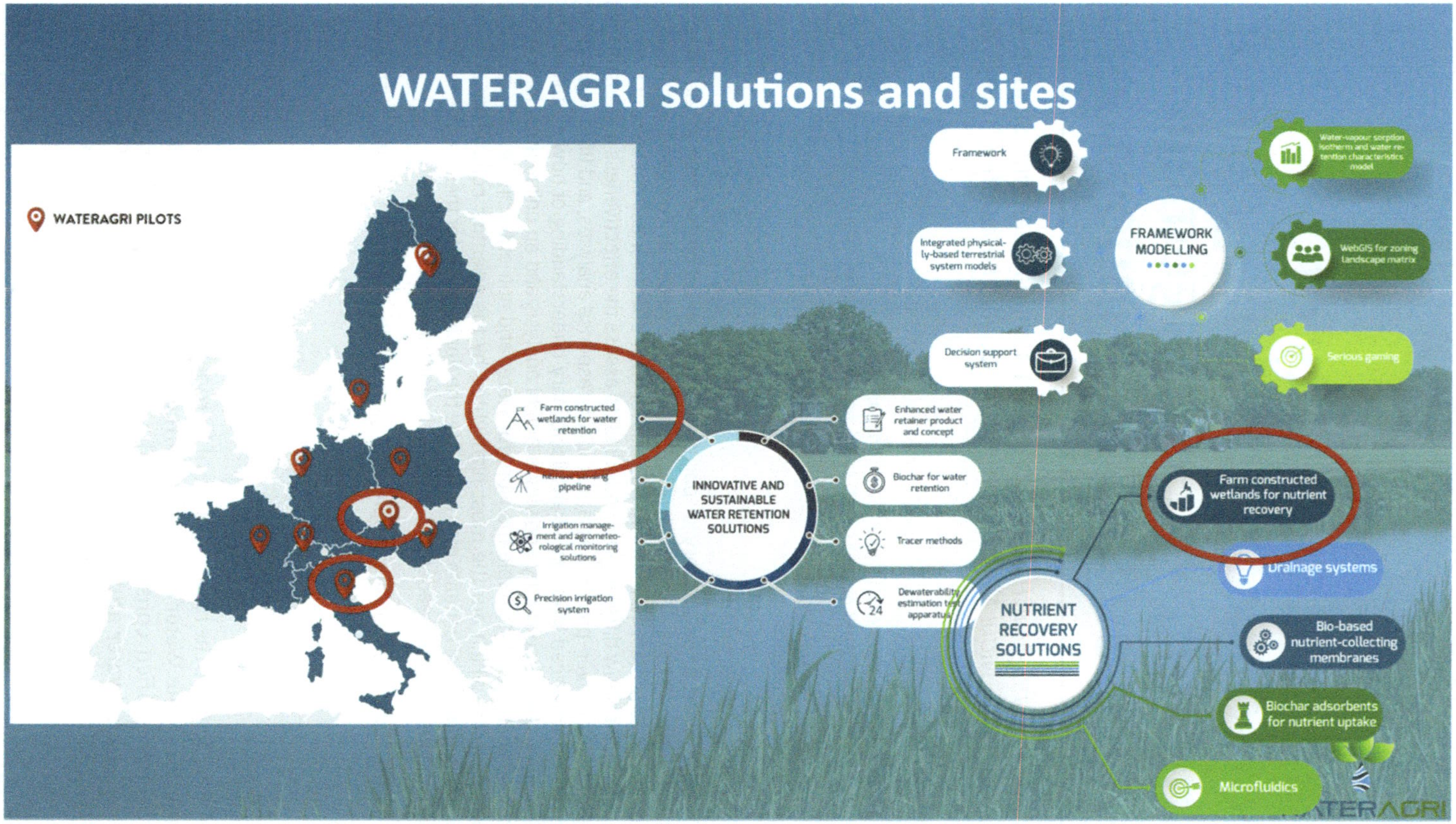

Figure 14.1 Overview of WATERAGRI sites and solutions with farm constructed wetlands for nutrient recovery and/or water retention being implemented in the cases in Austria and Italy. In the context of WATERAGRI, the term farm constructed wetlands was used to refer to FWS wetlands, also known as surface flow constructed wetlands.

numbers, can be traced back directly to the use of agrochemicals. Accordingly, soil biodiversity has suffered a significant toll as well, and soil structure has been altered leading to poor water absorption, retention, and release for plant uptake. Overall, there is thus a clear need to rethink agricultural practices not only to rescue the remaining healthy ecosystems, but also to restore the degraded ones for the sake of agricultural production itself and our own human livelihoods (Boix-Fayos & de Vente, 2023; Giannakis *et al.*, 2019).

There is thus a need for a holistic picture to the implications of constructed wetland (CW) implementation in agricultural and landscape contexts. This chapter collates the findings from the EU H2020 project WATERAGRI (https://wateragri.eu/), that intended to solve agricultural water management and soil fertilization challenges in a sustainable manner to secure affordable food production in Europe for the 21st century. This project was implemented through the development of a new framework for the use of small water retention approaches for managing excess and shortage of water as well as better recovery of nutrients from agricultural catchments applying a multiactor approach. In the context of European river basins, where diffuse pollution from agriculture poses a significant challenge, efficient water management becomes pivotal. Solutions include infrastructural changes such as the implementation of CWs towards better water or nutrient retention, modelling approaches towards better soil moisture forecasts, and more generic knowledge tools such as the development of a serious game (see all solutions in Figure 14.1). Integrating WATERAGRI's approaches also aligns with the European bio-economy strategy, aiming to reduce pollution, enhance resilience, and promote sustainable agricultural food production. The project's commitment to developing integrated water management approaches underscores its dedication to achieving sustainable agricultural production and food security. Operating in diverse climatic and socio-economic contexts, ranging from large farms to regions with numerous small farms, the project encompasses 10 distinct cases. Its implementation spans 10 case study sites strategically distributed across the Boreal, Continental, and Pannonian climatic zones.

Although WATERAGRI offered a number of different solutions for water and nutrient management in European agriculture, not all solutions are suitable nor sustainable in all contexts. This chapter delves into the complexities of identifying sustainable water and nutrient management practices for European agriculture, acknowledging that not all solutions offered by WATERAGRI are universally applicable or sustainable. The chapter begins by outlining methods to determine the suitability and sustainability of various solutions under different conditions. The chapter then provides a detailed sustainability assessment of a specific practice – a CW system in Italy – using life-cycle thinking. This assessment includes an exploration of how WATERAGRI stakeholders perceive sustainability in the context of CW systems. Concluding the chapter, policy recommendations aimed at promoting sustainable practices in agricultural nutrient and water management are presented, with a particular emphasis on the role of CWs.

14.2 MODALITIES TO FIND SUSTAINABLE SOLUTIONS IN AGRICULTURAL WATER AND NUTRIENT MANAGEMENT

14.2.1 Understanding which solutions addresses best the causes: driving forces, pressures, states, impacts, and responses

The driving force, pressure, state, impact, and response (DPSIR) framework (European Environmental Agency, 1999) is a simple systematic, rapid, and holistic method recommended by the European Commission as a causal framework for describing the interactions between society and the environment. This tool helps to outline the environmental problems at hand including the assessment of solutions and how they interact with the environmental problem. The framework divides a given environmental concern/problem into five compartments, namely: driving forces, pressures, states, impacts, and responses. The driving forces 'D' represent broader trends and changes in social, economic, and human activities such as agriculture. It can also consider changing climate conditions.

The drivers generate pressures 'P' that can be understood as processes that affect biological, chemical, and physical cycles in water systems. Pressures influence the state 'S' of water systems. State changes lead to impacts 'I', which is a multidimensional concept involving economic, social, health, and environmental aspects, but often with a focus on environmental impacts if environmental problems are to be solved. Responses 'R' represent actions by water managers, society, or policy makers to respond to solve the environmental issue under consideration. These actions consist in elaborating and implementing regulations, planning documents, or economic instruments. Actions also include the emergence of technological innovations and changes in behaviour.

Within WATERAGRI, the primary objective revolved around devising strategies for 'water retention and nutrient recovery'. The project thoroughly evaluated the DPSIR model, leveraging a comprehensive understanding of water resources and their interlinkage with the agriculture sector. To facilitate this assessment, the project relied on its case studies and factsheets that were developed for the different solutions (https://wateragri.eu/factsheets/). These resources offer a detailed perspective on the identified issues, serving as a foundation for the DPSIR evaluation. Figure 14.2 illustrates the overarching assessment, intricately connecting DPSIR compartments with the research efforts dedicated to developing and evaluating solutions. The devised solutions within WATERAGRI play a pivotal role by providing responses denoted as 'R' to one or multiple compartments within the DPSIR model. This integrative approach ensures a holistic understanding of the challenges related to water resources in agriculture, ultimately contributing to outlining effective and targeted solutions.

The WATERAGRI solutions affect the DPSI compartments in different ways. An assessment of where the effects can be seen is outlined in Table 14.1. Depending on where the solutions respond to the DPSI compartments, the solutions can have wide, synergetic, cross-benefit, and win–win effects. For example, water retention will decrease leaching as less water is required for irrigation, reducing the pressure on irrigation demand (P), which, in theory, can lead to more yields and less leaching. All solutions have the potential to influence compartments S and I. The serious game, AgriLemma, can lead to changes in agriculture and agricultural management itself, which influence all compartments.

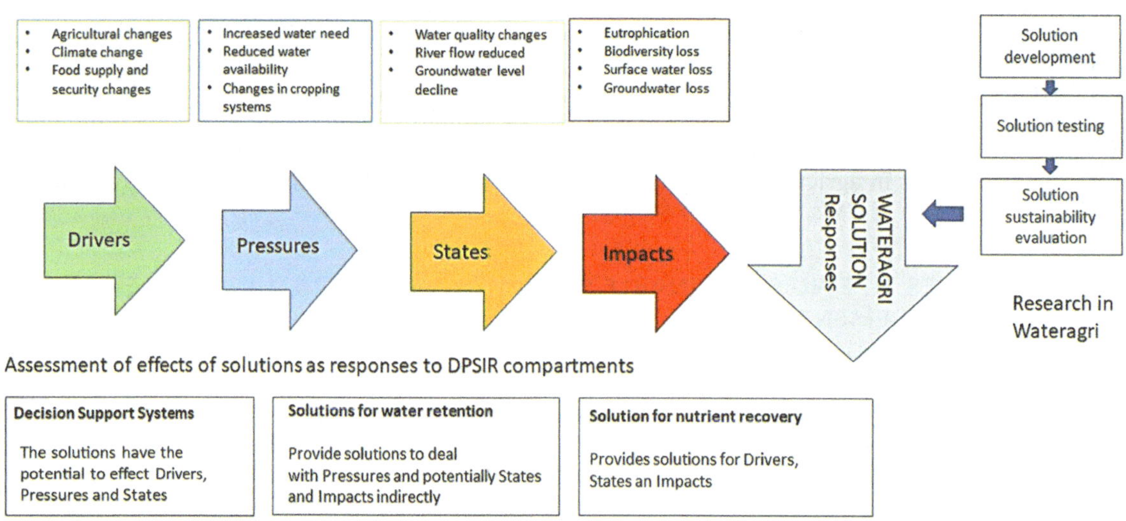

Figure 14.2 Overview of the DPSIR process showing the responses and their development in WATERAGRI. The DPSIR compartments are evaluated for the case of water use and pollution from agriculture. The effect and potential effect of the solutions on DPSIR compartments is drafted.

Table 14.1 Solution classified into the DPSI framework.

Solution Name	DPSI Compartments	More Information and Reasoning
AgriLemma serious game	DPSI	Potentially all DPSI compartments as the method is a game where different stakeholders participate
Water retainer	PSI	An organic liquid that can be sprayed on soil to increase water retention. This can reduce irrigation needs, thus limiting the leaching of nutrients and pesticides. Also, plants can take up more nutrients, which might limit leaching
Biomembranes	(P)SI	It works as an end-of-pipe solution that allows for nutrient recycling. Biomembranes tested, but solutions not ready for application yet
Filters at end of drain pipes	(P)SI	It works as an end-of-pipe solution that allows for nutrient recycling. Tested in field operation, but frequent replacement might limit usage
Multilayered filter system	(SI)	Can retain nutrients. Developed and tested in meso-scale field operation
Microfluidics	(SI)	Tested in lab scale set-up
Wetland for Water Retention	PSI	Water stored for irrigation which limits nutrient runoff losses
Wetland for Nutrient Retention	PSI	Nutrients recycled, which limits nutrient runoff losses

Solution names relate here to the ones used in the factsheets.

The DPSIR analysis allowed WATERAGRI to better understand which solutions might address best the underlying drivers and pressures to influence the state and cause beneficial impacts. However, it did not fully address sustainability.

14.2.2 What is 'sustainability' in agricultural water management

As a first step the vague concept of sustainability needed to be defined. In the past, sustainability has been defined in slightly different ways, reflecting institutional values. The Brundtland Commission first defined sustainable development in 1987 as 'development that meets the needs of the present without compromising the ability of future generations to meet their own needs' (WCED, 1987). This definition laid the foundation for sustainability's three-dimensional concept: social, environmental, and economical.

WATERAGRI is an inter- and transdisciplinary project and encompasses project partners from different scientific and professional backgrounds. These different backgrounds may lead to different understandings of key concepts, which in turn may lead to misunderstandings across the project members. To overcome this, Dahal *et al.* (2023) conducted a consensus-building approach through a Delphi survey. A Delphi survey is 'a method of structuring a group communication process so that the process effectively allows a group of individuals as a whole to deal with a complex problem' (Hugé *et al.*, 2010). Experts from the WATERAGRI consortium members (9 WATERAGRI case study owners, 11 WATERAGRI solution providers), local WATERAGRI case study stakeholders (10), and persons not related to the project in any way (9) were invited to participate in this survey. A two-round survey was conducted: the former focused on prioritizing keywords in the three dimensions of sustainability, whereas the latter intended to clarify remaining conflicts among the responses of the experts.

The findings showed that environmental sustainability emerges as a prime focus, with most experts emphasizing climate change, water quality, and quantity concerns. Additionally, they expressed worries about the impact of agricultural practices on environmental degradation. Notably, experts from the Global North emphasize water quality due to concerns about pollution from agricultural activities, whereas those from the Global South stressed water quantity due to water-scarcity issues. Agricultural sustainability remains a central concern, with experts advocating for efficient water use in farming practices. Social sustainability considerations encompassed stakeholder participation and capacity building, with differences observed between expert groups based on geographic regions. Economic sustainability priorities varied, with consortium members emphasizing economic growth, stakeholders focusing on affordability, and external experts prioritizing income and profit generation. Gender differences in perceptions highlighted varying concerns, suggesting the need for nuanced policy formulations tailored to diverse expert perspectives. A recent policy brief on gendered agricultural practices in Europe also evidenced gendered differences in perceptions, with female experts prioritizing water availability, and underscored the importance of inclusive decision-making and targeted interventions tailored to regional challenges and the voices of marginalized groups (UNU-CRIS, 2024).

Based on the most prominently featuring keywords the following definition was coined for 'sustainable management of water in agriculture' for the WATERAGRI project:

> Managing water in a way that ensures its quality and has minimal negative effects on the natural environment. It involves stakeholder satisfaction and wellbeing, ensuring equal participation, and improving competency by entailing effective governance and equitable laws. It promotes long-term economic growth and provisions of incentives and subsidies. In summary, it exemplifies the ideal level of water management incorporating climate change concerns.

Establishing a common definition within the WATERAGRI consortium facilitated a more cohesive understanding and communication among its members. This process not only helped in harmonizing perspectives but also uncovered the consortium's predisposition towards prioritizing the environmental aspects of sustainability. Additionally, it highlighted a male-centric viewpoint from the Global North, which tended to focus more on water quality rather than water quantity concerns.

14.2.3 Understanding which solutions work where: rapid assessment of sustainability

Before diving deeply into various sustainability assessments of solutions and sites, a rapid assessment of which solutions might be considered sustainably 'suitable' for which site was performed. For this purpose, several criteria for the three dimensions of sustainability, namely environmental, economic, and social, were identified by the research team at the University of Oulu and vetted by the solution and case study owners for general soundness (Figure 14.3). From the literature and the project

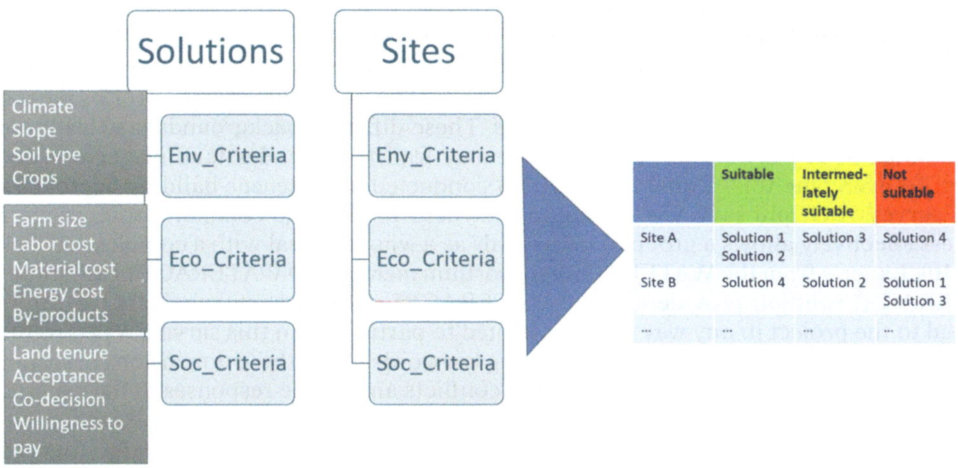

Figure 14.3 Overall schematic of the rapid assessment and criteria in each of the dimensions of sustainability used to determine the potential suitability of WATERAGRI solutions for WATERAGRI sites.

documentation for the respective solution and case study information data for each criterion was sought and solution providers and case study owners were asked to double-check and correct the data if needed. In this process, several solutions were removed as the evidence base was not considered to be strong enough for a sufficiently sound judgement/. The assessments were therefore only conducted for four WATERAGRI solutions that had seen implementation on sites, namely: farm constructed wetlands, precision irrigation systems, biochar application, and Drainage systems.

Finally, the 'fit' of the solutions for the respective sites was carried out by matching the information of the criteria through a traffic light system (highly matching, medium, not at all) for each of the dimensions of sustainability. Not matching at all meant that the solution was considered to be 'Not Suitable' in any of the three dimensions, whereas for a solution to be fully suitable for a site it had to be suitable in all dimensions. All other (mixed) scenarios were considered as moderately suitable. As such, across all sustainability dimensions, farm constructed wetlands and precision irrigation systems were deemed suitable solutions for all cases except for Finland. Biochar application and drainage systems were considered moderately to highly suitable solutions in some cases (see Table 14.2). It must be noted that in particular biochar is still being tested and that the data used for the rapid assessment

Table 14.2 Overview of the suitability of WATERAGRI solutions for WATERAGRI sites.

WATERAGRI Sites	Suitable WATERAGRI Solution	Potentially Suitable WATERAGRI Solution	Non-suitable WATERAGRI Solution (Hindrance)
Finland I	–	Farm constructed wetlands (climate, costs) Biochar (climate, costs) Drainage systems (costs)	Precision irrigation systems (farm size)
Finland II	–	Farm constructed wetlands (climate, costs) Biochar (climate, costs) Drainage systems (costs)	Precision irrigation systems (farm size)
Sweden	Farm constructed wetlands Precision irrigation systems	Biochar (costs) Drainage systems (costs)	–
France	Farm constructed wetlands Precision irrigation systems	Biochar (costs) Drainage (costs)	–
Germany	Farm constructed wetlands Precision irrigation systems	Biochar (costs) Drainage (costs)	–
Poland	Farm constructed wetlands Precision irrigation systems Biochar Drainage systems	–	–
Switzerland	Farm constructed wetlands Precision irrigation systems	Biochar (costs) Drainage (costs)	–
Austria	Farm constructed wetlands Precision irrigation systems	Biochar (costs) Drainage (costs)	–
Italy	Farm constructed wetlands Precision irrigation systems	Biochar (costs) Drainage (costs)	–
Hungary	Farm constructed wetlands Precision irrigation systems Biochar Drainage systems	–	–

Element in parenthesis marks the criteria that limit the suitability of the respective solution.

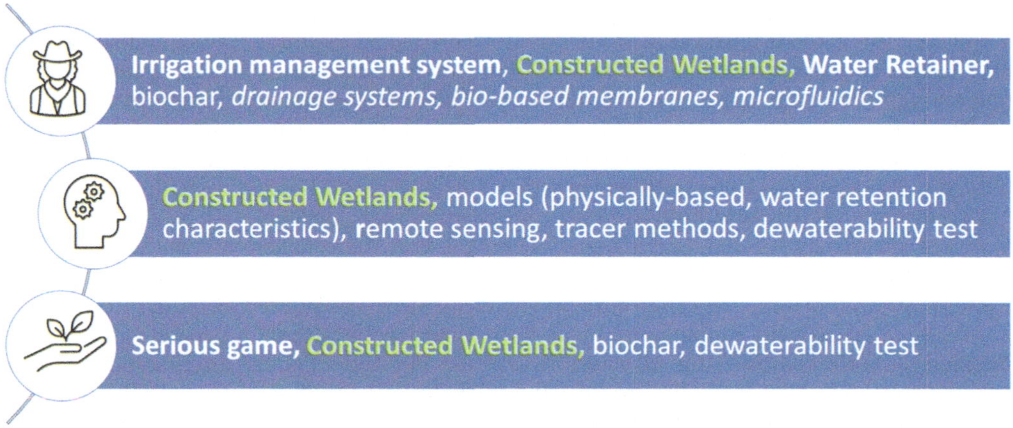

Figure 14.4 Overview of the maturity (bold) and usefulness of the WATERAGRI solutions by user groups (top to bottom: farmers, researchers, advisory services); solutions in italics were considered useful for 'future farmers'.

are mostly based on experimental information and not so much on field testing. Suitability here is therefore to be taken with caution.

The swift evaluation indicated that categorizing WATERAGRI solutions based on sustainability criteria present challenges. Primarily, this is due to the fact that numerous WATERAGRI solutions have not been applied at the farm level, exhibiting low technology readiness levels, and the scarcity of definitive data pertaining to economic and social criteria. Caution is essential when utilizing these findings, and expert consultation is vital for ensuring secure on-site application. For those interested in self-guided exploration and application of the rapid assessment, resources are available through the WATERAGRI website and its framework.

14.2.4 Sustainable solutions for water and nutrient management: preliminary considerations

The preliminary considerations showed that sustainability in the field of agricultural water (and nutrient) management is a challenging one. Being able to determine which solutions to assess in more detail was strongly limited by several factors, such as (1) the maturity of the solution to be used by end-users, (2) the intent of the solution to be used by end-users or not, and (3) the availability and access to data in all three dimensions of sustainability. Of the 19 solutions, some can be lumped as they represent one product package, and others are similar. Only four (groups of) solutions were actually destined for end-users, that is farmers, and of those, the maturity level of deployment in the field was only given for three, namely the family of precision irrigation through GIS, those of CWs, and the water retainer product and concept (Figure 14.4).

14.3 UNDERSTANDING THE SUSTAINABILITY OF FARM CONSTRUCTED WETLANDS

CWs for pollution control are nature-based solutions for wastewater purification, with a long-standing history of reducing organic and inorganic pollution loads from a variety of liquid and semi-liquid effluents, for example, domestic wastewater, acid mine drainage, stormwater runoff, animal manure, and human faecal sludge (Dotro *et al.*, 2017; Kadlec & Wallace, 2008; Vymazal, 2011). They not only reduce nutrients and organic matter concentrations in effluents, but are also known to support the reduction of medication-derived substances such as hormones, antibiotic resistance, and painkillers (e.g., Cavalheri *et al.*, 2022). The removal of contaminants is achieved by physical, chemical, and biological processes (Garcia *et al.*, 2010). In addition to pollution reduction, these systems offer various co-benefits such as enhanced biodiversity, aesthetics, water retention, and thus potential flood protection, the possibility to reuse the treated effluent, and the use of the biomass for multiple purposes, including bioenergy production, construction material, or artwork (Turcios *et al.*, 2021). CWs can, therefore, be used to reduce and compensate for the losses caused by intense agricultural purposes and have been used effectively in agricultural and landscape settings (Carty *et al.*, 2008).

CWs, particularly in agricultural settings, represent a significant investment. They require land that might otherwise be allocated for more immediately productive purposes. Moreover, the design, construction, and maintenance mandate the involvement of trained professionals, rendering them a lower priority for farmers and farm managers (Soldo *et al.*, 2022). Notably, the need for incentives, both in terms of subsidies and enforcement, deters their adoption and scaling potential. Therefore, although the return on investment may not be immediately attractive for farm owners or managers, it is important to consider the broader societal benefits and perspectives associated with these systems in short, medium, and long time frames.

14.3.1 Assessing sustainability of a farm constructed wetland

To better understand what drives sustainability of farm constructed wetlands, the CW of WATERAGRI's Italian case study located in the Emilia-Romagna region, specifically in Budrio, was chosen. The wetland treats the agricultural drainage water of an experimental farm of 12.4 hectares cultivated with cereals, vegetables, and orchards. The CW has a surface of 5557.5 m^2 with a water course length of 470 m composed of four meanders that are 8–10 m wide. The outlet is set at 0.4 m above the bed surface level, setting the overall CW capacity at 1500 m^3. However, the volume of drainage water treated and present in the system depends on the frequency and volume of the seasonal precipitation. Furthermore, the water outflow is discharged into a network of ditches from which farmers in the areas withdraw the water for irrigation purposes. The case study area is equipped with a hydraulic system and sensors which measure the water level inside the CW as well as a weather station. The vegetation is mainly composed of *Phragmites australis*, *Typha latifolia*, and *Carex* spp.

The experimental data for the sustainability assessments were collected in 2020 and 2021 in the context of the WATERAGRI project and complemented with other existing data collected in 2018 and 2019. During the project years the research team monitored and analysed the system functioning through operations for hydraulic system maintenance, sampling for water quality monitoring, and processes for meteorological data collection (Nan *et al.*, 2023). These operations were carried out by work and training activities of PhD students, master students, and researchers of the University of Bologna (Italy).

For this CW, a full sustainability analysis based on life-cycle thinking was carried out. Life-cycle thinking provides a comprehensive framework for assessing the impacts of processes or products throughout their entire life cycle, incorporating environmental, social, and economic dimensions of sustainability (Hunkeler *et al.*, 2008; ISO, 2006a, 2006b; UNEP *et al.*, 2020). Life-cycle assessment (LCA) specifically focuses on environmental impacts and is a standardized procedure for evaluating sustainability, aiding decision-making by identifying environmental hotspots and trade-offs (Corominas *et al.*, 2020). Life-cycle costing assesses the economic dimension of sustainability, considering costs over the entire life cycle of a product or service and aligning with LCA techniques. Social life-cycle assessment (sLCA) is a newer methodology focusing on social aspects, offering guidelines for studying the social performance of products or services. The integration of these analyses enables a more holistic understanding of sustainability across all dimensions.

By combining the results of each assessment, an integrated score was computed, offering an overarching perspective on the overall sustainability of the WATERAGRI case study. This score averages positive (environmental and social) and negative (economic) impacts of the WATERAGRI scenario compared to the business-as-usual scenario. In this instance, the integrated score indicated that implementing the WATERAGRI scenario results in a 32% increase in sustainability. Although standalone, this metric lacks external comparison, it serves as a starting point for understanding the role of CWs in sustainable land and water management strategies. Despite its limitations, an integrated score can aid advocacy and policy-making due to its communicative simplicity. Further research is necessary to refine scoring strategies, develop reference cases, and test scoring designs across multiple case studies.

14.3.2 Understanding the criteria that drive decision-making: asking the community

Although our findings indicated that the implementation of farm CW is sustainable for the one WATERAGRI case, the research team wanted to further understand the perceptions of WATERAGRI stakeholders about the sustainability of CW implementation. The research team pondered: What criteria drive actors towards implementing CWs?

CWs have been shown to serve as a cost-effective solution for the removal of pollutants from wastewater (García-Herrero *et al.*, 2022). Moreover, they contribute to the enhancement of biodiversity and play a significant role in ameliorating habitat and ecosystem conditions. They therefore foster environmental sustainability, an aspect that was already seen as prominent in the WATERAGRI community. Although CWs offer co-benefits, they also entail trade-offs, particularly concerning the allocation of costs for implementation and maintenance. Farmers operate with narrow margins and thus need to invest into new technology conservatively.

Various sustainability aspects were identified as pertinent for the implementation of CWs through a comprehensive review of the literature, supplemented by expert consultations. Subsequently, these identified aspects were subjected to a Delphi-like exercise to solicit expert opinions and achieve consensus. A survey, conducted from 7th July to 26th September 2023, was utilized to gather responses from relevant stakeholders. To ensure broad stakeholder participation, the survey was disseminated through multiple channels, including WATERAGRI project platforms and professional networking interfaces such as LinkedIn.

A total of 48 respondents were engaged in the survey, among which only 22 completed the survey and agreed to the treatment of personal data: the results presented in this study are henceforth referred to this subsample. As far as gender distribution is concerned, 59% of the respondents identified as men and 41% as women. The respondents' geographical distribution was mainly focused on the Mediterranean (38%) and Continental (27%) zones, followed by the Boreal zone (14%). The Pannonian zone (3%) was the least covered. In total, 18% of the respondents selected another country outside the European continent. Due to the sampling strategy and self-selection bias, stakeholder coverage was unbalanced towards researchers, accounting for 68% of the total respondents. The remaining 32% of non-researchers included advisory services (17%), decision-makers (5%), farmers (5%), and others (5%). This distribution represents a limit of the study that could be addressed with more refined stakeholder engagement strategies.

14.3.2.1 Ranking sustainability dimensions

Initially, participants were tasked with prioritizing the triple bottom-line sustainability dimensions (i.e., environmental, social, and economic sustainability) in the ascending order of relevance. As anticipated, a significant majority of respondents (72%) identified environmental sustainability as the most pertinent dimension in CW implementation, trailed by economic (22%) and social (6%) considerations (Figure 14.5). Notably, none of the researcher identified social sustainability as the foremost dimension of importance. Moreover, social sustainability received primary ranking only once, from a respondent affiliated with advisory services. Interestingly, one-fourth of the researchers identified economic sustainability as the primary dimension of importance.

In the second phase, stakeholders were asked to assign a score the relevance of selected sustainability aspects for the implementation of CW in the rural context. Respondents could choose to rank each

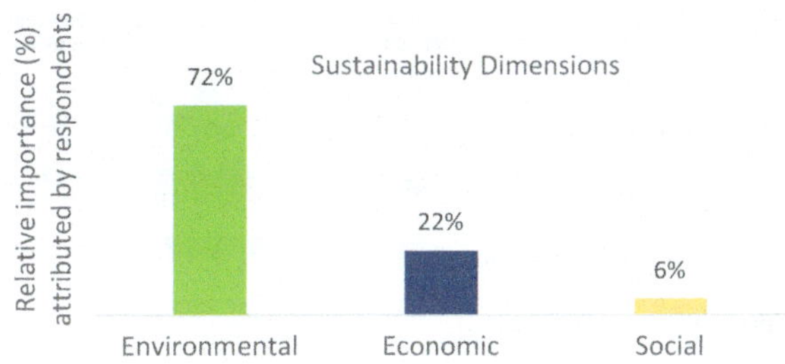

Figure 14.5 Relative importance (% of respondents ranked the dimension first) of sustainability dimensions.

sustainability aspect as having low (0–1), medium (1–2), or high relevance (2–3). Coherently with the literature evidence and the control question on the sustainability dimensions (see above), the respondents agreed on ranking the sustainability aspects related to environmental sustainability as the most relevant, while social aspects were overarchingly disregarded. However, as a general trend the sample showed a high degree of heterogeneity in the ranking (Figure 14.6).

14.3.2.1.1 Environmental aspects

The capacity to treat effluents is the only aspect that none of the stakeholder perceived as having a low relevance. More than 60% of the respondents agree on the high relevance of this aspect, and this percentage increases to 82% among researchers while dropping to 48% among non-researchers. Another important environmental aspect that was considered of high relevance is the water reuse potential. Although none of the researcher ranked water reuse relevance as low, around one-third of non-researchers attributed low relevance to this aspect. There are no significant differences among the two stakeholder groups regarding the perception of the relevance of the role of CW in ameliorating the ecosystem conditions, which is considered to be of low relevance by 28% of the respondents.

14.3.2.1.2 Economic aspects

As far as economic sustainability is concerned, more than 50% of the respondents agreed on ranking investment cost as a highly relevant aspect of CW sustainability. This aspect is considered highly relevant for researchers, while on average it was ranked as of low relevance by the rest of the respondents. This perspective is mirrored by the high relevance ranking of the need for a subsidy to implement a CW by non-researchers (48%), whereas only 18% of the researchers ranked this aspect's relevance as high. The lowest-ranked economic sustainability aspects are those related to the production of biomass from CW. In particular, the production of biomass itself is considered of low relevance by 67% of the respondents: this perspective is consistent with the low level of knowledge on CW biomass valorization opportunities evidenced by Avellán and Gremillion (2019).

14.3.2.1.3 Social aspects

Social aspects can be found at the lower end of the spectrum of relevance. As already hinted by the sustainability dimensions' control question, it was confirmed that these aspects are among the least known and considered. For instance, the improvement of the landscape quality is considered of low relevance by 57% of the non-researcher respondents, even if only 9% of researchers ranked this aspect low and most expressed a more balanced opinion. Similarly, potential social benefits are also perceived generally as of low relevance especially by non-researchers. All the non-researchers ranked the possibility to provide training to disadvantaged people as of low relevance, whereas 45% of researchers ranked this aspect as a highly or medium relevant aspect.

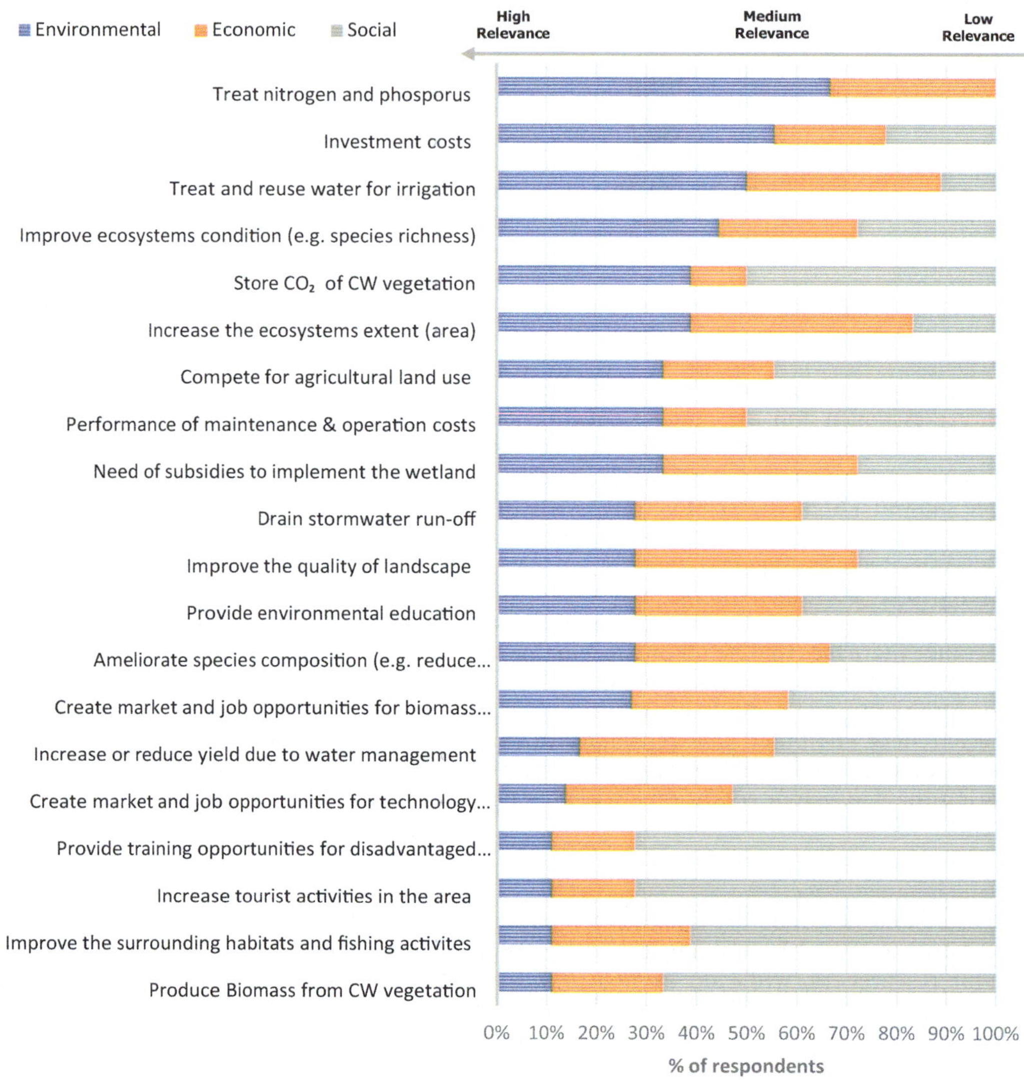

Figure 14.6 Gradient of the relevance of the perception of sustainability aspects according to respondents.

Testing the validity of these assumptions would require a more statistically significant sample, but this consultation offered a first overview of the possible trends, and some overarching conclusions can be drawn. Apart from the consensus on the nutrient retention capacity, there is a wide variability in responses, hinting that sustainability aspects might be perceived differently according to stakeholders' groups and context. Moreover, environmental aspects seem to be largely perceived as more relevant.

14.3.2.2 Impact of different sustainability aspects on different types of wetlands

In the last phase of the questionnaire, respondents were asked to rate their potential willingness to implement free water surface (FWS), horizontal flow (HF), and vertical flow (VF) CWs on a likelihood scale (ranging from very unlikely to very likely) considering the different sustainability aspects presented above. The top three sustainability aspects were then analysed and compared among the three different systems, as shown in Figure 14.7.

More than 60% of respondents responded that they would implement all three types of wetlands due to their nutrient retention capacity even though differences in the treatment capacity for different pollutants were not investigated in the survey. This perspective is mirrored by the likelihood of implementing the three systems for their water reuse potential: about 60% of the respondents would implement FWS and HF CWs for this function.

In the case of HF CWs, more than 50% of the respondents would implement this system considering maintenance and operation costs, which are generally lower than those of the other types of wetlands. Similarly, the investment costs would be an important theme in VF CWs, as more than 50% of the respondents would implement this infrastructure considering this aspect. Flood management scored among the lowest for HF CWs, consistently because these systems are feeble to drain stormwater runoff (Langergraber *et al.*, 2020).

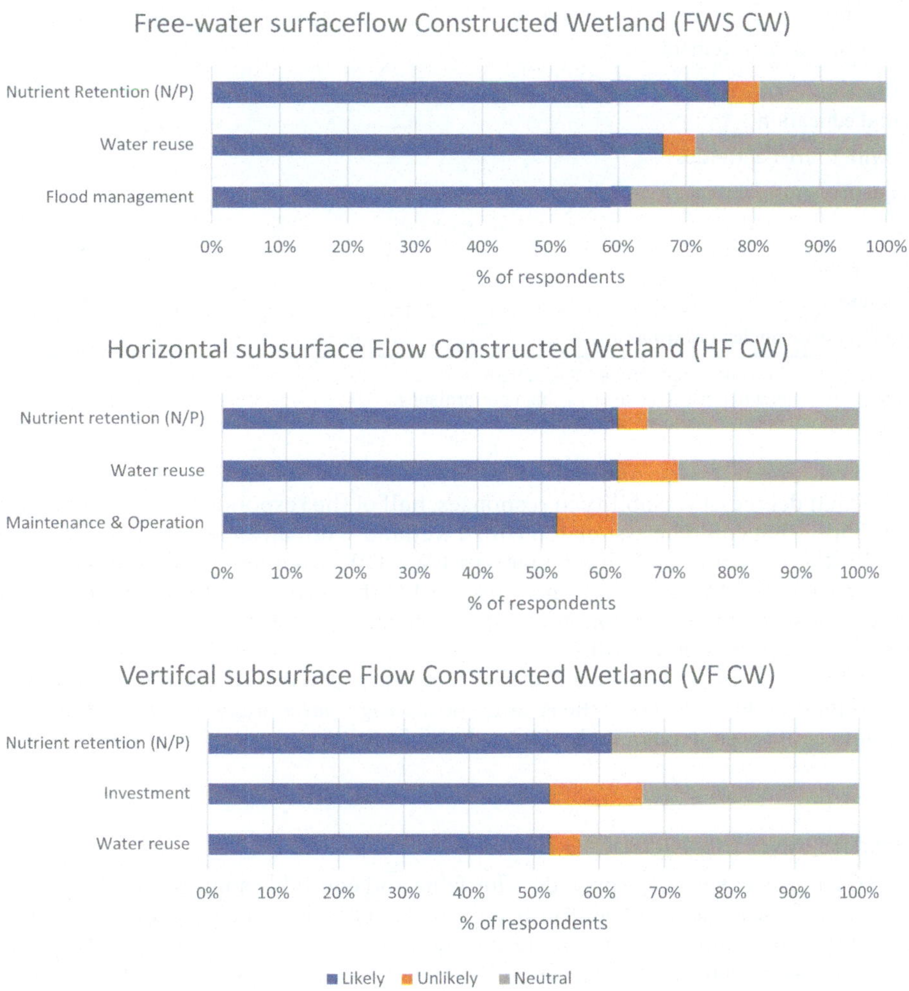

Figure 14.7 Three most relevant aspects to implement different type of wetlands (FWS, HF, VF).

Table 14.3 Comparison of requirements and performance of different types of CWs.

	FWS Wetlands	HF Wetlands	VF Wetlands
Required treatment area/land use	+++	++	+
O&M costs (OPEX)/maintenance & operation	+	++	++
Investment costs (CAPEX)	++	+	+
Removal of ammonia	++	+	+++
Removal of total nitrogen	++	++	+
Removal of phosphorus (long term)	+	+	+
Removal of dissolved organic matter	+	+	++
Removal of suspended solids	+	+	++
Removal of coliforms	+	++	++
Biomass production	+	++	++
Water retention/flood management	++	+	O
Water reuse	++	++	++
Environmental education	+++	+	+
Association with tourist activities	+++	+	+
Landscape quality	+++	++	++
CO_2 storage	++		
Ecosystem condition	+++	+	+
Ecosystem extent	+++	+	+
Enhance biodiversity/species composition	++	+	+

Source: From Dotro *et al.* (2017) and Langergraber *et al.* (2020).
+: low performance, ++: medium performance, +++: high performance.
O: not used for this aspect; blanks: unknown performance.

Despite the high degree of variability in responses, half of the respondents agree that an increase in the fishery activities that could be associated with a wetland is not an aspect they would consider when confronted with the opportunity to implement an FWS CW, whereas those percentages are around 40% for HF and VF CWs. As a general trend, one-third of the respondents would not implement these infrastructures considering socio-economic aspects such as the association with tourist activities, yield increase, and training opportunities.

As noted in Table 14.3, these aspects are also less investigated in the academic literature, except for biomass production. Furthermore, the results show a high percentage of neutral responses, hinting that these results could stem from a lack of knowledge in these sustainability aspects more than from an opposition.

14.4 SUMMARY AND CONCLUDING REMARKS

The WATERAGRI project set out to provide affordable and equitable solutions to the growing need to keep more water and nutrients in agricultural soils for better yields and lower environmental impacts. The sustainability assessments are one method to determine the sustainability of the WATERAGRI solutions to make sure that future sustainability is not impacted by their recommendations. Our analysis shows that many of the WATERAGRI solutions are not at a level of maturity to provide meaningful recommendations about their sustainability. Information about their environmental effects, their costs, and the perception of actors about their usefulness is largely missing.

Overall, solutions were found to be mostly sustainable in their environmental dimension, contributing either directly to increased yields, or indirectly to improved environmental conditions by reducing eutrophication and/or contributions to climate change. The solutions are also not sustainable in their economic dimension. They are currently too costly or do not exhibit a large enough return on investment either through additional benefits such as biomass or irrigation use from CWs. The social dimension is the weakest link in full sustainability assessments. The sLCA for CWs looked at an atypical CW which is not really used for farming conditions, but represents an experimental site.

It is interesting to note that the discourse in sustainability has been dominated by the environmental dimension of sustainability, and this consortium is no exception to this as we saw from the Delphi study by Dahal *et al.* (2023) and the stakeholder assessment on CWs. Our analyses show positive environmental effects for the solutions. Under traditional settings, this consortium may have gone ahead in recommending the use of these solutions. A more holistic picture that also includes the other dimensions of sustainability paints a different picture. Any of the solutions may be more sustainably suitable in other contexts and with more information and maturity.

Despite challenges for farmers due to high construction and operation costs, the analysis suggests that with initial subsidies and ongoing support, CW implementation could be financially viable. Additionally, integrating CWs into established value chains for bioenergy and bio-based products could enhance profitability. Upscaling CWs requires engagement strategies such as water-oriented living labs and water citizen science approaches to foster knowledge transfer and community involvement. Overall, although technical improvements do not enhance perceptions of elements for social innovation, economic support can enhance the economic sustainability of CW implementation, enabling effective utilization of the system's environmental and social benefits (Figure 14.8).

CWs, hence, represent nature-based solutions with significant potential, offering both challenges and opportunities within European agrarian systems. Challenges arise in integrating wetlands' functions into policies effectively, particularly within the agricultural sector, where a cohesive policy on water management is lacking. However, as classified under nature-based solutions, CWs provide green infrastructure that can be strategically integrated into regional policy frameworks, supporting environmental sustainability goals. Recognizing their role in improving water quality and offering diverse functions, CWs hold promise for providing sustainable solutions for wastewater treatment. By addressing challenges and embracing opportunities, policymakers can facilitate the integration of CWs into European agrarian systems, thereby enhancing environmental sustainability and agricultural landscape resilience.

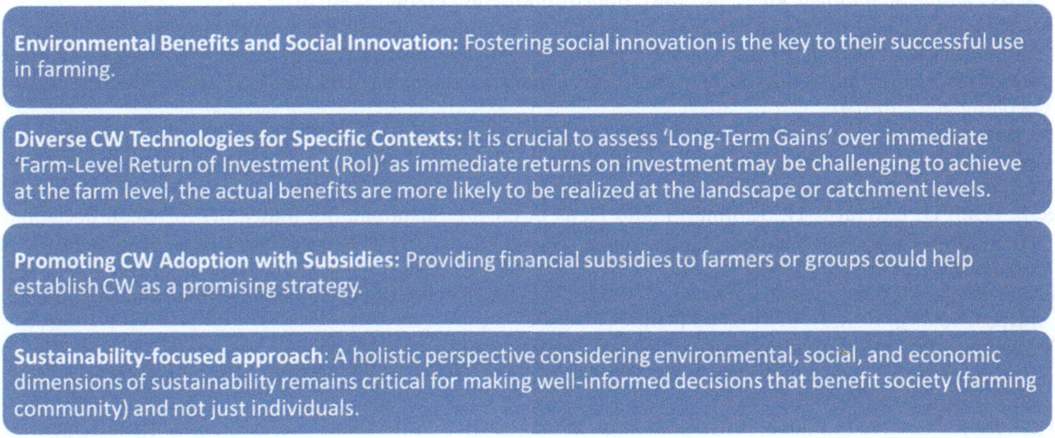

Figure 14.8 Main findings from the WATERAGRI studies on sustainable water and nutrient management practices in particular for CWs.

ACKNOWLEDGEMENTS

This project has received funding from the European Union's Horizon 2020 research and innovation programme under Grant Agreement No. 858375.

REFERENCES

Avellán T. and Gremillion P. (2019). Constructed wetlands for resource recovery in developing countries. *Renewable and Sustainable Energy Reviews*, **99**, 42–57, https://doi.org/10.1016/j.rser.2018.09.024

Bieroza M. Z., Bol R. and Glendell M. (2021). What is the deal with the green deal: will the new strategy help to improve European freshwater quality beyond the water framework directive? *Science of The Total Environment*, **791**, 148080, https://doi.org/10.1016/j.scitotenv.2021.148080

Boix-Fayos C. and de Vente J. (2023). Challenges and potential pathways towards sustainable agriculture within the European green deal. *Agricultural Systems*, **207**, 103634, https://doi.org/10.1016/j.agsy.2023.103634

Carty A., Scholz M., Heal K., Gouriveau F. and Mustafa A. (2008). The universal design, operation and maintenance guidelines for farm constructed wetlands (FCW) in temperate climates. *Bioresource Technology*, **99**(15), 6780–6792, https://doi.org/10.1016/j.biortech.2008.01.045

Cavalheri P. S., Mello M. A., Pereira A. L., Marques T. R., Moraes T. N., Cavazzana G. H. and Magalhães Filho F. J. C. (2022). Chapter 16 – Ecotoxicity evaluation of diclofenac potassium in vertical flow constructed wetlands as posttreatment of septic tank effluent. In: Circular Economy and Sustainability, A. Stefanakis and I. Nikolaou (eds), Elsevier, pp. 271–282, https://doi.org/10.1016/B978-0-12-821664-4.00028-5

Corominas L., Byrne D. M., Guest J. S., Hospido A., Roux P., Shaw A. and Short M. D. (2020). The application of life cycle assessment (LCA) to wastewater treatment: a best practice guide and critical review. *Water Research*, **184**, 116058, https://doi.org/10.1016/j.watres.2020.116058

Dahal B., Avellán T., Haghighi A. T. and Kløve B. (2023). Defining sustainability in agricultural water management using a Delphi survey technique. *Water Policy*, **25**(6), 597–621, https://doi.org/10.2166/wp.2023.057

Dotro G., Langergraber G., Molle P., Nivala J., Puigagut J., Stein O. and von Sperling M. (2017). Treatment Wetlands. Biological Wastewater Treatment Series. IWA Publishing, London, UK, Vol. **7**, p. 172. eISBN: 9781780408774, https://doi.org/10.2166/9781780408774

Dudley N. (2022). Impact of agriculture on insect species decline. In: Imperiled: The Encyclopedia of Conservation, A. Dominick, Sala D. and M. I. Goldstein (eds), Elsevier, Oxford, pp. 500–506, https://doi.org/10.1016/B978-0-12-821139-7.00238-5

European Environmental Agency (1999). The DPSIR Framework – European Environment Agency. https://www.eea.europa.eu/publications/92-9167-059-6-sum/page002.html (accessed 16 April 2024)

Garcia J., Rousseau D. P. L., Morato J., Lesage E. L. S., Matamoros V. and Bayona J. M. (2010). Contaminant removal processes in subsurface-flow constructed wetlands: a review. *Critical Reviews in Environmental Science and Technology*, **40**(7), 561–661, doi: 10.1080/10643380802471076

García-Herrero L., Lavrnić S., Guerrieri V., Toscano A., Milani M., Cirelli G. L. and Vittuari M. (2022). Cost-benefit of green infrastructures for water management: a sustainability assessment of full-scale constructed wetlands in northern and southern Italy. *Ecological Engineering*, **185**, 106797, https://doi.org/10.1016/j.ecoleng.2022.106797

Giannakis E., Kushta J., Giannadaki D., Georgiou G. K., Bruggeman A. and Lelieveld J. (2019). Exploring the economy-wide effects of agriculture on air quality and health: evidence from Europe. *Science of The Total Environment*, **663**, 889–900, https://doi.org/10.1016/j.scitotenv.2019.01.410

Hugé J., Trinh H. L., Hai P. H., Kuilman J. and Hens L. (2010). Sustainability indicators for clean development mechanism projects in Vietnam. *Environment, Development and Sustainability*, **12**(4), 561–571, https://doi.org/10.1007/s10668-009-9211-6

Hunkeler D., Lichtenvort K. and Rebitzer G. (2008). Environmental life cycle costing. Boca Raton: CRC Press, https://doi.org/10.1201/9781420054736

ISO (2006a). ISO 14040:2006, Environmental Management – Life Cycle Assessment – Principles and Framework. International Standards Organization. Geneva, Switzerland, https://www.iso.org/obp/ui/en/#iso:std:iso:14040:ed-2:v1:en

ISO (2006b). ISO 14044:2006b, Environmental Management – Life Cycle Assessment – Requirements and Guidelines. International Standard Organization. Geneva, Switzerland, https://www.iso.org/obp/ui/en/#iso:std:38498:en

Kadlec R. H. and Wallace S. (2008). Treatment Wetlands, 2nd edn, CRC Press, Boca Raton, USA.

Langergraber G., Dotro G., Nivala J., Rizzo A. and Stein O. (2020). Wetland Technology: Practical Information on the Design and Application of Treatment Wetlands. IWA Publishing, London, UK, https://doi.org/10.2166/9781789060171

Nan X., Lavrnić S., Mancuso G. and Toscano A. (2023). Effects of design and operational conditions on the performance of constructed wetlands for agricultural pollution control – critical review. *Water, Air, & Soil Pollution*, **234**(7), 434, https://doi.org/10.1007/s11270-023-06380-y

Schäfer R. B., Caquet T., Siimes K., Mueller R., Lagadic L. and Liess M. (2007). Effects of pesticides on community structure and ecosystem functions in agricultural streams of three biogeographical regions in Europe. *Science of The Total Environment*, **382**(2), 272–285, https://doi.org/10.1016/j.scitotenv.2007.04.040

Soldo C., Wilson R. S., Walpole H. and Shaffer-Morrison C. D. (2022). Farmer willingness to implement constructed wetlands in the Western Lake Erie Basin. *Journal of Environmental Management*, **321**, 115928, https://doi.org/10.1016/j.jenvman.2022.115928

Turcios A. E., Miglio R., Vela R., Sánchez G., Bergier T., Włodyka-Bergier A., Cifuentes J. I., Pignataro G., Avellan T. and Papenbrock J. (2021). From natural habitats to successful application – role of halophytes in the treatment of saline wastewater in constructed wetlands with a focus on Latin America. *Environmental and Experimental Botany*, **190**, 104583, https://doi.org/10.1016/j.envexpbot.2021.104583

UNEP, Benoît Norris C., Traverso M., Neugebauer S., Ekener E., Schaubroeck T., Russo Garrido S., Berger M., Valdivia S., Lehmann A., Finkbeiner M. and Arcese G. (2020). Guidelines for social life cycle assessment of products and organizations 2020. *United Nations Environment Programme (UNEP)*, p. 140, http://www.unep.fr/shared/publications/pdf/DTIx1164xPA-guidelines_sLCA.pdf

UNU-CRIS (2024). A Regional Overview of Gender Sensitive Dimensions in the European Agricultural Sector. https://cris.unu.edu/regional-overview-gender-sensitive-dimensions-european-agricultural-sector (accessed 16 April 2024)

Vymazal J. (2011). Constructed wetlands for wastewater treatment: five decades of experience. *Environmental Science & Technology*, **45**(1), 61–69, https://doi.org/10.1021/es101403q

WATERAGRI (2023). Deliverable 6.3: Assessment of Sustainability.

WCED (1987). Our Common Future. Report of the World Commission on Environment and Development (WCED). United Nations, New York, https://sustainabledevelopment.un.org/milestones/wced

doi: 10.2166/9781789065015_0289

Chapter 15

Life-cycle assessment framework applied to sustainable integrated aquaculture systems for urban living

*Azalea D. M. Satya[1], Awalina Satya[2], Ika A. Satya[3], Sara K. Yazdi[1], Pau-Loke Show[4], Tjandra Chrismadha[2] and Tjandra Setiadi[5]**

[1]Department of Chemical and Environmental Engineering, Faculty of Science and Engineering, University of Nottingham Malaysia, Jalan Broga, Semenyih 43500, Selangor Darul Ehsan, Malaysia
[2]Research Centre for Limnology and Water Resources, National Research and Innovation Agency-Rep. of Indonesia (BRIN), KST Soekarno, Jl. Raya Bogor-Jakarta km46, Cibinong, Bogor 16911, Indonesia
[3]Research Centre for Computation, National Research and Innovation Agency-Rep. of Indonesia (BRIN), Komplek Cisitu, Jl. Sangkuriang, Coblong, Bandung 40135, Indonesia
[4]Department of Chemical Engineering, Khalifa University, Shakhbout Bin Sultan St, Zone 1, Abu Dhabi, United Arab Emirates
[5]Department of Chemical Engineering, Faculty of Industrial Technology, Institut Teknologi Bandung, Jl. Ganesa No. 10, Bandung 40132, Indonesia
**Corresponding author: tjandra.setiadi@itb.ac.id*

ABSTRACT

The acceleration of economic and urban population growth leads to a higher demand for water due to the inefficiency of public water supply, sanitation, and climate change that exasperated management of water resources. The employment of low-cost and low environmental impact technologies is of great value as part of nature-based solutions (NBSs) applicable for the growing number of densely populated areas. This chapter includes recent case studies and trends of integrated multitrophic aquaculture in a scientific and methodical approach which will allow an overall framework for the application of these technologies as part of NBSs in urban wastewater solutions. In ensuring the position of the technological approaches as sustainable solutions, a life-cycle assessment for environmental impact must be conducted.

Keywords: Environmental assessment, IMTA, integrated multitrophic aquaculture systems, LCA, LCIA.

15.1 INTRODUCTION

Fishing has long been part of human's history in actively maintaining a resilient food supply system. The practice predates on two separate occasions as early as 4000 BC in China and 500 BC in Rome (Little, 2003). Both practiced aquaculture in its simplest form by utilizing natural resources such as ponds, lakes, and lagoons. Although different in nature, their practices remained largely unchanged until the 20th century. Unfortunately, due to climate change and fast growth in population foregoing the environmental impacts and unsustainable practices, global marine life has been reduced by half in the last 40 years with 90% of commonly captured fish species have disappeared from the ocean

© 2025 IWAP. This is an Open Access book chapter distributed under the terms of the Creative Commons Attribution License (CC BY-NC-ND 4.0) which permits copying and redistribution for non-commercial purposes with no derivatives, provided the work is properly cited (https://creativecommons.org/licenses/by-nc-nd/4.0/). The chapter is from the book *Nature-Based Solutions for Urban Sustainability*, Piet N.L. Lens and Xuan-Thanh Bui (Editors).

since 1950 (Nadarajah & Flaaten, 2017; WWF, 2021). In addition, environmental degradation due to poor management of waste has caused considerable changes that polluted wild fish with plastics and higher amounts of heavy metals due to its food source (plankton) being contaminated. This effect has also been reported to be a problem in freshwater fisheries which has reported a decline in biodiversity twice the rate of that in oceans (WWF, 2021). Aquaculture is the frontier build to overcome this biodiversity loss as well as food demand as fish are cultivated in a controlled manner. Aquaculture production has considerably grown over recent decades, showing that it has predominantly contributed to meeting food demands from the doubled global per capita fish consumption since 1960 (Mair *et al.*, 2023). Total aquaculture production in 2021 attained 126 million tonnes live weight, largely produced in Asia accounting to 91.6% of the global production (FAO, 2022). However, growth of aquaculture production is on a brink of slowing down due to the increasing feed costs and impact of policy changes in addressing some concerns of high fish producers (Naylor *et al.*, 2021). The concerns of aquaculture's negative impacts revolved around the irreversible damages it caused resulting from inadequate management such as conversion of mangrove forests or natural ecosystems for profitable fish types, abandonment of fish farms resulting in salinization, improper dealing with the generated waste causing deterioration of nearby water bodies resulting in unsafe water for consumption, eutrophication of effluent receiving ecosystems, ecological impacts caused by medication practices, ecological impacts on the natural ecosystems and of exotic species due to changes in landscape and hydrological patterns, and collapse of fisheries as fishmeal supplies for fish production increased in the area (Martinez-Porchas & Martinez-Cordova, 2012).

It is also noted that the traditional way of culturing is often carried out in a high culture density manner to keep up with major demand resulted in a build-up of waste (Stentiford *et al.*, 2017). This contributes to fast spreading or more virulent pathogens and occurrence of disease outbreaks especially with the limited gene pool (Reverter *et al.*, 2020). Yet addressing food supply problems for urban living requires aquaculture or food production systems to be developed near urban centres as transport of fresh food production has turned rather costly in recent decades. Local food production has also positive impact on the environment as it promotes the local economy and strengthens social development in cities. Freshwater aquaculture however requires large land areas, averaging around 0.25 ha necessary for direct support and another 3.51 ha for plant meals for feed or higher depending on the kind of fish feed chosen, that ensure high production and avoid fast spread of pathologic problems within the ponds that increase the environmental costs (Heimann & Delzeit, 2024; Reverter *et al.*, 2021).

In maintaining a sustainable aquaculture, valorization of waste must be integrated with fish cultivation. Integrated multitrophic aquaculture systems (commonly known as IMTA) are considered as current sustainable solution as it mimics the natural cycle of nutrients by including a minimum of two trophic cultures (see Chapter 11). The system relies on the symbiotic relationship between the two cultures involved with the most prevalent example being rice and fish farming. Integrated aquaculture has been employed for centuries, primarily in Asia by utilizing rice fields for carp farming. The rice fields function as a protective system to fish by filtering potential hazardous compounds as well as providing habitat or protection for the smaller fish added to a aquaculture–crop integrated system. Aquaculture can also be run with livestock farming where the incorporated livestock serves as a source of nutrients for aquaculture. Aquaculture can also be integrated with vegetable farming where each fish growth cycle's production of muck in ponds is utilized to fertilize plants (Hasimuna *et al.*, 2023). In all instances, a pond is used *de facto* and built relatively close to the house where domestic and cooking wastes can be drained into the pond. In urban living, the idea of integrated aquaculture systems is found to be the most prospective business model and reportedly secured economical needs during the COVID-19 pandemic (Haryanti *et al.*, 2023). It also adds benefits to the surrounding urban communities and provides a resilient food source. Building of IMTA, however, is not an easy fit and can be expensive especially in an urban setting. Furthermore, resilience and resource-efficiency become increasingly imperative but building one is not as easy as picking the cheapest options or betting on the available natural resources to serve their roles efficiently. Within this context, the study of environmental impacts, now known as life-cycle assessment (LCA), stands as a valuable tool by evaluating the environmental footprint of aquaculture operations from cradle-to-gate, offering insights into a holistic perspective in attaining a sustainable practice.

Ensuring quality of life with LCA only is not enough. For an LCA to be built, there must be a design or system to assess. Among all the aquaculture systems, IMTA could offer more than just securing food. Knowing the natural capability of nature in governing itself, adding ecosystem services (ESs) into IMTA can lead to a beneficial support not only to the environment but also to the surrounding communities (Le Gouvello *et al.*, 2022). This ESs encompassed into an action term known as nature-based solutions (NBSs). Essentially, IMTA is part of NBSs as it is the mean to return aquaculture to its most natural ecosystem by providing means of recycling waste or closing the loop to make a sustainable process. As the fastest growing livestock activity, aquaculture combining with hydroponic vegetable crops, a common type of IMTA, has long been run in one loop layout. This helps to enhance and improve urban resilience as well as diversifying resilience in food security. However, it is needed to emphasize that its function is not only to support food self-sufficiency of urban communities but also the multifunctional values ranging from cultural, regulation, and provisioning (Russo & Cirella, 2023). The practices of urban IMTA should help promote fish as an affordable alternative source of protein-promoting regimes that could help a productive harvest with better financial returns. Thus, in management practice, there should be a keen focus on developing better feeds, harvesting smaller fish, and preparing larger fingerlings for future cycles. Evaluating IMTA systems by using an LCA approach has been formally introduced in 2009 (Lazard *et al.*, 2009), however the term was not familiarized. The formal publication of LCA studies of IMTA is still quite limited.

According to the International Union for Conservation of Nature (IUCN), actions in protecting, managing, and restoring natural or modified ecosystems are needed to address societal challenges effectively and adaptively, while providing human well-being and biodiversity (IUCN, 2022). Although it started as an umbrella term, an NBS is formally considered as an active action by abiding to the eight criteria which include societal challenges, design at scale, biodiversity net-gain, economic feasibility, inclusive governance, balance trade-off, adaptive management, and mainstreaming sustainability (IUCN, 2020). An LCA in NBSs can provide detailed information on the land use and the land-use changes associated with different regions and functional units (FUs) for a production system. LCA can also be used to assess environmental benefits of different solutions or designs originating from applied NBSs in ensuring environmentally sustainable and effective ways to deal with urban challenges. This means LCA works beyond the traditional scope of urban NBSs in which the service is not only focused on providing citizens with supply of food, but also to include the generated products over the entire life cycle of NBSs starting from raw material extraction, through the processing, production, distribution, and use stages, to end-of-life management (Larrey-Lassalle *et al.*, 2022).

Drawing upon interdisciplinary insights from environmental science, aquaculture management, urban planning, and sustainability studies, this chapter endeavours to contribute to the ever-growing knowledge surrounding sustainable application of IMTA systems. Through the development of a robust LCA framework tailored specifically for urban aquaculture, this chapter aims to inform policymakers, industry stakeholders, and urban planners in their efforts to promote resilient, environmentally sound, and socially equitable food systems for urban living. In the subsequent sections, we will delve into the theoretical underpinnings of LCA, explore the current state of urban aquaculture initiatives, and present our proposed framework for assessing the sustainability of integrated aquaculture systems in urban environments. Through empirical case studies and comparative analyses, we will illustrate the practical application of the LCA framework, highlighting its utility in guiding decision-making processes and fostering the development of more sustainable urban food systems.

15.2 BASIC APPROACH OF LCA FOR TECHNOLOGICAL BENCHMARKING IN APPLIED NBSS AS IMTA

The extensive application of LCA in aquaculture and fishery systems has long been carried out typically to compare different production techniques of one species and to assess 'hotspots' or main contributing activities of the species' total impact. The strongest aspects of having an LCA applied for aquaculture are to identify environmental trade-offs among the alternatives of aquaculture systems as well as identifying the impact of shifting environmental effects of feeding wild caught fish to farmed fish in comparison to feeding agricultural products (Pelletier & Tyedmers, 2008). In building an LCA for NBS technology in IMTA, it is important to understand the current understanding of both concepts and technological benchmarking. Benchmarking, as defined in ISO 21678:2020, is a step-by-step approach to collecting, analysing, and relating performance data of comparable buildings or other types of construction works. This process has its origin in the business management field and its core is linked to comparative procedures and improvements from knowledge acquired (Galindro *et al.*, 2019). LCA is generally closely linked to ESs which are essential for life-sustaining benefits to the urban biosphere. According to Larrey-Lassalle *et al.* (2022) in applying LCA for NBSs, harmonized LCA guidelines for all NBSs are not feasible; however, the following common methodological challenges and research were identified:

(1) Defining the FU is extremely crucial to include all the systems involved. To define a multifunction FU, an innovative approach by shifting ESs into the scope phase or early on in the process rather than defining it in the inventory analysis part of an LCA.
(2) In approaching the multifunctionality of NBSs, LCA methodology to be used must compute either eco-efficiency ratios or allocate procedures based on a systems' capacity in delivering ESs.
(3) In improving current LCA studies, emissions from field application must be considered particularly on usage of organic/mineral fertilizers or even pesticides in systems. The mass balance on chemical elements, especially, must be added to avoid uncertainty in input or output flows. These chemical elements should particularly focus on nitrogen (N), carbon (C), and phosphorus (P).
(4) LCA for NBSs can be carried out at different scales and adapted to the decision maker. This means there should be different perspectives used between local and global impacts on the one hand and impacts on-site and off-site must be facilitated on the other.

There are different types of LCA that need to be considered. Attributional LCA (ALCA) is the most used approach in LCA studies. It involves quantifying the environmental impacts associated with a product or process throughout its life cycle. In the context of sustainable aquaculture, ALCA can be used to assess the environmental burdens of various aquaculture technologies, including fish farming systems, feed production, energy consumption, waste management, and transportation. By conducting ALCA, researchers can identify the key hotspots of environmental impact within aquaculture systems and compare the environmental performance of different technologies. Consequential LCA (CLCA) goes beyond the direct environmental impacts of a product or process to consider the broader system-wide effects of changes in demand or production. In the context of sustainable aquaculture, CLCA can be used to assess the indirect environmental consequences of adopting different aquaculture technologies. For example, CLCA can help evaluate the impacts of changes in fish farming practices on land and water use, habitat destruction, biodiversity loss, and greenhouse gas emissions. By incorporating CLCA into technological benchmarking studies, researchers can assess the full life-cycle environmental implications of different aquaculture technologies and inform decision-making processes accordingly. Prospective LCA, also known as anticipatory LCA or foresight LCA, involves assessing the potential environmental impacts of emerging technologies or scenarios that have not yet been implemented.

In the context of sustainable aquaculture, prospective LCA can be used to evaluate the environmental consequences of future technological advancements, policy interventions, or market trends. By conducting prospective LCA, researchers can anticipate the environmental implications

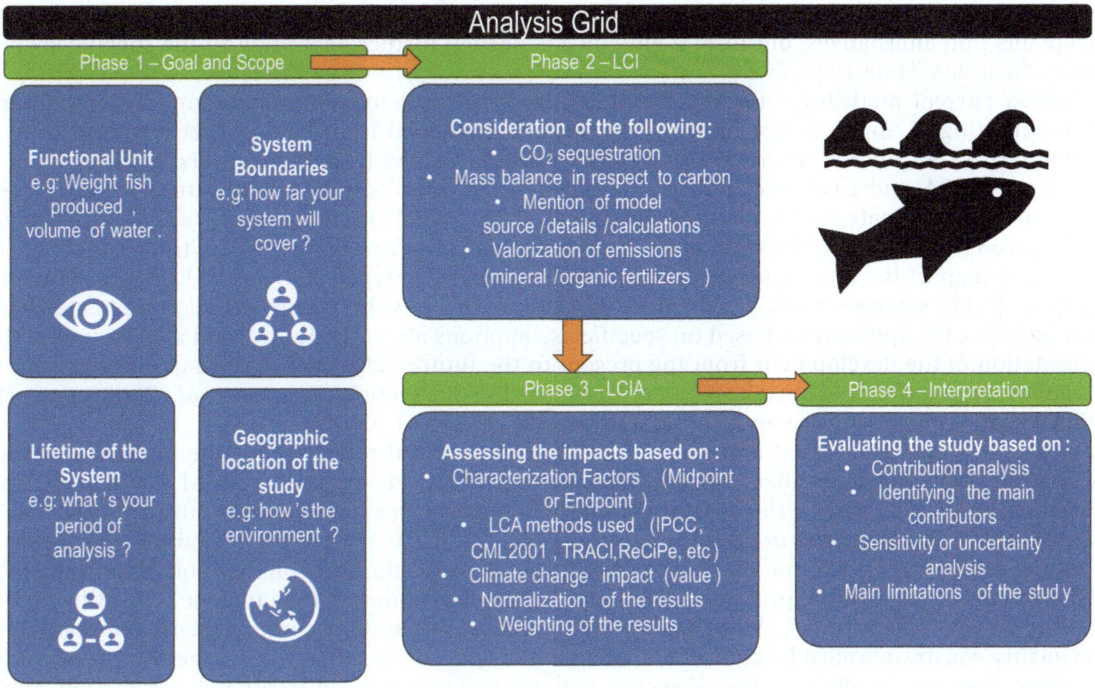

Figure 15.1 Criteria description and analysis steps necessary for building an LCA for NBSs.

of adopting new aquaculture technologies or implementing sustainability initiatives, thereby guiding research and development efforts towards more environmentally benign solutions (Steubing *et al.*, 2023). These basic approaches to LCA provide valuable tools for benchmarking the environmental performance of different aquaculture technologies and systems. By employing ALCA, CLCA, and prospective LCA methodologies, researchers can gain insights into the environmental implications of various technological choices and inform decision-making processes towards more sustainable aquaculture practices. As in selecting LCA, a set of criteria can be considered from the standard (ISO 14040:2006), ILCD guidelines, and Environmental footprints, and from specific challenges associated with NBS application as shown in Figure 15.1.

15.3 GOAL AND SCOPE OF LCA

The goal and scope definition of LCA is ultimately the most critical stage that determines the outcome of a study. The goal can be varied depending on what is specifically needed. The goal expresses what type of FU would be considered. As for scope, it must define the outreach of the studied system such as boundary, FU, and assumptions made in the study. In assessing IMTA's life cycle, the goal mostly focuses on either quantifying environmental impacts of the system, identifying hotspots in the process, and comparison studies of IMTA to monoculture production in multiple scenarios or settings. The scope of LCA in IMTA meanwhile is largely dominated with the cradle-to-gate approach where harvesting processes are the end gate, for example in farms. Generally, the process is depicted as a process flow diagram that can be utilized to identify hotspots. Hotspots are needed to determine the significant contributor(s) to the system's environmental impacts. In an LCA, there should be at least more than one impact assessed as stated in the respective ISO quality standards. From the assessment, the identified significant contributors can then be explored further by conducting scenario analysis. Scenario analysis helps decision-making by providing trade-offs amidst various potential viewpoints and alternatives, but formal guidance is needed further according to the specific scopes and applications (Hala *et al.*, 2024).

Amidst current modelling of scenarios, there are three archetypal combinations in preliminary LCA consisting of 'input' where future scenarios are developed from key aspects retrieved from a preliminary LCA, 'hybrid' where future scenarios are developed from key aspects resulting from a preliminary LCA and evaluated with a subsequent LCA, and 'output' where future scenarios are developed and evaluated with an LCA (Bisinella *et al.*, 2021). Predominantly, as an explorative, model-based technology, IMTA would often be conducted in a hybrid setting with a time fixed horizon which can support the return period of the applied system. The types of 'hybrids' LCA are shown in Figure 15.2. The scenario itself is defined as 'description of a possible future situation that is relevant for a specific LCA application, based on specific assumptions about the future, and also including the presentation of the development from the present to the future'. This does not mean guessing, but is an important question that must be explicitly described in this section (Pesonen *et al.*, 2000). A poorly defined goal, scope, and FUs can result in a low quality of the LCA.

In determining a clear goal and scope, a primary purpose of reference for the input and output of the system must be normalized mathematically. For IMTA, FU is closely related to the scope and allocation method as well as the critical comparison between species or across cultures (Beltran & Guinée, 2014; Henriksson *et al.*, 2014a, 2014b). In IMTA, FUs that have been used include: 1 tonne of mixed fish products, 1 tonne of protein content of the products, and combination of multiple FUs or the economic value of the products (Hala *et al.*, 2024). Depending on the type of LCA utilized, the FU utilized would be different. As for example, in an ALCA of a system, comprehensive temporal data and quality constraints must be considered. Meanwhile, in a comparative LCA, a specific FU might be a problem because, ideally, the studied system will need an extent of similarity to be compared. Table 15.1 shows some reviewed studies of LCA carried out on IMTA systems (Hala *et al.*, 2024).

A proposed uniform and consistent FU for aquaculture practices as nutritional criteria (protein and energy content) has been the consensus in the LCA food community. Nevertheless, the multifunctional nature of IMTA challenges this conformity as by-products of the system can be viewed much more impactful through the economical lenses. Allocation, meanwhile, stood for the partitioning of input and output processes or product system to reflect the environmental burden. Allocation in IMTA can be challenging due to its multifunctionality. According to ISO 14040:2006, LCA's guidelines

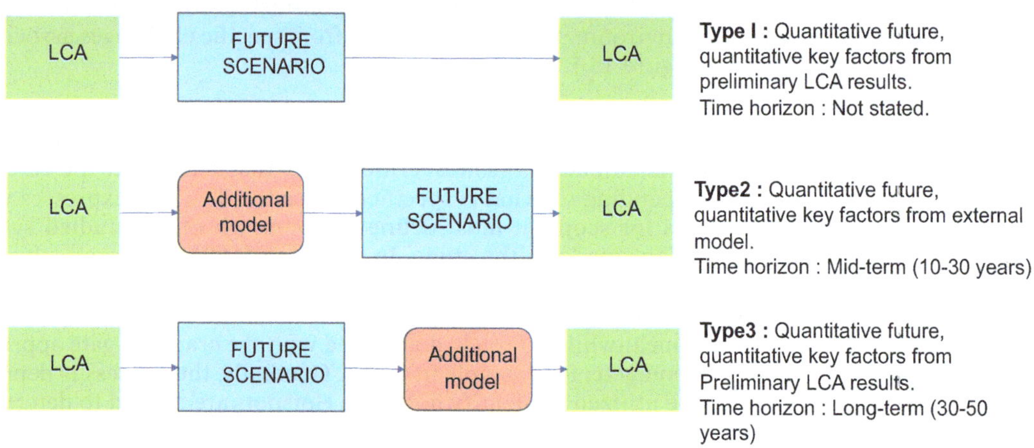

Figure 15.2 Explorative LCA modelling approaches based on explorative studies (*source*: Bisinella *et al.*, 2021).

Table 15.1 Pattern and relevance of LCA of IMTA practice applied at urban scale.

Study Location –Source	Scope	FU	Software	Database	Types of IMTA	Degree of Intensification
Mekong Delta – Phong *et al.* (2011)	Cradle-to-gate (farm)	1 kg farm product/ farm 1 kcal of farm products	Excel	Not mentioned	Agriculture–aquaculture (rice-based with high input fish, rice-based with medium input fish, and orchard-based with low input fish showed no difference for total land use and total energy used per kcal of farm product).	Extensive
Mekong Delta – Kluts *et al.* (2012)	Cradle-to-gate (farm)	1 kg/ tonne fish	SimaPRO	EcoInvent, LCA Food DK	Integrated aquaculture–agriculture (rice-based with high input fish culture and rice-based and medium input fish culture, rice input was pig manure).	Intensive
A Greenhouse in Switzerland – Cohen *et al.* (2018)	Cradle-to-gate (farm)	1 tonne of tilapia and 5 tonnes of lettuce	SimaPRO	EcoInvent, Agri-footprint, Industry Data, USLCI	Aquaponics (lettuce production with *Tilapia* fish grown in tank (closed system) vs pond system and lake system adapted from previous studies).	Intensive

Source: Hala *et al.* (2024).

for allocation are as follows: (1) to avoid allocation whenever possible by dividing the unit process into two or more sub-processes that are associated directly with multiproducts or by expanding the product system to include the additional functions related to the co-products; (2) performing allocations based on the physical relationship of the different products; and (3) the allocation should reflect other relationships between the different products if the physical allocation cannot be used as basis, such as opted to base on economic value of the product. For IMTA, the majority of LCA uses either mass, nutritional value, or economic value to allocate the system's impact. The summarized allocation factor on IMTA is shown in Tables 15.1 and 15.2. However, there are few who used the sub-division approach instead by separating the production processes between different topics. For high volume/low market value cases, mass and energy-based locations may not be the best representation as fishermen would exert greater effort economically, making economic allocation become more prevalent. In determining the proper allocation effort, it is best to conduct a sensitivity analysis with two or more allocation methods, thus providing a deeper insight. In this case, having several FUs to compare might be the best course in iterative LCA processes.

Table 15.2 Most commonly used bases for allocation and their characteristics (Henriksson *et al.*, 2013).

Allocation Factor	Accessibility	Relationship	Market Oriented
Mass	Good	Yes	No
Value	Average	No	Yes
Nutritional energy content	Average	Yes	No
System expansion	Poor	Sometimes	Sometimes

The scope consisted of identifying the system boundaries or how far the system will be assessed. In addition to scope, aquaculture can also benefit from the inclusion of temporal scope. Temporal scope provided by defining questions and FU together by considering evolution and comparing products and explicitly defining the validity of an LCA study because consumer behaviour and markets change rapidly. In making a dynamic LCA, there are data quality requirements (DQRs) which are provided in Table 15.3. This matrix is needed to deal with temporal issues in LCA-based studies.

Table 15.3 DQRs to consider in building a dynamic LCA.

	DQRs as Dynamic FUs		
	Definition	**Function**	**Example**
Age of data	Metadata of datasets informing age and minimum length of time for data collection.	Providing insights on temporal scope of a system model representing human activities.	Description of solar energy installations from 1990 would probably still be relevant for LCA of solar energy before 2000.
Technology coverage	Technology coverage that ensures qualitative assessment of temporal representativeness.	Although relative to the sector, this can be important to determine levels of complexity to apply in a prospective LCA study.	In EcoInvent, there are five levels (i.e., outdated, old, current, modern, and new) of technology to represent transforming activities.
Source of data	Choosing data and qualitative assessment provide indirect assessment of temporal scope for the modelled system. In EcoInvent, the inclusion of date of generation, date of review, and the period of validity with a start and end date for any dataset.	To avoid false sense of security from LCA by doing database updates that follow market shares or evolution of technology.	Taking precise temporal information from current market case in comparison to off-period journals.
Uncertainty description	Other indirect sources of information to clarify the temporal scope and validity period associated with flows.	Provides a quantitative assessment of the discrepancy between the time when data were acquired and the intended temporal scope for the dataset.	A study of product flow was indicated with temporal indicator of 3 which means their value was collected between 6 and 9 years before or after the target time frame for a dataset. This means the precision is low, limiting its applicability.

Source: Beloin-Saint-Pierre *et al.* (2020).

15.4 LIFE-CYCLE INVENTORY

The next step is to create a comprehensive inventory of the aquaculture's system; data gathering must be carried out on the system's operations. These operations include the inputs and outputs that affect the environmental performance in relation to the FU chosen. This includes the raw materials used, energy consumption, water usage, and waste generation of any type. In addition to inputs, outputs (e.g., fish biomass, waste, and emissions) associated with the aquaculture system must also be added into the flow. In building an inventory, the data consist of two types: primary and secondary. Primary data are usually collected from the location shortlisted for study for the systems studied, whereas secondary data originate from supported or available standard databases. Usually, standard datasets can be collected from EcoInvent (https://www.ecoinvent.ch/) but it is quite important to contact local authorities for obtaining national standards. If the aquaculture system is working on comparison and would like to include damage caused by the conventional system or a way to prevent problems in the future, it would be best to investigate eco-costs. Averaging of data is commonly carried out when building life-cycle inventory (LCI) thus resulting in quantitative uncertainties. Usually, this uncertainty at the unit process level is largely generated using a numerical unit spread assessment pedigree approach that disregards inaccurate measurements and variability around the means. A method of weighting was then used to estimate the characteristics of uncertainties for the secondary data. It is also important to average the unit process data using the horizontal averaging methodology that includes dispersion estimates around inventory movements with larger dispersion necessitating a more diverse approach (Koniyo, 2022). This adopts inherent uncertainty, spread, and unrepresentativeness as a fundamental choice for propagating uncertainties such as Monte Carlo analysis or Latin hypercube, later known as the Henriksson protocol which can be accessed as a spreadsheet at: https://www.universiteitleiden.nl/en/research/research-output/science/cml-quantitative-uncertainty-in-lci (Henriksson *et al.*, 2014a, 2014b). The type of data uncertainty in aquaculture is provided in Table 15.4.

In fully utilizing urban space, microalgae cultivation can be encouraged as an addition to IMTA as a sustainable approach of providing a circulatory system. This is due to aquaculture systems usually resulting in high load of nutrients. Microalgae biomass as a unit in circular bioeconomy is quite versatile from their nutrient recoverable as well as by-products perspective (Satya *et al.*, 2023). In addition, microalgae can also be cultivated in a relatively small space in the form of photobioreactors (Chrismadha *et al.*, 2022; Satya *et al.*, 2021, 2024). These by-products are often high in antioxidants

Table 15.4 Types of data uncertainty to be considered in aquaculture case studies as exemplified in *Pangasius* aquaculture.

	Uncertainty		
	Inherent	**Unrepresentativeness**	**Spread**
Definition	Inaccuracies in measurement	Mismatch between the representativeness and use of data	Variability resulting from horizontal averaging
Examples specifically for aquaculture	Inaccuracies in foreground data measurements such as quantities in water, feed, and chemicals used throughout the process	Mismatch in respect to completeness, reliability, temporal correlation, geographical correlation, and technological correlation	Variability in foreground data among the studied aquaculture farms
How is this covered in the foreground system?	Coefficient variation assumed (5% according to Henriksson's protocol)	By the pedigree matrix approach	Coefficient of variation was calculated based on-site foreground data

Source: Nhu *et al.* (2016).

Table 15.5 Some of the summarized LCA nutrient input data for production of 1 kg of dry algal biomass and the types of fertilizers applied and the produced lipid from the cultivation.

Species	N (g/kg DW)	P (g/kg DW)	K (g/kg DW)	Lipid (g/kg DW)	Energy Used (MJ/kg DW)
Chlorella vulgaris	Ammonia – 10.9	SSP – 2.4	Potassium chloride – 2	175	19.3
C. vulgaris	Calcium nitrate – 46	SSP – 9.9	Potassium chloride – 8.2	500	25.8
C. vulgaris	Ammonium nitrate – 24	TSP – 2.2	–	40	–
Generic freshwater algae	Urea – 73	SSP – 19.4	–	–	24
Nannochloropsis sp.	150	20	–	250	–
Nannochloropsis sp.	Ammonium nitrate – 82	Calcium phosphate – 7	–	29	14
Nannochloropsis sp.	70	10	10	20.1	80

that help promote growth as fish feed and even as a valuable nutrient (Trimbee & Prepas, 1987). Mayers *et al.* (2016) summarized the following LCA nutrient input data in producing 1 kg of dry algal biomass that can be incorporated in urban LCAs that include microalgae in the system (Table 15.5). Mayers *et al.*'s (2016) study was also equipped with CO_2 usage based on EcoInvent, thoroughly recommended as a future reference.

According to Henriksson *et al.* (2017), in obtaining primary data, there should be differed categories as follows: (i) timing of survey and (ii) the data characteristics of some important flows such as feed types, water input, and nutrient emissions. The groups then require separate analysis and are compared with certified group which are later horizontally averaged and quantified for uncertainties (Henriksson *et al.*, 2014a, 2014b). In evaluating the significance of innovation regarding the sustainable growth of aquaculture applied for fewer ecological impacts, secondary data must be consulted. The consulted sources must be available to collect data on resource use, energy consumption, material flows, and emissions throughout the entire life cycle of the system, from raw material extraction to product disposal as secondary data or, usually, this can be obtained from globally monitored data of several sources such as the United Nations Environment Program and World Conservation Monitoring Centre (UNEP-WCMC) 2019, Global Forest Watch 2018, and UNEP-WCMC 2014. Land utilization must also be considered to find the equivalent of emission. For example, for integration of the LCA framework in aquaculture growth in Indonesia, concurrent classification of the states of seagrass beds, corals, and mangrove forests were needed from the Sub Directorate of Environment Statistics (2020), Assembling Maps and BPS (Central Bureau of Statistic Indonesia). Through that, a lifetime assessment can be based for freshwater ponds consisting of different trophic levels.

For background data from IMTA LCA, it is most commonly obtained from Agribalyse and EcoInvent. Both databases are available online but EcoInvent has a much more broader sector ranging around 18,000 LCI datasets. Hala *et al.* (2024) have analysed a number of entries associated with aquaculture from both databases. It is shown that in both databases, the number of database entries for aquaculture had always been shown to be lower compared to the agricultural sector and other animal production, contributing only to 7% of total entries.

15.5 LIFE-CYCLE IMPACT ASSESSMENT

Life-cycle impact assessment (LCIA) is the most detailed and exhaustive part of an LCA. This can be carried out by qualitatively and quantitatively assessing environmental impact based on the resource, energy consumption data, and various emission data provided from the inventory. Generally, the impact categories are usually divided into three or as many as one identified in a study with the reference equivalent of the unit chosen. This part of building an LCA focused on categorizing impacts into different environmental categories such as climate change (global), eutrophication (local), and acidification (regional) in a quantified magnitude. In addition to those, it is recommended to include human toxicity and freshwater ecotoxicity which can be carried out by using the USEtox model. For freshwater use, it is important to include the 'blue water footprint' concept, as fresh water is required to supply FU throughout the supply chain.

After the process of categorizing, the LCIA phase then characterizes each factor by converting the inventory data into a common unit of interest, such as the equivalent of carbon dioxide emissions for greenhouse gas emissions. Although practitioners can carefully collect and validate the data manually, the modelling part of the system is usually aided by software such as OpenLCA (GreenDelta), SimaPRO (PRE Sustainability B.V.), GaBi (Sphera), CMLCA (Leiden University), OneClickLCA (One Click LCA Ltd.), and Ecodesign Studio (Altermaker). For Mediterranean case studies, the CML-IA database (CML-Department of Industrial Ecology 2016) is usually preferred. The advantages and disadvantages of these modelling systems are summarized in Table 15.6. It is to note, however, that we should only utilize one LCA throughout all the assessment processes as there are several differences in their normalization. For example, a comparison study of LCA programmes by Iswara *et al.* (2020) has shown the percentage contributor of environmental impact normalization in the entire process using the SimaPRO and OpenLCA. The study showed global warming was spotted as the largest normalization impact in SimaPRO followed by eutrophication, whereas OpenLCA presented acidification as the largest normalization impact followed by eutrophication.

Inventory analysis requires exhaustive data collection. For IMTA, this data collection must be focused in the details of infrastructure and equipment, resource consumption, transportation, productivity, and direct emissions from the studied system. These collected data would then be analysed as input to the models provided in the programme each to their own. The impact categories most common to IMTA are global warming potential (GWP), acidification, eutrophication, water use, biotic resources, land use, toxicity, resource use, and ozone depletion. Table 15.7 shows the common analysis tool used for these impact categories on IMTA as summarized by Hala *et al.* (2024).

In addition to the environmental impacts listed in Table 15.7, it is also quite common to assess the biotic resources used in IMTA by utilizing LCA. This is carried out by counting the net primary production use (NPPU). NPPU as a specific impact category for aquaculture and fisheries was proposed by Papatryphon *et al.* (2005) in their study of aquafeed with incorporation of high-quality plant ingredients and fishmeal. This impact category was calculated by determining the carbon content of the feed composition for plant-based ingredients and using the formula described by Pauly and Christensen (1995). Nevertheless, proper definition and calculation of NPPU still need to be improved. NPPU is, however, globally defined as the amount of carbon produced by primary producers per unit area and time that can be obtained through subtraction of the respiratory cost from gross primary productivity of total photosynthesis (Hala *et al.*, 2024):

$$P = \left(\frac{M}{9}\right) \times 10^{(T-1)} \tag{15.1}$$

with P as in required NPP with assumed C content of 111.1 g C/kg fish (wet weight), M is the mass on a wet basis of the organism of which NPP use is calculated, and T is the trophic level or organism (Papatryphon *et al.*, 2004). When it comes to IMTA's NPP, allocative efficiency must be understood in this sector due to the limited biotic resources. In addition, it is also challenging to implement this impact category because of it being overlaid onto fixing eutrophication as an environmental impact. This might have been the reason why NPP has not been fully implemented by common LCA tools and is mostly characterized manually. As the IMTA system and our understanding of aquaculture grows, it

Table 15.6 Advantages and disadvantages of some LCIA tools.

LCA Assessment Tools	Advantages	Disadvantages
SimaPRO	Suitable to a range of sectors (manufacturing, energy, construction, agriculture, transportation, and consumer goods). This application also has a regionalized database (geographically). The library is a combination of EcoInvent, USLCI, ELCD, Agri-footprint, and many other databases that is credited with a very comprehensive and extensive list of inputs and outputs. The outputs are categorized into emission to air, emissions to water, and emissions to soil.	There are some methods that are no longer supported. This program may load very slowly as the calculation engine is its memory usage as it uses the matrix inverse.
GaBi Sphera	Applicable to diverse sectors especially chemicals, manufacturing, energy, electronics, automotive, and packaging. Supporting study materials are available. Easy Graphical User Interface. GaBi also has its own database that can be accessed in the application itself and online. The GaBi data are also known to be far more detailed than ELCD and CPM databases and are updated often by the team annually.	Paid license service. GaBi also does not allow the same product multiple times in a process so user might need to do a workaround by simplifying in OpenLCA.
OpenLCA (GreenDelta)	Open-source and versatile for different sectors and applications such as product manufacturing, building construction, waste management, and renewable energy.	The method packs are outsourced and might need to be updated. Database must be downloaded in the NEXUS, not within the application.
Umberto	Highly recommended for circular economy initiatives. Robust. The software includes scenario tools. More professional and comprehensive as it has been used for more than 25 years by LCA experts ranging from industry, consulting, and research & education.	Paid service and not recommended for students who are not backed with long-term projects.
OneClickLCA	Specifically used for assessing the environmental impacts of buildings.	Expensive and the payment structure requires an upfront payment. References within the import wizard may be unclear when matching back to items within the model software. Difficulty in finding materials is also reported.

is important to address how medications or antimicrobial agents affect these factors onto production. There has yet to be an LCA that addresses this, but this could be good for future development of LCA tools.

It is important to remember that different methods of assessment in LCI may use different factors. Thus, it is additionally important to perform normalization and weighing that depends on the goal

Table 15.7 Common impact categories and characterization methods used in LCA studies on IMTA systems.

Methods	Climate Change/ GWP	Acidification	Eutrophication	Water Use	Land Use	Energy Use	Other Specified
ReCIPe 2016 midpoint	x	X	x	X	x	x	Fuel resource scarcity, mineral resource scarcity, stratospheric ozone depletion, human carcinogenic toxicity, human non-carcinogenic toxicity, ecotoxicity
ReCIPe 2016 endpoint	X	x	x	X	x	x	Stratospheric ozone depletion, ionizing radiation, ozone formation, fine particular matter formation, ecotoxicity (terrestrial, freshwater, marine, human carcinogenic/ non-carcinogenic), mineral resource scarcity, fossil resource scarcity, abiotic depletion
CML-IA Baseline	X	X	x	–	x	x	Abiotic depletion (elements and fossil fuels), ozone layer depletion, human toxicity potential
TCED	–	–	–	–	–	x	
TRACI	X	x	x	–	–		Fossil fuel depletion, ozone depletion, photochemical smog, human health carcinogenic, human health non-carcinogenic, respiratory effects
CML2 2000	X	x	x	–	x		
IPCC	X						
Environmental Footprint	X	X	x	x	x	x	Ozone depletion, ecotoxicity, resource use, mineral and metals
EDIP 2003	–	x	–	–	–	–	Ozone depletion
USEtox	–	–	–	–	–	–	Human toxicity, freshwater toxicity

and scope of the study. In a recent review for IMTA LCAs, Hala *et al.* (2024) reported that there have not been normalization studies specifically used that are relevant to most of the goals and scope of the study. In addition, there are arguments regarding how toxicity impacts categories in current models might not be reliable. There needs to be a thorough research when it comes to assigning the weights of impacts for normalization as geological location can also play a role, especially in the fishing sector.

15.6 INTERPRETATION

The final step in LCA is to interpret the resulting LCIA. The interpretation involves recognizing the most significant contributors to the impacted environment, evaluating the study as a complete work, analysing sensitivity, and checking consistency, and providing conclusions, limitations, and future recommendations. For current IMTA systems, interpretation usually states how IMTA performs in comparison to monoculture. However, it is generally supported that the impact of IMTA is lower in some categories such as eutrophication, land use, and net primary production (Chary *et al.*, 2019). This is also supported by a previous study in assessing combined fish and lettuce cultivation in aquaponics that showed a smaller impact than conventional greenhouse lettuce with lake and pond culture in normalized value of global warming and freshwater eutrophication according to end-value ReCiPe 2016 Endpoint (Cohen *et al.*, 2018). However, it is noted that in the same study by Chary *et al.* (2019), IMTA systems also presented high impacts on climate change, acidification, and cumulative energy demand. But this result cannot be definite as the systems compared in Chary *et al.*'s study have high uncertainties due to the massive difference in scale between fish and other organisms production in the system. In interpretation of IMTA systems, it is also important to see what are the environmental impacts mostly contributed by the feed. So far in the recent studies, biotic sources are certainly most linked to the feed thus future IMTA LCA should focus on analysing different feed alternatives (Papatryphon *et al.*, 2004). One alternative is to use plant-based feed through fish grow-out phase as it also significantly improves the system's environmental performances. It is also noted that local and global impacts found from the LCIA must be considered to facilitate decision-making processes (Larrey-Lassalle *et al.*, 2022). In addition to that, interpretation should also depict the current system's performance in a wider lens, providing another form of possible demonstration of sustainability in IMTA such as the use of differential drivers–pressures–states–impacts–responses (El-Serehy *et al.*, 2018; Hala *et al.*, 2024). This approach was carried out by incorporation of multiple aspects into the assessment through assigned cost measures from nutrient loading, water quality, condition of natural habitats, greenhouse gas emissions, and by-product harvesting. Although IMTA has been introduced as early as 1997 through Troell *et al.*'s work in utilizing integrated seaweed cultivation to treat fish farm waste, it has only been in 2009 that LCA was utilized to analyse the overall sustainability of the system (Lazard *et al.*, 2009). In interpreting an LCA, additional information might also be needed to compare with one product in mind when building LCA. As an example, an extensive review comparing *tilapia* aquaculture equipped with recirculation systems operated in low demand of water and effluent production has also reported a challenge in summarizing quantitative data on nutrient flows and balance in integrated systems (Zimmermann *et al.*, 2023).

15.7 FEATURING TRENDS AND METHODS OF LCA APPLIED FOR NBSS IN IMTA AROUND THE GLOBE

The application of IMTA is generally applied quite similarly in terms of the systems applied. There is no doubt, however, that the geographical location plays a huge role in how an LCA is determined or utilized in assessing the system. This is due to the nature of NBSs that is used to provide services to the community whether economically, financially, or both. According to the latest review among studies involving urban aquaculture in the northern countries, this system has been shown to be the best in securing food security while also serving in climate change, biodiversity and ESs, sustainable agriculture, resource efficiency, urban regeneration, land management, public health, social cohesion as well as economic growth (Artmann & Sartison, 2018). IMTA research, however, is still considered as under-researched globally. Arguments regarding the future of IMTA alongside other aquaculture technologies such as recirculating aquaculture systems and aquaponics for future global fish supply have shown the need to improve the technology to make end-products less competitive due to high production cost (Laktuka *et al.*, 2023).

The application of LCA for IMTA in the continent of Asia is mostly conducted by universities and research institutions. However, there are some studies originating from the private sector although quite limited. LCA for IMTA is mostly expected to play a significant role in meeting the region's growing demand for seafood, although not exclusively for urban areas. The LCA studies on IMTA systems in Asia are mostly focused on the environmental aspects such as their impacts on greenhouse gas emissions, eutrophication, and acidification. Studies also focused more on combining vegetables with fish production which notably profit from both biomasses produced (Nissar *et al.*, 2023). The main findings in LCA studies of IMTA in this region often focus on economic benefits through reduced costs from feed and energy used. Increased efficiency and improved water quality have also been the major findings, pursuing later into an enhanced biodiversity in the systems that leads to increased efficiency in terms of feed conversion ratios (Henriksson *et al.*, 2014a, 2014b; Peter *et al.*, 2022; Phong *et al.*, 2011). The addition of microalgae as feed has also been considered but the inclusion in an integrated approach has also been proposed but nothing has been published about LCA in the region. However, the utilization of seaweed has been existing earlier in Asia. In the Fisheries Research Institute of Ciguu Lagoon in southwest Taiwan, IMTA modules consisted of three ponds which ran along to assess water purification where the southern system would run cyanobacteria alongside seaweeds (Yeh *et al.*, 2017). The study did, however, not include LCA and focused more on observing water quality and the nutrient recycling in comparison between the systems.

In Europe and America, meanwhile, LCA has been utilized quite thoroughly and is the norm when it comes to sustainable living. Recent LCAs from this region on IMTA focuses more on valorization and highlighting the importance of designing and configuring IMTA systems for an optimized environmental performance (Laktuka *et al.*, 2023). Studies from this region are also completed with relevance and uncertainty analysis which presented how allocation methods can significantly influence results of an LCA. This is then followed by technical variables of different trophic levels, their integration, and the choice of species in determining their trade-offs. An innovative IMTA module has also recently been studied in Italy combining culture of extractive species, polychaetes, sponges, and macroalgae ran together in a somewhat simple settlement made of coconut fibre ropes, assessing the applicability of macroalgae – often run in Asia's IMTA (Giangrande *et al.*, 2020). With regards to waste recycling, the inclusion of microalgae into IMTA has long been considered (Herrador, 2016). It has also been well-documented in terms of CO_2 fixation, even available for an inventory-based LCA on photobioreactors ran in Arizona (Mayers *et al.*, 2016; Satya *et al.*, 2023). The utilization of IMTA specifically used for urban space is nevertheless still relatively scarce as IMTA is often run at large scale due to its complexity and cost to set up.

The application of LCA in Africa has also been reported, focusing on the circulatory aspect and biodiversity, resulting in more sustainable products that are utilized to up-keep food security in the region compared to other regions (Chalchisa, 2022). Checa *et al.* (2024) recently performed circularity assessment of IMTA systems of a semi-closed abalone and urchin system to that of IMTA's run in Brazil and Ireland. The study focused on the two pillars in circularity performance: use of resources and nutrient management. In the use of resources category, there are feed-linearity, water, energy, and infrastructure. The study showcased that the IMTA built in Africa has given a 90% water circularity performance which puts them right after the Brazilian's IMTA circularity performance in reusing water at 99.68%. The abalone–urchin system in South Africa also showed higher nitrogen bioremediation at 80% although nothing regarding phosphorus was recorded.

15.8 CONCLUSION

LCA is an important tool in assessing a sustainable technology. Its application in assessing IMTA systems has been largely utilized in verifying and showcasing their lower environmental impact in comparison to monoculture systems. However, its application in IMTA remains a challenge due to the varying degrees of how an LCA can be interpreted as it depends on how the goals and scope definition are set. Due to this nature, there is still a lack of specific definitions or framework when it comes to applying LCA to IMTA. Challenges also come from the numerous allocation methods as it should be catered towards the diverse uses of species and products. There also needs to be a thorough update in the inventory to properly represent the regions. Overall, IMTA has showcased the potential to lower environmental impacts when it comes to monoculture systems. Future LCA should be more focused on how to deal with uncertainties and promote collaborative research in providing inventory that is usable efficiently and effectively.

REFERENCES

Artmann M. and Sartison K. (2018). The role of urban agriculture as a nature-based solution: a review for developing a systemic assessment framework. *Sustainability*, MDPI, **10**(6), 1937 https://doi.org/10.3390/su10061937

Beloin-Saint-Pierre D., Albers A., Hélias A., Tiruta-Barna L., Fantke P., Levasseur A., Benetto E., Benoist A. and Collet P. (2020). Addressing temporal considerations in life cycle assessment. *Science of the Total Environment*, **743**, 1–16, https://doi.org/10.1016/j.scitotenv.2020.140700

Beltran M. A. and Guinée J. (2014). Goal and scope definition for life cycle assessment of integrated multi-trophic marine aquaculture systems. In: International Conference on Life Cycle Assessment in the Agri-Food Sector, R. Schenck and D. Huizenga (eds), ACLCA, San Francisco, pp. 817–822.

Bisinella V., Christensen T. H. and Astrup T. F. (2021). Future scenarios and life cycle assessment: systematic review and recommendations. *International Journal of Life Cycle Assessment*, Springer Science and Business Media Deutschland GmbH, **26**(11), pp. 2143–2170, https://doi.org/10.1007/s11367-021-01954-6

Chalchisa T. (2022). Benefits of integrated multi-trophic aquaculture systems, https://doi.org/10.35248/2150-3508.22.13.314

Chary K., Aubin J., Sadoul B., Fiandrino A., Covès D. and Callier M. D. (2019). Integrated multi-trophic aquaculture of red drum (*Sciaenops ocellatus*) and sea cucumber (*Holothuria scabra*): assessing bioremediation and life-cycle impacts, *Aquaculture*, **516**, 734621, https://doi.org/10.1016/j.aquaculture.2019.734621

Checa D., Macey B. M., Bolton J. J., Brink-Hull M., O'Donohoe P., Cardozo A., Poersch L. H. and Sánchez I. (2024). Circularity assessment in aquaculture: the case of integrated multi-trophic aquaculture (IMTA) systems. *Fishes*, **9**(5), 165, https://doi.org/10.3390/fishes9050165

Chrismadha T., Satya A., Satya I., Rosidah R., Satya A., Pangestuti R., Harimawan A., Setiadi T., Chew K. and Show P. (2022). Outdoor inclined plastic column photobioreactor: growth, and biochemicals response of *Arthrospira platensis* culture on daily solar irradiance in a tropical place. *Metabolites*, **12**(12), 1199, https://doi.org/10.3390/metabo12121199

Cohen A., Malone S., Morris Z., Weissburg M. and Bras B. (2018). Combined fish and lettuce cultivation: an aquaponics life cycle assessment. In: 25th CIRP Life Cycle Engineering (LCE) Conference, 30 April–2 May 2018, Procedia CIRP, Copenhagen, Denmark, pp. 551–556, https://doi.org/10.1016/j.procir.2017.11.029

El-Serehy H. A., Abdallah H. S., Al-Misned F. A., Al-Farraj S. A. and Al-Rasheid K. A. (2018). Assessing water quality and classifying trophic status for scientifically based managing the water resources of the Lake Timsah, the lake with salinity stratification along the Suez Canal. *Saudi Journal of Biological Sciences*, **25**(7), 1247–1256, https://doi.org/10.1016/j.sjbs.2018.05.022

FAO (2022). The State of World Fisheries and Aquaculture 2022. Towards Blue Transformation. Rome, FAO. https://doi.org/10.4060/cc0461en

Galindro B. M., Zanghelini G. M. and Soares S. R. (2019). Use of benchmarking techniques to improve communication in life cycle assessment: a general review. *Journal of Cleaner Production*, **213**, 143–157, https://doi.org/10.1016/j.jclepro.2018.12.147

Giangrande A., Pierri C., Arduini D., Borghese J., Licciano M., Trani R., Corriero G., Basile G., Cecere E. and Petrocelli A. (2020). An innovative IMTA system: polychaetes, sponges and macroalgae co-cultured in a southern Italian in-shore mariculture plant (Ionian Sea). *Journal of Marine Science and Engineering*, **8**(10), 733, https://doi.org/10.3390/jmse8100733

Hala A. F., Chougule K., Cunha M. E., Caria M. M., Oliveira I., Bradley T., Forbes J. and Galileu S. L. (2024). Life cycle assessment of integrated multi-trophic aquaculture: a review on methodology and challenges for its sustainability evaluation. *Aquaculture*, **590**, 741035, https://doi.org/10.1016/j.aquaculture.2024.741035

Haryanti I., Rizal A., Aliah R. S. and Sachoemar S. I. (2023). Urban farming aquaculture as an alternative business for food and economic security during the COVID-19 pandemic – case study in the sub-urban area of Jakarta, Indonesia. *Polish Journal of Environmental Studies*, **32**(5), 4023–4036, https://doi.org/10.15244/pjoes/166362

Hasimuna O. J., Maulu S., Nawanzi K., Lundu B., Mphande J., Phiri C. J., Kikamba E., Siankwilimba E., Siavwapa S. and Chibesa M. (2023). Integrated agriculture–aquaculture as an alternative to improving small-scale fish production in Zambia. *Frontiers in Sustainable Food Systems*, **7**, 1161121, https://doi.org/10.3389/fsufs.2023.1161121

Heimann T. and Delzeit R. (2024). Land for fish: quantifying the connection between the aquaculture sector and agricultural markets. *Ecological Economics*, **217**, 108090, https://doi.org/10.1016/j.ecolecon.2023.108090

Henriksson P. J. G., Pelletier N. L., Troell M. and Tyedmers P. H. (2013). Life Cycle Assessments and Their Applications to Aquaculture Production Systems. In *Sustainable Food Production*, Springer, New York, *pp.* 1050–1066, https://doi.org/10.1007/978-1-4614-5797-8_191

Henriksson P., Zhang W., Nahid S., Newton R., Phan L., Hai D., Zhang Z., Jaithiang J., Andong R., Chaimanuskul K., Vo N., Hua H., Haque M., Das R., Kruijssen F., Satapornvanit K., Nguyen P. T., Liu Q., Wahab M. and Rico A. (2014a). Final LCA case study report – results of LCA studies of Asian aquaculture systems for tilapia, catfish, shrimp, and freshwater prawn. SEAT Deliverable.

Henriksson P. J. G., Guinée J. B., Heijungs R., De Koning A. and Green D. M. (2014b). A protocol for horizontal averaging of unit process data – including estimates for uncertainty. *The International Journal of Life Cycle Assessment*, **19**(2), 429–436, https://doi.org/10.1007/s11367-013-0647-4

Henriksson P. J. G., Dickson M., Allah A. N., Al-Kenawy D. and Phillips M. (2017). Benchmarking the environmental performance of best management practice and genetic improvements in Egyptian aquaculture using life cycle assessment. *Aquaculture*, **468**, 53–59, https://doi.org/10.1016/j.aquaculture.2016.09.051

Herrador M. (2016). The microalgae/biomass industry in Japan – an assessment of cooperation and business potential with European companies. EU–Japan Centre for industrial Cooperation, pp. 1–184, https://www.eu-japan.eu/sites/default/files/publications/docs/microalgaebiomassiindustryinjapan-herrador-min16-1.pdf

Iswara A. P., Farahdiba A. U., Nadhifatin E. N., Pirade F., Andhikaputra G., Muflihah I. and Boedisantoso R. (2020). A comparative study of life cycle impact assessment using different software programs. *IOP Conference Series: Earth and Environmental Science*, **506**(1), https://doi.org/10.1088/1755-1315/506/1/012002

IUCN (2020). Guidance for Using the IUCN Global Standard for Nature-Based Solutions: First Editions. IUCN, International Union for Conservation of Nature, Gland, Switzerland, https://doi.org/10.2305/IUCN.CH.2020.09.en

IUCN (2022). Nature-Based Solutions in the Post-2020 Global Biodiversity Framework Targets. IUCN, Gland, Switzerland.

Kluts I. N., Potting J., Bosma R. H., Phong L. T. and Udo H. M. J. (2012). Environmental comparison of intensive and integrated agriculture–aquaculture systems for striped catfish production in the Mekong Delta, Vietnam, based on two existing case studies using life cycle assessment. *Reviews in Aquaculture*, **4**(4), 195–208, https://doi.org/10.1111/j.1753-5131.2012.01072.x

Koniyo Y. (2022). Role of innovations/interventions to bring sustainability in aquaculture growth in Indonesia: integration of life cycle assessment (LCA) framework. *FishTaxa*, **26**, 19–32.

Laktuka K., Kalnbalkite A., Sniega L., Logins K. and Lauka D. (2023). Towards the sustainable intensification of aquaculture: exploring possible ways forward. *Sustainability*, **15**(24), 16952, https://doi.org/10.3390/su152416952

Larrey-Lassalle P., Decker S. A., Perfido D., Naneci S. and Rugani B. (2022). Life cycle assessment applied to nature-based solutions: learnings, methodological challenges, and perspectives from a critical analysis of the literature. *Land*, ***11***(5), 649, https://doi.org/10.3390/land11050649

Lazard J., Baruthio A., Mathé S., Rey-Valette H., Chia E., Aubin J., Clément O., Morissens P., Mikolasek O. and Legendre M. (2009). Adaptation of aquaculture system typologies to the requirements of sustainable development. *Cahiers Agricultures*, **18**(3), 199–210, https://doi.org/10.1684/agr.2009.0297

Le Gouvello R., Brugère C. and Simard F. (2022). Aquaculture and Nature-Based Solutions: Identifying Synergies Between Sustainable Development of Coastal Communities, Aquaculture, and Marine and Coastal Conservation. IUCN, International Union for Conservation of Nature, Gland, Switzerland, https://doi.org/10.2305/IUCN.CH.2022.02.en

Little D. C. (2003). Integrated livestock fish farming systems. https://www.researchgate.net/publication/259999240

Mair G. C., Halwart M., Derun Y. and Costa-Pierce B. A. (2023). A decadal outlook for global aquaculture. *Journal of the World Aquaculture Society*, John Wiley and Sons Inc., **54**(2), 196–205, https://doi.org/10.1111/jwas.12977

Martinez-Porchas M. and Martinez-Cordova L. R. (2012). World aquaculture: environmental impacts and troubleshooting alternatives. *The Scientific World Journal*, **4**, 389623, https://doi.org/10.1100/2012/389623

Mayers J. J., Ekman Nilsson A., Svensson E. and Albers E. (2016). Integrating microalgal production with industrial outputs – reducing process inputs and quantifying the benefits. *Industrial Biotechnology*, Mary Ann Liebert Inc., **12**(4), 219–234, https://doi.org/10.1089/ind.2016.0006

Nadarajah S. and Flaaten O. (2017). Global aquaculture growth and institutional quality. *Marine Policy*, **84**, 142–151, https://doi.org/10.1016/j.marpol.2017.07.018

Naylor R. L., Hardy R. W., Buschmann A. H., Bush S. R., Cao L., Klinger D. H., Little D. C., Lubchenco J., Shumway S. E. and Troell M. (2021). A 20-year retrospective review of global aquaculture. *Nature*, **591**, 551–563, https://doi.org/10.1038/s41586-021-03308-6

Nhu T. T., Schaubroeck T., Henriksson P. J. G., Bosma R., Sorgeloos P. and Dewulf J. (2016). Environmental impact of non-certified versus certified (ASC) intensive *Pangasius* aquaculture in Vietnam, a comparison based on a statistically supported LCA. *Environmental Pollution*, **219**, 156–165, https://doi.org/10.1016/j.envpol.2016.10.006

Nissar S., Bakhtiyar Y., Arafat M. Y., Andrabi S., Mir Z. A., Khan N. A. and Langer S. (2023). The evolution of integrated multi-trophic aquaculture in context of its design and components paving way to valorization via optimization and diversification. *Aquaculture*, Elsevier B.V., **565**, 739074, https://doi.org/10.1016/j.aquaculture.2022.739074

Papatryphon E., Petit J., Kaushik S. J. and Van Der Werf H. M. G. (2004). Environmental impact assessment of salmonid feeds using life cycle assessment (LCA). *AMBIO: A Journal of the Human Environment*, **33**(6), 316–323, https://doi.org/10.1579/0044-7447-33.6.316

Papatryphon E., Petit J., Van Der Werf H. M. G., Sadasivam K. J. and Claver K. (2005). Nutrient-balance modeling as a tool for environmental management in aquaculture: the case of trout farming in France. *Environmental Management*, **35**(2), 161–174, https://doi.org/10.1007/s00267-004-4020-z

Pauly D. and Christensen V. (1995). Primary production required to sustain global fisheries. *Nature*, **374**(6519), 255–257, https://doi.org/10.1038/374255a0

Pelletier N. and Tyedmers P. (2008). Life cycle considerations for improving sustainability assessments in seafood awareness campaigns. *Environmental Management*, **42**, 918–931, https://doi.org/10.1007/s00267-008-9148-9

Pesonen H. L., Ekvall T., Fleischer G., Huppes G., Jahn C., Klos Z. S., Rebitzer G., Sonnemann G. W., Tintinelli A., Weidema B. P. and Wenzel H. (2000). Framework for scenario development in LCA. *The International Journal of Life Cycle Assessment*, **5**, 21–30, https://doi.org/10.1007/BF02978555

Peter A. P., Chew K. W., Koyande A. K., Munawaroh H. S. H., Bhatnagar A., Tao Y., Sun C., Sun F., Ma Z. and Show P. L. (2022). Integrated microalgae culture with food processing waste for wastewater remediation and enhanced biomass productivity. *Chinese Chemical Letters*, **34**(2), 107721, https://doi.org/10.1016/j.cclet.2022.08.001

Phong L. T., de Boer I. J. M. and Udo H. M. J. (2011). Life cycle assessment of food production in integrated agriculture–aquaculture systems of the Mekong Delta. *Livestock Science*, **139**(1), 80–90, https://doi.org/10.1016/j.livsci.2011.03.015

Reverter M., Sarter S., Caruso D., Avarre J.-C., Combe M., Pepey E., Pouyaud L., Vega-Heredía S., De Verdal H. and Gozlan R. E. (2020). Aquaculture at the crossroads of global warming and antimicrobial resistance. *Nature Communications*, **11**(1), 1870, https://doi.org/10.1038/s41467-020-15735-6

Reverter M., Tapissier-Bontemps N., Sarter S., Sasal P. and Caruso D. (2021). Moving towards more sustainable aquaculture practices: a meta-analysis on the potential of plant-enriched diets to improve fish growth, immunity and disease resistance. *Reviews in Aquaculture*, Wiley-Blackwell, **13**(1), 537–555, https://doi.org/10.1111/raq.12485

Russo A. and Cirella G. T. (2023). Urban ecosystem services: advancements in urban green development. Multidisciplinary Digital Publishing Institute (MDPI). *Land*, ***12***(3), 522, https://doi.org/10.3390/land12030522

Satya A., Chrismadha T. and Atman I. (2021). Irradiance optimation for growing *Spirulina fusiformis*: biomass, phycocyanin and protein production. *Indonesian Journal of Limnology*, **2**, 76–85, https://doi.org/10.51264/inajl.v2i1.7

Satya A. D. M., Cheah W. Y., Yazdi S. K., Cheng Y.-S., Khoo K. S., Vo D.-V. N., Bui X. D., Vithanage M. and Show P. L. (2023). Progress on microalgae cultivation in wastewater for bioremediation and circular bioeconomy. *Environmental Research*, **218**, 114948, https://doi.org/10.1016/j.envres.2022.114948

Satya I. A., Satya A., Chrismadha T., Rosadi R., Maysarah A. D., Harimawan A., Setiadi T., Tang D. Y. Y. and Show P. L. (2024). Cultivation of *Spirulina platensis* for carbon dioxide bio sequestration in hybrid photobioreactor with real-time monitoring system. *Journal of Environmental Chemical Engineering*, **12**(3), 112396, https://doi.org/10.1016/j.jece.2024.112396

Stentiford G. D., Sritunyalucksana K., Flegel T. W., Williams B. A. P., Withyachumnarnkul B., Itsathitphaisarn O. and Bass D. (2017). New paradigms to help solve the global aquaculture disease crisis. *PLoS Pathogens*, **13**(2), e1006160, https://doi.org/10.1371/journal.ppat.1006160

Steubing B., Mendoza Beltran A. and Sacchi R. (2023). Conditions for the broad application of prospective life cycle inventory databases. *The International Journal of Life Cycle Assessment*, Springer Science and Business Media Deutschland GmbH, **28**(9), 1092–1103, https://doi.org/10.1007/s11367-023-02192-8

Trimbee A. M. and Prepas E. E. (1987). Evaluation of total phosphorus as a predictor of the relative biomass of blue-green algae with emphasis on Alberta lakes. *Canadian Journal of Fisheries and Aquatic Sciences*, **44**, 1337–1342, https://doi.org/10.1139/f87-158

WWF (2021). The world's forgotten fishes. https://wwfint.awsassets.panda.org/downloads/world_s_forgotten_fishes__report_final__1.pdf

Yeh S.-L., Dahms H.-U., Chiu Y.-J., Chang S.-J. and Wang Y.-K. (2017). Increased production and water remediation by land-based farm-scale sequentially integrated multi-trophic aquaculture systems – an example from southern Taiwan. *Sustainability*, **9**, 2173, https://doi.org/10.3390/su9122173

Zimmermann S., Kiessling A. and Zhang J. (2023). The future of intensive tilapia production and the circular bioeconomy without effluents: biofloc technology, recirculation aquaculture systems, bio-RAS, partitioned aquaculture systems and integrated multitrophic aquaculture. *Reviews in Aquaculture*, John Wiley and Sons Inc., **15**(Suppl. 1), 22–31, https://doi.org/10.1111/raq.12744

Index

IWA Publishing's authorised EU representative for General Product Safety Regulations is Diane D'Arras, 15 rue Duret, 75116 Paris, France, e-mail: safety@iwap.co.uk.

Printed and bound by CPI Group (UK) Ltd, Croydon, CR0 4YY
20/07/2026
02171581-0001